PARTICLES, STRINGS AND SUPERNOVAE

Proceedings of the Theoretical Advanced Study Institute in
Elementary Particle Physics

PARTICLES, STRINGS AND SUPERNOVAE

Volume 1

Brown University, Providence
6 June — 2 July 1988

Editors

A Jevicki
C-I Tan

World Scientific
Singapore • New Jersey • London • Hong Kong

Published by

World Scientific Publishing Co. Pte. Ltd.
P O Box 128, Farrer Road, Singapore 9128

USA office: World Scientific Publishing Co., Inc.
687 Hartwell Street, Teaneck, NJ 07666, USA

UK office: World Scientific Publishing Co. Pte. Ltd.
73 Lynton Mead, Totteridge, London N20 8DH, England

PARTICLES, STRINGS AND SUPERNOVAE

Copyright © 1989 by World Scientific Publishing Co. Pte. Ltd.

All rights reserved. This book, or parts thereof, may not be reproduced in any form or by any means, electronic or mechanical, including photocopying, recording or any information storage and retrieval system now known or to be invented, without written permission from the Publisher.

ISBN 9971-50-792-7
      9971-50-793-5 (pbk)

Printed in Singapore by General Printing Services Pte. Ltd.

Preface

The 1988 Theoretical Advanced Study Institute in Elementary Particle Physics (TASI-88) was hosted by the Brown High-Energy theory Group during the four week period, June 5 to July 1, 1988. The format and aims of the TASI-88 were similar to those of previous TASI's, held at the University of Michigan (1984), Yale (1985), the University of California, Santa Cruz (1986), and the St. John's College Campus, Santa Fe (1987).

The objective of the Theoretical Advanced Study Institutes is to provide a series of lectures and seminars for advanced graduate students on topics of current research in theoretical elementary particle physics and related experimental subjects. Particle physics is an ever evolving discipline. In choosing the final program, we were guided by the following considerations: Since TASI is designed for thesis-level graduate students, the program should be pedagogical. Not all theoretical areas of current interest have to be covered in depth during a given year; on the contrary, a deliberate effort was made to include subjects not dealt with adequately during past TASI's. However, each TASI should always include a coupling to the current and future frontier of experimental physics.

The central theme of Brown TASI-88 was "**PARTICLES, STRINGS, AND SUPERNOVAE**", with primary focus on recent developments. In addition to introductory lectures on Standard Particle Physics and on Strings and Superstrings, special emphasis was placed on Non-perturbative Aspects of Quantum Field Theories, an area not covered in depth during past TASI's. It has become increasingly clear that non-accelerator events promise to emerge as an important branch of experimental particle physics. The rapidly increasing information about structure in the universe on very large scales can provide a clue to and a constraint on high energy particle theories. At TASI-88, lectures on Cosmology and Neutrino-Astrophysics were also provided.

This volume and the accompanying one consist mainly of lectures given at Brown TASI-88. Unfortunately, written versions of lectures by S. Coleman: "*Why There Is Nothing Rather Than Something: A Theory Of Cosmological Constant*", by R. Schwitter: "*Experimental Physics at Collider Energies*", and by

J. Cronin: "*Gamma-Ray Astronomy*", are not included. In order to make these volumes more useful to a wider audience, as supplements, we have also included a reprint of lectures on *Construction of Interacting String and Superstring Field Theory*, and a comment piece on *Statistical Mechanics of Strings at High Energy Densities*, a subject heatedly debated at TASI-88. We would like to thank all the lecturers for their contributions. Color photographs of the Brown campus and environs were taken by John Foraste of the Brown University News Bureau.

We would like to give special acknowledgement to members of the Local Organizing Committee, as well as other members of the Brown High-Energy Theory Group for their enthusiastic help in making TASI-88 a success. We are thankful to the TASI Scientific Advisory Board for their suggestions and encouragement. More details on the organization of the Institute can be found at the end of the Volume II under "Information on TASI-88." We are grateful to the National Science Foundation and the Department of Energy for their continuing support of TASI's. We would like to thank M. Glickman, Provost of Brown University, for his encouragement and financial support. Special thanks should also go to P. Stiles, Dean of Brown Graduate School and Reseach for his help, Ms. N. Catterall and Ms. S. Guillet for their handling of administrative details. Finally, we would like to thank the students of TASI-88 for their entusiastic participation.

Antal Jevicki
Chung-I Tan
Editors, TASI-88

TABLE OF CONTENTS

VOLUME I

Preface v

R.D. Peccei, Radiative Effects and Flavor Mixing in the Standard Model . 1

Jihn E. Kim, The Standard Model and Beyond: the Axion Physics . . 79

R.D. Ball, Decoupling Anomalies 147

D.N. Spergel, A Short Tour of the Universe: An Introduction to Supernovae, the Solar Neutrino Puzzle and the Missing Mass Problem . 171

Michael S. Turner, Cosmology and Particle Physics 219

Andrew Strominger, Baby Universes 315

Neil Turok, Phase Transitions as the Origin of Large Scale Structure in the Universe 393

Gregory G. Athanasiu and Joseph J. Atick, Remarks on Thermodynamics of Strings 495

Chung-I Tan, Statistical Mechanics of Strings at High Energy Densities 517

VOLUME 2

Don Weingarten, Lattice Quantum Chromodynamics 529

Ian Affleck, Field Theory Methods and Strongly Correlated Electrons . 581

N. Andrei, Integrable Models: 4 Introductory Lectures 615

Michael Dine, Supersymmetry: Microphysics and Macrophysics 653

Steven B. Giddings, Fundamental Strings 731

C. Callan and L. Thorlacius, Sigma Models and String Theory 795

J.-B. Zuber, Conformal Field Theory and Applications 879

Antal Jevicki, Construction of Interacting String and Superstring Field Theory 901

Cumrun Vafa, Conformal Algebra of Riemann Surfaces 967

Philip Nelson, Lectures on Supermanifolds and Strings 997

Courses and Lectures 1075

Information on TASI-88 1076

List of Participants 1077

Radiative Effects and Flavor Mixing in the Standard Model

R.D. Peccei
Deutsches Elektronen Synchrotron DESY
Hamburg, Fed. Rep. Germany

Abstract

After a brief introduction to the standard model and a summary of some salient experimental results, a fairly extensive, but at the same time elementary, discussion of electroweak radiative corrections is given. Special attention is paid to the effect of having a large top and/or Higgs mass. Experimental constraints, both direct and indirect, for the presence of top are discussed and the present bounds on the number of generations are reviewed. Assuming the existence of only three generations, a detailed discussion is given of the uncertainties associated with the determination of the elements of the quark mixing matrix. Several issues in particle-antiparticle mixing and CP violation are also broached.

1 A Précis of the Standard Model

The standard model for the electroweak interactions [1] is based on an $SU(2) \times U(1)$ gauge theory which is spontaneously broken to $U(1)_{em}$. The model has three distinct sectors:

1. The Gauge-Higgs sector

2. The Fermion-Gauge sector

3. The Fermion-Higgs sector

Experimentally, very little is known about the first sector. The only thing that has been established with some accuracy here is the interrelation between the W and the Z boson masses, which is predicted by Higgs doublet breaking. The Fermion-Gauge sector, on the contrary, is quite well checked by experiment. For instance, as I will discuss in some detail, it is already necessary to include electroweak radiative corrections, for a comparison of theory with experiment, testifying to the precision with which one understands this part of the standard model. The last sector above, has essentially no experimental support. The only evidence here is inferential. Quark mass generation, via the Fermion-Higgs Yukawa couplings, induces flavour mixing in the charged weak currents and such mixing has been observed. Indeed, the mixing matrix V plays a fundamental role in a variety of phenomena: CP violation, flavor oscillations, etc- topics which will occupy us for a fair fraction of these lectures.

1.1 The Gauge-Higgs Sector

The gauge Lagrangian for $SU(2) \times U(1)$ is augmented by including a scalar field Φ, whose self interactions lead to a breakdown of $SU(2) \times U(1)$ to $U(1)_{em}$. The simplest such scalar field is a complex doublet and the Gauge-Higgs Lagrangian reads

$$\mathcal{L}_{GH} = -\frac{1}{4} F_a^{\mu\nu} F_{a\mu\nu} - \frac{1}{4} F_Y^{\mu\nu} F_{Y\mu\nu} - (D_\mu \Phi)^\dagger (D^\mu \Phi) - V(\Phi) \tag{1}$$

In the above $F_a^{\mu\nu}$ and $F_Y^{\mu\nu}$ are the $SU(2)$ and $U(1)$ field strengths, respectively:

$$F_a^{\mu\nu} = \partial^\mu W_a^\nu - \partial^\nu W_a^\mu + ig\epsilon_{abc} W_b^\mu W_c^\nu \tag{2}$$

$$F_Y^{\mu\nu} = \partial^\mu Y^\nu - \partial^\nu Y^\mu \tag{3}$$

and $D_\mu \Phi$ is the $SU(2) \times U(1)$ covariant derivative [1]

$$D_\mu \Phi = [\partial_\mu - ig \frac{\tau_a}{2} W_{a\mu} + i \frac{g'}{2} Y_\mu] \Phi \tag{4}$$

Here g and g' are the $SU(2)$ and $U(1)$ coupling constants, respectively. The potential $V(\Phi)$ is such that the doublet Higgs field Φ acquires a vacuum expectation value, thereby causing the breakdown $SU(2) \times U(1) \to U(1)_{em}$. The vacuum expectation value

$$<\Phi> = \frac{v}{\sqrt{2}} \begin{pmatrix} 1 \\ 0 \end{pmatrix} \tag{5}$$

breaks $SU(2)$ and $U(1)$ but conserves $U(1)_{em}$, since the charge

$$Q = T_3 + Y \tag{6}$$

acting on $<\Phi>$ vanishes:

$$Q <\Phi> = \begin{pmatrix} 0 & 0 \\ 0 & -1 \end{pmatrix} \frac{v}{\sqrt{2}} \begin{pmatrix} 1 \\ 0 \end{pmatrix} = 0 \tag{7}$$

The breakdown of $SU(2) \times U(1) \to U(1)_{em}$ gives mass to three of the gauge fields. These mass terms are readily computed from the Higgs field kinetic energy term by replacing Φ by its vacuum expectation value $<\Phi>$:

$$\mathcal{L}_{mass} = -(D_\mu <\Phi>)^\dagger (D^\mu <\Phi>) \tag{8}$$

With $<\Phi>$ given by Eq(5), one sees readily that the charged fields

$$W_\pm^\mu = \frac{1}{\sqrt{2}} [W_1^\mu \mp i W_2^\mu], \tag{9}$$

as well as

$$[g \frac{\tau_3}{2} W_3^\mu - \frac{g'}{2} Y^\mu]_{11} = \frac{1}{2} [g W_3^\mu - g' Y^\mu] \tag{10}$$

---
[1] The field Φ has hypercharge Y equal to $-\frac{1}{2}$.

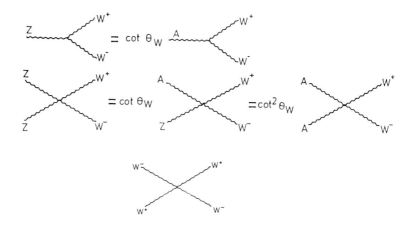

Figure 1: Non vanishing 3-gauge and 4-gauge vertices.

get mass. It is convenient to define two new neutral fields Z^μ and A^μ, related to W_3^μ and Y^μ by

$$\begin{pmatrix} W_3^\mu \\ Y^\mu \end{pmatrix} = \begin{pmatrix} cos\Theta_W & sin\Theta_W \\ -sin\Theta_W & cos\Theta_W \end{pmatrix} \begin{pmatrix} Z^\mu \\ A^\mu \end{pmatrix} \quad (11)$$

By picking the mixing angle Θ_W - the Weinberg angle - as

$$tan\Theta_W = \frac{g'}{g} \quad (12)$$

so that $[gW_3^\mu - g'Y^\mu] = \frac{g}{cos\Theta_W} Z^\mu$, one insures that Z^μ and A^μ are the mass eigenstates. A simple calculation then gives

$$\mathcal{L}_{mass} = -\frac{1}{2}[\frac{gv}{2cos\Theta_W}]^2 Z^\mu Z_\mu - [\frac{gv}{2}]^2 W_+^\mu W_{-\mu} \quad (13)$$

Since the A^μ field has no mass term, it is naturally identified as the photon field. Furthermore, one learns that for doublet breaking there is a simple relation between the W and Z masses. Namely,

$$M_W = M_Z cos\Theta_W = \frac{gv}{2} \quad (14)$$

The gauge interactions arise from the $F_a^{\mu\nu} F_{a\mu\nu}$ term in the Lagrangian (1). Because of this they obey particular symmetry properties, due to the ϵ_{abc} term in (2). Furthermore, the interactions of the photon and Z fields have fixed ratios, since $W_3^\mu = cos\Theta_W Z^\mu + sin\Theta_W A^\mu$. These interconnections, for the nonvanishing 3-gauge and 4-gauge vertices are shown in Fig 1.

The interactions of the Higgs field with the gauge fields are also fixed. The complex doublet Φ is composed of four real fields. Of these, three are "eaten" by the Higgs mechanism to give longitudinal components to the $W^\pm$ and Z fields. The remaining scalar field is the physical Higgs boson, H. It is possible to parametrize Φ so as to make its physical content manifest. One writes Φ as

$$\Phi = \frac{1}{\sqrt{2}} e^{i\vec{\tau}\cdot\vec{\xi}/v} \begin{pmatrix} v+H \\ 0 \end{pmatrix} \tag{15}$$

In the above the fields $\vec{\xi}$ are unphysical. Indeed, since the potential $V(\Phi)$ is really only a function of $\Phi^\dagger \Phi$, it is clear that $\vec{\xi}$ does not enter in V at all. Furthermore, $\vec{\xi}$ can be eliminated by a gauge choice from the covariant derivative $D_\mu \Phi$ [2]. Hence, effectively, one can replace Φ by

$$\Phi \equiv \frac{1}{\sqrt{2}} \begin{pmatrix} v+H \\ 0 \end{pmatrix} \tag{16}$$

The Gauge-Higgs interactions are immediately read off from the Higgs kinetic energy terms in (1), using the above representation. There are both trilinear and quadrilinear couplings. The trilinear couplings, for instance, are proportional to the mass of the gauge fields and one finds

$$\mathcal{L}_{trilinear} = -\frac{e}{2\sin\Theta_W \cos\Theta_W} M_Z Z^\mu Z_\mu H - \frac{e}{\sin\Theta_W} M_W W_+^\mu W_{-\mu} H \tag{17}$$

Note that there is no direct $HA^\mu A_\mu$ coupling.

1.2 The Fermion-Gauge Sector

The interactions of the fermions (quarks and leptons) with the $SU(2) \times U(1)$ fields are fixed, once their $SU(2) \times U(1)$ quantum numbers are specified. The Gauge-Fermion Lagrangian is simply

$$\mathcal{L}_{GF} = -\sum_i \bar{\psi}_i \gamma^\mu \frac{1}{i} D_\mu \psi_i \tag{18}$$

where $D_\mu \psi_i$ is the appropriate covariant derivative for the ith fermion:

$$D_\mu \psi_i = [\partial_\mu - ig(t_a)_i W_{a\mu} - ig'(y)_i Y_\mu] \psi_i \tag{19}$$

The known fermions have left handed helicity components [2] which transform as an $SU(2)$ doublet ($t_a = \frac{\tau_a}{2}$), while their right handed components are $SU(2)$ singlets ($t_a = 0$). Their hypercharge values follow directly from their charge, via Eq(6). Furthermore, fermions come in repetitive patterns (families, or generations) which have the same quantum numbers. I summarize the relevant $SU(2) \times U(1)$ properties of the first generation of quarks and leptons in Table 1 below [3].

---

[2] In my conventions $\psi_L = \frac{1}{2}(1-\gamma_5)\psi$; $\psi_R = \frac{1}{2}(1+\gamma_5)\psi$.
[3] The right handed neutrino is usually omitted, since it has no $SU(2) \times U(1)$ quantum numbers

Table 1: Fermion assignements in the standard model

Fermion	$SU(2)$	$U(1)$	$Q = T_3 + Y$
$\begin{pmatrix} \nu_e \\ e \end{pmatrix}_L$	2	$-\frac{1}{2}$	$\begin{pmatrix} 0 \\ -1 \end{pmatrix}$
e_R	1	-1	-1
$\begin{pmatrix} u \\ d \end{pmatrix}_L$	2	$\frac{1}{6}$	$\begin{pmatrix} \frac{2}{3} \\ -\frac{1}{3} \end{pmatrix}$
u_R	1	$\frac{2}{3}$	$\frac{2}{3}$
d_R	1	$-\frac{1}{3}$	$-\frac{1}{3}$

From Eq(18) it follows that the interactions between the gauge fields and the fermions can be written as

$$\mathcal{L} = g J_a^\mu W_{a\mu} + g' J_Y^\mu Y_\mu \tag{20}$$

where the $SU(2)$ and $U(1)$ fermionic currents are given by

$$J_a^\mu = \sum \bar{\psi}_i \gamma^\mu (t_a)_i \psi_i \tag{21}$$

$$J_Y^\mu = \sum \bar{\psi}_i \gamma^\mu (y)_i \psi_i \tag{22}$$

Using Eq.(12), Eq.(20) can be rewritten in terms of the physical fields Z^μ, A^μ and $W_\pm^\mu$ and their associated currents:

$$Z^\mu \leftrightarrow J_{NC}^\mu = 2(J_3^\mu - \sin^2 \Theta_W J_{em}^\mu) \tag{23}$$

$$A^\mu \leftrightarrow J_{em}^\mu = J_3^\mu + J_Y^\mu \tag{24}$$

$$W_\pm^\mu \leftrightarrow J_\mp^\mu = 2(J_1^\mu \pm i J_2^\mu) \tag{25}$$

One finds in this way for $\mathcal{L}_{int}$ the expression

$$\mathcal{L}_{int} = e J_{em}^\mu A_\mu + \frac{e}{2\sqrt{2}\sin\Theta_W}(W_+^\mu J_{-\mu} + W_-^\mu J_{+\mu}) + \frac{e}{2\cos\Theta_W \sin\Theta_W} Z_\mu J_{NC}^\mu \tag{26}$$

In the above, the identification of the electric charge e as the strenght of the coupling between the photon field and the electromagnetic current, has been used. This identification provides the coupling constant interrelation (unification condition)

$$e = g' \cos\Theta_W = g \sin\Theta_W \tag{27}$$

For weak interaction processes where the momentum transfer q^2 is such that $q^2 \ll M_W^2, M_Z^2$, the effects of $\mathcal{L}_{int}$ can be well approximated by a current-current effective Lagrangian

$$\mathcal{L}_{eff}^{weak} = \frac{i}{2!} \int \mathcal{L}_{int} \cdot \mathcal{L}_{int}$$

$$\simeq (\frac{e}{2\sqrt{2}\sin\Theta_W})^2 \frac{1}{M_W^2} J_+^\mu J_{-\mu} + \frac{1}{2}(\frac{e}{2\cos\Theta_W \sin\Theta_W})^2 \frac{1}{M_Z^2} J_{NC}^\mu J_{NC\mu} \tag{28}$$

In this limit the Glashow Salam Weinberg model reduces to the Fermi theory, plus an additional neutral current interaction contribution. This allows one to identify the Fermi constant G_F as

$$\frac{G_F}{\sqrt{2}} = \frac{e^2}{8 \sin^2 \Theta_W M_W^2} \tag{29}$$

It has become conventional to introduce a parameter

$$\rho = \frac{M_W^2}{M_Z^2 \cos^2 \Theta_W} \tag{30}$$

Then Eq(28) simplifies to

$$\mathcal{L}_{eff}^{weak} = \frac{G_F}{\sqrt{2}} [J_+^\mu J_{-\mu} + \rho J_{NC}^\mu J_{NC\mu}] \tag{31}$$

Note that, in the case of doublet Higgs breaking, Eq(14) gives $\rho = 1$.

1.3 The Fermion-Higgs Sector

Ordinary fermion mass terms, since they connect $\bar{\psi}_{iL}$ with ψ_{iR}, are forbidden by $SU(2)$. However, a Yukawa coupling of $\bar{\psi}_{iL}$ and ψ_{iR} to the doublet Higgs field Φ is allowed. After the breakdown of $SU(2) \times U(1)$ to $U(1)_{em}$, these interactions give rise to fermion mass terms. Thus a doublet Higgs field performs a dual role in the standard model: it gives mass to both gauge bosons and to the quarks and leptons. A general Yukawa interaction will mix fermions of different families, which have the same hypercharge. Thus, when Φ gets replaced by its vacuum expectation value $<\Phi>$, one obtains mass matrices for fermions of the same charge.

Let me discuss this point a bit more in detail. One considers besides Φ, the Higgs doublet field of hypercharge $-\frac{1}{2}$, also its charge conjugate $\tilde{\Phi}$, of hypercharge $+\frac{1}{2}$:

$$\tilde{\Phi} = i\tau_2 \Phi^* \tag{32}$$

If i and j are family indices then, in an obvious notation, the most general $SU(2) \times U(1)$ invariant Yukawa interaction is [4]

$$\mathcal{L}_{Yukawa} = -\Gamma_{ij}^u (\bar{u}_i \bar{d}_i)_L \Phi u_{jR} + \Gamma_{ij}^d (\bar{u}_i \bar{d}_i)_L \tilde{\Phi} d_{jR} + \Gamma_{ij}^l (\bar{v}_{l_i} \bar{l}_i)_L \tilde{\Phi} l_{jR} + h.c. \tag{33}$$

The effective replacement of Φ by Eq(16) and of $\tilde{\Phi}$ by

$$\tilde{\Phi} = -\frac{1}{\sqrt{2}} \begin{pmatrix} 0 \\ v + H \end{pmatrix} \tag{34}$$

yields mass terms for fermions of the same charge and couplings for the Higgs boson H with the fermions. Since the role of v and H in Eqs(6) and (34) are interchangeable, it suffices to focus on the mass terms, which read

$$\mathcal{L}_{Mass} = -\bar{u}_{iL} M_{ij}^u u_{jR} - \bar{d}_{iL} M_{ij}^d d_{jR} - \bar{l}_{iL} M_{ij}^l l_{jR} + h.c. \tag{35}$$

with

$$M_{ij}^f = \frac{1}{\sqrt{2}} \Gamma_{ij}^f v \quad (f = \{u, d, l\}) \tag{36}$$

---

[4] No Yukawa interactions containing right handed neutrinos are generally included, so as not to generate neutrino masses.

Because the above mass matrices are not necessarily diagonal or hermitean one needs to perform a basis charge to get to the physical fermion fields. It is this basis change which causes the family mixing of the states probed by the weak interactions.

Specifically, the matrix M^f can be diagonalized by the bi-unitary transformation [5]

$$(U_L^f)^\dagger M^f (U_R^f) = M_{diag}^f \tag{37}$$

and the fermion fields are replaced by

$$\psi_{L,R}^f \to U_{L,R}^f \psi_{L,R}^f \tag{38}$$

These replacements introduce a mixing matrix for the charged currents, but do not affect the neutral currents. This is easily checked. For the case of three generations, for example, before the basis change the charged current J_+^μ reads

$$J_+^\mu = 2(\bar{\nu}_e \bar{\nu}_\mu \bar{\nu}_\tau)_L \gamma^\mu \mathbf{1} \begin{pmatrix} e \\ \mu \\ \tau \end{pmatrix}_L + 2(\bar{u}\bar{c}\bar{t})_L \gamma^\mu \mathbf{1} \begin{pmatrix} d \\ s \\ b \end{pmatrix}_L \tag{39}$$

After the basis change Eq(38) this current becomes

$$J_+^\mu = 2(\bar{\nu}_e \bar{\nu}_\mu \bar{\nu}_\tau)_L \gamma^\mu \mathbf{U_L^l} \begin{pmatrix} e \\ \mu \\ \tau \end{pmatrix}_L + 2(\bar{u}\bar{c}\bar{t})_L \gamma^\mu (\mathbf{U_L^u})^\dagger \mathbf{U_L^d} \begin{pmatrix} d \\ s \\ b \end{pmatrix}_L \tag{40}$$

Because the neutrinos are massless, the matrix U_L^l above can be eliminated by a redefinition of the neutrino fields. The matrix $(U_L^u)^\dagger U_L^d$ appearing in the quark sector, on the other hand, remains. It is the 3×3 unitary mixing matrix of the charged current sector - the Cabibbo Kobayashi Maskawa matrix V [3]. It is clear that the basis change of Eq(38) will not change the form of the neutral or electromagnetic currents, since these currents always involve fermions of the same charge and $(U_L^f)^\dagger U_L^f = (U_R^f)^\dagger U_R^f = 1$.

The basis change (38) which diagonalizes the mass matrices M^f, obviously diagonalizes also the Yukawa couplings Γ^f. Clearly, therefore, after the basis change to physical fermion fields, the coupling of the Higgs boson H to the fermions is purely diagonal and the Higgs-Fermion Lagrangian is simply

$$\mathcal{L}_{HF} = -\sum [m_i \bar{\psi}_i \psi_i + \frac{m_i}{v} H \bar{\psi}_i \psi_i] \tag{41}$$

Note that here also the Higgs couplings are proportional to the mass of the particles with which it interacts. The only remnant of the mass generation mechanism coming from the Yukawa interactions Eq(33) in the mixing matrix V entering in the charged currents. As can be seen from Eq(40) V is a unitary matrix

$$V^\dagger V = V V^\dagger = 1 \tag{42}$$

For the case of n families, such a matrix is described by $\frac{1}{2}n(n-1)$ real angles and $\frac{1}{2}n(n+1)$ phases. However, by further diagonal phase redefinitions of the quark fields one can always remove $2n-1$ of the phases [6]. Thus physically the Cabibbo Kobayashi Maskawa mixing matrix V is described by $\frac{1}{2}n(n-1)$ real angles and $\frac{1}{2}(n-2)(n-1)$ phases. In particular, one needs at least three generations to have a phase in V. As we shall see later in these lectures, this phase can account for the CP violating phenomena seen in the Kaon system.

---
[5] If M^f is hermitean, then $U_L^f = U_R^f$.
[6] $2n-1$, not $2n$ because an overall phase has no physical meaning.

2 Testing the Standard Model

Since the standard model, by construction, reproduces the low q^2 charged current weak interactions [cf Eq(31)], tests of the standard model have necessarily focused on neutral current processes. These interactions, if one has no prejudice on the Higgs sector, depend on two parameters: $\sin^2 \Theta_W$ and ρ. On the other hand, if one assumes that the breakdown $SU(2) \times U(1) \to U(1)_{em}$ is effected by a Higgs doublet, then $\rho = 1$ and the only parameter to describe neutral current processes is the Weinberg angle. In a nutshell, the electroweak model of Glashow Salam and Weinberg [1] has become the standard model since **all** neutral current experiments [Neutrino deep inelastic scattering; polarized lepton deep inelastic scattering; W and Z properties; neutrino electron scattering; elastic neutrino nucleon scattering; parity violation in atoms] lead, within errors, to the same values of $\sin^2 \Theta_W$ and ρ.

2.1 Resumé of Salient Experimental Results

In this subsection, I want to summarize the present experimental status of the most accurate neutral current experiments to:

- show the "state of the art" testing to which the standard model has been subjected
- indicate the beginning importance which electroweak radiative corrections are playing in comparing standard model predictions with experiment

I will focus particularly, therefore, on deep inelastic experiments involving neutrinos (ν_μ) and antineutrinos ($\bar{\nu}_\mu$) and on the experimentally determined properties of the W and Z bosons.

The comparison of the neutral current processes $\nu_\mu N \to \nu_\mu X$ and $\bar{\nu}_\mu N \to \bar{\nu}_\mu X$ with the charged current processes $\nu_\mu N \to \mu^- X$ and $\bar{\nu}_\mu N \to \mu^+ X$ allows the determination of both ρ and $\sin^2 \Theta_W$. For the case of scattering on an isoscalar target, as provided by a nucleus with an equal number of neutrons and protons, so that $N = \frac{1}{2}(n+p)$, and in a world of only one generation [so that the quark "flavors" in a nucleon are only u, d and their antiparticles $\bar{u}, \bar{d}$] there is a simple connection between the charged and neutral current processes, shown in Fig 2. Defining as usual the kinematical variables x and y by

$$x = \frac{q^2}{-2P \cdot q} \quad ; \quad y = \frac{P \cdot q}{P \cdot l} \tag{43}$$

there are three possible ratios measured experimentally:

$$R_\nu = \frac{\int dx dy < \frac{d\sigma}{dx dy} >_{\nu_\mu N \to \nu_\mu X}}{\int dx dy < \frac{d\sigma}{dx dy} >_{\nu_\mu N \to \mu^- X}} \tag{44}$$

$$R_{\bar{\nu}} = \frac{\int dx dy < \frac{d\sigma}{dx dy} >_{\bar{\nu}_\mu N \to \bar{\nu}_\mu X}}{\int dx dy < \frac{d\sigma}{dx dy} >_{\bar{\nu}_\mu N \to \mu^+ X}} \tag{45}$$

and

$$r = \frac{\int dx dy < \frac{d\sigma}{dx dy} >_{\bar{\nu}_\mu N \to \mu^+ X}}{\int dx dy < \frac{d\sigma}{dx dy} >_{\nu_\mu N \to \mu^- X}} \tag{46}$$

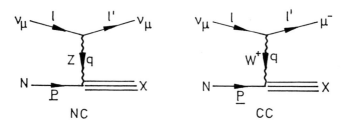

Figure 2: Deep inelastic neutral current and charged current processes involving neutrinos.

For $q^2 \ll M_W^2, M_Z^2$ and for scattering off isoscalar targets assumed to be composed only of quarks and antiquarks of the first family, one can readily establish the, so called, Llewellyn Smith relations [4] among these ratios:

$$R_\nu = \rho^2[(\frac{1}{2} - \sin^2\Theta_W) + \frac{5}{9}\sin^4\Theta_W(1+r)] \qquad (47)$$

$$R_{\bar\nu} = \rho^2[(\frac{1}{2} - \sin^2\Theta_W) + \frac{5}{9}\sin^4\Theta_W(1+r^{-1})] \qquad (48)$$

That is, for this simplified situation, one can altogether bypass the structure functions to compute the neutral current cross sections, provided one measures at the same time the charged current cross sections. Of course, in real life, scattering is done on not purely isoscalar targets. Furthermore, one must also take into account that nucleons have also a sea of strange quarks, charm quarks, etc. These real life effects, however, give rise to small corrections, which one can account for in a systematic way.

As an illustration, I indicate briefly the procedure followed by the CDHS collaboration [5] to extract $\sin^2\Theta_W$. Analogous corrections have been also applied in other deep inelastic experiments [6,7,8]. What was measured by the CDHS collaboration are the ratios R_ν and r, with the results

$$R_\nu = 0.3072 \pm 0.0025 \pm 0.0020 \qquad (49)$$

$$r = 0.39 \pm 0.01 \qquad (50)$$

where the two errors in Eq(49) are, respectively, statistical and systematic. Assuming $\rho = 1$, the Llewellyn Smith relation Eq(47) implies a value

$$(\sin^2\Theta_W)_{uncorr} = 0.236 \pm 0.005 \qquad (51)$$

To obtain a final value for the Weinberg angle, however, a number of corrections must be applied. The major "experimental" correction comes from the fact that the target, Fe, is not purely isoscalar. This lowers the value of $\sin^2\Theta_W$ by -0.009 [5]. There are, on the other hand, a number of smaller "experimental" corrections, connected with the presence of strange and charm quarks in the nucleon, which essentially cancel this shift. In addition, however, one must apply electroweak radiative corrections to the data and when this is done - in the manner which I shall explain below - one lowers $\sin^2\Theta_W$ by -0.011 [5]. In performing all of these corrections one incurs a certain theoretical error, because one needs to estimate many things which are not precisely known (e.g. the amount of strange sea). The biggest error in the final value one obtains for $\sin^2\Theta_W$ is due to the uncertainty in the value of the charm quark mass, which affects particularly the ratio r. Carrying through all these corrections CDHS obtains finally a value for $\sin^2\Theta_W$

$$\sin^2\Theta_W = 0.225 \pm 0.005 \pm [0.003 \pm 0.013(m_c - 1.5 GeV)] \qquad (52)$$

The error in the square bracket is an estimate of the theoretical uncertainty. Assuming, as is reasonable, that $m_c = 1.5 \pm 0.3 GeV$, then the full m_c uncertainty gives a theoretical error of [0.005]. The electroweak radiative effects included in Eq(52) are computed using a definition of $\sin^2\Theta_W$ in which

$$\sin^2\Theta_W = 1 - \frac{M_W^2}{M_Z^2} \qquad (53)$$

I will explain shortly why one needs to specify precisely what one means by $\sin^2\Theta_W$, when one computes radiative corrections. Here I note only that these radiative effects have **lowered** the raw experimental value for $\sin^2\Theta_W$ by about 4%.

The results of CDHS are in perfect agreement with results obtained at Fermilab [7] [8] and by the other large neutrino scattering experiment at CERN, CHARM [6]. Assuming that $\rho = 1$, the results of all these experiments, including CDHS, are reported in Table 2. The error in brackets in this Table is an estimate of the theoretical uncertainty.

Table 2: Results of Deep Inelastic Neutrino Experiments for $\sin^2\Theta_W$

$\sin^2\Theta_W = 0.235 \pm 0.005 \pm [0.005]$ CDHS[5]
$\sin^2\Theta_W = 0.236 \pm 0.005 \pm [0.005]$ CHARM[6]
$\sin^2\Theta_W = 0.239 \pm 0.010$ CCFRR[7]
$\sin^2\Theta_W = 0.246 \pm 0.016$ FHH[8]

Recently a global reanalysis of these inelastic experiments, and of all other neutral current experiments, has been performed by two groups [9] [10]. I will give below, for definitiveness, the results of Amaldi et al [9]. However, those of Costa et al [10] are quite similar. Correcting the raw experimental value of $\sin^2\Theta_W$ for all experimental effects, but not including radiative corrections, Amaldi et al [9] arrive at a "bare" average value for the Weinberg angle [9] of

$$(\sin^2\Theta_W^0)_{DIS} = 0.242 \pm 0.006 \qquad (54)$$

Applying radiative corrections, adopting the definitions Eq(53) for $\sin^2\Theta_W$ and using in addition the values $m_t = 45 GeV$ and $M_H = 100 GeV$, for the yet unknown top and Higgs masses, gives a downward shift

$$\delta s^2 = -0.009 \pm 0.001 \qquad (55)$$

The final value quoted by Amaldi et al [9] from their global fit of all neutrino deep inelastic data, assuming $\rho = 1$, is

$$(\sin^2 \Theta_W)_{DIS} = 0.233 \pm 0.003 \pm [0.005] \tag{56}$$

which is in perfect agreement with the values given by the individual experiments in Table 2.

To determine ρ in addition one needs to measure also $R_{\bar{\nu}}$. The most accurate values for $R_{\bar{\nu}}$ come from the CERN experiments [11] [12]

$$\begin{aligned} R_{\bar{\nu}} &= 0.363 \pm 0.015 \quad CDHS[11] \\ R_{\bar{\nu}} &= 0.377 \pm 0.020 \quad CHARM[12] \end{aligned} \tag{57}$$

Including also experimental information on $R_{\bar{\nu}}$, Amaldi et al [9] obtain values for both ρ and $\sin^2 \Theta_W$. A two parameter fit of all data yields:

$$\sin^2 \Theta_W = 0.232 \pm 0.014 \pm [0.008] \tag{58}$$

$$\rho = 0.999 \pm 0.013 \pm [0.008] \tag{59}$$

One sees that the above is quite consistent with the fit obtained by fixing $\rho = 1$. Furthermore, Eq(59) shows that the hypothesis of doublet Higgs breaking is well established experimentally.

A second set of precision measurements for testing the electroweak theory comes from the CERN collider measurements of the W and Z properties. The mass values for the $W^\pm$ and the Z bosons, measured by the $UA1$ and $UA2$ collaborations, have been summarized by Jenni [13] and are displayed in Table 3 below.

Table 3: Values of W and Z Masses [13]

	Average
$M_W = (82.7 \pm 1.0 \pm 2.7) GeV \quad UA1(e\nu)$	
$M_W = (80.2 \pm 0.6 \pm 1.7) GeV \quad UA2(e\nu)$	$M_W = (80.8 \pm 1.3) GeV$
$M_Z = (93.1 \pm 1.0 \pm 3.1) GeV \quad UA1(e^+e^-)$	Average
$M_Z = (91.5 \pm 1.2 \pm 1.7) GeV \quad UA2$	$M_Z = (92.0 \pm 1.8) GeV$

Using Eq(53) one can compute $\sin^2 \Theta_W$ directly from the W and Z mass values. In so doing, the large systematic errors in Table 3 largely cancel. However, the statistical accuracy is still not good enough to obtain a really significant number. One finds in this way that

$$\sin^2 \Theta_W = 1 - \frac{M_W^2}{M_Z^2} = \left\{ \begin{array}{c} 0.211 \pm 0.025 \; UA1 \; (e\nu) \\ 0.232 \pm 0.025 \pm 0.010 \; UA2 \end{array} \right\} \tag{60}$$

These values agree with those of Eq(56), but the errors are such that not much more can be said.

One can, however, do a much more accurate comparison between different determinations of $\sin^2 \Theta_W$ by using a radiatively corrected version of the formula [cf Eq(29)] relating the Fermi constant to M_W and $\sin^2 \Theta_W$. One can write this formula as an equation for $\sin^2 \Theta_W$:

$$\sin^2 \Theta_W = [\frac{\pi \alpha}{\sqrt{2} G_F M_W^2}][\frac{1}{1 - \Delta r}] = \frac{(37.281 GeV)^2}{M_W^2 [1 - \Delta r]} \qquad (61)$$

In the above α is the fine structure constant $\alpha = \frac{e^2}{4\pi}$, measured at zero momentum transfer. The quantity Δr contains the effects of the radiative corrections and we shall discuss it shortly in detail. For the definition Eq(53) of $\sin^2 \Theta_W$, Δr is rather large. Adopting again as canonical values $m_t = 45 GeV$ and $M_H = 100 GeV$, one finds [14] [9] [7]

$$\Delta r = 0.0713 \pm 0.0013 \qquad (62)$$

Using the UA1 and UA2 vector boson masses and Eq(61), one obtains

$$\sin^2 \Theta_W = 0.218 \pm 0.005 \pm 0.014 \quad UA1(e\nu)$$
$$\sin^2 \Theta_W = 0.232 \pm 0.003 \pm 0.008 \quad UA2 \qquad (63)$$

which lead to an average value - indicating also an estimate of possible theoretical uncertainties [9] - of

$$(\sin^2 \Theta_W)_{M_W/M_Z} = 0.228 \pm 0.007 \pm [0.002], \qquad (64)$$

The values for $\sin^2 \Theta_W$ obtained in deep inelastic scattering experiments, Eq(56), is in very good agreement with that of Eq(64), which follows from the measurements of the W and Z masses. I should note, however, that this comparison is very much aided by the inclusion of radiative corrections. If one had set in Eq(61) $\Delta r = 0$, then the "bare" value of $\sin^2 \Theta_W$ one would have obtained would have been

$$(\sin^2 \Theta_W^0)_{M_W/M_Z} = 0.212 \pm 0.007 \qquad (65)$$

The effect of radiative corrections is to **increase** this value by about 7%. In the case of deep inelastic scattering, however, the effect of the radiative corrections was to **decrease** the value of $\sin^2 \Theta_W$ by about 4%. So the bare values, Eqs(54) and (65), are actually quite far apart. This is summarized in Table 4.

Table 4: Radiative effects for Θ_W

	$\sin^2 \Theta_W^0$	$\sin^2 \Theta_W$
Deep Inelastic	0.242 ± 0.006	$0.233 \pm 0.003 \pm [0.005]$
W/Z Masses	0.212 ± 0.007	$0.228 \pm 0.007 \pm [0.002]$

As a final point I display in Fig 3 the results in the $\rho - \sin^2 \Theta_W$ plane of the global fit of Amaldi et al [9] of all neutral current experiments. As can be seen, the most accurate determinations for ρ and $\sin^2 \Theta_W$ come from the two processes discussed above. However, all neutral current experiments find a consistent value of ρ and $\sin^2 \Theta_W$, testifying to the validity of the standard model - at least to the level of accuracy to which it is probed today. Setting $\rho = 1$, this global fit yield for $\sin^2 \Theta_W$ the value

$$\sin^2 \Theta_W = 0.230 \pm 0.0048 \qquad (66)$$

---

[7] The error here is due to uncertainties in the cross section $e^+ e^- \to hadrons$ at low energy, needed for a full estimate of Δr.

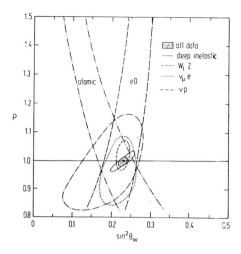

Figure 3: Allowed regions in the $\rho - \sin^2 \Theta_W$ plane at 90% C.L. for various neutral current processes. From [9]

2.2 Electroweak Radiative Corrections and the Renormalization Group

I have indicated briefly earlier than when considering radiative corrections one must specify how $\sin^2 \Theta_W$ is defined. In the 0th order discussion of Sec.1 we have defined $\sin^2 \Theta_W$ in a number of different - but to this order, equivalent - ways:

- Through the unification condition, Eq(27):

$$e = g' \cos \Theta_W = g \sin \Theta_W \tag{67}$$

- Via the interrelation of the W and Z masses, corresponding to doublet Higgs breaking, Eq(14):

$$\sin^2 \Theta_W = 1 - \frac{M_W^2}{M_Z^2} \tag{68}$$

- By comparing the low energy charged current interactions with the Fermi theory, Eq(29):

$$\frac{G_F}{\sqrt{2}} = \frac{e^2}{8 \sin^2 \Theta_W M_W^2} \tag{69}$$

- From the structure of the neutral current itself, Eq(23):

$$J_{NC}^\mu = 2(J_3^\mu - \sin^2 \Theta_W J_{em}^\mu) \tag{70}$$

Although these ways of specifying the Weinberg angle are equivalent in lowest order in α, each of these definitions will get different corrections in higher order, with the corrections depending precisely on how the Weinberg's angle is physically defined. Thus one must agree on some definition of $\sin^2 \Theta_W$, and then all other definitions will be related in a calculable way to this conventionally picked value.

Sirlin [15] suggested in 1980 that a particularly useful definition for $\sin^2 \Theta_W$ would tie this parameter directly with the measured values of the W and Z masses, also at $0(\alpha)$. That is, he suggested that one should adopt Eq(68) as **the** definition of $\sin^2 \Theta_W$ also in higher order. Once one defines $\sin^2 \Theta_W$ this way, then one fixes the way in which one absorbes infinities in higher order calculations and one can express all corrections to the lowest order results in terms of a well defined power series in α. The advantage of using Sirlin's definition of $\sin^2 \Theta_W$, which is the definition I shall use in these lectures, is that one can estimate in most instances the dominant electroweak radiative corrections by means of the renormalization group [16]. Radiative corrections in the $SU(2) \times U(1)$ theory can only be large if the electromagnetic coupling constant squared is accompanied by a large logarithm. A non logarithmic correction is proportional to $\frac{\alpha}{\pi} \sim \frac{1}{500}$ and is negligible. A logarithmic correction, on the other hand, is enhanced since $M_W^2 >>< q^2 >$, with $< q^2 >$ being the relevant momentum or energy transfer in the process, so that, typically,

$$\frac{\alpha}{\pi} \ln \frac{M_W^2}{< q^2 >} \sim (2-3) \times 10^{-2} \tag{71}$$

These logarithmic factors in the calculation can be tracked by means of the renormalization group.

To understand the idea behind this, it is useful to look at the formula I used earlier connecting G_F with $\sin^2 \Theta_W$, in which the effects of radiative corrections are included [Eq(61)]:

$$M_W^2 \sin^2 \Theta_W = \frac{\pi \alpha}{\sqrt{2} G_F} [\frac{1}{1-\Delta r}] \tag{72}$$

On the left-hand side of this equation, using Sirlin's [15] definition of $\sin^2 \Theta_W$, all the parameters are fixed at a large scale $(M_W, M_Z \sim 100 GeV)$. On the right-hand side of this equation, on the other hand, the parameters α and G_F are fixed by low energy processes (Thomson scattering for α and μ-decay for G_F). Thus it is not surprising the correction Δr is large. One should be able to account for the dominant electroweak radiative corrections Δr by replacing α and G_F by their "running" counterparts, evaluated at the large scale M_W. That is, the replacement of α by $\alpha(M_W)$ and G_F by $G_F(M_W)$ [the running electromagnetic coupling and the running Fermi constant, respectively] should remove the bulk of the electroweak radiative corrections, since no large logarithms can then appear. Thus one expects, approximately,

$$M_W^2 \sin^2 \Theta_W = \frac{\pi \alpha}{\sqrt{2} G_F} [\frac{1}{1-\Delta r}] \simeq \frac{\pi \alpha(M_W)}{\sqrt{2} G_F(M_W)} \tag{73}$$

which gives the estimate for Δr:

$$\Delta r \simeq 1 - (\frac{\alpha}{\alpha(M_W)})(\frac{G_F(M_W)}{G_F}) \tag{74}$$

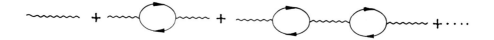

Figure 4: Graphs giving rise to the running of α

The computation of the running electromagnetic constant α proceeds through the summation of the usual QED vacuum polarization graphs, which include all the charged fermions f in the model (see Fig. 4). This yields the standard result [17]

$$\alpha(M_W) = \frac{\alpha}{1 - \frac{\alpha}{3\pi}\sum_f e_f^2 \ln \frac{M_W^2}{m_f^2}} \quad (75)$$

which, for the three generation electroweak model gives $\alpha(M_W) \simeq \frac{1}{128}$ [8]. The Fermi constant in μ-decay, as I will discuss immediately below, does not in fact run, so that $G_F(M_W) \simeq G_F$. Thus our considerations lead us to predict that

$$\Delta r \simeq 1 - \left(\frac{\alpha}{\alpha(M_W)}\right)\left(\frac{G_F(M_W)}{G_F}\right) \simeq 1 - \frac{128}{137} \simeq 0.066 \quad (76)$$

This number is very near the result of the exact $0(\alpha)$ calculation in [9,14] quoted in Eq(62). Thus one sees that via these renormalization group considerations one has got the main effect.

It remains to understand why the Fermi constant does not run. To see this, one must look at the photonic corrections to μ-decay in the Fermi theory. If these corrections are infinite in the Fermi theory - that is, proportional to $A \ln \Lambda$, where Λ is a high energy cutoff introduced to regularize the theory - then one expects in the $SU(2) \times U(1)$ model, which is less singular at short distances, that there will be corrections proportional to $A \ln M_W$. So if there are **no** infinite photonic corrections in the Fermi theory, one should also have no logarithmic dependences on M_W. So for G_F not to run, it must be that the photonic corrections to μ-decay in the Fermi theory, indicated in Fig. 5, including the contributions from the fermion wave function renormalization, are in fact finite.

---

[8]This value actually uses the experimental cross section for the process $e^+e^- \to hadrons$ to take more carefully into account the effects of the light quarks. For a careful discussion, see [18]

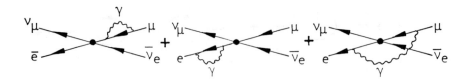

Figure 5: Photonic corrections to μ-decay in the Fermi theory

One can express this somewhat more formally, as follows. Higher order corrections modify the coefficient $\frac{G_F}{\sqrt{2}}$ of the 4 Fermi operator $\bar{\mu}\gamma_\mu(1-\gamma_5)\nu_\mu \bar{\nu}_e \gamma^\mu(1-\gamma_5)e$ in the effective current-current Lagrangian of Eq(31). To define this operator properly one must introduce a scale μ where the operator is normalized. The corresponding coefficient will also depend on μ and one has in general a power series expansion in powers of $\alpha \ln M_W^2/\mu^2$ for this coefficient

$$G_F(\mu^2) = G_F(M_W^2)\{1 + a\frac{\alpha}{\pi}\ln\frac{M_W^2}{\mu^2} + \cdots\} \tag{77}$$

where a is a numerical constant. This logarithmic series can be summed by making use of the renormalization group equation [19], so that one relates the Fermi constant at different scales

$$G_F(\mu^2) = G_F(M_W^2)[\frac{\alpha(\mu^2)}{\alpha(M_W^2)}]^{\bar{d}} \tag{78}$$

where $\bar{d}$ is the anomalous dimension of the operator entering in μ-decay. It is simple to see that this anomalous dimension vanishes - or equivalently that a in Eq(77) vanishes. Thus the Fermi constant does not run:

$$G_F(\mu^2) = G_F(M_W^2) \tag{79}$$

The above can be demonstrated in two ways, either by making use of a Fierz transformation trick, or by direct calculation. The operator entering in μ decay can be readily transformed to one where the neutrino fields are contracted with one another:

$$\bar{\mu}\gamma^\mu(1-\gamma_5)\nu_\mu \bar{\nu}_e \gamma_\mu(1-\gamma_5)e = \bar{\mu}\gamma^\mu(1-\gamma_5)e\bar{\nu}_e\gamma_\mu(1-\gamma_5)\nu_\mu \tag{80}$$

Only the $\bar{\mu}e$ term above is affected by photonic corrections. However, the anomalous dimension of this current, like that of the ordinary electromagnetic current, vanishes. Hence the anomalous dimension $\bar{d}$ in Eq(78) must also vanish.

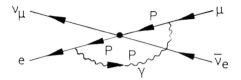

Figure 6: Momentum flow for the amplitude of Eq(81)

Although the above argument suffices to prove the desired non running of G_F, as I will need to perform similar computations again later on, it is useful to demonstrate this explicitly by evaluating directly the photonic diagrams of Fig 5. It is particularly convenient to perform this computation in the Landau gauge, where the photonic self energies are finite. In this way one needs not to worry about any wavefunction renormalization factors and the only diagram that needs to be examined for μ decay is the last one in Fig 5. What I want to show is that this diagram is not ultraviolet divergent. For these limited purposes I can simplify my task further by also setting all external momenta to zero and dropping all masses. The relevant diagram to evaluate then is that indicated in Fig 6. The amplitude associated with Fig 6, in Landau gauge and setting all masses to zero, is given by:

$$A = \int \frac{d^4p}{(2\pi)^4} \frac{-i}{p^2}(\eta^{\mu\nu} - \frac{p^\mu p^\nu}{p^2})[\gamma_\alpha(1-\gamma_5)(\frac{i\gamma p}{p^2})ie_e\gamma_\mu][ie_\epsilon\gamma_\nu(\frac{i\gamma p}{p^2})\gamma^\alpha(1-\gamma_5)] \quad (81)$$

This expression can be simplified by using the fact that under the integral sign

$$p_\alpha p_\beta = \frac{1}{4}p^2\eta_{\alpha\beta}, \quad (82)$$

so that

$$A = -ie_e e_\mu \int \frac{d^4p}{(2\pi)^4} \frac{1}{(p^2)^2} \{\frac{1}{4}[\gamma_\alpha\gamma_\beta\gamma_\mu(1-\gamma_5)][\gamma^\mu\gamma^\beta\gamma^\alpha(1-\gamma_5)] - [\gamma_\alpha(1-\gamma_5)][\gamma^\alpha(1-\gamma_5)]\} \quad (83)$$

The quantity in the curly bracket can be shown to vanish by making use of the γ-matrix identity [9]

$$\gamma_\alpha\gamma_\beta\gamma_\rho = -\eta_{\alpha\beta}\gamma_\rho + \eta_{\alpha\rho}\gamma_\beta - \eta_{\beta\rho}\gamma_\alpha - i\epsilon_{\alpha\beta\rho\delta}\gamma^\delta\gamma_5, \quad (84)$$

as well as the usual relations

$$\gamma^\alpha\gamma_\alpha = -4; \quad \gamma^\alpha\gamma_\mu\gamma_\alpha = 2\gamma_\mu; \quad \epsilon_{\alpha\beta\gamma\delta}\epsilon^{\alpha\beta\gamma\rho} = -6\eta_\delta^\rho \quad (85)$$

---
[9] My γ-matrices obey $\{\gamma_\alpha,\gamma_\beta\} = -2\eta_{\alpha\beta}$ and my metric is $-1+1+1+1$.

Thus the potentially logarithmic divergent piece of the photonic corrections to μ decay vanishes, which establishes, in a different way, the assertion that G_F does not run.

Having understood the magnitude of Δr, using these renormalization group techniques, I want to apply this same method to compute the electroweak corrections to the Llewellyn Smith formulas, Eqs (47) and (48). For definitiveness, I will consider the case where the symmetry breaking is caused by a Higgs doublet, so that $\rho = 1$. Thus it suffices to look only at R_ν, whose 0th order expression, reads

$$R_\nu = \frac{\int dx dy < \frac{d\sigma}{dx dy} >^{\nu_\mu N \to \nu_\mu X}}{\int dx dy < \frac{d\sigma}{dx dy} >^{\nu_\mu N \to \mu^- X}} = (\frac{1}{2} - \sin^2 \Theta_W) + \frac{5}{9}\sin^4 \Theta_W (1+r) \tag{86}$$

Now R_ν experimentally involves rather low momentum transfers ($q^2 << M_W^2, M_Z^2$). Thus one expects that if one had used for $\sin^2 \Theta_W$ a definition which related it to a low energy scale measurement - call it $\sin^2 \Theta_W (m^2)$ - then one could have just used Eq(86) also to $0(\alpha)$. However, since I want to use Sirlin's definition for $\sin^2 \Theta_W$, I need to derive a relation between this $\sin^2 \Theta_W$ and $\sin^2 \Theta_W (m^2)$.

There is a further subtlety that needs also to be accounted for. Eq(86) hides the fact that both the neutral current cross section and the charged current cross section are proportional to the Fermi constant. Now the neutral current Fermi constant, just like the Fermi constant for μ decay does not run. This is because the relevant operator $(\bar{\nu}_\mu \gamma_\alpha (1 - \gamma_5)\nu_\mu)(\bar{q}[v_q \gamma^\alpha + a_q \gamma^\alpha \gamma_5]q)$ also has zero anomalous dimension, for precisely the same reasons as were discussed in μ decay. On the other hand, the Fermi constant associated with the charged current interactions involving quarks **does** run. The relevant operator now contains three charged fields - the μ and two quarks - and the Fierz identity trick is not of any use anymore. Thus, apart from corrections arising out of reexpressing $\sin^2 \Theta_W (m^2)$ in terms of $\sin^2 \Theta_W$, there is a correction due to the fact that

$$R_\nu \sim \frac{(G_F^{NC})^2}{(G_F^{CC})^2} \neq 1 \tag{87}$$

Let me examine this last point first. For the charged current process $\nu_\mu + d \to \mu^- + u$, even in Landau gauge, there exist a divergent graph among the diagrams shown in Fig 7. The self energy diagrams and the vertex diagrams are finite in Landau gauge. Furthermore, the potentially divergent 5th diagram is also finite, since it is precisely the same as the one we considered for μ decay. However, the last diagram in Fig 7 is divergent. To calculate the running of G_F^{CC}:

$$G_F^{CC}(\mu^2) = G_F^{CC}(M_W^2)[1 + \frac{\alpha}{\pi} a_u \ln \frac{M_W^2}{\mu^2}] \tag{88}$$

it suffices to get the coefficient of the logarithmic divergence in this diagram, using as a cutoff $\Lambda \leftrightarrow M_W$. For the above computation it is enough again to set all external momenta to zero and drop all masses. The relevant diagram to evaluate is that indicated in Fig 8, whose amplitude, in the Landau gauge and setting all masses to zero, is given by

$$A^{CC} = \int \frac{d^4 p}{(2\pi)^4} \frac{-i}{p^2}(\eta^{\mu\nu} - \frac{p^\mu p^\nu}{p^2})[ie_\mu \gamma_\mu (\frac{i\gamma p}{p^2})\gamma_\alpha (1-\gamma_5)][ie_u \gamma_\nu (\frac{-i\gamma p}{p^2})\gamma^\alpha (1-\gamma_5)] \tag{89}$$

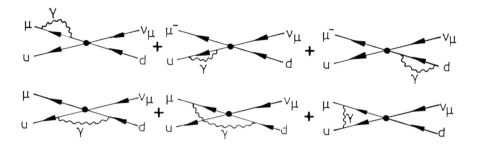

Figure 7: Photonic corrections to the process $\nu_\mu + d \to \mu^- + u$ in the Fermi theory

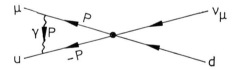

Figure 8: Momentum flow for the amplitude of Eq(89)

This expression can be simplified as before, but now the resulting integrand does not vanish identically. One has

$$A^{CC} = \frac{i\alpha}{4\pi^3}[e_\mu e_u]\int \frac{d^4p}{(p^2)^2}\{\frac{1}{4}[\gamma_\mu\gamma_\beta\gamma_\alpha(1-\gamma_5)][\gamma^\mu\gamma^\beta\gamma^\alpha(1-\gamma_5)] - [\gamma_\alpha(1-\gamma_5)][\gamma^\alpha(1-\gamma_5)]\}$$

$$= \frac{3\alpha i}{4\pi^3}[e_\mu e_u]\int \frac{d^4p}{(p^2)^2}[\gamma_\alpha(1-\gamma_5)][\gamma^\alpha(1-\gamma_5)] \qquad (90)$$

In the above e_μ and e_u are the charges of the μ and the u quarks in units of e. The divergent integral in (90), evaluated with a cutoff $\Lambda = M_W$ gives

$$\int \frac{d^4p}{(p^2)^2} \leftrightarrow i\pi^2 \ln M_W^2 \qquad (91)$$

Hence one identifies the coefficient a_u in Eq(88) as

$$a_u = -\frac{3}{4}e_\mu e_u = \frac{1}{2} \qquad (92)$$

and one has

$$G_F^{CC}(\mu^2) = G_F^{CC}(M_W^2)[1 + \frac{\alpha}{2\pi}\ln\frac{M_W^2}{\mu^2}] \equiv \rho^{CC}G_F^{CC}(M_W^2) \qquad (93)$$

It remains to compute the running of $\sin^2\Theta_W$. For these purposes it suffices to consider some convenient subprocess and look at the relevant photonic corrections. A particularly nice reaction to study is $\nu_\mu e \to \nu_\mu e$ scattering, where there are only two different types of corrections to consider, as depicted in Fig 9. The correction to the electron vertex - G_F correction - are identical to those in μ-decay (after the Fierz transformation) and obviously do not produce any logarithmic modifications. The, so called, Penguin contribution, on the other hand, modify $\sin^2\Theta_W$. This correction corresponds, essentially, to an electromagnetic neutrino charge radius, giving rise to a multiplicative modification of the local 4-Fermi interaction. The neutrino-neutrino-photon vertex of Fig 10, given that the neutrino charge vanishes, has the gauge invariant form.

$$\Gamma^\lambda(q) = \bar{\nu}_\mu[q^2\gamma^\lambda - q^\lambda \gamma q](1-\gamma_5)\nu_\mu F(q^2) \qquad (94)$$

However, the 2nd term above does not contribute, since $m_{\nu_\mu} = 0$. Hence, effectively, the Penguin contribution gives a correction to the $\nu_\mu e \to \nu_\mu e$ T-matrix of the form

$$T_{Penguin} = \bar{\nu}_\mu\gamma_\mu(1-\gamma_5)\nu_\mu \bar{e}\gamma^\mu e F(q^2) \qquad (95)$$

Clearly the above corresponds to a correction to $\sin^2\Theta_W$, since the lowest order contribution to the T-matrix contains a term of the same form as (95) but proportional to $\sin^2\Theta_W$:

$$T_{l.o.} = \bar{\nu}_\mu\gamma_\mu(1-\gamma_5)\nu_\mu \bar{e}\gamma^\mu e[G_F\sqrt{2}\sin^2\Theta_W] + \cdots \qquad (96)$$

The contributions to the Penguin vertex of Fig 10 are of two kinds, as indicated in Fig 11 Both these terms are technically divergent, but can be made convergent by subtracting at zero momentum transfer, since the Penguin vertex must vanish there [cf Eq(94)]. I will illustrate the relevant calculations by considering the CC contribution of Fig 11. Neglecting

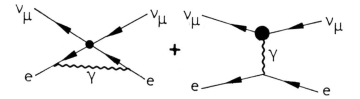

Figure 9: Photonic corrections to $\nu_\mu e \to \nu_\mu e$ scattering: (a) G_F correction; (b) Penguin correction

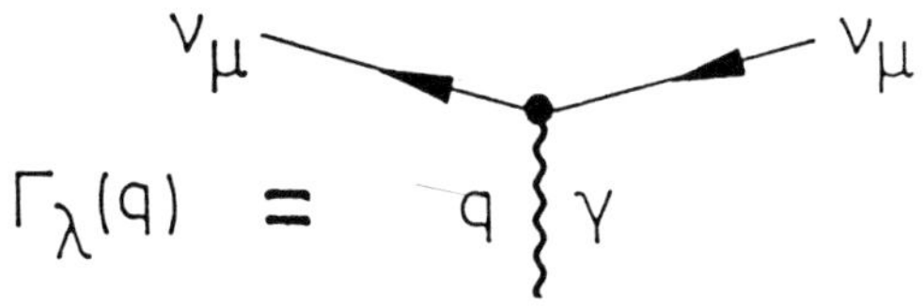

$$\Gamma_\lambda(q) =$$

Figure 10: The neutrino-neutrino-photon vertex

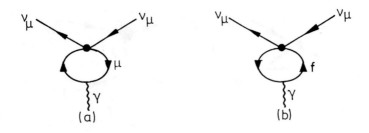

Figure 11: Contribution to the neutrino-neutrino photon vertex, to lowest order in the weak interactions: (a) CC contribution; (b) NC contribution

again mass terms, which are irrelevant for these purposes, the charged current vertex correction is

$$\Gamma_\lambda^{CC}(q) = \frac{G_F}{\sqrt{2}} \bar{\nu}_\mu \{ \int \frac{d^4p}{(2\pi)^4} \gamma_\alpha (1-\gamma_5) [\frac{i\gamma(p+q)}{(p+q)^2}] i e_\mu \gamma_\lambda [\frac{i\gamma p}{p^2}] \gamma^\alpha (1-\gamma_5) \} \nu_\mu - \{q=0\} \quad (97)$$

The quantity in the 1st curly bracket can be rewritten as

$$\begin{aligned} \{\}_\lambda &= -4ie_\mu \int \frac{d^4p}{(2\pi)^4} \frac{\gamma p \gamma_\lambda \gamma(p+q)(1-\gamma_5)}{p^2(p+q)^2} \\ &= -4ie_\mu \int \frac{d^4p}{(2\pi)^4} \int_0^1 dx \frac{[\gamma p - \gamma q(1-x)]\gamma_\lambda [\gamma p + \gamma qx](1-\gamma_5)}{[p^2 + q^2 x(1-x)]^2} \end{aligned} \quad (98)$$

Since all terms odd in p vanish and

$$\gamma q \gamma^\lambda \gamma q = q^2 \gamma^\lambda - 2q^\lambda \gamma q, \quad (99)$$

with the 2nd term above not contributing between the neutrino spinors, one has effectively

$$\{\}_\lambda^{eff} = -4ie_\mu \int_0^1 dx \int \frac{d^4p}{(2\pi)^4} \frac{\frac{1}{2}p^2 - q^2 x(1-x)}{[p^2 + q^2 x(1-x)]^2} \gamma_\lambda (1-\gamma_5) \quad (100)$$

The quadratically divergent piece arising from Eq(100) vanishes if one subtracts the $q^2 = 0$ contribution, but one still has a remaining logarithmic divergence. It is this divergence which details the running of $\sin^2 \Theta_W$. One has

$$\Gamma_\lambda^{CC}(q^2) = \bar{\nu}_\mu \gamma_\lambda (1-\gamma_5) q^2 \nu_\mu \{ 4ie_\mu \sqrt{2} G_F \int_0^1 dx x(1-x) \int \frac{d^4p}{(2\pi)^4} \frac{1}{[p^2 + q^2 x(1-x)]^2} \} \quad (101)$$

Cutting off the divergence as before with a cutoff $\Lambda = M_W$ gives for the integral over momentum in Eq(101) [cf Eq. (91)] $\frac{i}{16\pi}\ln M_W^2$. Hence, using that $e_\mu = -e$,

$$\Gamma_\lambda^{CC}(q^2) = \bar{\nu}_\mu \gamma_\lambda (1-\gamma_5) q^2 \nu_\mu [\sqrt{2} G_F \frac{e}{24\pi^2} \ln M_W^2] \tag{102}$$

This result can be used to compute the modification to the $\nu_\mu e \to \nu_\mu e$ scattering amplitude due to the CC Penguin contribution. One has

$$T_{Penguin}^{CC} = \bar{\nu}_\mu \gamma_\mu (1-\gamma_5) \nu_\mu \bar{e} \gamma^\mu e [G_F \sqrt{2} \sin^2 \Theta_W] \{-\frac{\alpha}{6\pi \sin^2 \Theta_W} \ln M_W^2\} \tag{103}$$

Whence one obtains immediately the form of the running $\sin^2 \Theta_W$, arising from these contributions[10]

$$\sin^2 \Theta_W(m^2) = \sin^2 \Theta_W \{1 - \frac{1}{\sin^2 \Theta_W} \frac{\alpha}{6\pi} \ln \frac{M_W^2}{m^2}\} \tag{104}$$

One can evaluate in an analogous way the NC contributions to the neutrino-neutrino-photon verex, with the result involving a sum over all the flavours f entering in the loop in Fig 11. Adding the result of this calculation in, one obtains finally the desired formula detailing the running of $\sin^2 \Theta_W$:

$$\sin^2 \Theta_W(m^2) = \sin^2 \Theta_W \{1 - \frac{\alpha}{6\pi \sin^2 \Theta_W}[1 + \sum_f (2e_f^2 \sin^2 \Theta_W - T_{3f} e_f)] \ln \frac{M_W^2}{m^2}\} \tag{105}$$

In the above T_{3f} is the T_3 value of the fermions in question. It turns out that there is a fairly large cancellation in the quantity in the square bracket in Eq(105), since

$$\sum (2e_f^2 \sin^2 \Theta_W - T_{3f} e_f) + 1 = Ng[\frac{16}{3} \sin^2 \Theta_W - 2] + 1 \simeq -1.32 \tag{106}$$

where the 2nd line applies for three generations ($Ng = 3$) and $\sin^2 \Theta_W = 0.23$. Thus, approximately,

$$\sin^2 \Theta_W(m^2) = \sin^2 \Theta_W [1 + 0.96 \frac{\alpha}{\pi} \ln \frac{M_W^2}{m^2}] \equiv \kappa \sin^2 \Theta_W \tag{107}$$

We have now all the elements to compute $\sin^2 \Theta_W$ from the measured value for R_ν, in the leading logarithmic approximation. Instead of the lowest order approximation of Eq(86), if we want to extract $\sin^2 \Theta_W$ - given by Sirlin's definition Eq(68) - from R_ν we must reexpress $\sin^2 \Theta_W(m^2)$ in terms of $\sin^2 \Theta_W$, using Eq(107), and correct for the mismatch between the neutral and charged current Fermi constants. That is, one has

$$R_\nu \simeq \frac{1}{(\rho^{CC})^2} \{[\frac{1}{2} - \kappa \sin^2 \Theta_W] + \frac{5}{9} \kappa^2 \sin^4 \Theta_W (1+r)\} \tag{108}$$

Comparing the above with Eq(68) one deduces for the shift δs^2 [cf Eq(55)] the formula

$$\delta s^2 \simeq -\sin^2 \Theta_W [\kappa - 1] - \frac{R_\nu [(\rho^{CC})^2 - 1]}{[1 - \frac{20(1+r)(R_\nu - \frac{1}{2})}{9}]^{\frac{1}{2}}} \simeq -0.23[\kappa - 1] - 0.48[(\rho^{CC})^2 - 1] \tag{109}$$

---

[10]Here the scale m^2 is defined so that, if $m = M_W$, then $\sin^2 \Theta_W(m^2) = \sin^2 \Theta_W$.

where to obtain the last line we have used the approximate numerical values of the various physical parameters. Using Eq(93) and Eq(107) one has, approximately,

$$\delta s^2 \simeq -0.7\frac{\alpha}{\pi}\ln\frac{M_Z^2}{\mu^2} \simeq -0.007 \tag{110}$$

where the last result uses for $\mu^2 \simeq 100 GeV^2$, which is a typical momentum transfer for the deep inelastic neutrino scattering experiments. The value just computed in the leading log approximation, Eq(110), is in excellent agreement with the result of the exact calculation, quoted in Eq(55). This reasserts the contention that the dominant effects of the electroweak radiative corrections can be computed by retaining just the logarithmic corrections.

2.3 Effects of m_t and M_H

The actual numbers for δs^2 and Δr used to compare theory and experiment took, as canonical values, $m_t = 45 GeV$ and $M_H = 100 GeV$ for the two still unknown parameters in the standard model. It is interesting to ask how sensitively does the good agreement obtained with experiment depend on these choices. We shall see that both δs^2 and Δr are quite sensitive to m_t, once it becomes much bigger than M_W, but they are rather insensitive to the Higgs mass. Indeed, one finds

$$\delta s^2, \ \Delta r \sim \begin{cases} (\frac{m_t}{M_W})^2 \\ \ln M_H \end{cases} \tag{111}$$

Furthermore, it turns out that the correction to δs^2 is positive as m_t increases, while it is negative for Δr. This gives rise to a bound on m_t, since as this parameter grows so does the discrepancy between the values of $sin^2\Theta_W$ obtained in deep inelastic scattering and that obtained through the values of the W and Z masses [9]. This bound on m_t lies in the range of 200 GeV, with the actual value depending slightly on the Higgs mass.

The quadratic behaviour in m_t [11] shown in Eq(111) comes from the failure of a total cancellation between the W and Z propagator corrections. This is perhaps most simply demonstrated by considering the effect of fermion loops to the neutral current (NC) to charged current (CC) ratio in R_ν. The relevant graphs in both cases are shown in Fig 12. For $q^2 \ll M_W^2, M_Z^2$ these graphs modify the NC and CC amplitudes as follows

$$A^{NC} \rightarrow \frac{e^2}{8\sin^2\Theta_W\cos^2\Theta_W}[\frac{1}{M_Z^2} + \frac{1}{M_Z^2}\pi_f^Z(0)]$$

$$A^{CC} \rightarrow \frac{e^2}{8\sin^2\Theta_W}[\frac{1}{M_W^2} + \frac{1}{M_W^2}\pi_f^W(0)] \tag{112}$$

where $\pi_f^Z(0)$ and $\pi_f^W(0)$ are the Z and W self energy contributions, due to the fermion(s) in question, at zero momentum transfer [12]. Obviously, these modifications change the ratio of these amplitudes from unity in lowest order ($\rho_0^{NC} = 1$) to:

$$\rho^{NC} = 1 + [\frac{\pi_f^Z(0)}{M_Z^2} - \frac{\pi_f^W(0)}{M_W^2}] \tag{113}$$

---

[11] More correctly, this behaviour concerns the top-bottom mass squared difference: $m_t^2 - m_b^2$. However, since $m_b \ll M_W$, I shall consistently set $m_b = 0$ in our discussion.

[12] More precisely, they are the coefficients of the $\eta^{\mu\nu}$ piece of the weak boson vacuum polarization tensors

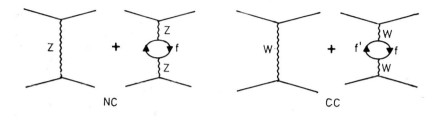

Figure 12: Fermion loop modifications to the NC and CC amplitudes

Let me consider first the modification to the Z propagator. What enters at each vertex in the fermion loop of Fig 12 is the expectation for this fermion of $2(J_3^\mu - sin^2\Theta_W J_{em}^\mu)$. However, since purely electromagnetic contributions do not give rise to terms proportional to fermion masses, one needs only concern oneself with the J_3^μ pieces [13]. These contributions are quite divergent individually and one must be a bit careful. In particular, it is important to consider **simultaneously** both the top and bottom corrections to the Z^0 propagator. Thus the $\pi_f^Z(0)$ contribution of Eq(113) which needs to be studied contains both the t and b quark loops.

Using only the expectation of $(2J_3^\mu)$ at each vertex the Z^0 self energy due to both t and b quark loops is given by

$$[\pi_f^Z(0)]^{\mu\nu} = -i(-1)3\int\frac{d^4p}{(2\pi)^4}\ \{\ Tr\frac{ie\gamma^\mu(1-\gamma_5)}{4\sin\Theta_W\cos\Theta_W}[\frac{i(\gamma p - m_t)}{p^2+m_t^2}]\frac{ie\gamma^\nu(1-\gamma_5)}{4\sin\Theta_W\cos\Theta_W}[\frac{i(\gamma p - m_t)}{p^2+m_t^2}]\} \\ +\ \{m_t = 0\} \quad (114)$$

In the above the $(-i)$ factor is the one appropriate to the definition of π_f^Z used in Eq(112), while the factors of (-1) and 3 are Pauli principle and color factors of the fermion loops in questions. A little algebra reduces Eq(114) to

$$[\pi_f^Z(0)]^{\mu\nu} = \frac{3i\alpha}{32\pi^3\sin^2\Theta_W\cos^2\Theta_W}\int d^4p[\frac{1}{(p^2+m_t^2)^2}+\frac{1}{(p^2)^2}]Tr\gamma^\mu\gamma p\gamma^\nu\gamma p(1-\gamma_5) \quad (115)$$

A similar calculation must be performed for the corrections to the W propagator. Here what enter are the $t-b$, $t-s$ and $t-d$ loops. However, if one neglects the masses of the

---

[13] The cross term between J_3^μ and J_{em}^μ also does not contribute.

bottom - like quarks, the unitarity of the mixing matrix V implies that the sum of these three contributions is equivalent to that of one loop with unit strength:

$$V_{ti}(V^\dagger)_{it} = 1 \tag{116}$$

Hence, effectively one has

$$\begin{aligned}[\pi_f^W(0)]^{\mu\nu} &= -i(-1)3\int \frac{d^4p}{(2\pi)^4}\{Tr\frac{ie\gamma^\mu(1-\gamma_5)}{2\sqrt{2}\sin\Theta_W}[\frac{i(\gamma p - m_t)}{p^2+m_t^2}]\frac{ie\gamma^\nu(1-\gamma_5)}{2\sqrt{2}\sin\Theta_W}[\frac{i\gamma p}{p^2}]\} \\ &= \frac{3i\alpha}{32\pi^3\sin^2\Theta_W}\int d^4p[\frac{2}{p^2(p^2+m_t^2)}]Tr\gamma^\mu\gamma p\gamma^\nu\gamma p(1-\gamma_5)\end{aligned} \tag{117}$$

Although Eqs(115) and (117) are individually divergent, the difference appearing in Eq(113) for ρ^{NC} is perfectly finite. One has

$$\begin{aligned}\frac{1}{M_Z^2}[\pi_f^Z(0)]^{\mu\nu} - \frac{1}{M_W^2}[\pi_f^W(0)]^{\mu\nu} &= \frac{3i\alpha}{32\pi^3 M_W^2 \sin^2\Theta_W}\int d^4p\{\frac{1}{(p^2+m_t^2)^2} + \frac{1}{(p^2)^2} - \frac{2}{p^2(p^2+m_t^2)}\} \\ &\quad Tr\gamma^\mu\gamma p\gamma^\nu\gamma p(1-\gamma_5) \\ &= \frac{3i\alpha}{32\pi^3 M_W^2\sin^2\Theta_W}\int d^4p\frac{Tr\gamma^\mu\gamma p\gamma^\nu\gamma p(1-\gamma_5)}{(p^2)^2(p^2+m_t^2)^2}\end{aligned} \tag{118}$$

The resulting integral is clearly convergent and a simple calculation secures

$$\int d^4p \frac{Tr\gamma^\mu\gamma p\gamma^\nu\gamma p(1-\gamma_5)}{(p^2)^2(p^2+m_t^2)^2} = \frac{2\pi^2}{im_t^2}\eta^{\mu\nu} \tag{119}$$

Hence one finds [20]

$$[\rho^{NC}]_{top} = 1 + \frac{3\alpha}{16\pi\sin^2\Theta_W}(\frac{m_t}{M_W})^2 \tag{120}$$

Eq(120) shows that the ρ parameter departs strongly from unity if the t-quark is very heavy - or more precisely if the mass squared difference between t and b quarks is much greater than M_W^2.

Quite similar considerations give the t-dependence of Δr and κ [21]. To trace the sensitivity of these corrections to m_t, let me look again at the radiative corrected formula, Eq(61), for M_W^2:

$$M_W^2 = \frac{\pi\alpha}{\sqrt{2}G_F\sin^2\Theta_W}[\frac{1}{1-\Delta r}] \tag{121}$$

This formula can be obtained from the "bare" expression

$$(M_W^0)^2 = \frac{\pi\alpha^0}{\sqrt{2}G_F\sin^2\theta_W^0} \tag{122}$$

by shifting all the bare values by a certain amount:

$$M_W^2 = (M_W^0)^2 + \delta M_W^2; \quad G_F = G_F^0 + \delta G_F; \quad etc \tag{123}$$

Hence the correction Δr can be written as

$$\Delta r = -\frac{\delta\alpha}{\alpha} + \frac{\delta G_F}{G_F} + \frac{\delta\sin^2\Theta_W}{\sin^2\Theta_W} + \frac{\delta M_W^2}{M_W^2} \tag{124}$$

On the other hand, by the definition of $\sin^2 \Theta_W$, one has also that

$$M_Z^2 = \frac{\pi \alpha}{\sqrt{2} G_F \sin^2 \Theta_W \cos^2 \Theta_W} [\frac{1}{1 - \Delta r}] \qquad (125)$$

Thus one can compute an alternative expression for Δr:

$$\Delta r = -\frac{\delta \alpha}{\alpha} + \frac{\delta G_F}{G_F} + \frac{\cos^2 \Theta_W - \sin^2 \Theta_W}{\cos^2 \Theta_W \sin^2 \Theta_W} \delta \sin^2 \Theta_W + \frac{\delta M_Z^2}{M_Z^2} \qquad (126)$$

Comparison of Eqs(124) and (126) yields a connection between the shift needed for $\sin^2 \Theta_W$ and the shifts associated with the W and Z masses:

$$\delta \sin^2 \Theta_W = \cos^2 \Theta_W [\frac{\delta M_Z^2}{M_Z^2} - \frac{\delta M_W^2}{M_W^2}] \qquad (127)$$

We have already computed the shift of the Fermi constant, defined through the CC process, due to the presence of a heavy t-quark loops [c.f. Fig 12]

$$\frac{\delta G_F}{G_F} = \frac{\pi_f^W(0)}{M_W^2} \qquad (128)$$

The mass shifts δM_W^2 and δM_Z^2, on the other hand, are related to the relevant self energy functions on their mass shells

$$\delta M_W^2 = -\pi^W(M_W^2); \qquad \delta M_Z^2 = -\pi^Z(M_Z^2) \qquad (129)$$

However, in the limit when $m_t >> M_W^2$, the self energy contribution due to heavy t-quarks on shell, is approximately equal to that evaluated at zero momentum transfer. Thus for the shifts due to heavy t-quarks one has also, approximately

$$\delta M_W^2 \simeq -\pi_f^W(0); \qquad \delta M_Z^2 \simeq -\pi_f^Z(0) \qquad (130)$$

Using Eqs(128) and (130) the large m_t dependence of Δr simplifies to

$$[\Delta r]_{top} \simeq -\frac{\delta \alpha}{\alpha} + \frac{\delta \sin^2 \Theta_W}{\sin^2 \Theta_W} \qquad (131)$$

However, as I remarked earlier, purely electromagnetic corrections never give rise to quadratic mass shifts. Thus using Eq(127) and Eq(130) one has [21]

$$[\Delta r]_{top} \simeq \frac{\delta \sin^2 \Theta_W}{\sin^2 \Theta_W} \simeq -\frac{\cos^2 \Theta_W}{\sin^2 \Theta_W} [\frac{\pi_f^Z(0)}{M_Z^2} - \frac{\pi_f^W(0)}{M_W^2}] \simeq -\frac{3\alpha \cos^2 \Theta_W}{16\pi \sin^4 \Theta_W} (\frac{m_t}{M_W})^2, \qquad (132)$$

where to obtain the last line, I have used the result previously obtained for ρ^{NC}, Eq(120). Furthermore, recalling the definition of κ, Eq(107), one sees that for κ the large m_t contribution produces a shift opposite to the shift in Δr [21]

$$[\kappa]_{top} = -\frac{\delta \sin^2 \theta_W}{\sin^2 \Theta_W} = \frac{3\alpha \cos^2 \Theta_W}{16\pi \sin^4 \Theta_W} (\frac{m_t}{M_W})^2 \qquad (133)$$

We have now all the relevant formulas to explore the effects of having a large t-quark mass in the standard model. Recall that the change in $\sin^2\Theta_W$ for deep inelastic scattering, δs^2, was given by the approximate formula Eq(109). Since $R_\nu \sim [\rho^{NC}/\rho^{CC}]^2$ it follows therefore also that

$$\delta s^2 \simeq -\sin^2\theta_W[\kappa - 1] + 0.48[(\rho^{NC})^2 - 1] \tag{134}$$

Using our results the large m_t shift is given by

$$[\delta s^2]_{top} \simeq \frac{3\alpha}{16\pi \sin^2\Theta_W}\{-\cos^2\theta_W + 2(0.48)^2\}(\frac{m_t}{M_W})^2 \simeq 4.4 \times 10^{-4}(\frac{m_t}{M_W})^2 \tag{135}$$

Note that the net effect is not particularly big, because of the large cancellation among the two terms in the curly bracket above. Recall that for $m_t = 45 GeV$, $\delta s^2 = -0.009$ [9]. Using Eq(135), this shift is partially cancelled for $m_t = 200 GeV$, where one would obtain only $\delta s^2 = -0.006$.

The effects of a large top mass are more severe for Δr. Using Eq(132) one has, numerically,

$$[\Delta r]_{top} \simeq -6.3 \times 10^{-3}(\frac{m_t}{M_W})^2 \tag{136}$$

Hence the value for Δr, obtained with the canonical value of $m_t = 45 GeV$, $\Delta r = 0.071$, [9] is more than halved at $m_t = 200 GeV$ where, using Eq(136), one obtains only $\Delta r = 0.031$. The good agreement between $(\sin^2\Theta_W)_{DIS}$ and $(\sin^2\Theta_W)_{M_W/M_Z}$ of Table 4 is spoiled for large m_t, since the former goes up and the latter goes down as m_t grows. This is illustrated in Table 5

Table 5: Effects of m_t on $\sin^2\Theta_W$

	$m_t = 45 GeV$	$m_t = 200 GeV$
Deep Inelastic	$0.233 \pm 0.003 \pm [0.005]$	$0.236 \pm 0.003 \pm [0.005]$
W/Z Masses	$0.228 \pm 0.007 \pm [0.002]$	$0.219 \pm 0.007 + [0.002]$

This phenomena has been analyzed in detail by Amaldi et al [9], who have studied the variations of $\sin^2\Theta_W$ as a function of m_t for a variety of neutral current processes besides those discussed here [14]. They obtain in this way a 90 % C.L. region for $\sin^2\Theta_W$ and m_t, as shown in Fig 13. The actual precise limits on m_t depent slightly on the choice one takes for the Higgs mass, but one sees that, roughly speaking, values of m_t above $200 GeV$ are not allowed.

The weak dependence of the bound on m_t on the Higgs mass follows since, as I have previously noted [c.f. Eq(111)], the corrections δs^2 and Δr only depend logarithmically on this parameter. As a last point in this already long section, I will demonstrate this explicitly for Δr. Similar considerations apply for δs^2. Using Eqs(127) - (129) one can rewrite Δr entirely in terms of differences of W and Z self energies, plus the purely electromagnetic contribution $-\frac{\delta\alpha}{\alpha}$. However, this last term gives no M_H dependence and I will drop it entirely. Hence, effectively, one has

$$(\Delta r)_{Higgs} = -\left\{[\frac{\pi^W(M_W^2)}{M_W^2} - \frac{\pi^W(0)}{M_W^2}] + \frac{\cos^2\Theta_W}{\sin^2\Theta_W}[\frac{\pi^Z(M_Z^2)}{M_Z^2} - \frac{\pi^W(M_W^2)}{M_W^2}]\right\} \tag{137}$$

Because the formula for $(\Delta r)_{Higgs}$ involves differences in self energies, there are some simplifications. In particular, the seagull self energy graphs of Fig 14 do not contribute

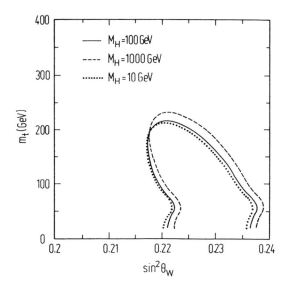

Figure 13: Allowed 90 % C.L. region for $\sin^2 \Theta_W$, for Higgs boson masses of 10, 100 and 1000 GeV. From [9].

Figure 14: Seagull self energy graphs

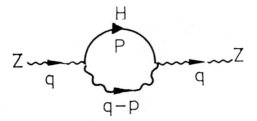

Figure 15: Higgs contribution to the Z self energy

at all. This is certainly clear for the first difference in Eq(137), since these contributions are q^2 independent. The second difference in Eq(137) also vanishes because the Z and W terms, divided by their respective masses, cancel. As is the case with the gauge mass terms [c.f. Eq(13)], the $WWHH$ and $ZZHH$ vertices just differ by a factor of $\cos^2\Theta_W$. This factor is compensated by M_Z^2 in the difference appearing in Eq(137).

Hence one needs only consider the $H - W/Z$ loop contributions to the self energies, arising out of the trilinear interactions detailed in Eq(17). The answer for Δr is gauge invariant and so can be evaluated in any gauge. I will do the computation in the unitary gauge, which is the physical gauge where the W and Z propagators contain a projection $(\eta^{\mu\nu} + q^\mu q^\nu/M^2)$, with M^2 being the appropriate mass. The integrals which enter in the computation diverge and must be regularized. It is convenient to use dimensional regularization, where the divergent contributions are segregated in terms proportional to $(n-4)^{-1}$, with n being the continued dimension. When one adds up all the contributions in Δr, of course, these terms cancel. In our limited calculation, we shall show that there are no singular terms depending on M_H and shall extract the finite contributions proportional to $\ln M_H$.

The contribution to the Z self energy due to the graph shown in Fig 15 is given by, in n-dimensions,

$$[\pi^Z(-q^2)]_{\mu\nu} = -i \int \frac{d^n p}{(2\pi)^4} [\frac{ieM_Z}{\sin\Theta_W \cos\Theta_W}]^2 \frac{-i}{p^2 + M_H^2} \frac{-i\{\eta_{\mu\nu} + \frac{(q-p)_\mu(q-p)_\nu}{M_Z^2}\}}{(q-p)^2 + M_Z^2} \qquad (138)$$

What is needed is the $\eta_{\mu\nu}$ coefficient in the above. To proceed we Feynman parametrize the denominator in Eq(138) and shift to a new integration variable $p' = p - qx$. Then the denominator becomes

$$Den = \int_0^1 dx [p'^2 + D_Z]^{-2} \qquad (139)$$

---

[14] The processes I have discussed, however, are the most significant, since they are the most precise

where
$$D_Z = M_Z^2 x + M_H^2(1-x) + q^2 x(1-x) \tag{140}$$

In the numerator, since we need only the $\eta_{\mu\nu}$ coefficient, effectively one can replace

$$[q(1-x) - p']_\mu [q(1-x) - p']_\nu \to p'_\mu p'_\nu \equiv \frac{p'^2}{n}\eta_{\mu\nu} \tag{141}$$

Thus, going to Euclidean space $[-i \int d^n p' \to \int d^n p'_E]$ and doing the trivial angular integral $[\Omega_n = \frac{2\pi^{n/2}}{\Gamma(n/2)}]$ yields

$$\pi^Z(-q^2) = \frac{\alpha M_Z^2}{4\pi^3 \sin^2\Theta_W \cos^2\Theta_W} \left\{ \Omega_n \int_0^1 dx \int_0^\infty dp\, p^{n-1} \frac{1 + \frac{p^2}{nM_Z^2}}{[p^2 + D_Z]^2} \right\} \tag{142}$$

Using
$$\int_0^\infty dp \frac{p^{2\beta+n-1}}{[p^2 + D]^\alpha} = \frac{\Gamma(\beta + \frac{n}{2})\Gamma(\alpha - \beta - \frac{n}{2})}{2\Gamma(\alpha)D^{\alpha-\beta-\frac{n}{2}}} \tag{143}$$

the curly bracket in Eq(142) reduces to

$$\{\}^Z = \frac{\pi^{n/2}\Gamma(2 - \frac{n}{2})}{D_Z^{2-\frac{n}{2}}} \int_0^1 dx [1 - \frac{D_Z}{M_Z^2(n-2)}] \tag{144}$$

The singularity is isolated in $\Gamma(2 - \frac{n}{2})$. Using

$$\frac{\Gamma(2 - \frac{n}{2})}{D_Z^{2-\frac{n}{2}}} = -\frac{2}{(n-4)} - \ln D_Z \tag{145}$$

one has, as $n \to 4$:

$$\{\}^Z = -\frac{2\pi^2}{(n-4)} \int_0^1 dx[1 - \frac{D_Z}{2M_Z^2}] - \pi^2 \int_0^1 dx[1 - \frac{D_Z}{2M_Z^2}]\ln D_Z \tag{146}$$

Since D_Z contains M_H there is a potentially, Higgs mass dependent, divergent term. However, it is easy to see that in Δr these divergent terms cancel. An analogous calculation for the Higgs correction to the W propagator secures a formula akin to Eq(142)

$$\pi^W(-q^2) = \frac{\alpha M_W^2}{4\pi^3 \sin^2\Theta_W}\{\}^W \tag{147}$$

where $\{\}^W$ is identical to $\{\}^Z$, except that $M_W \leftrightarrow M_Z$. Thus the singular contributions for π^W and π^Z depending on M_H are

$$\begin{aligned}\pi^W(-q^2)|_{M_H}^{sing} &= \frac{\alpha M_H^2}{8\pi \sin^2\Theta_W} \frac{1}{(n-4)} \\ \pi^Z(-q^2)|_{M_H}^{sing} &= \frac{\alpha M_H^2}{8\pi \sin^2\Theta_W \cos^2\Theta_W} \frac{1}{(n-4)}\end{aligned} \tag{148}$$

Obviously, given the mass weighting in Eq(137), the Higgs mass dependent singular terms in Δr cancel.

It remains to detail the finite M_H dependence in Δr. This can be extracted from the formulas just obtained for π^W and π^Z, using Eq(137). Collecting terms, one finds

$$[\pi^W(-q^2)]_{Higgs} = -\frac{\alpha M_W^2}{4\pi \sin^2 \Theta_W} \int_0^1 dx \left[1 - \frac{M_W^2 x + M_H^2(1-x) + q^2 x(1-x)}{M_W^2}\right]$$
$$\cdot \ln[M_W^2 x + M_H^2(1-x) + q^2 x(1-x)] \quad (149)$$

$$[\pi^Z(-q^2)]_{Higgs} = -\frac{\alpha M_W^2}{4\pi \sin^2 \Theta_W \cos^2 \Theta_W} \int_0^1 dx \left[1 - \frac{M_Z^2 x + M_H^2(1-x) + q^2 x(1-x)}{M_Z^2}\right]$$
$$\cdot \ln[M_Z^2 x + M_H^2(1-x) + q^2 x(1-x)] \quad (150)$$

Using the above, the Higgs mass dependence of Δr is readily computable. One finds, in particular, for large M_H [21]

$$(\Delta r)_{Higgs} = \frac{11\alpha}{48\pi \sin^2 \Theta_W} \ln \frac{M_H^2}{M_Z^2} \simeq 0.0023 \ln \frac{M_H^2}{M_Z^2} \quad (151)$$

From Eq(151) one sees that a $1TeV$ Higgs would increase, approximately, Δr by 0.01. Such a shift is clearly well below the present experimental accuracy. Using the value of $\sin^2 \Theta_W$ obtained in deep inelastic scattering, Eq(56), and the present experimental value for M_W, given in Table 3, Eq(61) implies for Δr:

$$(\Delta r)_{exp} = 0.077 \pm 0.037 \quad (152)$$

This value is in perfect agreement with the theoretical prediction given in Eq(62), which uses $m_t = 45 GeV$ and $M_H = 100 GeV$. However, it is too inaccurate to be able to test Eq(151), even if one were to know m_t precisely. Only when the new e^+e^- colliders at the Z energy, LEP and SLC, come into operation will one hope to achieve the accuracy needed ($\delta \Delta r < 0.01$) to be able to test for virtual Higgs boson effects [22].

3 Counting Flavors

I want to begin discussing a different aspect of the standard model, connected with the observed family repetition of the fundamental fermions. I shall start by looking at issues connected with the top quark.

3.1 Why there is Top

Demanding agreement between the standard model predictions and experiment is a sophisticated way to infer that top must exist. Indeed, as we saw in the last section, $m_t < 200 GeV$ was needed to make sure that the radiatively corrected values of $\sin^2 \Theta_W$ measured in different experiments agree, within errors, with each other. Langacker [23] has pointed out a much more direct and simpler way by which one can tell from experiment that top exists, even though top has not been directly observed! His observation makes use of the measurement of the axial charge of b-quarks in e^+e^- annihilation experiments.

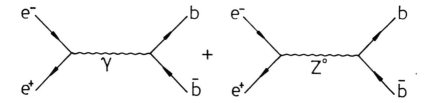

Figure 16: Feynman graph contributions to $e^+e^- \to b\bar{b}$

In the standard model, where the left handed components of b and t quarks sit together in a doublet, the neutral current of b-quarks has an axial piece:

$$[J^\mu_{NC}]_b = 2[J^\mu_3 - \sin^2\theta_W J^\mu_{em}]_b = \bar{b}(-\frac{1}{2})\gamma^\mu\gamma_5 b + ... \quad (153)$$

Conventionally, the axial charge a_b of the b-quark is defined as twice the coefficient of $\gamma^\mu\gamma_5$ in the neutral current. Thus in the standard model

$$a_b = 2[-\frac{1}{2}] = -1 \quad (154)$$

Obviously, if the b-quark had a different assignment in the standard model, the value of a_b would change. In particular, if the top quark did not exist and b-quarks were $SU(2)$ singlets then $a_b = 0$.

The axial charge a_b has been measured experimentally by studying the forward-backward asymmetry in the process $e^+e^- \to b\bar{b}$. Events originating from b-quarks are identified from the high transverse momentum leptons produced in semileptonic decays $b \to c l \nu_l$. Experimentally, a large p_T cut serves to discriminate between the b and the c semileptonic decay sample. In the process $e^+e^- \to b\bar{b}$, the existence of a Z exchange contribution, besides the usual photon annihilation term, (see Fig 16) gives an asymmetry in the angular distribution of the produced b quarks. Let Θ be the angle in the CM system between the direction of the incoming electron and the outgoing b-quark, as shown in Fig 17. Then the differential angular distribution has the form

$$\frac{d\sigma}{d\cos\Theta} = \frac{\pi\alpha^2}{2s}[A(1+\cos^2\Theta) + B\cos\Theta] \quad (155)$$

It is the presence of the B coefficient, due to the Z graph, which causes a forward-backward asymmetry

$$A_{FB}(s) = \frac{\int_0^1 d\cos\Theta \left(\frac{d\sigma}{d\cos\Theta}\right) - \int_{-1}^0 d\cos\Theta \left(\frac{d\sigma}{d\cos\Theta}\right)}{\int_0^1 d\cos\Theta \left(\frac{d\sigma}{d\cos\Theta}\right) + \int_{-1}^0 d\cos\Theta \left(\frac{d\sigma}{d\cos\Theta}\right)} = \frac{3B}{8A} \quad (156)$$

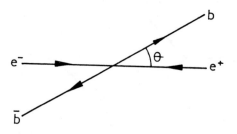

Figure 17: Definition of the angle Θ for $e^+e^- \to b\bar{b}$

In the PEP/PETRA energy range, where this asymmetry has been measured, since $s \ll M_Z^2$, B is determined by the $Z - \gamma$ interference term, as the $|Z|^2$ term is negligible. A straightforward calculation of the graphs in Fig 16 gives

$$A_{FB}^b(s) \simeq \frac{9G_F}{16\pi\sqrt{2}\alpha}[\frac{sM_W^2}{M_Z^2 - s}]a_b \simeq 0.29a_b \qquad (157)$$

The numerical value above is that appropriate for $\sqrt{s} = 35 GeV$, which is the energy where most of the data on this asymmetry has been gathered. Clearly, at this energy, the asymmetry expected in the standard model is sizable.

In Fig 18, I show the data of the JADE collaboration [24], which are the most accurate. The measured distribution shows a substantial asymmetry and the data is backward peaked, implying $a_b < 0$. The result of JADE for a_b ($a_b = -0.90 \pm 0.24 \pm 0.10$), combined with those of the other PEP and PETRA experiments [25] yields an average experimental value for the axial charge of the b-quark

$$a_b = -0.84 \pm 0.21 \qquad (158)$$

Actually, this raw number needs to be corrected for $B - \bar{B}$ mixing - a subject which I'll discuss, in detail in Sec. 5. When this is done, the final central value for a_b is slightly higher, but the error has also increased. One finds [25]

$$a_b = -1.08 \pm 0.29 \qquad (159)$$

Clearly Eq(159) is in perfect agreement with the expectations of the standard model, in which the left handed b quark is part of an $SU(2)$ doublet. So top must exist, but where is it?

3.2 Where is Top?

The safest lower bound for top is provided by the study of the process $e^+e^- \to hadrons$ at high energy at TRISTAN. The ratio of the cross section for this process to that for

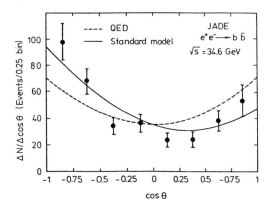

Figure 18: The angular distribution for the process $e^+e^- \to b\bar{b}$, measured by JADE [24].

producing μ-pairs is proportional to the charged squared of the quarks being produced:

$$R = \frac{\sigma(e^+e^- \to hadrons)}{\sigma(e^+e^- \to \mu^+\mu^-)} \simeq 3\sum_q e_q^2 \qquad (160)$$

where the factor of 3 above is a color factor. Thus one expects a step in R of $(\Delta R)_{top} = \frac{4}{3}$, after one goes above the threshold in energy for producing $t - \bar{t}$ pairs. Although R rises due to the tail of the Z pole, no sign of this threshold jump can be seen at the highest TRISTAN energies [26], as Fig 11 demonstrates. This allows the various collaborations working at TRISTAN to set a lower bound on m_t of about 28 GeV. More precisely, the 95% C.L. TRISTAN bounds are [27]:

$$m_t \geq \begin{cases} 27.4 \; GeV & TOPAZ \\ 28.0 \; GeV & VENUS \\ 27.6 \; GeV & AMY \end{cases} \qquad (161)$$

The strongest experimental bound on top comes from the UA1 collaboration, working at the Sp$\bar{p}$S collider [28]. In $p\bar{p}$ collisions top can be produced either as a byproduct of W decay, if its mass is below the $W \to t\bar{b}$ threshold, or in association with a $\bar{t}$, by gluon-gluon fusion, as shown in Fig. 20. The first process has a well determined rate, which can be inferred from the experimentally measured W production and the known $W \to t\bar{b}$ branching fraction in the standard model. The rate for the second process, however, is more uncertain, since it depends on the knowledge of the gluon structure functions and of higher order QCD corrections. The signal for both these sources of top is an isolated muon with large transverse momentum, from the semileptonic decay of top, plus jets resulting from

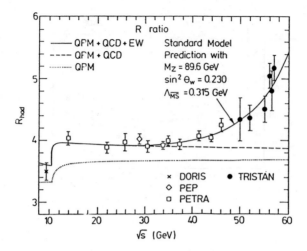

Figure 19: R values measured by the VENUS, AMY and TOPAZ detectors at TRISTAN, from[26]. Data from DESY and SLAC is also plotted. The solid curve is the prediction of the standard model, in which the tail of the Z propagator is beginning to be noticeable.

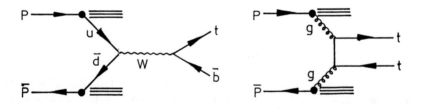

Figure 20: Mechanisms for producing top in $p\bar{p}$ collisions: (a) through W decay; (b) by gluon-gluon fusion.

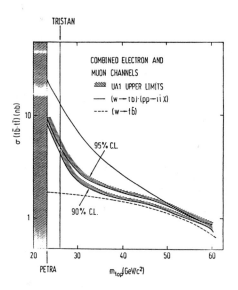

Figure 21: Expected top signal for the UA1 cuts [28], as a function of m_t. The dashed line is the signal from $W \to t\bar{b}$.

the accompanying debris. This signal, although characteristic, is not without background and a careful Monte Carlo study is needed to ascertain the presence of top.

Fig. 21, taken from [28], shows the expected signal, as a function of m_t, for the cuts imposed by the UA1 collaboration. As can be seen, for $m_t \leq 50$ GeV the main source of the signal is the gluon-gluon fusion process. The signal in Fig. 21 lies above the calculated background for $m_t \geq 56$ GeV (at 90% confidence) and this allows the setting of a lower bound on m_t.

Because the predicted rate is dominated by the (somewhat) theoretically uncertain QCD process, this bound is not the most conservative one can set. UA1 sets a more cautious lower bound of $m_t \geq 44$ GeV [28], by varying the QCD inputs for the theoretically predicted top cross section and selecting the lowest value among these. A similar result has also been obtained recently by Altarelli et al. [29], who incorporated the effects of higher order QCD corrections for heavy quark production [30] and obtained $m_t \geq 41$ GeV. In summary, from the nonobservation of a clear signal in high energy $p\bar{p}$ scattering one can deduce - subject to the above mentioned theoretical ambiguities - an experimental lower bound for the top mass, in the range

$$m_t \geq (41 - 56) \; GeV \qquad (162)$$

The measurement of a_b gives us the reasonable certitude that top exists and so that there are three generations of quarks and leptons. The above lower bound on m_t, along with the upper bound provided by the electroweak radiative corrections, tell us that there is a range of about $150 GeV$ open for top-hunting: $50 GeV \leq m_t \leq 200 GeV$. The interesting and still open question, however, is if there are more generations? Direct evidence against more generations is rather scant, with the best bounds again being those of UA1 [28] of $m'_b \geq 32\ GeV$ (90% C.L.) and $m'_l \geq 41\ GeV$ (90% C.L.) for a fourth generation charge $-\frac{1}{3}$ quark and a fourth generation lepton, respectively. On the other hand, rather good limits are beginning to emerge from e^+e^- experiments and from the CERN collider on the number of light neutrinos. These limits, which are becoming of comparable quality to the cosmological bound from nucleosynthesis [31] ($N_\nu \leq 4$), can be taken to be limits on the number of families, under the reasonable assumption that all neutrinos species are either massless, or at least very much lighter than the corresponding charged fermions. I discuss these interesting limits below.

3.3 Neutrino Counting Experiments

The idea of neutrino counting experiments is very simple. In the standard model the coupling of the Z boson to each neutrino species is universal [c.f. Eq(26)]

$$\mathcal{L}_{Z\nu\bar{\nu}} = \frac{e}{4\sin\Theta_W \cos\Theta_W} Z^\mu \sum_i \bar{\nu}_i \gamma_\mu (1-\gamma_5)\nu_i \tag{163}$$

Thus if the neutrinos are light, $m_{\nu_i} \simeq 0$, all types of neutrinos are able to be produced via their coupling to the Z. Two experimental particle physics bounds exist on the number of neutrino species. One of these, the PEP/PETRA limit, arising from the study of the process $e^+e^- \to \gamma\ Nothing$, makes use of the coupling of a virtual Z to neutrinos. The other, coming from the CERN collider, measures indirectly the total Z width, and hence can bound in this way the number of neutrinos. I want to describe briefly both of these bounds, since they also illustrate nicely various general features of the standard model.

3.3.1 $e^+e^- \to \gamma\ Nothing$

At PEP and PETRA the production of photons unaccompanied by other charged or neutral interacting particles - the process $e^+e^- \to \gamma\ Nothing$ - has been studied. This reaction in the standard model can occur by pair producing neutrinos, along with some bremsstrahlung photon: $e^+e^- \to \nu_i\bar{\nu}_i\gamma$. Clearly, the more neutrino species there are, the larger the contribution one expects from this process to the total cross section for $e^+e^- \to \gamma\ Nothing$. Experimentally, however, the situation is complicated, since the signal can also be faked by radiative e^+e^- production, in which the final charged particles never come out of the beam pipe. To minimize this background one must require that the produced photon comes out at a rather large angle, with respect to the incident e^+e^- direction. This cut, however, also reduces the expected signal and, typically, one is left with just a handful of events.

The two graphs that contribute to the process $e^+e^- \to \nu_i\bar{\nu}_i\gamma$ are shown in Fig 22. Since the coupling of the Z is universal, it is obvious that the total rate will be the same for each species. There is, however, a complication for the case in which ν_i is an electron neutrino ($\nu_i = \nu_e$). For the process $e^+e^- \to \nu_e\bar{\nu}_e\gamma$, in addition to the s-channel NC contribution, there is also a t-channel CC contribution, as Fig 23 shows. For the energies of PEP and PETRA, where $\sqrt{s} << M_Z$, one can compute all processes in the Fermi approximation, where all gauge boson propagators are just replaced by constants. This has two consequences:

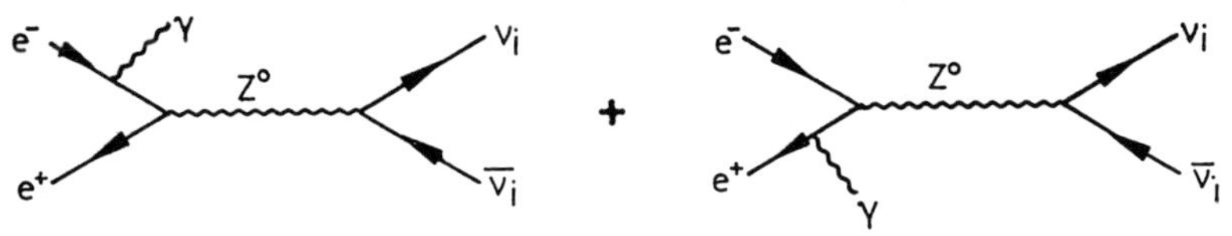

Figure 22: Graphs contributing to the process $e^+e^- \to \nu_i\bar{\nu}_i\gamma$, with $\nu_i \neq \nu_e$.

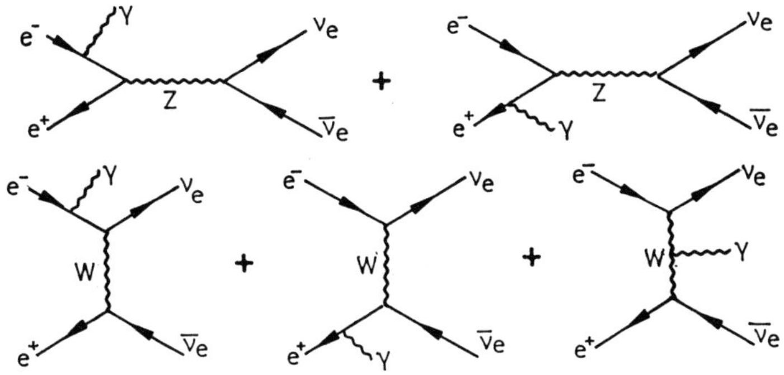

Figure 23: Graphs contributing to the process $e^+e^- \to \nu_e\bar{\nu}_e\gamma$.

1. The internal photon emission graph for the process $e^+e^- \to \nu_e \bar{\nu}_e \gamma$ is totally negligible, since it involves two weak propagators, and can be omitted.

2. By means of a Fierz transformation one can recast the ν_e calculation in a similar form as that appropriate for arbitrary neutrinos.

Let me expand a bit on the last point. In the Fermi approximation, the effective neutral current Lagrangian describing $e^+e^- \to \nu_i \bar{\nu}_i$ is given by [c.f. Eq(31)] [15]

$$\mathcal{L}_{NC}^{eff} = \frac{G_F}{\sqrt{2}} [\bar{\nu}_i \gamma^\mu (1-\gamma_5) \nu_i][\bar{e}(Q_L^e \gamma_\mu (1-\gamma_5) + Q_R^e \gamma_\mu (1+\gamma_5))e] \tag{164}$$

where the chiral charges $Q_{L,R}^e$ are

$$Q_L^e = -\frac{1}{2} + \sin^2 \Theta_W; \quad Q_R^e = \sin^2 \Theta_W \tag{165}$$

The charged current Lagrangian, corresponding to the process $e^+e^- \to \nu_e \bar{\nu}_e$ on the other hand, is given by

$$\mathcal{L}_{CC}^{eff} = \frac{G_F}{\sqrt{2}} [\bar{\nu}_e \gamma^\mu (1-\gamma_5) e][\bar{e} \gamma_\mu (1-\gamma_5) \nu_e] \tag{166}$$

This Lagrangian, however, can be rewritten as that appearing in Eq(164), after a Fierz transformation:

$$[\mathcal{L}_{CC}^{eff}]_{Fierz} = \frac{G_F}{\sqrt{2}} [\bar{\nu}_e \gamma^\mu (1-\gamma_5) \nu_e][\bar{e} \gamma_\mu (1-\gamma_5) e] \tag{167}$$

Clearly, therefore, the result of the computation for producing electron neutrinos is just like that for producing any other neutrino, but with the replacement

$$Q_L^e \to (Q_L^e)_{eff} = Q_L^e + 1; \quad Q_R^e \to (Q_R^e)_{eff} = Q_R^e \tag{168}$$

The calculation of the process $e^+e^- \to \nu\bar{\nu}\gamma$ was done originally by Ma and Okada [32] and it was repeated soon after by Gaemers, Gastmans and Renard [33], who corrected some errors in the first calculation. If one denotes by x the ratio of the photon energy to the incident electron energy in the CM system: $x = \frac{2E_\gamma}{\sqrt{s}}$ and by y the cosine of the angle that the photon makes with respect to the e^- direction: $y = \cos \Theta_\gamma$ (see Fig 24), one finds [33]

$$\frac{d\sigma}{dxdy} = \frac{\alpha G_F^2 s(1-x)}{3\pi^2 x(1-y)^2} [(1-\frac{1}{2}x)^2 + \frac{1}{4}x^2 y^2] F \tag{169}$$

One sees that this cross section has the typical x^{-1} behaviour of a bremsstrahlung process. The factor F, in the Fermi approximation, is simply given by

$$\begin{aligned} F &= (N_\nu - 1)[(Q_L^e)^2 + (Q_R^e)^2] + [(Q_L^e)_{eff}^2 + (Q_R^e)_{eff}^2] \\ &= N_\nu [\frac{1}{4} - \sin^2 \Theta_W + 2\sin^4 \Theta_W] + 2\sin^2 \Theta_W \simeq 0.126 N_\nu + 0.46 \end{aligned} \tag{170}$$

That is, all neutrinos contribute equally, except for the electron neutrino which has just some different effective chiral charges. Unfortunately, as can be seen from Eq(170), the

---
[15] I have taken $\rho = 1$ here, as that is what experiment indicates.

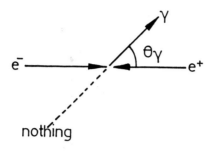

Figure 24: Definition of the angle Θ_γ

part in F not proportional to N_ν is rather large. Thus the difference between having three ($F \simeq 0.84$) or four ($F \simeq 0.96$) families is only about 15%.

Because of the cuts on y imposed to get rid of background at PEP and PETRA [16], very few events of the type $e^+e^- \to \gamma$ $Nothing$ are expected and very few are observed. Nevertheless, one has been able to set rather good limits on the number of neutrino species. The most sensitive results come from ASP at PEP [34] and CELLO at PETRA [35], giving 90 % C.L. of $N_\nu \leq 7.5$ and $N_\nu \leq 8.7$, respectively. An analysis of all e^+e^- data, done by the CELLO collaboration [35], gives a stronger combined limit

$$N_\nu < 4.6 \quad (90\% \ C.L.) \tag{171}$$

3.3.2 Neutrino Counting in Collider Experiments

A bound on the number of neutrino species has also been obtained at the $Sp\bar{p}S$ collider through the measurement of the ratio of the production of W bosons, decaying into an electron and a neutrino $[W \to e\nu_e]$, to that of Z bosons, decaying into e^+e^- pairs $[Z \to e^+e^-]$:

$$R = \frac{\sigma_W B(W \to e\nu_e)}{\sigma_Z B(Z \to e^+e^-)} \tag{172}$$

This ratio has been measured by both the UA1 and UA2 collaborations with the result

$$R = \begin{cases} 9.1 \ ^{+1.7}_{-1.2} & UA1[36] \\ 7.2 \ ^{+1.7}_{-1.2} & UA2[37] \end{cases} \tag{173}$$

---

[16]Typically $\Theta_{gamma} > 20°$.

This leads to an average value [38] of R:

$$<R> = 8.4\,^{+1.2}_{-0.9} \qquad (174)$$

and a 90 % confidence limit

$$R < 10.1 \quad (90\% C.L.) \qquad (175)$$

The ratio R depends on the number of neutrinos N_ν and on the top mass, m_t. This is easily seen by rewriting Eq(172) as a function of the total Z and W widths:

$$R = [\frac{\sigma_W}{\sigma_Z}][\frac{\Gamma(W \to e\nu_e)}{\Gamma(Z \to e^+e^-)}][\frac{\Gamma^Z_{tot}}{\Gamma^W_{tot}}] \qquad (176)$$

The first two factors in square brackets above are fixed by our present theoretical knowledge. The ratio of the W and Z production cross sections is calculable in QCD, with reasonably good accuracy, while the ratio of the partial widths of $W \to e\nu_e$ to $Z \to e^+e^-$ is fixed by the $SU(2) \times U(1)$ model. The total width ratio, however, is affected by the unknown parameters in the standard model. Since Γ^Z_{tot} goes up the more neutrino species there are, it is clear that the bound on R of Eq(175) also gives a bound on N_ν. However, since the total widths depend on whether the decays $Z \to t\bar{t}$ and $W \to t\bar{b}$ are possible - and on how big their respective contributions are- the bound on N_ν depends also on the value of the t-quark mass.

To make the above remarks more quantitative, one needs to calculate the three quantitives in square brackets in Eq(176). The widths $\Gamma(W \to e\nu_e)$ and $\Gamma(Z^0 \to e^+e^-)$ are straightforward to compute from the interaction Lagrangian of Eq(26). Neglecting the electron mass, a simple calculation gives

$$\Gamma(W \to e\nu_e) = \frac{\sqrt{2}G_F M_W^3}{12\pi} \qquad (177)$$

$$\Gamma(Z^0 \to e^+e^-) = \frac{\sqrt{2}G_F M_Z^3}{6\pi}[(Q_L^e)^2 + (Q_R^e)^2] \qquad (178)$$

In the above, I have replaced the factor of $\frac{\alpha}{\sin^2\Theta_W}$ which comes from a direct calculation using the Lagrangian of Eq(26), by $\frac{\sqrt{2}G_F M_W^2}{\pi}$. This replacement, effectively, takes care of the leading logarithmic radiative effects because, as can be seen in Eqs(177) and (178), both sides contain only quantities at large scales, or quantities like G_F which do not run. Using Eqs(177) and (178) one has

$$\frac{\Gamma(W \to e\nu_e)}{\Gamma(Z \to e^+e^-)} = \frac{2M_W^3}{M_Z^3[1 - 4\sin^2\Theta_W + 8\sin^4\theta_W]} = 2.68 \qquad (179)$$

The numerical value above uses a set of values for the parameters in the standard model which correspond to the central values presently measured, consistent with Eq(53): $\sin^2\Theta_W = 0.23$; $M_Z = 92\ GeV$; $M_W = 80.7\ GeV$.

To calculate the total Z width, in the standard model, one just needs to sum over the individual fermionic channels:

$$\Gamma^Z_{tot} = \sum_f \Gamma(Z \to f\bar{f}) \qquad (180)$$

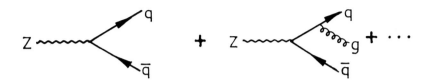

Figure 25: QCD corrected decays of the Z boson, including gluon emission.

From the UA1 bound on m_t, Eq(162), presumably the decay $Z \to t\bar{t}$ is not kinematically allowed. Thus the fermions f in Eq(180) include all those in the first three generations, except top, plus any extra (light) neutrinos of subsequent generations. Neglecting fermion masses again, which is an excellent approximation, and using Eqs(23) and (26) one finds

$$\Gamma(Z \to f\bar{f}) = \frac{\sqrt{2}G_F M_Z^2}{6\pi}[(Q_L^f)^2 + (Q_R^f)^2] \begin{cases} 1 & leptons \\ 3(1+\frac{\alpha_s}{\pi}) & quarks \end{cases} \quad (181)$$

In the above the chiral charges Q_L^f, Q_R^f are given by

$$Q_L^f = T_{3f} - e_f \sin^2\Theta_W; \quad Q_R^f = -e_f \sin^2\Theta_W \quad (182)$$

where T_{3f} are the $SU(2)$ quantum numbers of the left-handed fermions and e_f are the fermion charges. The extra factor entering in Eq(181) for the quarks is a color factor of 3, times a QCD correction which accounts for the fact that, as shown in Fig 25, a Z boson not only can decay into a $q\bar{q}$ pair but it can also have decays involving gluon emission. Using the estimate [39] $\frac{\alpha_s(M_Z)}{\pi} = 0.04 \pm 0.01$ and the standard values of M_Z and $\sin^2\Theta_W$ given above, one finds:

$$\Gamma(Z \to \nu\bar{\nu}) = 170 \; MeV \quad (183)$$

$$\Gamma(Z \to e^+e^-) = 86 \; MeV \quad (184)$$

$$\Gamma(Z \to u\bar{u}) = 306 \; MeV \quad (185)$$

$$\Gamma(Z \to d\bar{d}) = 393 \; MeV \quad (186)$$

Using these results, the total Z width is

$$\Gamma_{tot}^Z = 2560 \; MeV + (N_\nu - 3)170 \; MeV \quad (187)$$

I note that the error on this prediction - apart from the uncertainty in M_Z which is irrelevant for our considerations, since the value of M_Z does **not** affect the branching ratio

$B(Z \to e^+e^-)$ - is actually very small: the α_s uncertainty implies $\delta\Gamma^Z_{tot} = \pm 20\ MeV$, while the uncertainty in $\sin^2\Theta_W$ of Eq(66), only gives $\delta\Gamma^Z_{tot} = \pm 12\ MeV$.

A similar analysis yields Γ^W_{tot}. If one again neglects all fermion masses, except m_t, then Γ^W_{tot} is independent of the Cabibbo Kobayashi Maskawa matrix V. A simple calculation, using Eqs(25) and (26), yields

$$\Gamma^W_{tot} = \Gamma(W \to e\nu_e)\{3 + 2[3(1 + \frac{\alpha_s}{\pi})]\} + \Gamma(W \to t\bar{b}) \tag{188}$$

Using the standard model parameters adopted earlier, Eq(177) implies

$$\Gamma(W \to e\nu_e) = 229 MeV \tag{189}$$

For $W \to t\bar{b}$, on the other hand, one cannot forget phase space effects which kinematically suppress the rate. Neglecting again m_b, one finds

$$\Gamma(W \to t\bar{b}) = 3(1 + \frac{\alpha_s}{\pi})\Gamma(W \to e\nu)\{(1 + \frac{m_t^2}{2M_W^2})(1 - \frac{m_t^2}{M_W^2})^2\} \tag{190}$$

The above formula implies $\Gamma(W \to t\bar{b}) = 395\ MeV$ for $m_t = 45 GeV$, but $\Gamma(W \to t\bar{b}) \to 0$ as $m_t \to M_W$. For the standard model parameters adopted one has

$$\Gamma^W_{tot} = 2120\ MeV + \Gamma(W \to t\bar{b}) \tag{191}$$

with again little error - except for the error on M_W, but this error is irrelevant in the branching ratio that enters in R.

Collecting all the results one finds, in the standard model:

$$\frac{B(W \to e\nu_e)}{B(Z \to e^+e^-)} = \frac{\Gamma(W \to e\nu_e)}{\Gamma(Z \to e^+e^-)} \frac{\Gamma^Z_{tot}}{\Gamma^W_{tot}} = 2.68\left\{\frac{2560 + 170(N_\nu - 3)}{2120 + \Gamma(W \to t\bar{b})(MeV)}\right\} \tag{192}$$

It is clear from the above that this ratio increases as N_ν increases and that the ratio also grows as $m_t \to M_W$, since then $\Gamma(W \to t\bar{b}) \to 0$. This behaviour is depicted in Fig 26

To complete the calculation of R one needs an estimate of the ratio $\frac{\sigma_W}{\sigma_Z}$. This is the place where the theoretical uncertainty is the largest. There are basically two sources of error in this ratio:

1. The actual estimates for σ_W and σ_Z need values for the probability of finding u and d quarks in the proton. These probabilities are related to the corresponding structure functions and these, in turn, have some uncertainty.

2. In addition, the production cross sections for W and Z bosons are affected by QCD corrections, arising from either real or virtual emission of gluons- the so called K-factor.

Because one is dealing with a ratio of cross sections, it turns out that $\frac{\sigma_W}{\sigma_Z}$ is not very much affected by QCD corrections, since the W and Z K-factors tend to cancel out. On the other hand, the actual value for $\frac{\sigma_W}{\sigma_Z}$ depends to some extent on what structure functions one uses [17].

---

[17] σ_W requires a knowledge of $f_d \otimes f_u$, while σ_Z is proportional to $(f_d \otimes f_d) \oplus (f_u \otimes f_u)$, so there is not a complete cancellation in the ratio.

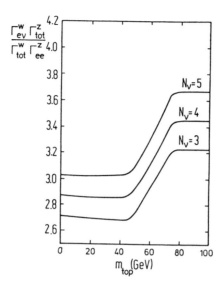

Figure 26: The ratio of the $W \to e\nu_e$ to $Z \to e^+e^-$ branching ratios as a function of m_t, for various values of N_ν.

Recently, Colas, Denegri and Stubenrauch [38] have done an extensive reanalysis of existing deep inelastic data from the BCDMS and EMC collaborations to try to determine $\frac{\sigma_W}{\sigma_Z}$ accurately. Their results, which are shown pictorially in Fig 27 along with some other determinations, lead to a value

$$\frac{\sigma_W}{\sigma_Z} = 3.25 \pm 0.10 \qquad (193)$$

Using Eqs(192) and (193) and the experimental bound on R of Eq(175), the bound on N_ν, as a function of the top quark mass, follows immediately. In effect, for the bound, what is important in the lowest value for $\frac{\sigma_W}{\sigma_Z}$ assumed. Using the 1σ value of $\frac{\sigma_W}{\sigma_Z}$ of Eq(193), $\frac{\sigma_W}{\sigma_Z} = 3.15$, Colas, Denegri and Stubenrauch obtain the results displayed in Fig 28. One sees that for $m_t > M_W$ one has rather tight limits. These limits are displayed in a different form in Fig 29. From this figure, one sees that one can allow a few extra families if $m_t \simeq 50 GeV$, but there is essentially no freedom left if $m_t > 80 GeV$. To keep the above limits in perspective, however, one should note that if the minimum value assumed for $\frac{\sigma_W}{\sigma_Z}$ goes down by 5% (i.e. $\frac{\sigma_W}{\sigma_Z} = 3$ and not 3.15) then the allowed value for N_ν rises by one unit.

I close this Section with a remark on the future. Namely, that very accurate neutrino counting can be expected at LEP and SLC. First of all, at these colliders operating at the Z energy, one is expected to measure M_Z to an accuracy of better than 50 MeV [40]. With this measurement the actual error on Γ^Z_{tot}, due to the uncertainty in the Z mass is tiny. Hence, one can really hope to keep

$$(\delta\Gamma^Z_{tot})^{LEP/SLC}_{theory} \leq 30 \; MeV \qquad (194)$$

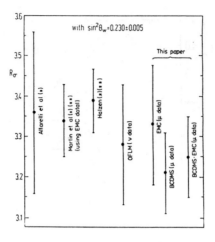

Figure 27: Compilation of the results on $\frac{\sigma_W}{\sigma_Z}$, from [38].

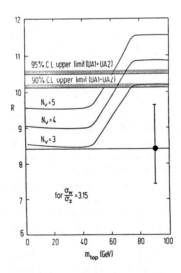

Figure 28: Values of N_ν allowed by the combined UA1 and UA2 data on R, as a function of m_t, from [38].

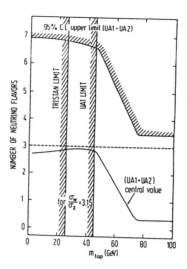

Figure 29: Bounds on N_ν versus m_t, from [38].

Experimentally, furthermore, one believes that one can measure Γ^Z_{tot} very accurately, with perhaps [40]

$$(\delta\Gamma^Z_{tot})^{LEP/SLC}_{experiment} \leq 10 \; MeV \oplus 15 \; MeV \tag{195}$$

where the first error is statistical and the second is an estimate of the expected systematic error. Given Eqs(194) and (195), one should have no trouble establishing if there are any extra neutrino species, since each extra neutrino type contributes:

$$\Gamma(Z \to \nu_i \bar{\nu}_i) = 170 \; MeV >> (\delta\Gamma^Z_{tot})^{LEP/SLC}_{theory} \oplus (\delta\Gamma^Z_{tot})^{LEP/SLC}_{experiment} \tag{196}$$

4 The Flavor Mixing Matrix

In this section I want to discuss in some detail one of the crucial elements that enter in the standard model: the flavor mixing matrix V. I will consider here only the case of 3 generations, where V is described by 3 angles and one phase. Many equivalent parametrization of the Cabibbo Kobayashi Maskawa matrix exist in the literature. I shall use a parametrization which is related to the one suggested by Maiani [41] and is the form of the CKM matrix now adopted by the Particle Data Group:

$$V = \begin{vmatrix} V_{ud} & V_{us} & V_{ub} \\ V_{cd} & V_{cs} & V_{cb} \\ V_{td} & V_{ts} & V_{tb} \end{vmatrix} = \begin{vmatrix} c_1 c_3 & s_1 c_3 & s_3 e^{-i\delta} \\ -s_1 c_2 - c_1 s_2 s_3 e^{i\delta} & c_1 c_2 - s_1 s_2 s_3 e^{i\delta} & s_2 c_3 \\ s_1 s_2 - c_1 c_2 s_3 e^{i\delta} & -c_1 s_2 - s_1 c_2 s_3 e^{i\delta} & c_2 c_3 \end{vmatrix} \tag{197}$$

Here $c_i = \cos\theta_i$, $s_i = \sin\theta_i$.

The advantages of the above parametrization for the mixing matrix V is that, experimentally, the mixing angles θ_i appear to have a natural hierarchy of values:

$$\theta_1 \gg \theta_2 \simeq \theta_1^2 \gg \theta_3 \simeq \theta_1^3 \tag{198}$$

Thus, it is possible to write Eq(197) in a convenient approximate form [42], which is easy to remember. For this purpose one defines

$$\sin\theta_1 = \lambda, \quad \sin\theta_2 = A\lambda^2 \quad \sin\theta_3 = A\rho\lambda^3 \tag{199}$$

Then to $O(\lambda^4)$, but keeping all the phase information, Eq(197) becomes

$$V = \begin{vmatrix} 1 - \frac{\lambda^2}{2} & \lambda & A\rho\lambda^3 e^{-i\delta} \\ -\lambda(1 + A^2\rho\lambda^4 e^{i\delta}) & 1 - \frac{\lambda^2}{2} - A^2\rho\lambda^6 e^{i\delta} & A\lambda^2 \\ A\lambda^3(1 - \rho e^{i\delta}) & -A\lambda^2(1 + \rho\lambda^2 e^{i\delta}) & 1 \end{vmatrix} \tag{200}$$

Experiment provides information on the parameters λ, A, ρ and δ. I shall discuss how their values, or allowed ranges, are determined below.

4.1 V_{ud} and Radiative Effects

The matrix element V_{ud} is the one which is known with the greatest precision. It is determined by comparing the rate for β-decay in nuclei to that for μ-decay. The ratio of these decay rates is proportional to $|V_{ud}|^2$. However, to extract a precise value for $|V_{ud}|$ one must:

1. make sure that in the comparison between theory and experiment, electroweak radiative corrections are taken into account;

2. pick appropriate decays, where hadronic and nuclear effects do not obscure the situation.

Recall from our discussion in Sec. 2 that electroweak radiative corrections affect, at leading logarithmic level, charged current processes involving hadrons but not μ-decay. The relevant correction for the rate of $d \to u + e^- + \bar{\nu}_e$, relative to that for $\mu^- \to \nu_\mu + e^- + \bar{\nu}_e$, is the CC coefficient ρ^{CC} of Eq(93):

$$\frac{\Gamma(d \to u + e^- + \bar{\nu}_e)}{\Gamma(\mu^- \to \nu_\mu + e^- + \bar{\nu}_e)} \sim |V_{ud}|^2 |\rho^{CC}|^2 \tag{201}$$

Now

$$|\rho^{CC}|^2 = |1 + \frac{3\alpha}{4\pi} e_u \ln\frac{M_W^2}{\mu^2}|^2 \simeq 1 + \frac{2\alpha}{\pi}\ln\frac{M_W}{\mu} \tag{202}$$

If μ is taken of the order of the typical energies involved in nuclear decays ($\mu \sim$ few MeV's), one sees that the correction of Eq(202) is of the order of $3-4\%$. As Sirlin[43] has emphasized, neglecting these effects would lead to a violation of unitarity of the CKM matrix. Hence, here one has another example where the physical importance of the electroweak radiative corrections is manifest.

The quark level ratio of Eq(201), characterized by V_{ud} and ρ^{CC}, can be obscured when one is dealing with a transition in a nucleus. The current

$$J_+^\mu = V_{ud}\bar{u}\gamma^\mu(1-\gamma_5)d, \qquad (203)$$

written in terms of quarks, receives different corrections for its vector and axial pieces in a hadronic matrix element. In general, between nucleon states, one can write

$$<p;p_1|\bar{u}\gamma_\mu d|n;p_2> = \bar{u}(p_1)[\gamma_\mu F_1(q^2) + \frac{i\sigma_{\mu\nu}q^\nu}{2M_N}F_2(q^2)]u(p_2) \qquad (204)$$

$$<p;p_1|\bar{u}\gamma_\mu\gamma_5 d|n;p_2> = \bar{u}(p_1)[\gamma_\mu\gamma_5 G_A(q^2) + \frac{iq_\mu\gamma_5}{M_N}G_P(q^2)]u(p_2) \qquad (205)$$

where $q^\mu = (p_1 - p_2)^\mu$ is the momentum transfer and F_1, F_2, G_A and G_P are form factors. Since the momentum transfers in nuclear decays are small compared to M_N, the F_2 and G_P form factors are not very important. Of the two remaining form factors, the axial form factor $G_A(0)$ is affected by the strong interactions and one finds experimentally that, at zero momentum transfer $G_A(0) \simeq 1.25$. The vector form factor F_1, on the other hand, is not affected by the strong interactions and one has

$$F_1(0) = 1 \qquad (206)$$

The result Eq(206), which follows from the conserved vector current hypothesis (CVC) [44], is easy to understand. The weak current J_+^μ, of Eq(203), apart from V_{ud}, is **also** the $1+i2$ component of the strong isospin current for the u and d quarks. Indeed, at the quark level, since u and d form an isospin doublet, the strong isospin current is just

$$J_i^\mu = (\bar{u}\bar{d})\gamma^\mu\frac{\tau_i}{2}\begin{pmatrix}u\\d\end{pmatrix} \qquad (207)$$

To the extent that strong isospin is a good symmetry, so that the current Eq(207) is conserved

$$\partial_\mu J_i^\mu = 0 \qquad (208)$$

it follows that $F_1(0) = 1$. This is easily seen, by computing the matrix element of the isospin raising operator, between a neutron and proton state, using Eq(204). By definition one has that [18]

$$<p;p_1|I_{1+i2}|n;p_2> = (2\pi)^3 2p_1^0 \delta^3(\vec{p}_1 - \vec{p}_2) \equiv <p;p_1|p;p_2> \qquad (209)$$

However, one has also that

$$<p;p_1|I_{1+i2}|n;p_2> = <p;p_1|\int d^3x J_{1+i2}^0|n;p_2> = F_1(0)(2\pi)^3 2p_1^0 \delta^3(\vec{p}_1 - \vec{p}_2) \qquad (210)$$

One can show, and I will sketch the proof of this in the next subsection, that corrections to the CVC result, Eq(206), only occur at 2nd order in isospin breaking [45]. This result,

---

[18] I use a covariant normalization for my states.

known as the Behrends, Sirlin, Ademollo, Gatto (BSAG) theorem, means that for nuclear transitions the corrections for $F_1(0)$ are really negligible

$$F_1(0) = 1 + 0[(\frac{\Delta M_N}{M_N})^2] \qquad (211)$$

So one can essentially eliminate all the hadronic uncertainties, if one can focus on transitions where only F_1 enters. Experimentally, to get rid of the axial vector contributions one studies Fermi transitions, where parity is conserved. Furthermore, one can also get rid of the contributions from F_2 - the, so called, weak magnetism term - by studying transitions from a spin zero state to another spin zero state (0-0 transitions).

The ideal prototype for a 0-0 transition is the β-decay of the π^+. Unfortunately, the three body decay mode $\pi^+ \to \pi^0 e^+ \bar{\nu}_e$, is so rare that there is not enough experimental precision yet to match that which can be obtained by studying nuclear transitions. The nuclear decays studied are the superallowed Fermi $0^+ \to 0^+$ transitions, of which 8 have been measured experimentally [47] and analyzed theoretically. In particular, Sirlin [46] has made a full comparison between theory and experiment, including the effects of the electroweak corrections. He finds, using only the best measured $^{14}0$ transition, $V_{ud} = 0.9739 \pm 0.0015$. Using all 8 decays, the value quoted by Sirlin is slightly more accurate:

$$V_{ud} = 0.9744 \pm 0.0010 \qquad (212)$$

I remark that, using the unitarity of the CKM matrix, the above allows one to determine λ, since $V_{ud} \simeq 1 - \frac{\lambda^2}{2}$. The value for λ obtained this way

$$\lambda = 0.226 \pm 0.004 \qquad (213)$$

is in very good agreement with the direct determation of V_{us}, which I will discuss below. If one had **not** applied any electroweak radiative corrections, then one would have found $V_{ud} \simeq 1$ and $|V_{ud}|^2 + |V_{us}|^2 > 1$, which would have violated the unitarity of the Cabibbo Kobayashi Maskawa matrix!

4.2 Fixing the Cabibbo Angle

The comparison of the rate for strangeness changing weak decays to that for μ-decay allows a direct determination of $V_{us} \simeq \lambda$. This parameter, for the case of two generations, is just the sine of the Cabibbo angle, $\sin \theta_C$ - introduced long ago by Cabbibo [48] to restore universality between hadronic and leptonic weak interactions. For analogous reasons to those discussed above, it is again safer to consider $\Delta S = 1$ decays where only the vector currents enter $[K \to \pi l \nu_l]$, rather than Hyperon β-decays.

The most comprehensive and careful analysis to date on how to extract λ from physical data on K_{e3} decays is due to Leutwyler and Roos [49] and I will base my discussion mostly on their work. In the decay $K \to \pi l \nu_l$, shown in Fig 30, only the vector piece of the $\Delta S = 1$ charged current contributes. The matrix element of this current, between a Kaon and a pion, is described by two form factors:

$$< K, p_1 |\bar{u}\gamma^\mu s|\pi, p_2> = C[(p_1 + p_2)^\mu f_+(q^2) + (p_1 - p_2)^\mu f_-(q^2)] \qquad (214)$$

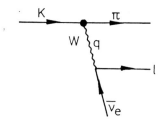

Figure 30: The decay $K \to \pi l \nu_l$

Here again q^μ is the 4-momentum transfer and C is a Clebsch Gordan factor, which takes the values

$$C = \begin{cases} \sqrt{\frac{1}{2}} & (K^+ \to \pi^0 l^+ \nu_l) \\ 1 & (K_L^0 \to \pi^- l^+ \nu_l) \end{cases} \tag{215}$$

Since $(p_1 - p_2)^\mu$, when contracted with the leptonic tensor coming from the $l\nu_l$ vertex in Fig 30, gives a factor of the lepton mass, for the electron decay mode only the f_+ form factor is important. For this reason one considers only K_{e3} decays. For the momentum transfers involved in K_{e3} decays, the variation of $f_+(q^2)$ with q^2 is well fit by a linear formula

$$f_+(q^2) = f_+(0)[1 + \lambda_+ q^2] \tag{216}$$

with the slope parameter λ_+ fixed from the data [50].

Using Eq(214) and (216) the K_{e3} decay rate is straightforwardly computed [49]. One finds

$$\Gamma = [\frac{G_F^2 M_K^5}{192\pi^3}][C^2 f_+^2(0) I_{ps}][1 + \frac{2\alpha}{\pi} \ln \frac{M_Z}{M_N} + \delta]|V_{us}|^2 \tag{217}$$

The 1st factor in square brackets above is the usual 3-body $(m = 0)$ β-decay rate. The 2nd factor contains hadronic effects - through $f_+(0)$ - and phase space corrections, coming from the non zero mass of the decay by-products. Finally, the 3rd factor contains the electroweak radiative corrections, where δ are the remaining corrections beyond the leading logarithmic ones. The phase space integrals reduce the $(m = 0)$ β-decay rate by almost an order of magnitude and Leutwyler and Roos [49] find

$$I_{ps} = \begin{cases} 0.1605 \pm 0.0009 & K^+ \to \pi^0 e^+ \nu_e \\ 0.1561 \pm 0.0009 & K_L^0 \to \pi^- e^+ \nu_e \end{cases} \tag{218}$$

where the errors above are reflections of the experimental uncertainties in λ_+. Using the branching ratios [50]

$$\Gamma(K^+ \to \pi^0 e^+ \nu_e) = (2.565 \pm 0.282) \times 10^{-15} \; MeV \tag{219}$$

$$\Gamma(K_L^0 \to \pi^- e^+ \nu_e) = (4.915 \pm 0.733) \times 10^{-15} \ MeV \tag{220}$$

Leutwyler and Roos [49] deduce that

$$f_+^{K^+}(0)|V_{us}| = 0.2161 \pm 0.0017 \tag{221}$$

$$f_+^{K^0}(0)|V_{us}| = 0.2103 \pm 0.0020 \tag{222}$$

In obtaining these results, the authors of [49] ignored the, difficult to calculate and model dependent, correction δ and just augmented the error brackets in Eq(221) and (222).

To proceed one must be able to estimate $f_+(0)$. Of course, since the $\Delta S = 1$ current $J_{\Delta S=1}^\mu = \bar{u}\gamma^\mu s$ is conserved in the $SU(3)$ limit $m_s = m_u = m_d$, one expects that $f_+(0) \simeq 1$. However, the correction to this relation, coming from an $SU(3)$ version of the BSAG theorem, naively appear to be quite large, since $\frac{(m_s-m_u)^2}{(m_s+m_u)^2} \sim 0(1)$! Thus a more careful analysis is warranted. I will briefly go into this analysis here as it will serve to demonstrate the difficulties inherent in extracting weak interaction information in hadronic processes. Furthermore, it will give me a chance to indicate how the BSAG theorem comes about, although, for the case of pions and Kaons, in fact, the theorem is violated by mass singularities! When all the theoretical smoke clears, however, we shall see that indeed $f_+(0)$ is very near unity, so that the value of $|V_{us}|$ is close to that indicated on the right hand side of Eqs(221) and (222).

Corrections to the $SU(3)$ limit result, $f_+(0) = 1$, are more complicated for the 0^- mesons, since these particles are quasi Goldstone bosons and one expects that their masses vanish, as the quark masses vanish. Thus, in calculating corrections to $f_+(0)$, one can encounter factors of m_q^{-1}, coming from virtual exchanges of these mesons, which alter the ordinary mass power counting. Indeed, these mass singularities will also give some logarithmic, and therefore non analytical, dependence on m_q. Because of these circumstances, the correct expansion for $f_+(0)$ is of the form [51]

$$f_+(0) = 1 + 0(m_q \ln m_q) + 0(m_q) + \cdots \tag{223}$$

Given Eq(223), it is obvious that for mesons the BSAG theorem [45] must be modified. Furthermore, at first sight, things look even worse than one would have naively expected. However, an actual computation shows that the coefficient of the $0(m_q \ln m_q)$ term is of $0(10^{-2})$ and thus these corrections are small [49].

The $0(m_q)$ corrections above have a rather trivial origin. If $m_u \neq m_d$ then there is some $\pi^0 - \eta$ mixing. That is, the π^0 is not purely the third component of an octet, but one has [52]

$$|\pi^0> \simeq |3> + \frac{\sqrt{3}(m_d - m_u)}{2(2m_s - m_d - m_u)}|8> \equiv |3> +\epsilon|8> \tag{224}$$

Then a simple calculation gives [49]

$$f_+^{K^0}(0) = 1; \quad f_+^{K^+}(0) = 1 + \sqrt{3}\epsilon \tag{225}$$

Note that using the present estimates of the quark masses [52] $\epsilon \simeq 2 \times 10^{-2}$.

To understand the origin of the mass singularities - and to derive the BSAG theorem in ordinary circumstances! - one can proceed as follows [49]. The $\Delta S = 1$ currents $J_+^\mu = \bar{u}\gamma^\mu s$, $J_-^\mu = \bar{s}\gamma^\mu u$ and $J_3^\mu = \frac{1}{2}(\bar{u}\gamma^\mu u - \bar{s}\gamma^\mu s)$, have an isospin charge algebra

$$[Q_+, Q_-] = 2Q_3, \tag{226}$$

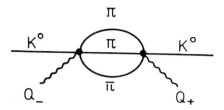

Figure 31: Dominant 3π correction to $f_+^{K^0}(0)$.

where $Q_i = \int d^3x J_i^0$. This algebra, upon sandwiching it with $|K^0>$ states and inserting a complete set of states $|n>$, yields a sum rule [53]

$$\sum_n |<K^0|Q_+|n>|^2 - \sum_m |<K^0|Q_-|m>|^2 = |<K^0|K^0>|^2 \qquad (227)$$

The states $|n>$ above have zero strangeness, while the states $|m>$ must have strangeness 2. These latter states are exotic and there is good reason to drop altogether their contribution from the sum rule. Furthermore, if $|n>$ is a $|\pi^->$ what enters in Eq(227) is precisely $f_+^{K^0}(0)$. Hence one finds

$$|f_+^{K^0}(0)|^2 = 1 - \sum_{|n>\neq|\pi^->} |\frac{<K^0|Q_+|n>}{<K^0|K^0>}|^2 \qquad (228)$$

The above formula, which could also be derived for other states besides Kaons is, essentially, the BSAG theorem [45] for SU(3). If one does not worry about mass singularities, then the matrix element on the RHS vanishes when $m_s = m_u$, so that the corrections are indeed of $0((\frac{m_s-m_u}{m_s+m_u})^2)$. Actually, in the case of Kaons, the corrections to $f_+(0)$ are only of $0((m_s - m_u)^2)$ near $m_s = m_u$. Otherwise, there are logarithmic modifications. The dominant contribution in the sum in Eq(228) is due to the state $|n>=|3\pi>$. Since as $m_u, m_d \to 0$, M_π vanishes, there are actually mass singularities when one performs the integrals over the $|3\pi>$ intermediate states, depicted in Fig 31. These integrals give rise to terms proportional to $M_K^2 \ln \frac{M_\pi^2}{M_K^2}$, which are not analytic in the quark masses. Fortunately, numerically these contributions are rather small, because the three body phase space of Fig 31 provides suppression factors of $\frac{1}{192\pi^3}$!

Leutwyler and Roos [49], make use of the chiral Lagrangian estimate of Gasser and Leutwyler [54] of the 3π corrections to $f_+(0)$, plus the $\pi - \eta$ mixing corrections of Eq(225), to calculate the departure of the K_{e3} form factors from unity. They find

$$\frac{f_+^{K^+}(0)}{f_+^{K^0}(0)} = 1.022 \qquad (229)$$

$$f_+^{K^0}(0) = 0.961 \pm 0.008 \tag{230}$$

where the first result is dominated by the mixing correction, while the second is purely a $m_q \ln m_q$ correction. Armed with these numbers, Leutwyler and Roos [49] use Eqs(221) and (222) to extract a combined value for $|V_{us}|$

$$|V_{us}|_{K_{e3}} = 0.2196 \pm 0.0023 \tag{231}$$

In addition, these authors also analyze Hyperon β-decay. Here there are more theoretical uncertainties, since also axial current matrix elements are present. The value obtained for $|V_{us}|$ by Leutwyler and Roos [49] in this case is somewhat larger than that given in Eq(231):

$$|V_{us}|_{Hyperon} = 0.231 \pm 0.005 \tag{232}$$

Taking both analyses into account, with an appropriate weighting, leads to a final value for V_{us}

$$|V_{us}| = 0.221 \pm 0.002 \tag{233}$$

This value of V_{us} - or equivalently λ - is in excellent agreement with the one I quoted earlier using the unitarity of the CKM matrix and the value for V_{ud}, Eq(213). So the unitarity of the CKM matrix is actually checked experimentally, at least for the first row!

4.3 Heavy Quark Matrix Elements

The matrix elements V_{cd} and V_{cs} of the CKM matrix are much more accurately fixed by unitarity than by direct measurements. To $0(\lambda^4)$ unitarity implies [19]

$$V_{cs} = V_{ud} = 0.9744 \pm 0.0010 \tag{234}$$

$$V_{cd} = V_{us} = 0.221 \pm 0.002 \tag{235}$$

Direct estimates for V_{cd} and V_{cs} come from neutrino deep inelastic experiments, in particular dimuon production $\nu_\mu N \to \mu^+ \mu^- X$. Opposite sign muons can arise from the production and subsequent semileptonic decay of charmed states. As is shown in Fig 32, there are two possible ways to produce charm in neutrino scattering. Both these contributions turn out to be comparable, since the probability of finding a valence d quark is much greater than that of finding a sea s quark, thereby compensating the disparity between V_{cd} and V_{cs}. Furthermore, since the sea and valence distributions have a different x-dependence, one can extract V_{cd} and V_{cs} separately. This analysis has been performed by the CDHS collaboration [55] and they obtain:

$$V_{cd} = 0.207 \pm 0.024 \tag{236}$$

$$V_{cs} = 0.95 \pm 0.14 \tag{237}$$

Clearly these numbers are consistent with those obtained through unitarity, but they are much less precise.

The remaining parameters of the mixing matrix, except for the phase δ, are fixed by B meson decays. For δ, CP violation in the Kaon system is of primary importance but, as

---

[19]If one does include λ^4 terms, then unitarity implies $V_{cs} = 0.973 \pm 0.001$.

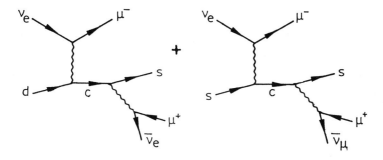

Figure 32: Processes giving rise to opposite sign dimuons in deep inelastic neutrino scattering.

I'll discuss in the next Section, also $B - \bar{B}$ mixing imposes constraints. The ratio of $\frac{|V_{ub}|}{|V_{cb}|}$, which meausures ρ, is determined by semileptonic B decays. The two contributions to these decays, coming from $b \to cl^-\bar{\nu}_l$ and $b \to ul^-\bar{\nu}_l$, can be separated out experimentally, in principle, by the shape of the lepton spectrum near the end point. This is shown in Fig 33, where the various contributions to the lepton spectrum observed by the ARGUS collaboration [56] are depicted. The relative strength of the $b \to u$ and $b \to c$ transitions, which is denoted by R, is related to the ratio of V_{ub} to V_{cb}.

To actually extract a value for $\frac{|V_{ub}|}{|V_{cb}|}$ necessitates a calculation of the end point spectrum which, in turn, depends (somewhat) on the bound state model one adopts for the B mesons. The ratio R, on the other hand, can be computed in a spectator model where, as shown in Fig 34, B decays proceeds as a free quark decay. The ratio of total rates, is after all, a very inclusive quantity and the spectator model, which is a partonic approximation, should work well in these circumstances. One has then

$$R = \frac{\Gamma(b \to u)}{\Gamma(b \to c)} = \frac{|V_{ub}|^2}{|V_{cb}|^2} \frac{\Phi(\frac{m_u}{m_b})}{\Phi(\frac{m_c}{m_b})} \tag{238}$$

Here $\Phi(x)$ is the usual three body phase space factor for the decay of a massive particle, of mass M, into a light particle, of mass m, and two massless particles, with $x = \frac{m}{M}$:

$$\Phi(x) = 1 - 8x^2 + 8x^6 - 24x^4 \ln x \tag{239}$$

For the case at hand, using $\frac{m_c}{m_b} \simeq 0.3$, the ratio of phase space factors in Eq(238) is nearly

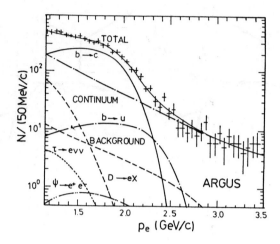

Figure 33: The electron spectrum at the $\Upsilon(4s)$ observed by the ARGUS collaboration [56].

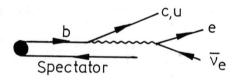

Figure 34: B decays in the spectator model.

2, so that
$$\frac{|V_{ub}|}{|V_{cb}|} \equiv \lambda\rho \simeq \sqrt{\frac{R}{2}} \tag{240}$$

Present experimental data shows no evidence for a $b \to u$ component. This, in turn, implies a bound on R. This bound is model dependent since to calculate the spectral shape of electrons near the end point:

$$\frac{d\Gamma}{dE_e} = \Gamma(b \to u)[\frac{1}{\Gamma(b \to u)}\frac{d\Gamma(b \to u)}{dE_e}] + \Gamma(b \to c)[\frac{1}{\Gamma(b \to c)}\frac{d\Gamma(b \to c)}{dE_e}] \tag{241}$$

one has to make some assumptions on the B-meson wavefunction, to calculate the individual factors in the square brackets above. What is usually assumed [57], is that the decay rate is no longer totally independent of the spectator in Fig 34. Rather, this particle has some momentum distribution $\Phi(p) \sim exp[-\frac{p^2}{p_F^2}]$, where p_F is a free parameter. What weak decays then is a virtual b quark, whose invariant mass depends on p and the mass of the spectator (another free parameter):

$$W^2 = M_B^2 + m_{sp}^2 - 2M_B\sqrt{p^2 + m_{sp}^2} \tag{242}$$

The free parameters m_{sp} and p_F are then fixed by fitting the spectrum of the decay electrons

$$\frac{d\Gamma_B}{dE_e} = \int p^2 dp \Phi(p) \frac{d\Gamma_b}{dE_e} \tag{243}$$

By following this procedure, or analogous steps depending on the detail of the model chosen, the various experimental collaborations arrive at a bound on R, based on the absence of a $b \to u$ signal. The present, rather conservative, bound on R [58], is

$$R < 0.08 \tag{244}$$

which, using Eq(240), implies

$$\frac{|V_{ub}|}{|V_{cb}|} < 0.20 \tag{245}$$

and

$$\rho < 0.9 \tag{246}$$

Smaller values of ρ, however, have also been given in the literature. For instance, a recent analysis of the ARGUS data [59] gives $\frac{|V_{ub}|}{|V_{cb}|} < 0.16$ and $\rho < 0.73$.

A lower bound on $\frac{|V_{ub}|}{|V_{cb}|}$, and therefore on ρ, can be inferred from the recent ARGUS observation of charmless B decays [60]:

$$\begin{aligned} B(B^+ \to p\bar{p}\pi^+) &= (3.7 \pm 1.3 \pm 1.4) \times 10^{-4} \\ B(B_d^0 \to p\bar{p}\pi^+\pi^-) &= (6.0 \pm 2.0 \pm 2.2) \times 10^{-4} \end{aligned} \tag{247}$$

These decay modes, however, have been looked for by CLEO and have not been found, at a level of perhaps a factor of 2-3 below that of the ARGUS claim [61]. Thus one must take the result in Eq(247) with a certain degree of caution. Apart from this controversy, which must be settled, it appears that Eq(247) provides the first evidence for a nonzero

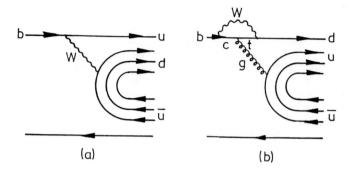

Figure 35: Diagrams contributing to charmless B decays.

V_{ub}. This assertion is not trivial, since one has to make sure that the results in Eq(247) are not induced by hadronic Penguin diagrams.

There are two sources for the charmless B decays of Eq(247), as depicted in Fig 35. Either they proceed directly through a $b \to u$ transition, or they are induced by a $b \to d$ Penguin diagram. The Penguin contribution is (slightly) suppressed, by being a factor of $0(\frac{\alpha_s}{\pi} \ln \frac{m_t}{m_c})$ down relative to the direct diagram, but is of the same order in electroweak parameters: $0(A\rho\lambda^3)$. Thus one must have some additional evidence that the decays in Eq(247) are not Penguin induced. This evidence fortunately exists, since if Penguin effects were dominant one would expect a very large signal for $B \to \Lambda \bar{p} \pi(\pi)$. Such a signal has not been observed. Indeed, ARGUS [60] quotes a limit

$$\frac{B(B \to \Lambda \bar{p} \pi(\pi))}{B(B \to p\bar{p}\pi(\pi))} < 0.5 \qquad (248)$$

If Penguin effects were to have dominated, since the rate associated with the $b \to s$ Penguin is enhanced by $0(\lambda^{-2})$ relative to the $b \to d$ Penguin, one would have expected Eq(248) to be of order 20.

It is not easy to translate the ARGUS results into a value for $\frac{|V_{ub}|}{|V_{cb}|}$. The collaboration itself gives a conservative bound, as follows. They write

$$B(B \to p\bar{p}\pi(\pi)) = \frac{|V_{ub}|^2}{|V_{cb}|^2} B(B \to \text{ charmed Baryons})[Ps \cdot r_{vis} \cdot r_B] \qquad (249)$$

and then ask that the "fudge factors" in the square bracket be maximized. These terms correspond to:

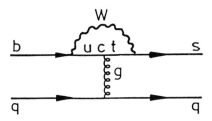

Figure 36: Penguin diagram entering in the $b \to sq\bar{q}$ decay.

- A factor, Ps, which takes into account of the fact that the phase space for the $b \to u$ decay is bigger. Using a quark model estimate, $Ps \simeq 2$ as we have argued earlier [cf. Eq(240)].

- A factor r_{vis} to account for the percentage of baryonic decay modes that are actually seen. ARGUS, assumes, that this is at most 10 %.

- A recombination factor, which takes into account the increase in probability that baryons are made after a $b \to u$ transition, relative to what happens after a $b \to c$ transition. Here what is assumed in [60] is that $r_B \simeq 2$.

In addition, one must make some assumption on the what is the biggest total charmed baryon rate that one can reasonably allow. ARGUS takes this as 25 %. Putting all these numbers together, then a branching ratio $B(B \to p\bar{p}\pi(\pi)) \simeq 5 \times 10^{-4}$ implies lower bounds on $\frac{|V_{ub}|}{|V_{cb}|}$ and ρ of [60]

$$\frac{|V_{ub}|}{|V_{cb}|} > 0.07; \quad \rho > 0.3 \qquad (250)$$

One can obtain roughly the same kind of estimate for ρ_{min}, by making use of the Penguin bound of Eq(248) [62]. The Penguin graph of Fig 36 gives rise to an effective Lagrangian which is analogous to the one which I computed when considering the photonic contribution to $\nu_\mu e$ scattering. The only difference here is that the electric charge is now replaced by the color charge. In Fig 36 the u-quark loop is irrelevant, since $V_{ub}V_{us}^\dagger \sim 0(\lambda^4)$. Furthermore, all dependence on $\ln M_W$ disappears since the contribution of the t-quark loop $[\sim V_{tb}V_{ts}^\dagger = -A\lambda^2]$ cancels against that of the c-quark loop $[\sim V_{cb}V_{cs}^\dagger = A\lambda^2]$. By straightforward calculation, one finds

$$\mathcal{L}_{eff} = \frac{G_F}{\sqrt{2}}[\frac{\alpha_s}{12\pi}A\lambda^2 \ln \frac{m_t^2}{m_c^2}]\left\{[\bar{s}\gamma^\mu(1-\gamma_5)\lambda_a b][\sum_q \bar{q}\gamma_\mu \lambda_a q]\right\} \qquad (251)$$

In the above λ_a are the color matrices.

To get a lower bound on ρ, one needs an estimate of the contribution of the operator in curly brackets in Eq(251) to the decay matrix element for $B \to \Lambda \bar{p} \pi(\pi)$. At the most naive level, one can assume that this operator contributes the same as that appearing in the CC Lagrangian responsible for the decay of $B \to p\bar{p}\pi(\pi)$:

$$\mathcal{L}^{CC} = \frac{G_F}{\sqrt{2}}[A\rho\lambda^3 e^{i\delta}]\left\{[\bar{u}\gamma^\mu(1-\gamma_5)b][\bar{d}\gamma_\mu(1-\gamma_5)u]\right\} \tag{252}$$

Then the ARGUS bound of Eq(248) implies a lower bound on ρ. One has

$$\frac{B(B \to \Lambda\bar{p}\pi(\pi))}{B(B \to p\bar{p}\pi(\pi))} \simeq \frac{|\frac{\alpha_s}{12\pi}A\lambda^2 \ln\frac{m_t^2}{m_c^2}|^2}{|A\lambda^3\rho|^2} \tag{253}$$

which, using $\alpha_s(m_b) \simeq 0.2$, $m_t = 45 GeV$, requires

$$\rho > 0.25, \tag{254}$$

which is comparable to the ARGUS limit of Eq(250).

To summarize the situation on ρ is as follows. Conservative limits on ρ put in the range:

$$0.3 < \rho < 0.9 \tag{255}$$

However, the lower limit is soft. If one believes in the ARGUS result of Eq(247), then probably ρ_{min} could be bigger. On the other hand, the experimental result of ARGUS [60] is not supported by CLEO data [61]. Thus ρ_{min} could also be smaller! The upper limit in eq(255) is probably too generous. Indeed, as we have indicated earlier, different analyses [59] arrive at stronger bounds for the values of $\frac{|V_{ub}|}{|V_{cb}|}$. I shall, nevertheless, in what follows imagine that ρ lies in the range of Eq(255), with my own working guesstimate being that $\rho \simeq 0.6$.

The lifetime of the B mesons, along with a measurement of the semileptonic branching ratio, can be used to estimate the parameter A which enters in CKM matrix. By definition one has

$$\tau_B^{-1} B(B \to l\nu_l X) = \Gamma(B \to l\nu_l X) \equiv \Gamma(b \to c)[1 + R] \tag{256}$$

The left hand side of this equation contains quantities that are measured and $\Gamma(b \to c) \sim |V_{cb}|^2 \sim A^2$. However, to extract a precise value for A one needs a good estimate of $\Gamma(b \to c)$. I have argued earlier that the spectator model should provide a dynamically adequate framework for calculating this rate. Unfortunately, the formula for $\Gamma(b \to c)$ in the spectator model has **another** unknown beside A, namely the mass of the decaying b-quark:

$$\Gamma(b \to c)_{spectator} = \frac{G_F^2 m_b^5}{192\pi^3}\Phi(\frac{m_c}{m_b})|V_{cb}|^2 \tag{257}$$

From Eq(257) one sees that a 10% uncertainty in m_b implies a 25 % error in V_{cb} and in A.

To avoid this large uncertainty associated with m_b, one can go back to the simple bound state model I discussed earlier [57]. In this model, after one fixes the free parameters m_{sp} and p_F by fitting the electron spectrum, then $\Gamma(b \to c)$ is very much better determined, since it depends only on M_B. For example, a recent analysis of Altarelli and Franzini [63] along these lines, finds

$$\Gamma(b \to c) = (4 \pm 0.6) \times 10^{13} sec^{-1}|V_{cb}|^2 \qquad (258)$$

which corresponds to an effective m_b mass, with a tiny error: $m_b^{eff} = 4.70 \pm 0.13\ GeV$. Using the average value of τ_B [64] [$\tau_B = (1.11 \pm 0.16) \times 10^{-12} sec$] and of the semileptonic branching ratio [50] [$B = 0.117 \pm 0.006$], Altarelli and Franzini [63] find [20]:

$$A = 1.05 \pm 0.17 \qquad (259)$$

5 Particle-Antiparticle Mixing and CP Violation

All the real parameters λ, ρ and A in the CKM matrix are now fixed, as best as it is presently possible [cf. Eqs(233), (255) and (259)]. It remains to constrain the phase δ. This parameter enters in the CP violating phenomena observed in Kaon decay. Furthermore, since $V_{td} \simeq A\lambda^3(1 - \rho e^{i\delta})$, a precise measurement of V_{td} has implications for δ. Such a measurement is afforded by the observation of $B_d - \bar{B}_d$ oscillations. In this section, therefore, I want to discuss both these topics.

5.1 Two State Formalism

Before I enter into these topics proper, it is useful to briefly discuss the formalism associated to a particle-antiparticle system (e.g. $K^0 - \bar{K}^0$; $B_d - \bar{B}_d$; etc) which are connected by the weak interactions. I shall denote the particle and antiparticle states, generically, by P and $\bar{P}$. Because, by assumption, there exist transition and decay channels which connect P and $\bar{P}$, this two state system can be described by an effective Hamiltonian which is complex

$$H \begin{pmatrix} P \\ \bar{P} \end{pmatrix} = (M - i\Gamma) \begin{pmatrix} P \\ \bar{P} \end{pmatrix} \qquad (260)$$

In the above, the mass matrix, M, and the decay matrix, Γ, are Hermitian

$$M = M^\dagger; \quad \Gamma = \Gamma^\dagger \qquad (261)$$

Furthermore, the CPT theorem [66] implies that the diagonal elements of M and Γ are equal. Thus

$$M = \begin{vmatrix} m & M_{12} \\ M_{12}^* & m \end{vmatrix}; \quad \Gamma = \begin{vmatrix} \gamma & \Gamma_{12} \\ \Gamma_{12}^* & \gamma \end{vmatrix} \qquad (262)$$

The presence of non vanishing phases in the off diagonal elements of M and Γ signifies CP violation [66].

Diagonalizing the 2×2 Hamiltonian of Eq(260) one obtains the eigenstates $|P_\pm >$ of the two state system

$$|P_\pm> = \frac{1}{[2(1 + |\epsilon_P|^2)]^{1/2}}[(1 + \epsilon_P)|P> \pm (1 - \epsilon_P)|\bar{P}>] \qquad (263)$$

---

[20] This number is consistent with that obtained in a new analysis of C.S. Kim [65], who, however, uses the new smaller experimental value of $B = 0.10 \pm 0.1$ [59].

where

$$\frac{1-\epsilon_P}{1+\epsilon_P} = [\frac{M_{12}^* - \frac{i}{2}\Gamma_{12}^*}{M_{12} - \frac{i}{2}\Gamma_{12}}]^{1/2} \qquad (264)$$

Clearly $\epsilon_P \to 0$ if M_{12} and Γ_{12} are real, so that ϵ_P is a measure of CP violation in the system. In the limit of good CP, then the states $|P_\pm>$ are CP eigenstates:

$$CP|P_\pm> = \mp|P_\pm> \qquad (265)$$

The masses and widths of the eigenstate $|P_\pm>$ involve a quantity

$$Q = [(M_{12} - \frac{1}{2}\Gamma_{12})(M_{12}^* - \frac{i}{2}\Gamma_{12}^*)]^{1/2} \qquad (266)$$

and one has

$$m_\pm = m \pm Re\ Q; \qquad \Gamma_\pm = \gamma \mp 2Im\ Q \qquad (267)$$

The physical states $|P_\pm>$ have a well defined time dependence:

$$|P_\pm(t)> = e^{-im_\pm t} e^{-\frac{\Gamma_\pm}{2}t}|P_\pm(0)> \qquad (268)$$

However, a state produced at $t = 0$ as $|P>$ (or as $|\bar{P}>$), will be a superposition of $|P>$ and $|\bar{P}>$ at later times. That is, a state $|P>$ will **oscillate** into a state $|\bar{P}>$. For a state born at $t = 0$ as a $|P>$ one finds, at later times,

$$|P_{phys}(t)> = f_+(t)|P> + f_-(t)(\frac{1-\epsilon_P}{1+\epsilon_P})|\bar{P}> \qquad (269)$$

where

$$f_+(t) = e^{-imt} e^{-\frac{\gamma}{2}t} \cos[(\frac{\Delta m}{2} - i\frac{\Delta\Gamma}{4})t] \qquad (270)$$

and

$$f_-(t) = ie^{-imt} e^{-\frac{\gamma}{2}t} \sin[(\frac{\Delta m}{2} - i\frac{\Delta\Gamma}{4})t] \qquad (271)$$

Here $\Delta m = m_+ - m_-$; $\Delta\Gamma = \Gamma_+ - \Gamma_-$.

5.2 CP violation and ϵ

In the Kaon system, $|P_+> \leftrightarrow |K_L>$ and $|P_-> \leftrightarrow |K_S>$. If CP were absolutely conserved, then the decay $K_L \to \pi^+\pi^-$ would be entirely forbidden, since $CP|\pi^+\pi^-> = |\pi^+\pi^->$. In fact this decay, along with the $\pi^0\pi^0$ decay mode of the K_L, has been seen. The two measured CP violating parameters η_{+-} and η_{00}, or ϵ and ϵ', characterize the ratio of the decay amplitudes for K_L and K_S mesons [67]:

$$\eta_{+-} = \frac{A(K_L \to \pi^+\pi^-)}{A(K_S \to \pi^+\pi^-)} \simeq \epsilon + \epsilon' \qquad (272)$$

$$\eta_{00} = \frac{A(K_L \to \pi^0\pi^0)}{A(K_S \to \pi^0\pi^0)} \simeq \epsilon - 2\epsilon' \qquad (273)$$

Experimentally, it has been known for a long time that the magnitude of the above ratios are very nearly the same and [50]

$$|\eta_{+-}| \simeq |\eta_{00}| \simeq |\epsilon| = (2.28 \pm 0.02) \times 10^{-3} \tag{274}$$

Very recently the NA31 experiment at CERN has obtained the first indications of a nonzero value for ϵ' [68].

$$\frac{\epsilon'}{\epsilon} = (3.5 \pm 1.1) \times 10^{-3} \tag{275}$$

The latest results from the Chicago Saclay experiment at Fermilab [69] are also consistent with a nonzero ϵ', but their result is not yet significant.

It is straightforward to relate the measured quantities ϵ and ϵ' to the CP violating contributions arising from the CKM model [70]. It is usual to define the decay amplitude for a K^0 into a 2π state of definite isospin I ($I = 0, 2$) as

$$< \pi\pi; I|H_{weak}|K^0> = A_I e^{i\delta_I} \tag{276}$$

Here δ_I is the strong interaction phase shift of the outgoing pions. The decay amplitude for a $\bar{K}^0$ would be the same, if CP were conserved, while if CP is not conserved, then $A_I \to A_I^*$:

$$< \pi\pi; I|H_{weak}|\bar{K}^0> = A_I^* e^{i\delta_I} \tag{277}$$

Using the above, along with the identification of $|K_L>$ and $|K_S>$ with $|P_+>$ and $|P_->$ and the definitions of ϵ and ϵ' of Eq(272) and (273), results in the following formulas for ϵ and ϵ' [70] [21]

$$\epsilon = \epsilon_K + i\frac{Im\, A_0}{Re\, A_0} \tag{278}$$

$$\epsilon' = \frac{1}{\sqrt{2}} e^{i(\delta_2 - \delta_0 + \frac{\pi}{2})} \frac{Re\, A_2}{Re\, A_0} \left\{ \frac{Im\, A_2}{Re\, A_2} - \frac{Im\, A_0}{Re\, A_0} \right\} \tag{279}$$

In the Kaon system one finds experimentally that $(\Delta m)_K \simeq -\frac{1}{2}(\Delta \Gamma)_K$ [50]. This implies that the quantity Q_K, defined in Eq(266) is just

$$Q_K = \frac{1}{2}(\Delta m - \frac{i}{2}\Delta\Gamma)_K \simeq \frac{e^{i\frac{\pi}{4}}}{\sqrt{2}}(\Delta m)_K \tag{280}$$

Furthermore, since the CP violating effects are small, one can approximate

$$(\Delta m)_K \simeq 2Re\,(M_{12}) \tag{281}$$

Using the above two equations, and working only to lowest order in significant quantites, a simple calculation starting from the definition of Eq(264), yields

$$\epsilon_K \simeq \frac{e^{i\frac{\pi}{4}}}{\sqrt{2}} \frac{Im\, M_{12}}{(\Delta m)_K} - \frac{e^{-i\frac{\pi}{4}}}{\sqrt{2}} \frac{Im\, \Gamma_{12}}{(\Delta \Gamma)_K} \tag{282}$$

However, saturating the decay rates by the 2π intermediate state [70] one has

$$\frac{Im\, \Gamma_{12}}{(\Delta\Gamma)_K} = -\frac{Im\, A_0}{Re\, A_0} \tag{283}$$

---

[21] I have dropped, in these formulas, nonleading $\Delta I = 1/2$ suppressed terms of $0(\frac{Re\, A_2}{Re\, A_0})$.

Hence, putting everything together, one arrives at the following formula for ϵ

$$\epsilon \simeq \frac{1}{\sqrt{2}} e^{\frac{i\pi}{4}} \{ \frac{Im\ M_{12}}{(\Delta m)_K} + \frac{Im\ A_0}{Re\ A_0} \} \tag{284}$$

Although the expressions for ϵ and ϵ', Eqs(284) and (279), are convention independent, one can pick phase conventions to eliminate individual pieces in these formulas. For instance, in the Wu-Yang convention [71], one choses $Im\ A_0 = 0$, thereby making the formula for ϵ depend only on the mass matrix. In the standard model, however, this convention is not so natural, since at the quark level it is the $\Delta I = 1/2$ amplitude A_0 which has an imaginary part. Thus in what follows I shall adopt the quark phase convention, where $Im\ A_2 = 0$, and shall denote by ξ the ratio

$$\xi = \frac{Im\ A_0}{Re\ A_0} \tag{285}$$

Using the experimental ratio for the $\Delta I = 3/2$ to $\Delta I = 1/2$ amplitudes, $\frac{Re\ A_2}{Re\ A_0} \simeq 0.05$, Eqs(284) and (279) simplify to

$$\epsilon \simeq \frac{e^{\frac{i\pi}{4}}}{\sqrt{2}} \{ \frac{Im\ M_{12}}{2Re\ M_{12}} + \xi \} \tag{286}$$

$$\epsilon' \simeq \frac{e^{i(\frac{\pi}{2}+\delta_2-\delta_0)}}{\sqrt{2}} \{ -\frac{1}{20} \xi \} \tag{287}$$

Remarkably, using the values of the $\pi\pi$ phase shifts of [72] the phase factor in Eq(287) is also very close to $\frac{\pi}{4}$, so that the phase of ϵ' is nearly the same as that of ϵ. Using the value obtained by the NA31 collaboration for $\frac{\epsilon'}{\epsilon}$, Eq(275), implies $|\xi| \simeq 0.1|\epsilon|$. Hence, the second contribution in Eq(286) is a negligible correction and I shall neglect it, henceforth.

To calculate ϵ, one needs to calculate the imaginary part of M_{12}. At the quark level, this is given by the imaginary part of the box diagram shown in Fig 37, which graphically depicts how by a 2nd order weak process one can transit from a K^0 to a $\bar{K}^0$. I shall show below that the box graph contribution is equivalently described by an effective 4-Fermi interaction Lagrangian for the $\Delta S = 2$ transition of the form [73]

$$\mathcal{L}_{eff}^{\Delta S=2} = A_{box} [\bar{d}\gamma^\mu (1-\gamma_5)s][\bar{d}\gamma_\mu (1-\gamma_5)s] \tag{288}$$

The coefficient A_{box}, does not contain any dominant $\ln M_W$ terms, because of the unitarity of the CKM matrix. However, it is a function of the parameters entering in this matrix.

To demonstrate that the box graph of Fig 37 is equivalent to having the interaction Lagrangian of Eq(288), it suffices to evaluate the box graph in the figure at zero external momentum. The associated amplitude, evaluated in the physical gauge for the vector bosons, is:

$$A^{eff} = -\frac{i}{2} [\frac{ie}{2\sqrt{2}\sin\Theta_W}]^4 [\frac{1}{i}]^4 \int \frac{d^4q}{(2\pi)^4} \left\{ \frac{[\eta_{\mu\alpha}+\frac{q_\mu q_\alpha}{M_W^2}][\eta_{\nu\beta}+\frac{q_\nu q_\beta}{M_W^2}]}{(q^2+M_W^2)^2} \right\}$$

$$\left\{ \sum_{ij} \lambda_i [\gamma^\mu (1-\gamma_5) \frac{(-\gamma q + m_i)}{q^2+m_i^2} \gamma^\nu (1-\gamma_5)] \cdot \lambda_j [\gamma^\beta (1-\gamma_5) \frac{(-\gamma q + m_j)}{q^2+m_j^2} \gamma^\alpha (1-\gamma_5)] \right\} \tag{289}$$

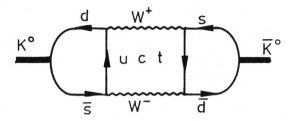

Figure 37: Box graph contributing to the 2nd order weak transition from a K^0 to a $\bar{K}^0$.

In the above the factor of $\frac{1}{2}$ is a combinatoric factor, while the factors in the first square brackets are associated with the vertices and propagators in the graph. Furthermore, λ_i is short hand for the product of CKM matrix elements:

$$\lambda_i = V_{is}V_{id}^\dagger \qquad (290)$$

Obviously, the unitarity of the CKM matrix implies that

$$\lambda_u + \lambda_c + \lambda_t = 0 \qquad (291)$$

Using this constraint, one can eliminate λ_u entirely from the above. Thus one can rewrite the 2nd curly bracket in Eq(289) as [22]:

$$\begin{aligned}\{2\}^{\mu\nu\beta\alpha} &= 4[\gamma^\mu \gamma q \gamma^\nu (1-\gamma_5)][\gamma^\beta \gamma q \gamma^\alpha (1-\gamma_5)] \\ &\quad \left\{\lambda_t^2\left[\frac{1}{(q^2+m_t^2)^2} - \frac{2}{q^2(q^2+m_t^2)} + \frac{1}{(q^2)^2}\right] + \lambda_c^2\left[\frac{1}{(q^2+m_c^2)^2} - \frac{2}{q^2(q^2+m_c^2)} + \frac{1}{(q^2)^2}\right]\right. \\ &\quad + 2\lambda_c\lambda_t\left[\frac{1}{(q^2+m_c^2)(q^2+m_t^2)} - \frac{1}{q^2(q^2+m_t^2)} - \frac{1}{q^2(q^2+m_c^2)} + \frac{1}{(q^2)^2}\right]\left.\right\} \qquad (292)\end{aligned}$$

Because of the unitarity of the CKM matrix, the divergent contributions of the individual graphs are now turned into very convergent factors. A little algebra transforms the product of the two curly brackets in the integrand in Eq(289) to

$$\begin{aligned}\{1\}\cdot\{2\} &= 4\frac{[\gamma^\mu(1-\gamma_5)][\gamma_\mu(1-\gamma_5)]}{M_W^4}\{1-\frac{3(q^2)^2}{4(q^2+M_W^2)^2}\}\left\{\lambda_t^2\left[\frac{m_t^4}{q^2(q^2+m_t^2)^2}\right]\right. \\ &\quad + \lambda_c^2\left[\frac{m_c^4}{q^2(q^2+m_c^2)^2}\right] + 2\lambda_c\lambda_t\left[\frac{m_c^2 m_t^2}{q^2(q^2+m_c^2)(q^2+m_t^2)}\right]\left.\right\} \qquad (293)\end{aligned}$$

---

[22] We have set $m_u^2 = 0$ here.

The factor above which contains a $(q^2 + M_W^2)^2$ contribution, for $m_q \ll M_W$ will give corrections of $O(\frac{m_q^2}{M_W^2})$. Thus it is irrelevant except for top. I will ignore it in the discussion below, but shall reinstate the corrections due to this term in the final answer. Using Eq(293), it is clear that the integral over d^4q in Eq(289) is very convergent. Evaluating this integral yields a result for the amplitude A^{eff} which is precisely of the form expected from the effective Lagrangian of Eq(288), and one identifies:

$$A_{box} = \frac{G_F^2}{16\pi^2}[\lambda_c^2 m_c^2 + \lambda_t^2 m_t^2 + 2\lambda_c \lambda_t m_c^2 \ln \frac{m_t^2}{m_c^2}] \tag{294}$$

For ϵ, since it is proportional to $Im\ M_{12}$, we need to consider the imaginary part of the effective Lagrangian of Eq(288). Using the form of the Cabibbo Kobayashi Maskawa matrix given in Eq(200) one has

$$\begin{aligned} Im\lambda_c^2 &= -2A^2\lambda^6\rho\sin\delta \\ Im\lambda_t^2 &= 2A^4\lambda^{10}\rho\sin\delta(1-\rho\cos\delta) \\ 2Im\lambda_c\lambda_t &= 2A^2\lambda^6\rho\sin\delta \end{aligned} \tag{295}$$

Hence

$$\mathcal{L}_{eff}^{CP\ viol} = \frac{G_F^2}{8\pi^2}[A^2\lambda^6\rho\sin\delta][m_c^2(-1+\ln\frac{m_t^2}{m_c^2}) + A^2\lambda^4(1-\rho\cos\delta)m_t^2]$$
$$[(\bar{d}\gamma^\mu(1-\gamma_5)s)(\bar{d}\gamma_\mu(1-\gamma_5)s)] \tag{296}$$

We note that:

- The CP violating contribution of Eq(296) is suppressed by the high power of λ entering in its common coefficient $[A^2\rho\lambda^6\sin\delta]$. This already "explains" why ϵ is so small.

- The two terms in the second square bracket are comparable numerically if $m_t > 45 GeV$, even though the t-quark contribution is quite Cabibbo suppressed $[\lambda^4]$.

To calculate ϵ we need to compute the contribution of $\mathcal{L}_{eff}^{CP\ viol}$ between a $|K^0>$ and a $|\bar{K}^0>$ state. One has:

$$e^{-\frac{i\pi}{4}}\epsilon = \frac{1}{\sqrt{2}(\Delta m)_K} < K^0|Im\ M_{12}|\bar{K}^0> = -\frac{1}{\sqrt{2}(\Delta m)_K}\frac{< K^0|\mathcal{L}_{eff}^{CP\ viol}|\bar{K}^0 >}{2M_K} \tag{297}$$

Unfortunately, it is not possible yet to compute the above hadronic matrix element accurately. What is done, usually, is to replace the quark matrix element in Eq(297) by

$$< K^0|[\bar{d}\gamma^\mu(1-\gamma_5)s][\bar{d}\gamma_\mu(1-\gamma_5)s]|\bar{K}^0 > = -\frac{8}{3}f_K^2 M_K^2 B_K \tag{298}$$

Apart from B_K, this is the result one would have obtained by inserting $|0><0|$ in the above, using the definition of the Kaon decay constant ($f_K \simeq 160 MeV$):

$$< K^0|\bar{d}\gamma_\mu\gamma_5 s|0> = if_K p_\mu \tag{299}$$

and taking into account of color factors [74]. The uncertainty in the hadronic matrix element resides then in B_K, with $B_K = 1$ - obviously - representing the vacuum insertion approximation. Although B_K is not really calculable, present day estimates for B_K [75] put it in the range

$$0.3 \leq B_K \leq 1 \tag{300}$$

I quote below the final formula for ϵ, which uses the results derived above, but includes also some kinematical correction factors connected with terms of $0(y_t = \frac{m_t^2}{M_W^2})$, which may not be negligible if top is not light [76]. This formula, in addition, contains parameters η_i which are short distance QCD corrections to the effective Lagrangian of Eq(296) [77] ($\eta_1 \simeq 0.7$; $\eta_2 \simeq 0.6$; $\eta_3 \simeq 0.4$),

$$e^{-\frac{i\pi}{4}}\epsilon = \frac{G_F^2 f_K^2 M_K}{6\sqrt{2}\pi^2 (\Delta m)_K} B_K [A^2 \rho \lambda^6 \sin\delta] \left\{ m_c^2 [-\eta_1 + \eta_3 (\ln\frac{m_t^2}{m_c^2} + f_3(y_t))] \right.$$
$$\left. + m_t^2 [\eta_2 f_2(y_t) A^2 \lambda^4 (1 - \rho\cos\delta)] \right\} \tag{301}$$

Here f_2 and f_3 are weakly dependent functions of the top quark mass

$$f_2(y_t) = 1 - \frac{3y_t(1 + y_t)}{4(1 - y_t)^2}\left[1 + \frac{2y_t}{1 - y_t^2}\ln y_t\right], \tag{302}$$

$$f_3(y_t) = -\frac{3y_t}{4(1 - y_t)}\left[1 + \frac{y_t}{1 - y_t}\ln y_t\right] \tag{303}$$

5.3 Penguin Operators and ϵ'

Before I analyze the formula for ϵ, I want to discuss the origin for ϵ' in the standard model. I mentioned earlier, that in the standard model-adopting the natural quark phase convention - there is only an imaginary part for A_0 and not A_2. That is, there is a complex weak contribution to transitions involving a change in strong isospin equal to $\frac{1}{2}$, but not $\frac{3}{2}$. Since

$$\epsilon' \sim Im \; < 2\pi | \mathcal{L}_{weak}^{\Delta S=1} | K > \tag{304}$$

one must find complex contributions to the $\Delta S = 1$ weak Lagrangian and show that, indeed, they affect only A_0. This is actually very easy to see, since the leading contribution in Eq(304) is that provided by the Penguin operator of Fig 38. Since this is a pure $d \to s$ transition, one has obviously a pure $\Delta I = \frac{1}{2}$ contribution.

Most of the work needed for the computation of the Penguin diagrams has already been done. We know already the $\ln\frac{M_W^2}{m_q^2}$ behaviour of the individual graphs. Furthermore, I have given an example earlier of the cancellations that ensue, due to CKM unitarity, for the case of Penguins in B decay [cf Eq(251)]. For the Penguin diagrams of Fig 38, the relevant logarithmic terms are given by

$$P \sim \sum_i \lambda_i \ln\frac{M_W^2}{m_i^2} = \lambda_t \ln\frac{m_c^2}{m_t^2} + \lambda_u \ln\frac{m_c^2}{m_u^2} \tag{305}$$

where I have used Eq(291) to eliminate λ_c. Since λ_u is purely real, the contribution to ϵ' arises only from the imaginary part of λ_t:

$$\lambda_t \simeq -A\lambda^5(1 - \rho e^{-i\delta}) \leftrightarrow Im\lambda_t = -A\lambda^5 \rho \sin\delta \tag{306}$$

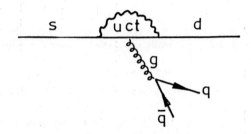

Figure 38: Penguin diagram giving rise to a $\Delta I = \frac{1}{2}$ CP violating contribution.

One finds, for the CP violating effective Lagrangian [78]

$$\mathcal{L}_{Penguin}^{CP\ viol} = \frac{G_F}{\sqrt{2}}[A^2\rho\lambda^5 \sin\delta][\frac{\alpha_s}{12\pi}\ln\frac{m_t^2}{m_c^2}]\{(\bar{s}\gamma^\mu(1-\gamma_5)\lambda_a d)\sum_q(\bar{q}\gamma_\mu\lambda_a q)\} \quad (307)$$

This expression is slightly modified by QCD corrections [79] and by $y_t = \frac{m_t^2}{M_W^2}$ effects. The net result of incorporating these modifications is to replace the 2nd function in the square brackets in Eq(307) by an effective coefficient C_P:

$$\frac{\alpha_s}{12\pi}\ln\frac{m_t^2}{m_c^2} \rightarrow C_P = 0.085 \pm 0.035 \quad (308)$$

The numerical value above is that estimated by a recent analysis of these effects by Altarelli and Franzini [80].

To calculate ϵ', according to Eq(287), one needs to calculate the ratio ξ of $Im\ A_0$ to $Re\ A_0$. Calling the operator in the curly brackets in Eq(307) O_P, one has

$$Im\ A_0 = -<\pi\pi; I=0|\mathcal{L}_{Penguin}^{CP\ viol}|K^0> = -[A^2\rho\lambda^4\sin\delta]<\pi\pi; I=0|\frac{G_F}{\sqrt{2}}\lambda C_P O_P|K^0> \quad (309)$$

On the other hand, $Re\ A_0$ is just given by the matrix element of the $\Delta S = 1$ current-current Lagrangian:

$$Re\ A_0 = -<\pi\pi; I=0|\frac{G_F}{\sqrt{2}}\lambda\bar{s}\gamma^\mu(1-\gamma_5)u\bar{u}\gamma_\mu(1-\gamma_5)d)|K^0> \quad (310)$$

Using Eqs(309) and (310) and Eq(287), one finds

$$\begin{aligned}\epsilon' &= -\frac{1}{\sqrt{2}}[\frac{1}{20}]e^{i(\frac{\pi}{2}+\delta_2-\delta_0)}\xi \\ &= -0.032 e^{i(\frac{\pi}{2}+\delta_2-\delta_0)}[A^2\rho\lambda^4\sin\delta]H\end{aligned} \quad (311)$$

Here H contains the ratio of the hadronic matrix elements entering in Eq(309) and (310):

$$H = \frac{<\pi\pi; I=0|C_P O_P|K^0>}{<\pi\pi; I=0|\bar{s}\gamma^\mu(1-\gamma_5)u\bar{u}\gamma_\mu(1-\gamma_5)d|K^0>} \quad (312)$$

Several remarks are in order:

1. Independent of a precise evaluation of H, one sees that ϵ' is numerically small, both because of the $\Delta I = \frac{1}{2}$ suppression factor of 0.032 and as a result of the high powers of λ entering in its coefficient $[A^2\rho\lambda^4\sin\delta]$.

2. The hadronic matrix element ratio H is slightly dependent on m_t through C_P. However, here the dependence is only logarithmic. Unfortunately this ratio is theoretically quite poorly estimated. For instance, the recent analysis of Altarelli and Franzini [80] gives the range

$$-0.43 \leq H \leq -0.03 \quad (313)$$

3. The sign of Eq(313) implies that the quantity multiplying the phase factor in ϵ' is positive, so that $\frac{\epsilon'}{\epsilon} > 0$ [23]. This sign agrees with that determined experimentally by the NA31 collaboration [68]. It is also in accord with some general arguments of Hagelin [81].

Unfortunately, given the large uncertainty in H, in my opinion the beautiful measurement of ϵ' does little to determine the phase δ. However, the NA31 result [68], Eq(275), is **perfectly consistent** with the standard model. For instance, if one uses the central values for λ, A, ρ and H [$\lambda = 0.22$, $A = 1.05$, $\rho = 0.6, H = 0.23$] one finds

$$|\epsilon'| = 1.9 \times 10^{-5}\sin\delta \quad (314)$$

while, experimentally, using Eqs(274) and (275) one has

$$|\epsilon'|_{exp} = (0.8 \pm 0.3) \times 10^{-5} \quad (315)$$

To get an estimate of the phase δ, it is much better to use ϵ and the recent new information on $B_d - \bar{B}_d$ oscillations, to which I now turn.

5.4 $B_d - \bar{B}_d$ Mixing

I discussed earlier how through $2nd$ order weak effects a particle-antiparticle transmutation can occur. A state which starts originally as a B^0 can, via mixing, become in time a superposition of B^0 and $\bar{B}^0$ states [cf. Eq. (269)]:

$$|B^0_{phys}(t)> = f_+(t)|B^0> + f_-(t)(\frac{1-\epsilon_B}{1+\epsilon_B})|\bar{B}^0> \quad (316)$$

---
[23] Recall that the phases in ϵ' and ϵ are approximately the same.

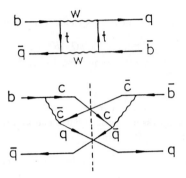

Figure 39: Mass and decay mixing in the B-$\bar{B}$ complex.

where $f_{\pm}(t)$ are given in Eqs(270) and (271). For B mesons the mass matrix M_{12} dominates over the decay matrix Γ_{12}. Hence the function $f_{\pm}(t)$ can be approximated by

$$f_+(t) = e^{-imt}e^{-\frac{\gamma}{2}t}\cos\frac{\Delta m}{2}t$$
$$f_-(t) = ie^{-imt}e^{-\frac{\gamma}{2}t}\sin\frac{\Delta m}{2}t \qquad (317)$$

with the characteristic parameter of the oscillation being $x = \frac{\Delta m}{\gamma}$.

The fact that $M_{12} \gg \Gamma_{12}$ for the B system is easily understood by comparing the graphs contributing to B-$\bar{B}$ transitions shown in Fig. 39. Clearly one sees from the figure that for the $B_q - \bar{B}_q$ system

$$\Delta m \sim m_t^2 |V_{tb}V_{tq}^*|^2 \qquad (318)$$

while

$$\Delta\Gamma \sim m_b^2 |V_{cb}V_{cq}^*|^2 \qquad (319)$$

Although the Kobayashi Maskawa matrix elements are comparable, one can neglect $\Delta\Gamma$ relative to Δm because of the phase space difference: m_b^2 versus m_t^2.

Normally a $B_d^0 \sim (\bar{b}d)$ state has a semileptonic decay into an l^+: $B_d^0 \to l^+\nu_l X$. However, with $B_d^0 - \bar{B}_d^0$ mixing one expects also semileptonic decays into an l^-,

$$B_d^0 \stackrel{mix}{\to} \bar{B}_d^0 \to l^- \bar{\nu}_l X \qquad (320)$$

The ratio [24]

$$r_d = \frac{\Gamma(B_d^0 \to l^-\bar{\nu}_l X)}{\Gamma(B_d^0 \to l^+\nu_l X)} = \frac{\int_0^\infty |f_-(t)|^2}{\int_0^\infty |f_+(t)|^2} = \frac{x_d^2}{2+x_d^2} \qquad (321)$$

measures the amount of $B_d^0 - \bar{B}_d^0$ mixing, with the parameter x_d

$$x_d = \frac{(\Delta m)_{B_d}}{(\gamma)_{B_d}} = \tau_{B_d}(\Delta m)_{B_d} \qquad (322)$$

---

[24] For the B_d system, ϵ_{B_d} is negligible in magnitude and can be neglected.

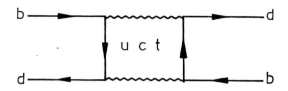

Figure 40: Box graph contributing to x_d.

being a function of the CKM matrix elements. The ARGUS collaboration in 1987 discovered that there was substantial mixing in the B_d system [82]. They had the good fortune to actually detect an event with two fully reconstructed B_d's, thereby obtaining explicit evidence for mixing. To actually extract the mixing parameter quantitatively, they studied their data sample for the presence of same sign leptons and of reconstructed B_d's associated with leptons of the wrong sign. From both these signals ARGUS deduced that [82]:

$$r_d = \frac{\Gamma(B_d^0 \to l^- \bar{\nu}_l X)}{\Gamma(B_d^0 \to l^+ \nu_l X)} = 0.21 \pm 0.08 \tag{323}$$

which implies that $x_d = 0.73 \pm 0.18$. Very recently the ARGUS result has been confirmed by CLEO [83], with the combined results leading to a mixing parameter

$$x_d = 0.70 \pm 0.13 \tag{324}$$

To calculate x_d theoretically, one is invited to calculate the box graph of Fig. 40. This calculation proceeds in an analogous fashion to the one in the Kaon system, done earlier in these lectures. Here, however, after having made use of the unitarity of the CKM matrix, one need only retain the t loop contribution, since both c and t loops are of $O(\lambda^6)$ and $m_t^2 >> m_c^2$. One finds [84]

$$x_d = \tau_{B_d} \frac{G_F^2 M_B}{6\pi^2} [f_{B_d}^2 B_{B_d} \eta_B][m_t^2 f_2(y_t)]|V_{td}|^2 \tag{325}$$

In the above $\eta_B \simeq 0.85$ [84] is a QCD correction factor, while $f_{B_d}^2 B_{B_d}$ is the analogue for the B_d system of $f_K^2 B_K$. A reasonable range for this hadronic uncertainty is [85]:

$$100 \; MeV << (f_{B_d}^2 B_{B_d})^{\frac{1}{2}} << 200 \; MeV \tag{326}$$

Because the ARGUS result for x_d [82] turned out to be bigger than theorists expected prior to the meausurment, one needed to push upwards the various factors appearing in the formula for x_d. In particular, the large mixing observed implies that top cannot be too light. Although one cannot make an absolute pronouncement on this, the results of very many detailed studies [86] suggest that

$$(m_t)_{theory} > 50 \ GeV, \tag{327}$$

which is comparable to the UA1 bound of Eq(162). Furthermore, at least in the 3 generation CKM model, also V_{td} cannot be too small. Now

$$|V_{td}|^2 = A^2 \lambda^6 \left[1 + \rho^2 - 2\rho \cos \delta \right], \tag{328}$$

so the mixing is largest when ρ is large and $\delta \to \pi$. However, δ is not really allowed to get to π, because then all CP violation disappears in the CKM model!

As the last point in these lectures, I want to give the results of a combined fit, made in collaboration with Krawczyk, London and Steger [85], of the x_d and ϵ formulas to determine the allowed area in the $\rho - \delta$ plane, for given values of m_t and of the hadronic uncertainties B_K and $f_{B_d}^2 B_{B_d}$. In the analysis of [85] $f_{B_d}^2 B_{B_d}$ was taken in the range of Eq(326), B_K in the range of Eq(300) and we assumed $40 \ GeV < m_t < 180 \ GeV$. Our results are shown in Fig. 41. As one can see from the figure, the allowed regions are "moons" in the $\rho - \delta$ plane, which fatten up as one increases the theoretical uncertainty. Knowledge of the top mass would greatly reduce the allowed region in $\rho - \delta$ space, even in the case depicted in Fig. 42 in which the whole range of theoretical uncertainty assumed is allowed.

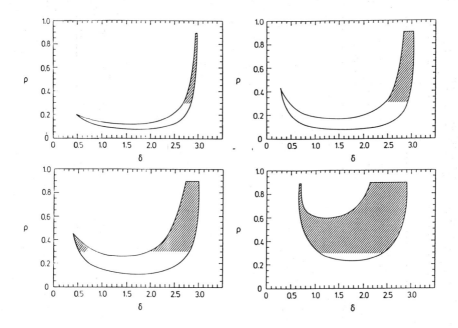

Figure 41: Summary of the present information on the CP violating phase δ. In all curves the whole m_t uncertainty is allowed. In the top left curve $f_{B_d}^2 B_{B_d} = 150\ MeV$ and $B_K = 1$. In the other curves $f_{B_d}^2 B_{B_d}$ was taken in the full range of uncertainty and $B_K = 1,\ \frac{2}{3},\ \frac{1}{3}$. From [85].

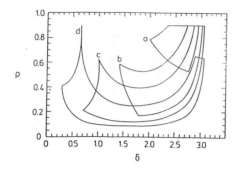

Figure 42: Allowed regions in the $\rho - \delta$ plane for fixed values of m_t, but for the full theoretical uncertainty. Regions $a - d$ correspond to $m_t = 60, 90, 120, 180\ GeV$, respectively. From [85].

References

[1] S.L. Glashow, Nucl. Phys. 22 (1961) 579; A. Salam in **Elementary Particle Theory**, ed. N. Svartholm (Almqvist and Wiksell, Stockholm, 1968); S. Weinberg, Phys. Rev. Lett. 19 (1967) 1264

[2] See for example, E. Abers and B.W. Lee, Phys. Rept. 9C (1973) 1

[3] N. Cabibbo, Phys. Rev. Lett. 10 (1963) 531; M. Kobayashi and T. Maskawa, Prog. Theo. Phys. 49 (1973) 652

[4] C. Llewellyn Smith, Nucl. Phys. B228 (1983) 205

[5] CDHS Collaboration H. Abramowicz et al., Phys. Rev. Lett. 57 (1986) 298

[6] CHARM Collaboration J.V. Allaby et al., Phys. Lett. 177B (1986) 446

[7] CCFRR Collaboration P.G. Reutens et al, Phys. Lett. 152B (1985)404

[8] FMM Collaboration D. Bogert et al., Phys. Rev. Lett. 55 (1985) 1969

[9] U. Amaldi et al., Phys. Rev. D36 (1987) 1385

[10] G. Costa et al., Nucl. Phys. B297 (1988) 244

[11] CDHS Collaboration H. Abramowicz et al., Z. Phys. C 28 (1985) 51

[12] CHARM Collaboration M. Jonker et al., Phys. Lett. 99B (1981) 265; 103B (1981) 469(E)

[13] P. Jenni, in Proceedings of the 1987 International Symposium on Lepton and Photon Interactions at High Energy, Hamburg, ed. W. Bartel and R. Ruckl, Nucl. Phys. B (Proc. Suppl.) 3 (1988) 341

[14] W. Marciano and A. Sirlin, Phys. Rev. D12 (1980) 2695; D29 (1984) 945; M. Consoli, S. Lo Presti and L. Maiani, Nuc. Phys. B223 (1983) 474; Z. Hioki, Prog. Theor. Phys. 68 (1982) 2134; Nuc. Phys. B229 (1983) 284

[15] A. Sirlin, Phys. Rev. D22 (1980) 971

[16] W. Marciano, Phys. Rev. D20 (1979) 274; F. Antonelli and L. Maiani, Nucl. Phys. B 186 (1981) 269; For a very nice discussion, see also, L. Maiani in the TASI Lectures 1984, Univ. of Michigan, Ann Arbor, Michigan.

[17] See for example, C. Itzykson and J.B. Zuber **Quantum Field Theory** (McGraw Hill, N.Y. 1980)

[18] W. Marciano and A. Sirlin, Phys.Rev. D29 (1984) 945

[19] For a discussion of renormalization group methods to sum QCD corrections to weak process, see for example, G. Altarelli, R.K. Ellis, L. Maiani and R. Petronzio Nucl. Phys. B 88 (1975) 215

[20] M. Veltman, Nucl. Phys. B 123 (1977) 89

[21] W. Marciano and A. Sirlin, Phys. Rev. D22 (1980) 2695

[22] See for example, P. Baillon et al. in **Physics at LEP** ed. J. Ellis and R.D. Peccei CERN Yellow Report CERN 86-02

[23] P. Langacker in **Neutrino Physics** ed. H.V. Klapdor (Springer, Berlin, 1988)

[24] JADE Collaboration, W. Bartel et al., Phys. Lett. 146B (1984) 437

[25] S.L. Wu in Proceedings of the 1987 International Symposium on Lepton and Photon Interactions at High Energy, Hamburg, ed. W. Bartel and R. Ruckl, Nucl. Phys. B (Proc. Suppl.) 3 (1988) 39

[26] T. Kamae, to appear in the Proceedings of the XXIV Int. Conference on High Energy Physics, Munich, West Germany, Aug. 1988

[27] H. Kichimi, to appear in the Proceedings of the 8th Physics in Collision Conference, Capri, Italy, Oct. 1988

[28] UA1 Collaboration, C. Albajar et al., Zeit. Phys. 37C (1988) 505

[29] G. Altarelli, M. Diemoz, G. Martinelli and P. Nason, Nucl. Phys. B 308 (1988) 724

[30] P. Nason, S. Dawson and R.K. Ellis, Nucl. Phys. B 303 (1988) 607

[31] J. Yang, M.S. Turner, G. Steigman, D.N. Schramm and K.A. Olive, Astrophys. J. 281 (1984) 443

[32] E. Ma and J. Okada, Phys. Rev. Lett. 41 (1978) 287; See also G. Barbiellini, B. Richter and J.L. Siegrist, Phys. Lett. 106B (1981) 414

[33] K.J.F. Gaemers, R. Gastmans and F.M. Renard, Phys. Rev. D19 (1979) 1605

[34] ASP Collaboration, C. Hearty et al., Phys. Rev. Lett. 58 (1987) 1711

[35] CELLO Collaboration, H.J. Behrend et al., Phys. Lett. 215B (1988) 179

[36] UA1 Collaboration, C. Albajar et al., Phys. Lett. 98B (1987) 271

[37] UA2 Collaboration, R. Ansari et al., Phys. Lett. 186B (1987) 440

[38] P. Colas, D. Denegri and C. Stubenrauch, Zeit. Phys. C40 (1988) 527

[39] See, for example, the recent global analysis by A. Ali and F. Barreiro, DESY preprint DESY 88-075

[40] A. Blondel et al., in **Physics at LEP** ed. by J. Ellis and R.D. Peccei, CERN Yellow Report CERN 86-02

[41] L. Maiani, Phys. Lett. B62 (1976) 183

[42] L. Wolfenstein, Phys. Rev. Lett. 51 (1983) 1945

[43] A. Sirlin, in Proceedings of the 1987 International Symposium on Lepton and Photon Interactions at High Energy, Hamburg, ed. W. Bartel and R. Ruckl, Nucl. Phys. B (Proc. Suppl.) 417 (1988)

[44] S.S. Gershtein and Ya. B. Zeldovich, JETP 29, 698 (1955); Soviet Physics JETP 2, 576 (1958), translation; E.C.G. Sudarshan and R.E. Marshak, Proc. Padua-Venice Conf. on Mesons and Recently Discovered Particles (1957) V-14. Reprinted in P.K. Kabir, **Development of the Weak Interaction Theory** (Gordon and Breach, New York 1963) p. 118; R.P. Feynman and M. Gell-Mann, Phys. Rev. 109, 193 (1958)

[45] R. Behrends and A. Sirlin, Phys. Rev. Lett. 4 (1960) 186; M. Ademollo and R. Gatto, Phys. Rev. Lett. 13 (1964) 264

[46] A. Sirlin, Phys. Rev. D35 (1987) 3423

[47] V.T. Koslowsky et al., in Proceedings of the Seventh International Conference on Atomic Masses and Fundamental Constants, ed. O. Klepper (GSI, Darmstadt, 1984)

[48] N. Cabibbo, Phys. Rev. Lett. 10 (1963) 531

[49] H. Leutwyler and M. Roos, Z. Phys. C25 (1984) 91

[50] Particle Data Book, Phys. Lett. 170B (1986) 1

[51] For a review, see for example H. Pagels, Phys. Rept. 16C (1975) 219; See also, J. Gasser and H. Leutwyler, Ann. Phys, (N.Y.) 158 (1984) 142

[52] See for example, J. Gasser and H. Leutwyler, Phys. Rept. 87 (1982) 77

[53] G. Furlan, F. G. Lannoy, C. Rossetti, G. Segre, Nuovo Cim. 38 (1965) 1747

[54] J. Gasser and H. Leutwyler, Nucl. Phys. B250 (1984) 465, 517, 539

[55] CDHS Collaboration H. Abramowicz et al, Zeit. Phys. C15 (1982) 19

[56] ARGUS Collaboration H. Schröder, Proceedings of the International Conference on Hadron Spectroscopy, ed. S. Oneda (AIP, New York, 1985)

[57] G. Altarelli, N. Cabibbo, G. Corbo, L. Maiani and G. Martinelli, Nucl.Phys. B 208 (1982) 365

[58] M. Gilchriese, in Proceedings of the XXIII International Conference on High Energy Physics, Berkeley, USA (1986)

[59] ARGUS Collaboration H. Schröder, to appear in the Proceedings of the XXIV International Conference on High Energy Physics, Munich, West Germany, August 1988; See also, CLEO Collaboration R. Fulton et al, preprint CBX-88-13 (1988)

[60] ARGUS Collaboration H. Albrecht et al., Phys. Lett. 209B (1988) 119

[61] CLEO Collaboration D. Kreinek, to appear in the Proceedings of the XXIV International Conference on High Energy Physics, Munich, West Germany, August 1988

[62] For a similar discussion, see also, M. Gronau and J. Rosner, Phys. Rev. D37 (1988) 688

[63] G. Altarelli and P. Franzini, Z. Phys. C37 (1988) 271

[64] D.M. Ritson, in Proceedings of the XXIII International Conference on High Energy Physics, Berkeley, ed. S. Loken (World Scientific, Singapore, 1987)

[65] G. S. Kim, private communication. See also, G. Belanger et al, to appear in the Proceedings of the 1988 Snowmass Summer Study on High Energy Physics in the 1990's, Snowmass, Colorado, USA, July 1988

[66] T. D. Lee and C. S. Wu, Ann. Rev. Nucl. Sci. 16 (1966) 511

[67] K. Kleinknecht, Ann. Rev. Nucl. Sci. 26 (1976) 1

[68] NA31 Collaboration H. Burkhardt et al., Phys. Lett. 206B (1988) 169

[69] M. Woods et al, Phys. Rev. Lett. 60 (1988) 1695

[70] L. Wolfenstein, Ann. Rev. Nucl. Sci. 36 (1986) 137

[71] T. T. Wu and C. N. Yang, Phys. Rev. Lett. 13 (1964) 380

[72] T. T. Devlin and J. O. Dickey, Rev. Mod. Phys. 51 (1979) 237

[73] B. W. Lee and M. K. Gaillard, Phys. Rev. D10 (1974) 897

[74] R.E. Shrock and S. B. Treiman, Phys. Rev. D19 (1979) 2148

[75] For a discussion of various determinations of B_K, see for example, L. Wolfenstein, Ann. Rev. Nucl. Sci. 36 (1986) 137

[76] A. J. Buras, W. Slominski and H. Steger, Nucl. Phys. B 238 (1984) 529

[77] F. Gilman and M. Wise, Phys. Rev. D27 (1983) 1128

[78] F. Gilman and M. Wise, Phys. Lett. 83B (1979) 83

[79] B. Guberina and R. D. Peccei, Nucl. Phys. B 163 (1980) 289; F. Gilman and M. Wise, Phys. Rev. D20 (1979) 2392

[80] G. Altarelli and P. Franzini, Cern Th 4914/87, to appear in the Proceedings of the Symposium on Present Trends, Concepts and Instruments of Particle Physics, Rome, Nov 1987

[81] J. Hagelin, in Proc. Moriond Workshop on Electroweak Interactions and Unified Theories, ed J. Tran Than Van (Editions Frontiers, 1984); L. Wolfenstein, Comments Nucl. Part. Phys. 14 (1985) 135

[82] ARGUS Collaboration H. Albrecht et al, Phys. Lett. 192B (1987) 245

[83] CLEO Collaboration A. Jawahery, to appear in the Proceedings of the XXIV International Conference on High Energy Physics, Munich, West Germany, August 1988

[84] A. J. Buras, W. Slominski and H. Steger, Nucl. Phys. B 245 (1984) 369; J. Hagelin, Nucl. Phys. B 193 (1981) 123

[85] P. Krawczyk, D. London, R.D. Peccei and H. Steger, Nucl. Phys. B 307 (1988) 19

[86] G. Altarelli and P. Franzini, Z. Phys. C37 (1988) 271; J. Donoghue, T. Nakada, E. Paschos and D. Wyler, Phys. Lett. 195B (1987) 285; J. Ellis, J. Hagelin and S. Rudaz, Phys. Lett. 192B (1987) 201; I. I. Bigi and A. I. Sanda, Phys. Lett. 194B (1987) 307; J. Maalampi and M. Roos, Phys. Lett. 195B (1987) 489; V. Barger, T. Han, D. V. Nanopoulos and R. J. N. Phillips, Phys. Lett. 194B (1987) 312

THE STANDARD MODEL AND BEYOND: THE AXION PHYSICS

JIHN E. KIM
Department of Physics*
Seoul National University
Seoul 151-742, Korea
and
Department of Physics
Brown University
Providence, RI 02912

ABSTRACT

I review the prototype grand unified theory SU(5) briefly, and discuss the theoretical aspects of axion physics.

---

* Permanent address with bitnet:JEKIM@KRSNUCC1

1. Introduction and Parameters of the Standard Model

One of the most successful theories in particle physics is the standard model[1] based on $SU(3) \times SU(2) \times U(1)$. The minimal particle content in the standard model requires the left-right asymmetric fermions of 3 families and 1 Higgs doublet,

$$\begin{pmatrix} u \\ d' \end{pmatrix}_L \quad \begin{pmatrix} c \\ s' \end{pmatrix}_L \quad \begin{pmatrix} t \\ b' \end{pmatrix}_L \quad u_R \quad c_R \quad t_R \quad d_R \quad s_R \quad b_R \quad (1.a)$$

$$\begin{pmatrix} \nu_e \\ e \end{pmatrix}_L \quad \begin{pmatrix} \nu_\mu \\ \mu \end{pmatrix}_L \quad \begin{pmatrix} \nu_\tau \\ \tau \end{pmatrix}_L \quad e_R \quad \mu_R \quad \tau_R$$

$$\phi = \begin{pmatrix} \phi^+ \\ \phi^\circ \end{pmatrix} \quad (1.b)$$

where d', s' and b' are the states in the weak interaction basis as suggested by Kobayashi and Maskawa.[2] The success [3] of the standard model can be seen from the predicted and verified neutral currents and the W/Z physics.[4] It is remarkable that the neutral current physics over the 10-11 orders in the energy scale ($10^{-9} GeV$ for the atomic parity violation and $100\ GeV$ for the W/Z physics) are successfully described by the minimal standard model. The essentially untested ingredients are t quark and the nature of $SU(2) \times U(1)$ symmetry breaking.

Just because of this remarkable success of the standard model, we take the standard model as the electroweak theory below $100\ GeV$ and can think of "beyond". "Beyond" here means any particle contents (gauge bosons, fermions, or scalars) in addition to those of the standard model. Otherwise, we may be still trying to find a reasonable theory at the electroweak scale. Because of the renormalizability, however, it is also possible to extend the standard model, without "beyond", from $100\ GeV$ scale up to $10^{19}\ GeV$ scale. But this dull extension does not answer the *parameter problem* of the standard model. We count 20 standard model parameters which arise as follows

Gauge bosons : α_3, α_2, α_1, M_W, M_Z

Higgs potential (1 Higgs doublet) : λ, M_H^2

Leptons (and Yukawa couplings) : m_e, m_μ, m_τ

Quarks (and Yukawa couplings) : m_u, m_d, m_c, m_s, m_t, m_b, θ_1, θ_2, θ_3, δ

There are two relations between these parameters,

$$M_H^2 = 2\mu^2 = 2\lambda v^2 = \lambda(2/\pi)(M_W^2/\alpha_2) \;,\; M_W^2 = M_Z^2/\cos^2\theta_W$$

where θ_W can be expressed in terms of α_2 and α_1. In addition, there is a θ parameter for each nonabelian gauge group,e.g.

$$\theta_{SU(3)}, \theta_{SU(2)}$$

Table 1.

The standard model parameters and some suggested theories to understand these. GUTs and axions are are most successful in this respect.

Parameters	*Theories to understand why*
Gauge couplings	**Grand unification**
Fermion masses and mixing angles	Horizontal symmetry, Extended GUTs, Composite models
Higgs masses (or electroweak symmetry breaking)	Technicolor, Supersymmetry (supergravity)
$\theta(\equiv \theta_{SU(3)})$ parameter	**Axions**
(Gravity)	(Superstring)

Thus the standard model contains 20 free parameters. The *parameter problem* is a theoretical problem of *why* these 20 parameters in the minimal model take the observed values, even though we know *how* the standard model describes the electroweak phenomena correctly with these parameters. In the last decade, we witnessed major theoretical achievements in attempts to understand the *parameter problem* as shown in Table 1.

Theoretical extensions of the standard model can be achieved in many various ways. But the most attractive track seems to be the one toward the understanding of *parameter problem*. In this lecture, I will concentrate on the QCD θ parameter and the axion physics after briefly introducing the simplest grand unified theory $SU(5)$.

2. Grand Unification

There are many excellent reviews on the grand unified theories(GUTs). [5] Here, I will be very brief on GUTs in general, but try to focus on the gauge coupling unification which elegantly solves some of the *parameter problem* of the standard model. For this I will discuss the prototype GUT $SU(5)$ which is the simplest one and contains basically most features encountered in the other GUT models.

The fermion representation, Eq.(1a), in the standard model is chiral, which will become clear in the following discussion of the Lorentz group. The standard caveat for the group theory discussion by particle physicists is really studying the *algebra* instead of the *group*. This means that we use the information of the group elements close to identity. So when I say *groups*, sometimes I really mean *algebras*.

The Lorentz transformation is a homogeneous linear transformation of the space-time coordinate

$$x^\mu \to x'^\mu = \Lambda^\mu_\nu x^\nu$$

The invariant distance is

$$\eta_{\mu\nu}x'^{\mu}x'^{\nu} = \eta_{\mu\nu}x^{\mu}x^{\nu}$$

where $\eta_{\mu\nu} = (1, -1, -1, -1)$ is the invariant metric for the flat space-time. Thus we have

$$\Lambda^{\mu}_{\rho}\Lambda^{\nu}_{\sigma}\eta^{\rho\sigma} = \eta^{\mu\nu} \tag{2}$$

The Lorentz algebra is obtained by considering an infinitesimal transformation,

$$\Lambda^{\mu}_{\nu} = \delta^{\mu}_{\nu} + \delta\omega^{\mu}_{\nu} \tag{3}$$

where $\omega_{\mu\nu}$ is derived to be antisymmetric

$$\delta\omega_{\mu\nu} = -\delta\omega_{\nu\mu} \tag{4}$$

The infinitesimal matrix $\delta\omega_{\mu\nu}$ can be split into the infinitesimal parameters $\delta\omega^{\alpha\beta}$ times matrices (= generators) $S_{\alpha\beta}$. Thus

$$\Lambda^{\mu}{}_{\nu} = \delta^{\mu}{}_{\nu} + \frac{i}{2}\delta\omega^{\alpha\beta}(S_{\alpha\beta})^{\mu}{}_{\nu} \tag{5}$$

For the vector representation Eq.(3), we have

$$(S_{\alpha\beta})^{\mu}{}_{\nu} = -i(\delta^{\mu}_{\alpha}\eta_{\beta\nu} - \delta^{\mu}_{\beta}\eta_{\alpha\nu}) \tag{6}$$

Obviously, S satisfies $(S^*_{\alpha\beta})_{\mu\nu} = (S_{\alpha\beta})_{\nu\mu}$. We can easily check that S satisfies the following algebra,

$$[S_{\alpha\beta}, S_{\gamma\delta}] = i(\eta_{\alpha\gamma}S_{\beta\delta} + \eta_{\beta\delta}S_{\alpha\gamma} - \eta_{\beta\gamma}S_{\alpha\delta} - \eta_{\alpha\delta}S_{\beta\gamma}) \tag{7}$$

The algebra Eq.(6) is the Lorentz algebra. For any representations, which we denote as $M_{\alpha\beta}$, of the Lorentz group, the same algebra is satisfied. Also it satisfies the conditions

$$M_{\alpha\beta} = -M_{\beta\alpha}$$
$$(M^*_{\alpha\beta})_{\mu\nu} = (M_{\alpha\beta})_{\nu\mu} \qquad (8)$$

The physical meaning of these Lorentz generators is clearer after spliting them into the rotation(J) and boost(K) generators

$$J_i = \frac{1}{2}\epsilon_{ijk}M_{jk} \quad (\text{or } M_{jk} = \epsilon_{jki}J_i)$$
$$K_i = M_{0i} \qquad (9)$$

In terms of J and K, the algebra can be written as

$$[J_i, J_j] = i\epsilon_{ijk}J_k$$
$$[J_i, K_j] = i\epsilon_{ijk}K_k \qquad (10)$$
$$[K_i, K_j] = -i\epsilon_{ijk}J_k$$

Both J and K are hermitian operators. J is the well-known angular momentum generator for the $SO(3)$ subgroup.

For the discussion of the chiral representations, it is more convenient to construct another nonhermitian generators out of J and K,

$$N_i = \frac{1}{2}(J_i + iK_i)$$
$$N_i^\dagger = \frac{1}{2}(J_i - iK_i) \qquad (11)$$

Then the Lorentz algebra is expressed in terms of N and $N^\dagger$ as

$$[N_i, N_j] = i\epsilon_{ijk}N_k$$
$$[N_i^\dagger, N_j^\dagger] = i\epsilon_{ijk}N_k^\dagger \qquad (12)$$
$$[N_i, N_j^\dagger] = 0$$

Because N and $N^\dagger$ commute and each set satisfies separately the $SU(2)$ algebra, we can see that the Lorentz algebra is isomorphic to the $SU(2) \times SU(2)$

algebra. Thus a representation of the Lorentz group can also be represented by a representation of $SU(2)_L \times SU(2)_R$ where we distinguished two different $SU(2)$'s by subscripts L and R. Thus let (p,q) be a representation of the Lorentz group where p and q denote the $SU(2)_L$ and $SU(2)_R$ representations, respectively. Since $N_i^\dagger = (N_i)^\dagger$, we have

$$(p,q)^* \to (q,p) \tag{13}$$

There are four frequently used Lorentz representations

(1 , 1) :   scalar
(2 , 1) :   left-handed spinor
(1 , 2) :   right-handed spinor
(2 , 2) :   vector (x_μ)

For the left-handed spinors we use [6] indices $a, b... (a = 1,2)$. For the right-handed spinors we use dotted ones $\dot{a}, \dot{b}, ... (\dot{a} = 1,2)$.

It is very useful to consider the tensor products to obtain the Lorentz invariant couplings. For example,

$$(2,1) \otimes (2,1) = (1,1) + (3,1) \tag{14}$$

from which we know that an invariant coupling is possible from a tensor product of two $(2,1)$'s. In fact the singlet arises from the antisymmetric combination,

$$\epsilon^{ab}\psi_a \psi'_b$$

where $\epsilon^{12} = 1$, $\epsilon^{21} = -1$, $\epsilon^{11} = \epsilon^{22} = 0$. Similarly, we can construct an invariant from the right-handed spinors,

$$\epsilon^{\dot{a}\dot{b}}\psi_{\dot{a}}\psi'_{\dot{b}}$$

There are more important invariant symbols,

$$(2,1) \otimes (1,2) \otimes (2,2) = 1(\sigma^\mu_{a\dot{a}}) \oplus \ldots$$
$$[(2,2) \otimes (2,2)]_S = 1(\eta_{\mu\nu}) \oplus \ldots \quad (15)$$
$$[(2,2) \otimes (2,2) \otimes (2,2) \otimes (2,2)]_A = 1(\epsilon_{\mu\nu\rho\sigma}) \oplus \ldots$$

Using the above invariant symbols, we can easily construct the Lorentz singlets. Let ψ be a left-handed Weyl field,

$$\begin{aligned}\epsilon^{ab}\psi_b\psi_a &= \psi\psi = -\epsilon^{ba}\psi_b\psi_a \\ (\psi\psi)^\dagger &= \epsilon^{\dot{a}\dot{b}}\psi^\dagger_{\dot{a}}\psi^\dagger_{\dot{b}} = \psi^\dagger\psi^\dagger\end{aligned} \quad (16)$$

For a Majorana fermion, we can therefore easily write the kinetic energy and mass terms,

$$i\psi^\dagger_{\dot{a}}\partial_\mu(\bar{\sigma}^\mu)^{\dot{a}a}\psi_a - \frac{m}{2}\psi\psi - \frac{m^*}{2}\psi^\dagger\psi^\dagger \quad (17)$$

The above Lagrangian is for a two-component spinor $\psi_M = (\psi, \psi^\dagger)^T$.

The Dirac γ matrices in the present basis are

$$\gamma^\mu \equiv \begin{pmatrix} 0 & (\sigma^\mu)_{a\dot{b}} \\ (\bar{\sigma}^\mu)^{\dot{a}b} & 0 \end{pmatrix}$$
$$\gamma_5 = -i\gamma^0\gamma^1\gamma^2\gamma^3 = \begin{pmatrix} \delta_a^b & 0 \\ 0 & -\delta^{\dot{a}}_{\dot{b}} \end{pmatrix} \quad (18)$$

where

$$(\sigma^\mu)_{a\dot{a}} = (\sigma^0, \sigma^1, \sigma^2, \sigma^3)$$
$$(\bar{\sigma}^\mu)^{\dot{a}a} = (\sigma^0, -\sigma^1, -\sigma^2, -\sigma^3)$$

which satisfy the Dirac commutation relation, $\{\gamma^\mu, \gamma^\nu\} = 2\eta^{\mu\nu}$. $\bar{\sigma}$ and σ are related by

$$(\bar{\sigma}^\mu)^{\dot{a}a} = \epsilon^{ab}\epsilon^{\dot{a}\dot{b}}(\sigma^\mu)_{b\dot{b}} \tag{19}$$

Therefore, Eq.(17) is equivalent to the familiar form*

$$\frac{1}{2}i\bar{\psi}_M\gamma^\mu\partial_\mu\psi_M - \frac{1}{2}m\bar{\psi}_M\psi_M \tag{17}'$$

A four-component Dirac field can be written in terms of two two-component Weyl fields, ξ and χ,

$$\psi_D \equiv \begin{pmatrix} \xi \\ \chi^\dagger \end{pmatrix} \tag{20}$$

where ξ is left-handed and $\chi^\dagger$ is right-handed,

$$\psi_{DL} = \frac{1+\gamma_5}{2}\psi_D = \begin{pmatrix} \xi \\ 0 \end{pmatrix}$$

$$\psi_{DR} = \frac{1-\gamma_5}{2}\psi_D = \begin{pmatrix} 0 \\ \chi^\dagger \end{pmatrix}$$

The Lagrangian for the Dirac field can be written as

$$\begin{aligned}\mathcal{L} &= i\bar{\psi}_D\gamma^\mu\partial_\mu\psi_D - m\bar{\psi}_D\psi_D \\ &= +i\xi^\dagger\bar{\sigma}^\mu\partial_\mu\xi - i\chi\sigma^\mu\partial_\mu\chi^\dagger - m(\xi^\dagger\chi^\dagger + \chi\xi) \\ &= +i\xi^\dagger\bar{\sigma}^\mu\partial_\mu\xi + i\chi^\dagger\bar{\sigma}^\mu\partial_\mu\chi - m(\xi\chi + \chi^\dagger\xi^\dagger)\end{aligned} \tag{21}$$

Note that ξ and χ are left-handed fields. Thus we have represented every field in terms of the left-handed field only (ξ and χ), which is called as *chiral representation*.(One could use the right-handed fields also.)

Now we can write a *chiral representation* for the fermions given in Eq.(1a),

---

* Note that $\{(\psi_a, \psi^{\dagger\dot{a}})^T\}^\dagger = (\psi^{\dagger a}, \psi_{\dot{a}})$.

$$\begin{pmatrix} u \\ d \end{pmatrix}_L \qquad \bar{u}_L \qquad \bar{d}_L \qquad \begin{pmatrix} \nu_e \\ e \end{pmatrix}_L \qquad \bar{e}_L \qquad (22)$$
$$(3,2,1/6) \quad (3^*,1,-2/3) \quad (3^*,1,1/3) \quad (1,2,-1/2) \quad (1,1,1)$$

where $\bar{u}_L$, $\bar{d}_L$ and $\bar{e}_L$ are just names for the left-handed chiral fields (anti-particles) as χ in Eq.(20). The second and the third families can be written in the same way. With the above *chiral representation*, Georgi and Glashow found the famous $SU(5)$ GUT.[7] (Also, there is another interesting class of GUTs due to Pati and Salam[8].) Let us note that the representation Eq.(22) is complex. The complexity of the GUT representations is very important in that it does not allow mass terms at the GUT scale, which can be a reason that light fermions exist in nature. Georgi and Glashow observed that the rank of the standard model group is 4 and hence the rank of a GUT group must be at least 4. The minimal GUT must have the minimum rank possible. So let us consider rank 4 groups. There are two candidates which allow complex representations, $SU(5)$ and $SU(3)_1 \otimes SU(3)_2$. ($SU(3)_1 \otimes SU(3)_2$ can be a GUT group by a discrete symmetry of exchanging two $SU(3)$'s.) However, for the latter group one needs at least 90 chiral fields to assign correctly the fifteen chiral fields of Eq. (22). This cannot be a minimal group. This leaves us with $SU(5)$. Indeed, the representation $10 + \bar{5}$ of $SU(5)$ can hold the fifteen chiral fields correctly. Thus in $SU(5)$ we need not introduce any other undiscovered fermions in the spectrum to fill the GUT representation.

Before discussing the gauge coupling unification in $SU(5)$, let us briefly summarize the group theory on $SU(N)$. Some of the $SU(N)$ representations are denoted as

fundamental representation (N) : $\psi_\alpha (\alpha = 1, 2, ..., N)$
anti-fundamental representation $(\bar{N})$: $(\psi_\alpha)^\dagger \equiv \psi^{\dagger \alpha}$
adjoint representation $(N^2 - 1)$: $\psi_\alpha{}^\beta$ (with $\text{Tr}\psi = 0$)
2nd rank antisymmetric $(\frac{N(N-1)}{2})$: $\psi_{[\alpha\beta]}$
2nd rank symmetric $(\frac{N(N+1)}{2})$: $\psi_{\{\alpha\beta\}}$

As usual, singlets are formed by contracting the $SU(N)$ indices, e.g. $\psi^{\dagger\alpha}\psi_\alpha$. The generators for an $SU(N)$ transformation of N are T_N^a ($a = 1, 2, ..., N^2 - 1$), since

$$N \otimes \bar{N} \otimes (N^2 - 1) = 1\{(T_N^a)_\alpha{}^\beta\} \oplus \cdots \quad (23)$$

Similarly, the generators for $\bar{N}$ are represented by $T_{\bar{N}}^a$. T_N^a and $T_{\bar{N}}^a$ are hermitian. The generators of the $SU(N)$ fundamental representations satisfy

$$Tr(T_N^a T_N^b) = Tr(T_{\bar{N}}^a T_{\bar{N}}^b) = \frac{1}{2}\delta^{ab}$$

from which the index of the fundamental of $SU(N)$ is known to be $1/2$. Taking a complex conjugation of the gauge transformed N, we obtain a gauge transformed $\bar{N}$,

$$e^{i\theta^a T^a}\psi(N) \to e^{-i\theta^a T^{a*}}\psi^*(\bar{N})$$

Thus we see that $-T_N^*$ corresponds to the gauge generator for $\bar{N}$. And from the hermiticity of T_N^a, we obtain

$$(T_{\bar{N}}^a)^\alpha{}_\beta = -(T_N^{a*})^\alpha{}_\beta = -(T_N^a)_\beta{}^\alpha \quad (24)$$

The generators satisfy the group algebra

$$[T_N^a, T_N^b] = if^{abc}T_N^c \quad (25)$$

It is easy to check that $T_{\bar{N}}^a$ given in Eq. (24) also satisfies the same algebra. If N is a real representation, T_N^a and $T_{\bar{N}}^a$ belong to the same equivalence class, i.e. by a gauge transformation of G,

$$-GT_N^{a*}G^{-1} = T_N^a \quad : \text{real representation} \quad (26)$$

Otherwise the representation is said to be *complex*. For example, 2 of $SU(2)$ is real in the above sense since $G = i\sigma_2$ satisfies Eq. (26). For $N \geq 3$, representation N of $SU(N)$ is always complex.

Similarly as the Lorentz invariants, one can construct $SU(N)$ invariant symbols,

$$\epsilon^{\alpha_1 \alpha_2 \ldots \alpha_N} : [N \otimes N \otimes \ldots \otimes N]_A = 1 \oplus \ldots$$
$$\delta^\alpha_\beta : \bar{N} \otimes N = 1 \oplus \ldots \tag{27}$$

which are frequently used in constructing $SU(N)$ invariants.

Let us now discuss the $SU(5)$ model. Let us denote $\alpha, \beta (= 1, 2, \ldots 5)$ as the $SU(5)$ indices, $a, b (= 1, 2, 3)$ as the color indices and $i, j (= 4, 5)$ as the weak $SU(2)$ indices. The electroweak hypercharge Y must be embedded in $SU(5)$,

$$Y = \begin{pmatrix} -\frac{1}{3} & 0 & 0 & 0 & 0 \\ 0 & -\frac{1}{3} & 0 & 0 & 0 \\ 0 & 0 & -\frac{1}{3} & 0 & 0 \\ 0 & 0 & 0 & \frac{1}{2} & 0 \\ 0 & 0 & 0 & 0 & \frac{1}{2} \end{pmatrix} = \sqrt{\frac{5}{3}} Y_5$$

$$Y_5 = \begin{pmatrix} -\sqrt{\frac{1}{15}} & 0 & 0 & 0 & 0 \\ 0 & -\sqrt{\frac{1}{15}} & 0 & 0 & 0 \\ 0 & 0 & -\sqrt{\frac{1}{15}} & 0 & 0 \\ 0 & 0 & 0 & \sqrt{\frac{3}{20}} & 0 \\ 0 & 0 & 0 & 0 & \sqrt{\frac{3}{20}} \end{pmatrix}$$

(28)

where the properly normalized Y_5 which belongs to $SU(5)$ satisfies $\text{Tr} Y_5^2 = \frac{1}{2}$. Also Q_{em} is given by

$$Q_{em} = \begin{pmatrix} -\frac{1}{3} & 0 & 0 & 0 & 0 \\ 0 & -\frac{1}{3} & 0 & 0 & 0 \\ 0 & 0 & -\frac{1}{3} & 0 & 0 \\ 0 & 0 & 0 & 1 & 0 \\ 0 & 0 & 0 & 0 & 0 \end{pmatrix} \tag{28}'$$

The fermions are embedded in 10_F and $\bar{5}_F$ of $SU(5)$,

$$10_F = \psi_{\alpha\beta} = \{\psi_{ab} = \epsilon_{abc}\psi^c (= \bar{u}); \psi_{ai}(= q); \psi_{ij}(= \bar{e})\}$$
$$= (\bar{3}, 1, -\frac{2}{3}) \oplus (3, 2, \frac{1}{6}) \oplus (1, 1, 1)$$
$$\bar{5}_F = \psi^\alpha = \{\psi^a(= \bar{d}); \psi^i(= \ell)\} \quad (29)$$
$$= (\bar{3}, 1, \frac{1}{3}) \oplus (1, 2, -\frac{1}{2})$$

where a, b, c are the color indices, and i, j are the electroweak indices. The representations are for $SU(3) \otimes SU(2) \otimes U(1)_Y$ where the $U(1)_Y$ charges are calculated from Eq.(28). Note that this embedding of the fifteen chiral fermions reproduce the standard model charges given in Eq.(22). No extra fermions are needed to fill up the representation, which is the reason that the $SU(5)$ model is minimal. For the breaking of $SU(5)$ down to $SU(3) \otimes SU(2) \otimes U(1)$, we need at least one real adjoint representation $(\Sigma), 24_H$, for the Higgs field. For the fermion masses, we need at least one fundamental representation $(H^\alpha), 5_H$. $< 5_H >$ breaks the $SU(2) \otimes U(1)_Y$ and gives fermions masses.

The $SU(5)$ model with fermion families of $(10_F + \bar{5}_F)$ is anomaly free[9], which must be the case because of the fact that we successfully embedded the fifteen chiral fermions of the standard model without any extra fermions to fill the $SU(5)$ representation, and the standard model is anomaly-free. The triangle anomaly $A(\mathbf{R})$ for a spin $\frac{1}{2}$ fermion transforming as $\mathbf{R}$ of the gauge group is defined as

$$Tr(\{T^a, T^b\}T^c) \equiv A(\mathbf{R})d^{abc} \quad (30)$$

From Eq.(24), we see that

$$A(\mathbf{R}) = -A(\bar{\mathbf{R}}) \quad (31)$$

It is also clear that

$$A(\mathbf{R}_1 \oplus \mathbf{R}_2 \oplus ... \oplus \mathbf{R}_n) = A(\mathbf{R}_1) \oplus ... \oplus A(\mathbf{R}_n) \quad (32)$$

since in the calculation we add all spin $\frac{1}{2}$ fermions circulating the triangle loop. From the physical ground we can also show that the $SU(5)$ model is anomaly-free.[10] For the representation 5, we note $\text{Tr} Y^3 \neq 0$ and $\text{Tr} Q_{em}^3 \neq 0$. Therefore, $d^{YYY} \neq 0$ and $d^{Q_{em} Q_{em} Q_{em}} \neq 0$. But, in the standard model the electromagnetic charges form a real representation, i.e. the nonzero Q_{em} values are paired with those of the same magnitude and opposite signs. Thus, $\text{Tr} Q_{em}^3 = 0$. From Eq.(29) and $d^{Q_{em} Q_{em} Q_{em}} \neq 0$, we must have $A(10_F) + A(\bar{5}_F) = 0$.

The covariant derivative acting on N of $SU(5)$ is $D_\mu = \partial_\mu - ig_5 T_N^a A_\mu^a$ where T_N^a contains $SU(3)_c$ generators $\lambda^a/2$, $SU(2)$ generators $\sigma^i/2$, and the $U(1)_{SU(5)}$ generator $Y_5 = \sqrt{\frac{3}{5}} Y$. Thus at the GUT scale, the color and electroweak couplings are related as

$$g_3 = g_2 = \sqrt{\frac{5}{3}} g' \equiv g_5 \tag{33}$$

Here we achieved the gauge coupling unification at the GUT scale. Therefore, the $SU(5)$ prediction of $\sin^2 \theta_W^0$ at the GUT scale is

$$\sin^2 \theta_W^0 = \frac{g'^2}{g^2 + g'^2} = \frac{3}{8} \tag{34}$$

Below the GUT scale, the gauge couplings α_3, α_2 and α_1 run differently and at the elecroweak scale one obtains[11,5]

$$\sin^2 \theta_W(M_W) = 0.214 - 0.004 \log_2 \left(\frac{\Lambda_{\overline{MS}}}{200 \text{ MeV}} \right) \tag{35}$$

which is in spectacular agreement with the global fit to the observed values in deep inelastic ν_μ neutral current experiments and W/Z physics[12,3]

$$\sin^2 \theta_W(\exp) = 0.230 \pm 0.0048 \tag{36}$$

As the final topic on GUTs, let us consider global symmetries. For every complex field which does not form a representation of a nonabelian global group, a priori there is a phase symmetry, i.e. a global $U(1)$. Thus we focus on $10_F + \bar{5}_F$

and Higgs quintet 5_H. All the other fields we introduced are real. Thus there can be $U(1)^3$ global symmetry. But the Yukawa couplings relate two of them

$$-\mathcal{L}_Y = f_d \psi_{\alpha\beta} \psi^\beta H^\alpha + f_u \psi_{\alpha\beta} \psi_{\gamma\delta} H_\epsilon \epsilon^{\alpha\beta\gamma\delta\epsilon} + h.c. \tag{37}$$

Therefore the minimal $SU(5)$ model has one global $U(1)$ which we call $U(1)_X$. The X charges of the complex fields are

$$X = \begin{array}{cccc} \psi_{\alpha\beta} & \psi^\alpha & H_\alpha & H^\alpha \\ \frac{1}{5} & -\frac{3}{5} & -\frac{2}{5} & \frac{2}{5} \end{array} \tag{38}$$

Table 2. $\Gamma(=B-L)$ quantum numbers in $SU(5)$.

Fermions	B	L	Γ	X	Y
$\bar{d}$	$-\frac{1}{3}$	0	$-\frac{1}{3}$	$-\frac{3}{5}$	$\frac{1}{3}$
ℓ	0	1	-1	$-\frac{3}{5}$	$-\frac{1}{2}$
$\bar{u}$	$-\frac{1}{3}$	0	$-\frac{1}{3}$	$\frac{1}{5}$	$-\frac{2}{3}$
q	$\frac{1}{3}$	0	$\frac{1}{3}$	$\frac{1}{5}$	$\frac{1}{6}$
$\bar{e}$	0	-1	1	$\frac{1}{5}$	1

Note that in the standard model we have two global symmetries (B and L), neglecting the nonperturbative effects.[13] When $<H^4>$ breaks $SU(2) \otimes U(1)_Y$ and $U(1)_X$, a combination of the gauge symmetry and the global symmetry is unbroken. This is known as the 't Hooft mechanism.[14] The unbroken global charge which we call Γ annihilate $<H^4>$: $\Gamma <H^4> = 0$. Thus, if $\Gamma \equiv X + aY$, we can determine a by

$$0 = (X + aY)_{<H^4>} = \frac{2}{5} - \frac{1}{2}a \quad \text{or} \quad a = \frac{4}{5}$$

In fact, Γ turns out to be $B - L$,

$$\Gamma = B - L = X + \frac{4}{5}Y \tag{39}$$

In Table 2, we calculate Γ of the complex fields. Because B is not separately conserved, proton is allowed to decay. The dominant decay process occurs through the super-heavy gauge boson (X and Y) exchange.[15] The predicted partial lifetime to $e^+\pi^0$ mode is taken as[16]

$$\tau(p \to e^+\pi^0)_{SU(5)} = 2.3 \times 10^{-28\pm0.7} \left(\frac{M_X}{\text{GeV}}\right)^4 \text{ yr} \tag{40}$$

while the observed lower bound is[17]

$$\tau(p \to e^+\pi^0) > 2.5 \times 10^{32} \text{ yr (90 per cent C.L.)} \tag{41}$$

But the theoretical prediction of τ_p using M_X derived from the successful $\sin^2\theta_W$ fit is not consistent with the experimental value.

3. QCD and the θ Vacuum

3.1. THE INSTANTON SOLUTION

Consider the following $1 + 1$ dimensional nonlinear field equation,

$$\partial^2\left(\frac{a}{f}\right) + m^2 \sin\left(\frac{a}{f}\right) = 0 \tag{42.a}$$

The time independent solution satisfies

$$\frac{d^2}{dx^2}\left(\frac{a}{f}\right) = m^2 \sin\frac{a}{f} \tag{42.b}$$

The x independent or trivial solutions are $a = 0, 2\pi f, 4\pi f, \cdots$. Also there is a class of nontrivial solutions

$$a = 4f \tan^{-1} e^{mz} + 2n\pi f \qquad n = \text{integer} \qquad (43)$$

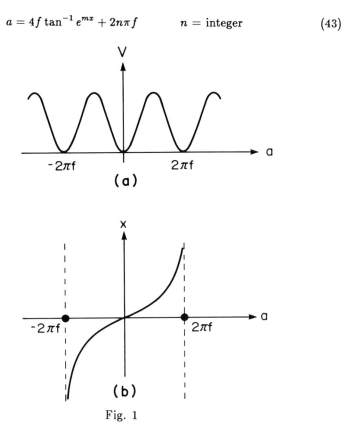

Fig. 1

This solution has a nontrivial topology. In Fig. 1, we show the potential of the a field and the solution for $n = 0$. We note that from Fig.1(a) the vacua are $<a> = 2n\pi f$, and from Fig.1(b) the nontrivial solutions connect these vacua. As in the above domain wall case, nonlinear differential equations can provide nontrivial classical solutions. In this vein, we anticipate to have more complex topological solutions in higher dimensions. They have the names

Domain walls : nontrivial $\pi_0(M)$
Strings : nontrivial $\pi_1(M)$

Monopoles : nontrivial $\pi_2(M)$

Instantons : nontrivial $\pi_3(M)$

QCD in 4 dimensions is expected to have nontrivial classical solutions, and the nontrivial topology solution was first discovered by Belavin et al (BPST)[18]. QCD is described by the Lagrangian

$$\mathcal{L} = -\frac{1}{2g^2} Tr F_{\mu\nu} F^{\mu\nu} + \bar{q}(i\gamma^\mu D_\mu - M_q)q \tag{44}$$

where

$$F_{\mu\nu} = \partial_\mu A_\nu - \partial_\nu A_\mu - i[A_\mu, A_\nu] \tag{45.a}$$

$$A_\mu = \sum A_\mu^a(\lambda^a/2) \;, \quad \lambda^a = \text{Gell-Mann matrices} \tag{45.b}$$

$$M_q = \begin{pmatrix} m_u & 0 & \cdots \\ 0 & m_d & \cdots \\ \cdots & \cdots & \cdots \end{pmatrix} \tag{45.c}$$

$$q = \begin{pmatrix} u \\ d \\ \cdot \\ \cdot \end{pmatrix} \tag{45.d}$$

Let us consider an $SU(2)$ gauge theory instead of the full QCD. The equation of motion is

$$D_\mu F_{\mu\nu} = 0 \tag{46}$$

which is expressed in the Euclidian space. Consider the quantity

$$\int d^4x \, Tr(F_{\mu\nu} \mp \tilde{F}_{\mu\nu})^2 \geq 0$$

$$\tilde{F}_{\mu\nu} = \frac{1}{2}\epsilon_{\mu\nu\rho\sigma} F_{\rho\sigma} \tag{47}$$

which is equivalent to

$$\frac{1}{2}\int d^4x Tr F_{\mu\nu}F_{\mu\nu} \geq \frac{1}{2}\int d^4x |F_{\mu\nu}\tilde{F}_{\mu\nu}|$$

The equality sign holds when

$$F_{\mu\nu} = \pm \tilde{F}_{\mu\nu} \tag{48}$$

Thus the self-dual fields satisfy the field equation which is a minimum action configuration. Independently from the above action argument, we can also see that the self dual fields are solutions of the equation

$$D_\mu F_{\mu\nu} = D_\mu \tilde{F}_{\mu\nu} = 0 \tag{49}$$

where the second equality comes from the Bianchi identity. Here the self-duality condition transformed the second order differential equation to the first order one which is easier to solve. Following BPST, it is easy to write the spherically symmetric solution, satisfying the boundary condition, $F_{\mu\nu} = 0$ at $r = \infty$,

$$\begin{aligned}
A_\mu &= if(r)g(x)^{-1}\partial_\mu g(x) \\
g(x) &= \frac{x_4 - ix_a\sigma_a}{r} = -\frac{i}{r}x_\mu\sigma_\mu \\
g(x)^{-1} &= \frac{i}{r}x_\mu\bar{\sigma}_\mu \\
r^2 = \sum_i x_i^2 \ , \ \sigma_\mu &= (\sigma_a, i), \text{ and } \bar{\sigma}_\mu = (\sigma_a, -i)
\end{aligned} \tag{50}$$

where $a = 1, 2, 3$. Then the self-duality condition for the positive sign in Eq.(48) reduces the equation to

$$\begin{aligned}
r\frac{d}{dr}f &= 2f(1-f) \\
f &= \frac{r^2}{r^2 + \rho^2}
\end{aligned} \tag{51}$$

which corresponds to the *instanton* solution where ρ is the integration constant called the instanton size. If we took the minus sign in Eq.(49), we would obtain the *anti − instanton* solution. Explicitly, the instanton solution is

$$A_\mu = \frac{ir^2}{r^2 + \rho^2} g(x)^{-1} \partial_\mu g(x)$$

$$F_{\mu\nu} = \frac{4\rho^2}{(r^2 + \rho^2)^2} \sigma_{\mu\nu} \qquad (52)$$

$$\sigma_{ab} = \frac{i}{4}[\sigma_a, \sigma_b] \;, \quad \sigma_{a4} = -\frac{1}{2}\sigma_a = -\sigma_{4a}$$

where $F_{\mu\nu}$ is shown to decrease sufficiently fast at $r = \infty$. The above classical solution obtained from $SU(2)$ is a solution of any nonabelian gauge group which contains $SU(2)$ as a subgroup. Since it is so, nonabelian gauge theories have *instanton* solutions.

The vacua we consider in Euclidian space have gauge field configurations like shown in Fig.2.

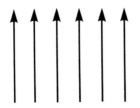   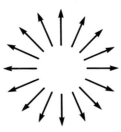

$g(x) = I$ (=trivial and $A_\mu = 0$) $g(x) \neq I$ (BPST type solution)

Fig. 2

So we can see that the instanton solution arises from the nontrivial topology, $\pi_3(SU(2)) = Z$ (nontrivial). In fact, in Eq.(52) we identified the $SU(2)$ gauge fields at $r = \infty(S_3)$. This nontrivial topology is classified by the Pontryagin index

$$q = \frac{1}{16\pi^2} \int d^4x Tr F_{\mu\nu} \tilde{F}_{\mu\nu} = \frac{1}{32\pi^2} \int d^4x F^a_{\mu\nu} \tilde{F}^a_{\mu\nu} \qquad (53)$$

Inserting the solution given in Eq.(52) into Eq.(53), we can show that $q = 1$. The anti-instanton solution would have given $q = -1$. q is always an integer.

3.2. THE θ VACUUM IN QCD

As we have seen in the previous section, gauge field configurations define different vacua labeled by the integer q: no instanton solution ($q = 0$), one instanton solution ($q = 1$), two instanton solution ($q = 2$), etc. Thus QCD vacuum is classified by the classical VEV of the gauge fields. Noting that the nonabelian gauge transformation is

$$A_\mu \to A'_\mu = \Omega A_\mu \Omega^{-1} + i\Omega \partial_\mu \Omega^{-1} \tag{54}$$

we know that our instanton solution Eq.(52) is of pure gauge solution at $r = \infty$. Therefore, we can gauge-transform from the vacuum with $A_\mu = 0$ to a vacuum with $A_\mu \neq 0$. (As in the domain wall case, the instanton solution connects the degenerate vacua.) For example, by using $g(x)$ for Ω, we can change q by one unit. Let us classify the gauge fields by their Pontryagin indices, $\mathcal{A}_i$ ($i = 0, \pm 1, \pm 2, ...$). Also G_i denotes the operation of gauge transformation where $i(= 0, \pm 1, \pm 2, \cdots)$ is the change of the Pontryagin index by G_i. Typically,

$$\begin{aligned} G_1 \mathcal{A}_0 G_1^{-1} &: A_{0\mu} \to \Omega A_{0\mu} \Omega^{-1} - i\Omega \partial_\mu \Omega^{-1} \quad \text{or} \\ G_1 \mathcal{A}_0 G_1^{-1} &: \mathcal{A}_0 \to \mathcal{A}_1 \text{ (BPST solution)} \end{aligned} \tag{55.a}$$

where $\Omega = g(x)$. In general,

$$G_1 \mathcal{A}_n G_1^{-1} : \mathcal{A}_n \to \mathcal{A}_{n+1} \tag{55.b}$$

So G_1 is known to be the operation of gauge transformation on vacua. Let us define $|n>$ vacua by the gauge field configurations with the Pontryagin index n,

$$\cdots, |-2>, |-1>, |0>, |1>, |2>, \cdots \tag{56}$$

G_1 shift these vacua to the right by one unit. Thus the classical solution in Euclidian space corresponds to tunneling.

As we have seen above the $|n>$ vacuum is not invariant under all possible gauge transformations.[19] In analogy with the problem of periodic potential, one can guess a gauge invaiant vacuum, $\Psi(A)$ = a linear combination of $|n>$. In obtaining $\Psi(A)$, therefore, we use the gauge invariance of the vacuum,

$$G_0\Psi(A) = \Psi(A)$$
$$G_1\Psi(A) = e^{-i\theta}\Psi(A) \qquad (57.a)$$

where we used the fact that G_0 belongs to the equivalence class of I, and G_1 is a unitary operator with one parameter (corresponding to $U(1)$) and its eigenvalue is simply $e^{-i\theta}$. Since $G_n \sim G_1^n$,

$$G_n\Psi(A) = e^{-in\theta}\Psi(A) \qquad (57.b)$$

From Eqs.(55) and (57), we determine the θ-vacuum of QCD, $|\theta> \equiv \Psi(A)$,

$$|\theta> = \sum_{n=-\infty}^{\infty} e^{in\theta}|n> \qquad (58)$$

which is gauge invariant and physically meaningful. But in this process we have introduced additional free parameter θ. Because of the definition of $|\theta>$ in Eq.(58), we note that θ is a periodic variable with periodicity of 2π,

$$|\theta> = |\theta + 2\pi(\text{integer})> \qquad (59)$$

To see the effects of θ-vacuum, we calculate the vacuum-to- vacuum transition amplitude. Note that

$$<n|e^{-HT}|m> = \int [dA_\mu]_{n-m} e^{-\int d^4x \mathcal{L}} \qquad (60)$$

where the pure gauge fields with the Pontryagin index $n-m$ contribute to the

$|m>$ to $|n>$ vacuum transition amplitude. In the θ-vacuum the vacuum transition amplitude is

$$<\theta'|e^{-HT}|\theta> = \sum_{n=-\infty}^{\infty}\sum_{n'=-\infty}^{\infty} e^{i(n\theta-n'\theta')} \int [dA_\mu]_{n'-n} e^{-I} \qquad (61)$$

The Jacobian for the transformation from $\{n,n'\}$ to $\{n,q\}$ is 1 where

$$q = n' - n \qquad (62)$$

Thus Eq.(61) becomes

$$\sum_{n'=-\infty}^{\infty} e^{in'(\theta-\theta')} \sum_{q=-\infty}^{\infty} \int [dA_\mu]_q \, e^{-iq\theta - \int d^4x \mathcal{L}}$$

Since

$$\sum_{n=-\infty}^{\infty} e^{-in\theta+in\theta'} = 2\pi\delta(\theta-\theta') \propto \delta(\theta-\theta')$$

we obtain

$$<\theta'|e^{-HT}|\theta> = \delta(\theta-\theta') \int [dA_\mu] e^{-\int d^4x(\mathcal{L}+\mathcal{L}_\theta^E)} \qquad (63)$$

where we have neglected the unimportant overall constant. Eq.(63) shows[19] the *superselection rule* that a θ–vacuum is stable and the effective Lagrangian of the θ–vacuum, $\mathcal{L}_\theta^E = iq\theta$. In Minkowski space

$$\mathcal{L}_\theta = \theta q = \frac{\theta}{32\pi^2} F^a_{\mu\nu} \tilde{F}^{a\mu\nu} \qquad (64)$$

So far we have considered only the gauge fields. Introduction of massive quarks does not change the above conclusion.

The effective Lagrangian $\mathcal{L}_\theta$ is a total derivative, but we have seen that nonabelian gauge theories can have instanton solutions or nontrivial gauge field configurations at infinity such that $e^{iq\theta} \neq I$ if $\theta \neq 2n\pi$.

In $U(1)$ gauge theories, one may consider $\theta F\tilde{F}$ term. But it is vanishing because it is a total derivative with a constant θ and there is no nontrivial $U(1)$ gauge field at infinity.

Thus we must include θ_{QCD} and θ_{EW} in the set of the standard model parameters. But θ_{EW} is not interesting phenomenologically because $SU(2)$ is broken at 100 GeV scale before $SU(2)$ has a chance to become strong.

4. The Strong CP Problem

The so-called *strong CP problem* is a parameter problem, consisting of two parts:"how does a theory determine the parameter θ ?", and "is the predicted θ consistent with observations ?" The first question is answered in the so-called *'naturalness'* framework.[20,21] At present and probably forever, the electric dipole moment of neutron ($\equiv d_n$) constrains the phenomenological bound on θ related to the second question. As we will see in the following section θ is physical, and can be in conflict with the observed d_n in a world with no massless quark and no axion. The electroweak[22] or GUT[23] solutions discussed in the literature in the framework of calculability are not aethetically very beautiful and will not be discussed here. (For reviews, see Ref.[21,24].)

It is easy to see that $\mathcal{L}_\theta$ violates both P and T discrete symmetries but concerves C. Thus in quatum field theory of QCD conserving CPT, it violates also CP. The violation of CP can be seen looking at the electric-type (**E**) and magnetic-type (**B**) gluon fields. Both of these change sign under C, but **E** changes sign while **B** does not change sign under P. Since $\mathcal{L}_\theta$ is proportional to $\mathbf{E} \cdot \mathbf{B}$, it violates the CP invariance of the strong interactions. In the following section, we estimate the magnitude of CP violation by this interaction.

4.1. Estimate of the Neutron Electric Dipole Moment

In late 1970's, there has been a great deal of interest in estimating the neutron electric dipole moment from the θ term[25]. Here we follow Fujikawa's technique relating an anomaly[26] to a noninvariant fermion measure under the continuous transformation.[27]

Let us consider the path integral

$$\int [dA_\mu][d\psi][d\bar{\psi}] \, e^{i \int d^4x \, \mathcal{L}} \tag{65}$$

where the fermion measure is

$$d\mu = [d\psi][d\bar{\psi}] \tag{66}$$

Fujikawa has shown that under a chiral transformation,

$$\begin{aligned}\psi'(x) &= e^{i\alpha\gamma_5} \, \psi(x) \\ \bar{\psi}'(x) &= \bar{\psi}(x) \, e^{i\alpha\gamma_5}\end{aligned} \tag{67}$$

the fermion measure changes to (since the Jacobian is not identity)

$$d\mu' = d\mu \exp\left(i \int d^4x \frac{1}{8\pi^2} \alpha(x) Tr F_{\mu\nu} \tilde{F}_{\mu\nu} \right) \tag{68}$$

This calculation of anomaly has been extensively used in the discussion of axion physics in Ref.[21].

QCD with three light quarks is of our interest in the following. The path integral is*

$$\int [\prod_f dq_f d\bar{q}_f][dA_\mu] e^{-S} \tag{69.a}$$

$$S = \int d^4x \{ \frac{1}{4g^2} F^a_{\mu\nu} F^a_{\mu\nu} + \frac{1}{4e^2} F^{em}_{\mu\nu} F^{em}_{\mu\nu} + \bar{q}(i\gamma^\mu D_\mu + M_q)q + \frac{i\theta}{32\pi^2} F^a_{\mu\nu} \tilde{F}^a_{\mu\nu} \} \tag{69.b}$$

where f runs over all the flavors of interest. Using the Fujikawa measure, we

---

⋆ When the indices are contracted in the subscripts only, it is performed in the Euclidian space.

want to express it in terms of the chirally rotated fields q'_f,

$$q_f \to q'_f = e^{i\alpha_f \gamma_5/2} q_f \quad \text{or} \quad q_f = e^{-i\alpha_f \gamma_5/2} q'$$
$$\bar{q}_f \to \bar{q}'_f = \bar{q}_f e^{i\alpha_f \gamma_5/2} \quad \text{or} \quad \bar{q}_f = \bar{q}'_f e^{-i\alpha_f \gamma_5/2} \tag{70}$$

Then the path integral becomes

$$\int [\prod_f dq'_f d\bar{q}'_f][dA_\mu] \exp(-) \int d^4x \left(i\frac{\sum_f \alpha_f(x)}{2} \frac{1}{16\pi^2} \{F^a_{\mu\nu}\tilde{F}^a_{\mu\nu} + cF^{em}_{\mu\nu}\tilde{F}^{em}_{\mu\nu}\} + \mathcal{L} \right) \tag{71}$$

where c is determined from the electromagnetic charges of the quarks and $\mathcal{L}$ is given in Eq.(69). In terms of q' and $\bar{q}'$ fields, the effective Lagrangian gets a contribution from the fermion measure. After q' is called q again and the sign of α is changed, we obtain

$$\mathcal{L} = \frac{1}{4g^2} F^a_{\mu\nu} F^a_{\mu\nu} + \frac{1}{4e^2} F^{em}_{\mu\nu} F^{em}_{\mu\nu} + \sum_f \bar{q}_f e^{i\alpha_f \gamma_5/2}(i\gamma^\mu D_\mu + m_f) e^{i\alpha_f \gamma_5/2} q_f$$
$$+ \frac{i}{32\pi^2}(\theta - \alpha) F^a_{\mu\nu} \tilde{F}^a_{\mu\nu} + \frac{i}{32\pi^2} c\alpha F^{em}_{\mu\nu} \tilde{F}^{em}_{\mu\nu} \tag{72}$$

where

$$\alpha = \sum_i \alpha_i$$
$$c\alpha = 2 \sum_{i=u,d,s} 3\alpha_i (Q^i_{em})^2 = 3 \sum_{i=u_L,u_R,\cdots} \alpha_i (Q^i_{em})^2 \tag{73}$$

where the factors 2 and 3 in the second equation are the ratio of the trace of $U(1)_{em}$ and nonabelian gauge group generators and the counting of color degrees, respectively. θ is given in the theory. We can choose α for our convenience. Since the expectation value of the gluon field operators between hadronic states is difficult to estimate, we choose α such that the gluon anomaly term is absent, i.e. $\alpha = \theta$. However, note that any choices of α_i's are equivalent at the Lagrangian level. Thus we have a master formula on the partition function

$$Z\left[e^{i\alpha_i\gamma_5/2} q_i \; ; \; \theta - \sum_i \alpha_i \; ; \; -3 \sum_{i=u_L,u_R,\cdots} \alpha_i (Q^i_{em})^2\right] = Z\left[q_i \; ; \; \theta \; ; \; 0\right] \quad (74)$$

where L and R refer to the left-handed and right-handed quarks, respectively. Note, however, that we count 4-component Dirac fields for the gluon anomaly. One should remember that both sides of Eq.(74) represent the same θ-vacuum. The path integral using ψ or ψ' should give the same physics. The term $F_{em}\tilde{F}_{em}$ which is a total derivative can be neglected in a constant-θ vacuum, but in axion physics it should be kept since θ is basically a dynamical field.

To remove $F\tilde{F}$ term, we choose the fermion basis such that

$$\alpha = \sum_i \alpha_i = \theta \quad (75)$$

Determination of individual α_i must be consistent with the stability of the QCD vacuum.[28,29] Rather than following the lengthy discussion of the current algebra technique, here we take a short cut. Note that the mass term becomes

$$-\mathcal{L}_m = m_u \cos\alpha_u \bar{u}u + m_d \cos\alpha_d \bar{d}d + m_s \cos\alpha_s \bar{s}s$$
$$+ m_u \sin\alpha_u \bar{u}i\gamma_5 u + m_d \sin\alpha_d \bar{d}i\gamma_5 d + m_s \sin\alpha_s \bar{s}i\gamma_5 s$$

Next we note that the original $F\tilde{F}$ term is a flavor singlet. So the first guess on the CP violating terms of $\bar{q}i\gamma_5 q$ is of the flavor singlet form,

$$\sim i(\bar{u}\gamma_5 u + \bar{d}\gamma_5 d + \bar{s}\gamma_5 s)$$

One may argue that this argument is not plausible, because it arises from the interrelation of the flavor singlet gluon anomaly and the flavor nonsinglet quark mass terms. But this must be the case since $m_u = 0$ should not include any CP violation effect originating from the θ term. Namely, trials with flavor nonsinglet mass coefficients in the above equation verifies this statement. Assuming that α_i are small, we can solve

$$m_u \alpha_u = m_d \alpha_d = m_s \alpha_s = x$$

$$\alpha_u + \alpha_d + \alpha_s = \theta = x(\frac{1}{m_u} + \frac{1}{m_d} + \frac{1}{m_s})$$

or

$$x = \frac{m_u m_d m_s}{(m_u + m_d)m_s + m_u m_d} \theta \simeq \frac{m_u m_d}{m_u + m_d} \theta$$

The exact solution for two flavors is

$$x = \frac{\sin\theta \sqrt{m_u m_d}}{\sqrt{2\cos\theta + (Z^2 + 1)/Z}} \tag{76}$$

where $Z = m_u/m_d$. Thus we have traded the $\theta F\tilde{F}$ term for a manageable Lagrangian,

$$\Delta \mathcal{L} = \frac{m_u m_d m_s \theta}{(m_u + m_d)m_s + m_u m_d}(\bar{u}i\gamma_5 u + \bar{d}i\gamma_5 d + \bar{s}i\gamma_5 s) \tag{77}$$

which was first derived by Baluni.[29] The Baluni Lagrangian is given in terms of quark fields with the flavor singlet form and shows that $m_u = 0$ does not contribute. From the expression on a finite θ, Eq. (76), $\theta = 0, \pi, ..., n\pi$ do not lead to the CP violating action. Because of the periodicity 2π of θ, we can consider only $\theta = 0$ and π.

The neutron electric dipole moment is estimated from the pion-nucleon interaction,

$$\mathcal{L} = \vec{\pi} \cdot \bar{N}(i\gamma_5 g_{NN\pi} + g^\theta_{NN\pi})\vec{\sigma} N \tag{78}$$

where the $g^\theta_{NN\pi}$ term comes from Eq.(77) evaluated between N and $N\pi$ states.

In the estimation one uses the soft-pion theorem

$$\lim_{q \to 0} f_\pi <\beta, \pi^a|\mathbf{O}|\alpha> = i <\beta|[Q_{\pi^a}, \mathbf{O}]|\alpha> \quad (79)$$

where Q_{π^a} is the spontaneously broken generator corresponding to the Goldstone boson π^a. Thus in the $q \simeq 0$ limit,

$$<N'\ \pi^a|\Delta\mathcal{L}|N> \simeq -\frac{\theta}{f_\pi}\frac{m_u m_d}{m_u + m_d} <N_f|\bar{q}\sigma_a q|N_i> \quad (80)$$

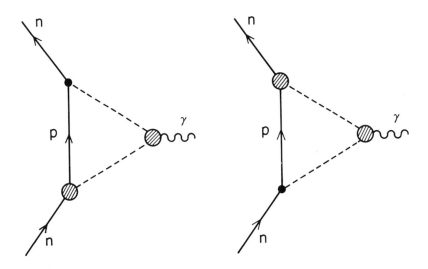

Fig. 3

The flavor $SU(3)$ symmetry relates the above matrix element to a mass ratio[25], and we obtain

$$g^\theta_{NN\pi} = -\theta \frac{m_u m_d}{m_u + m_d}\frac{1}{f_\pi}\frac{M_\Xi - M_N}{2m_s - m_u - m_d} \quad (81)$$

Now we can use the effective Lagrangian Eq.(78) to estimate the neutron electric dipole moment. The lowest order contribution comes from Fig.3, Crewther et al [25] estimate

$$d_n^{th} = 5.2 \times 10^{-16} \, \theta \, e \, \text{cm} \qquad (82.a)$$

which can be compared to the experimental bound[28]

$$d_n^{exp} < 0.6 \times 10^{-24} \, e \, \text{cm} \qquad (82.b)$$

Therefore we have a phenomenological bound on $\bar{\theta}^\star$

$$|\bar{\theta}| \text{ or } |\bar{\theta} - \pi| \le 10^{-9} \qquad (83)$$

The meson mass ratio $(M_{K^0}^2 - M_{K^+}^2 - M_{\pi^0}^2 + M_{\pi^+}^2)/M_\pi^2 = 0.27$ chooses $\bar{\theta}$ near 0. At $\bar{\theta} = 0$ this ratio is $(1 - m_u/m_d)/(1 + m_u/m_d)$ in the chiral perturbation theory, while at $\bar{\theta} = \pi$ it can be shown to be $(1 + |m_u/m_d|)/(1 - |m_u/m_d|)$ after choosing the correct vacuum.[32] Thus $\bar{\theta} = 0$ is prefered. From the bound (82), we note that the free parameter $\bar{\theta}$ of the standard model is extremely small.

4.2. THE STRONG CP PROBLEM

When we discussed the effective Baluni Lagrangian, we took the basis where the quark mass matrix is diagonalized and all the masses are positive, real and γ_5-free. Let the quark mass matrix is diagonalized but let us assume for a moment that the eigenvalues can be complex,

$$-\mathcal{L}_m = \bar{q}_R M_q q_L + \bar{q}_L M_q^\dagger q_R \qquad (84.a)$$

where

$$M_q = \begin{pmatrix} m_u e^{i\delta_1} & 0 & 0 \\ 0 & m_d e^{i\delta_2} & 0 \\ 0 & 0 & m_s e^{i\delta_3} \end{pmatrix} \qquad (84.b)$$

which is equivalent to writing $-\mathcal{L}_m$ as $\bar{u}_R e^{i\delta_1\gamma_5/2} m_u e^{i\delta_1\gamma_5/2} u_L + \cdots$. From our master formula, we then note that $\{\delta, \theta\}$ is equivalent to $\{0, \theta + \delta\}$ where $\delta =$

$\star$ $\bar{\theta}$ defined below is an effective θ at QCD scale and the one bounded by the experiment.

$\delta_1 + \delta_2 + \delta_3$. Then after making the mass matrix real, positive and γ_5-free, the coefficient of $F\tilde{F}$ becomes $(\delta_1 = \alpha_u, \text{etc})$

$$\begin{aligned}\bar{\theta} &= \theta + \alpha_u + \alpha_d + \cdots \\ &= \theta \text{ Arg Det } M_q\end{aligned} \quad (85)$$

Since the determinant is invariant under unitary transformations, the second line applies also to the bases where the quark mass matrices are not diagonal. The quark mass dependent term is called θ_{QFD}, since it gets contribution from the electroweak symmetry breaking by which quarks get masses. In comparison, the initial θ determined above the electroweak scale is called θ_{QCD}. Thus

$$\bar{\theta} = \theta_{QCD} + \theta_{QFD} \quad (86)$$

The phenomenological strong CP problem is the problem on $\bar{\theta}$. From now on we will denote $\bar{\theta}$ simply as θ.

The *strong CP problem* consists in two parts,

(i) *How is θ determined?*

(ii) *Why is the determined θ so small?*

The weak interactions violate the CP invariance, which has an effect on θ through θ_{QFD}. Hence the solutions of the strong CP problem by the calculability criteria relate the CP problem with the weak CP violation.[22] These solutions assume $\theta_{QCD} = 0$ by imposing a CP symmetry in the $d = 4$ terms in the effective electroweak Lagrangian. Thus the weak CP violations are introduced as '*spontaneous*'[30] or '*soft*'[33] CP violation. Here we will not discuss these solutions any more.

Theoretically, the attractive solutions are the cases with $m_u = 0$ and axions. Since we will discuss the axion solution of the strong CP problem extensively in next chapter, here we just mention the other attractive solution, $m_u = 0$.

Let us note that $m_u = 0$ solves the strong CP problem. Through the master formula, in the θ- vacuum we can choose some of the relevant terms as

$$\bar{u} \ e^{i\alpha\gamma_5/2} \left(i\gamma^\mu D_\mu - m_u\right) e^{i\alpha\gamma_5/2} u + \frac{\theta - \alpha}{32\pi^2} F^a_{\mu\nu} \tilde{F}^{a\mu\nu} \tag{87}$$

Note that α can be anything even for nonzero m_u. But if $m_u = 0$, we note that Eq.(87) becomes

$$\bar{u} \ i \ \gamma^\mu D_\mu \ u + \frac{\theta - \alpha}{32\pi^2} F\tilde{F}$$

Since α is a phase which we can choose, it implies that different θ-vacua give the same physical result. Namely, physically observed quantities do not depend on θ. θ is unobservable. This fact is usually said that we can rotate away θ by the chiral transformation,

$$u \rightarrow e^{i\theta\gamma_5/2} u$$

For a massless u quark, the above chiral symmetry holds at the effective Lagrangian level. However, the vacuum condensation $< \bar{u}u >$ will break this chiral symmetry and there will result a Goldstone boson. In fact, $\eta\prime$ is the one, and acts as an axion below the chiral symmetry breaking scale. The question on the massless u quark solution is really a phenomenological one whether $m_u = 0$ is consistent with the hadron mass spectrum. This possibility of massless u quark has been studied by several groups. Earlier phenomenological studies concluded that it is not possible to have $m_u = 0$.[34] However, a few years ago, Kaplan and Manohar[35] pointed out that in the chiral perturbation theory there is a room to allow the $m_u = 0$ possibility.

In addition we note that the determinental instanton interaction of 't Hooft breaks the flavor $U(1)$ symmetry and solves[36] the $U(1)$ problem.[37] One can generate a nonvanishing u quark mass by closing the quark lines except u in the Feynman diagram of the determinental interaction. But in earlier days it was thought that it is not enough to generate $m_u/m_d \sim 0.5$. However, Choi, Kim and

Sze[36] have shown recently that the 't Hooft interaction contributes significantly to light quark masses. In terms of quark fields, the determinental interaction in the θ vacuum is

$$e^{-i\theta}\left(K\,Det\Sigma + H\,(Det\Sigma)TrM\Sigma^\dagger + I\,(DetM)TrM^{-1}\Sigma + J\,DetM\right) + \text{h.c.} \quad (88)$$

where $K, H, I,$ and J are coupling constants, M is the quark mass matrix, $\Sigma = q_R \bar{q}_L$, and the quark mass term is $\bar{q}_R M q_L + \text{h.c.}$ The $U(3)_L \otimes U(3)_R$ symmetry

$$\begin{aligned}
\Sigma &\to R\,\Sigma\,L^\dagger \\
M &\to R\,M\,L^\dagger \\
q_L &\to L\,q_L \\
q_R &\to R\,q_R \\
\theta &\to \theta - Arg\,Det\,LR^\dagger
\end{aligned} \quad (89)$$

is consistent with Eq.(88). Choi, Kim and Sze argue that there exists a reasonable range of K to allow a phenomenologically acceptable m_u.

5. The Axion Physics

5.1. THE AXION SOLUTION TO THE STRONG CP PROBLEM

The axion solution interprets θ as a dynamical field instead of a free parameter. Integrating the partition function Eq.(69), we will obtain

$$Z \propto e^{-\int V[\theta,\cdots]d^4x} \quad (90)$$

If θ is a free parameter, $V[\theta]$ does not mean any since the *superselection* rule says that any choice of θ is stable. Nevertheless let us draw the shape of $V[\theta]$

for $m_u \neq 0$ in Fig.4 where we have represented the 2π period of θ and shown a vacuum as a black dot which does not roll down due to the *superselection* rule,

$$< \theta'|e^{-HT}|\theta > \propto \delta(\theta' - \theta)\{\cdots\} \tag{91}$$

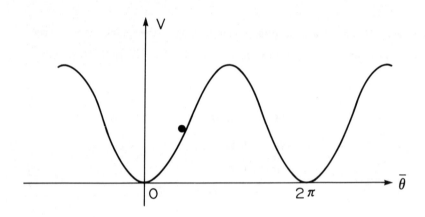

Fig. 4

The basic idea of the axion solution is

Different θ's do not describe different theories

but different vacua in the same theory.

In other words, θ is dynamical, and one needs a kinetic energy term for θ,

$$\theta \equiv \frac{a}{F_a} \tag{92}$$

$$\mathcal{L} = -\frac{1}{4g^2}F^a_{\mu\nu}F^a_{\mu\nu} + \frac{1}{2}\partial_\mu a\, \partial^\mu a + \sum_{i=u,d,s} \bar{q}_i(i\gamma^\mu D_\mu - m_i)q_i \\ + \frac{\theta_0}{32\pi^2}F^a_{\mu\nu}\tilde{F}^a_{\mu\nu} + \frac{a}{32\pi^2 F_a}F^a_{\mu\nu}\tilde{F}^a_{\mu\nu} \tag{93}$$

where θ_0 is the parameter given initially. The last term of Eq.(93) is a nonrenormalizable interaction term. Therefore it must appear from a more fundamental Lagrangian through the spontaneous symmetry breaking of a global symmetry or from a higher dimensional theories. This dynamical θ is called *axion* which can appear as the phase of some Higgs field[39-42], as a composite field of fermions[43] or is present in the string theory.[44]

In the axion solution of the strong CP problem, one has to show that $\theta = 0$ is the minimum of $V[\theta]$ as drawn in Fig.4. Peccei and Quinn[39] have shown that indeed this is the case. But here we follow Vafa and Witten's argument[45] which is elegant. We will consider a CP invariant theory, i.e. neglect weak CP violation. Inclusion of the weak CP violation will shift our ground state value of θ by a tiny amount. In a CP invariant world, one can choose the phases such that every coupling constant is real. Then let us consider Eq.(90),

$$e^{-\int d^4 z V[a]} \equiv |\int [dA_\mu \prod_i Det(\gamma_\mu D_\mu + m_i)$$
$$\exp\{-\int d^4 x (\frac{1}{4g^2} F^a_{\mu\nu} F^a_{\mu\nu} - \frac{ia}{32\pi^2 F_a} F^a_{\mu\nu} \tilde{F}^a_{\mu\nu})\}| \quad (94)$$

which is shown in Euclidian space. We have shifted a such that the θ_0 and a terms combine to remove the θ_0 term $(a \to a - F_a \theta_0)$. Note that the $F\tilde{F}$ term becomes pure imaginary. Using the γ-algebra*,

$$\gamma_4^\dagger = \gamma_4, \quad \gamma_i^\dagger = -\gamma_i$$
$$(i\gamma_\mu D_\mu)^\dagger = i\gamma_\mu D_\mu$$

one can show that $i\gamma_\mu D_\mu$ in Euclidian space is hermitian and has real eigenvalues. Hence, if λ is a nonzero eigenvalue of $i\gamma_\mu D_\mu$, so is $-\lambda$,

$$i\gamma_\mu D_\mu \psi = \lambda \psi$$
$$i\gamma_\mu D_\mu (\gamma_5 \psi) = -\lambda (\gamma_5 \psi)$$

$\star$ $\{\gamma_\mu, \gamma_\nu\} = 2\delta_{\mu\nu}$, $\gamma_4 = \gamma^0_{BD}$, $\gamma_i = i\gamma^i_{BD}$, where BD refers to the Bjorken-Drell γ-matrices.

Thus we can show that

$$Det(\gamma_\mu D_\mu + m_i) = \prod_\lambda(-i\lambda + m_i) = m_i^{N_0} \prod_{\lambda>0}(m_i^2 + \lambda^2) > 0 \tag{95}$$

Use the Schwartz inequality and Eq.(95) to show that Eq.(94) is bounded,

$$\begin{aligned}
e^{-\int d^4z V[a(z)]} &\equiv \\
&|\int [dA_\mu] \prod_i Det(\gamma_\mu D_\mu + m_i) \cdot e^{-\int d^4z(\frac{1}{4g^2}F^a_{\mu\nu}F^a_{\mu\nu} - \frac{ia}{32\pi^2 F_a}F^a_{\mu\nu}\tilde{F}^a_{\mu\nu})}| \\
&\leq \int [dA_\mu] \prod_i Det(\gamma_\mu D_\mu + m_i) \cdot |e^{-\int d^4z(\frac{1}{4g^2}F^a_{\mu\nu}F^a_{\mu\nu} - \frac{ia}{32\pi^2 F_a}F^a_{\mu\nu}\tilde{F}^a_{\mu\nu})}| \\
&= |\int [dA_\mu] \prod_i Det(\gamma_\mu D_\mu + m_i) e^{-\int d^4z \frac{1}{4g^2}F^a_{\mu\nu}F^a_{\mu\nu}}| \\
&= e^{-\int d^4z V[0]}
\end{aligned} \tag{96}$$

from which we conclude

$$V[0] \leq V[a] \tag{97}$$

showing that $a = 0$ is the minimum of the potential.

In the proof, we do not assume any a–dependent term in the potential other than that coming from the $aF\tilde{F}$ term. This feature must be present in any axion solution of the strong CP problem. As mentioned before we do not consider the weak CP violation. Consideration of the weak CP violation will shift the vacuum value of θ, which was first shown by Georgi[46]. In this case the axion potential contains both CP-conserving and CP-violating pieces,

$$f_\pi^2 m_\pi^2 \{CP \ conserving \ part \ + \ CP \ violating \ part\}$$

where the CP- violating part comes from the Feynman graphes containing two

W boson exchanges. Thus we can roughly estimate the CP-violating part as

$$\theta G_F^2 (1\,\text{GeV})^4 \cdot (CP\text{ phase}) f_\pi^2 m_\pi^2$$

Thus θ is determined to be $\sim 10^{-13}$ which is much smaller than the phenomenological bound 10^{-9}. More reliable estimate of 10^{-17} has been given in the chiral perturbation theory.[47] In any axion models with the Kobayashi-Maskawa type weak CP violation, the above estimate applies. Thus it is not likely that θ in the axion models will be determined experimentally in a near future.

Now it is clear that axion solution relies on the nonrenormalizable term $\frac{a}{32\pi^2 F_a} F\tilde{F}$. Thus this term must occur as an effective Lagrangian at low energy from a more fundamental theory. Firstly, it can be a Goldstone boson, but the classical global symmetry must be explicitly broken by the $aF\tilde{F}$ term to realize the scheme. Of course, a true Goldstone boson cannot have such a coupling, and the axion is a pseudo-Goldstone boson. Secondly, in higher dimensional consistent theories (such as some superstring theories) an axion may be present in the low energy effective Lagrangian. In both cases the appearance of the nonrenormalizable interaction is explained. In the literature, there appeared all these possibilities as shown below,

Electroweak axions:
    PQWW axion[39,40]   uses $\phi_1(\text{SM}) + \phi_2$
    Variant axions[48]   use $\phi_1(\text{SM}) + \phi_2 + (\cdots)$
Invisible axions:
    KSVZ axion[41]   uses $\phi_1(\text{SM}) + Q + \sigma$
    DFSZ axion[42]   uses $\phi_1(\text{SM}) + \phi_2 + \sigma$
Composite axions[43]: introduces an additional nonabelian gauge group
Superstring axions[44]: present in some superstring models

where ϕ_1 is the Higgs doublet of the standard model, and σ is an $SU(2) \otimes U(1)$ singlet complex scalar field. Since TASI students may like an education from the simplest model, let us start with the heavy quark axion model(KSVZ).[*] Then I will briefly discuss the PQWW and DFSZ models.

5.2. Heavy Quark Axion Model (or KSVZ Model)

As discussed in the previous section, the axion solution requires a global axial symmetry which is called the Peccei-Quinn symmetry(PQ symmetry). The global symmetry must be axial to have the gluon anomaly. It must be spontaneously broken to have the $aF\tilde{F}$ coupling. Both of these conditions are fulfilled by introducing a quark $Q(=SU(3)_c$ triplet) and a scalar σ which are $SU(2) \times U(1)$ singlets. The scalar potential is given such that σ gets a VEV. These are implemented by the Lagrangian,

$$\mathcal{L} = \bar{Q}i\gamma^\mu D_\mu Q + (\partial_\mu \sigma)^*(\partial^\mu \sigma) + \mathcal{L}_y - V \qquad (98.a)$$

$$\mathcal{L}_y = -f\bar{Q}_L \sigma Q_R - f^*\bar{Q}_R \sigma^* Q_L \qquad (98.b)$$

$$V = -\mu_\sigma^2 \sigma^*\sigma + \lambda_\sigma (\sigma^*\sigma)^2 + \lambda_{\phi\sigma}\sigma^*\sigma\phi^\dagger\phi + \cdots \qquad (98.c)$$

which has the following PQ symmetry,

$$\begin{aligned} \sigma &\to e^{i\alpha}\,\sigma \\ Q_L &\to e^{i\alpha/2}\,Q_L \\ Q_R &\to e^{-i\alpha/2}\,Q_R \end{aligned} \qquad (99.a)$$

This $U(1)_{PQ}$ is a good classical symmetry. If it is good to all orders, the spontaneous breaking of it will generate a real Goldstone boson, which we do not

---

[*] Some people may like others as the simple ones. But the axion is the pseudo-Goldstone boson arising from the breaking of the PQ symmetry. In this regard, the heavy quark axion model is the simplest, not mixing the global symmetry with $U(1)_Z$ gauge symmetry.

want to happen. But the PQ symmetry is broken at one loop order. The PQ symmetry is in fact chiral,

$$\begin{aligned} \sigma &\to e^{i\alpha}\sigma \\ Q &\to e^{i\alpha\gamma_5/2}Q \end{aligned} \qquad (99.b)$$

which shows that α couples to the anomaly through the master formula,

$$Z[e^{i\alpha\gamma_5/2}Q\,;\,\theta-\alpha\,;\,6\alpha Q_{em}(Q)^2] \;=\; Z[Q\,;\,\theta\,;\,0] \qquad (100)$$

σ gets a VEV v which can be complex but let us choose the phase such that it is real. Let us choose a gauge such that

$$\sigma \;=\; \frac{\tilde{v}+\rho}{\sqrt{2}}\,e^{ia/\tilde{v}} \qquad (101)$$

where ρ and a are real scalar and pseudoscalar fields, respectively. Through Eq.(100), we have chosen the phase of the heavy quark Q such that $(ia/\tilde{v})\bar{Q}\gamma_5 Q$ coupling is absent. Then the coupling will reappear as $(\partial^\mu a/\tilde{v})\bar{Q}\gamma_\mu Q$. But this term will not generate any potential of a when internal heavy Q lines are integrated, which follows from the property of the Goldstone boson nature. Then the effective Lagrangian of a below the symmetry breaking scale is

$$\mathcal{L}^a = \frac{1}{2}(1+\frac{\rho}{\tilde{v}})^2(\partial_\mu a)^2 - \frac{1}{32\pi^2}\frac{a}{\tilde{v}}F^a_{\mu\nu}\tilde{F}^a_{\mu\nu} \qquad (102)$$

and solves the strong CP problem as discussed before.

The axion mixes with the mesons, π^0 and η^0; thus we have to diagonalize the mass matrix. Here we follow Bardeen and Tye's current algebra method for a reliable estimate of the mass.[49] The PQ current of Eq.(98) is

$$J^{PQ}_\mu \;=\; \tilde{v}\partial_\mu a \;-\; \frac{1}{2}\bar{Q}\gamma_\mu\gamma_5 Q \qquad (103)$$

But we must also include the light quark currents to take care of the mixing. Let us note that divergence of J_μ^{PQ} has the anomaly term only

$$\partial^\mu J_\mu^{PQ} = 0 \text{ (from classical symmetry)} - \frac{1}{2}\frac{2}{32\pi^2}F_{\mu\nu}^a\tilde{F}^{a\mu\nu} \text{ (from loop)} \quad (104)$$

To show the first term of RHS of Eq.(104), we note that it is $\tilde{v}\partial^2 a - im_Q\bar{Q}\gamma_5 Q$ which can be shown to be zero using the equation of motion of a at $a = 0$,

$$\partial^2 a = \frac{m_Q}{\tilde{v}}\sin(\frac{a}{\tilde{v}})\bar{Q}Q + i\frac{m_Q}{\tilde{v}}\cos(\frac{a}{\tilde{v}})\bar{Q}\gamma_5 Q$$

If it were not for the anomaly term, the current conservation would be exact and a would be massless. Let us mix J_μ^{PQ} with the u and d quark currents and neglect the heavier quarks (i.e. consider a and π^0 mixing only) such that the divergence does not have the gluon anomaly $F\tilde{F}$,

$$J_\mu^a = \tilde{v}\partial_\mu a - \frac{1}{2}\bar{Q}\gamma_\mu\gamma_5 Q + \frac{1}{2(1+Z)}(\bar{u}\gamma_\mu\gamma_5 u + Z\bar{d}\gamma_\mu\gamma_5 d) \quad (105)$$

where

$$Z = \frac{m_u}{m_d} \quad (106)$$

Eq. (105) is obtained by using the diagonalization condition

$$[Q^3, \partial^\mu J_\mu^a] = 0 \quad (107)$$

where Q^3 is the generator corresponding to π^0. In this diagonalized basis, J_μ^a is constructed such that a does not mix with π^0 and η', but has a divergence

$$\partial^\mu J_\mu^a = \frac{im_u}{1+Z}(\bar{u}\gamma_5 u + \bar{d}\gamma_5 d) \quad (108)$$

Now we can relate the axion and pion masses through the following manipulation with the identification of $F_a = \tilde{v}$,

$$\begin{aligned}
F_a^2 m_a^2 &= <0|[-iQ_5^a, \partial^\mu J_{5\mu}^a]|0> \\
&= -\frac{m_u}{(1+Z)^2}(<\bar{u}u> + Z<\bar{d}d>) \\
&= -\frac{Z}{(1+Z)^2}(m_d<\bar{u}u> + m_u<\bar{d}d>) \\
&= \frac{Z}{(1+Z)^2} f_\pi^2 m_{\pi^0}^2
\end{aligned} \qquad (109)$$

where in the last line we used the well-known relation

$$f_\pi^2 m_{\pi^0}^2 = <0|[-iQ_5^3, \partial^\mu J_{5\mu}^3]|0> = -(m_u + m_d)<\bar{q}q> \qquad (110)$$

where $<\bar{q}q> \equiv <\bar{u}u> = <\bar{d}d>$. Thus the axion mass is given by

$$m_a = \frac{f_\pi m_{\pi^0}}{F_a} \frac{\sqrt{Z}}{1+Z} = 0.61 \times 10^7 \left(\frac{[\text{GeV}]}{F_a}\right) \text{ [eV]} \qquad (111)$$

The couplings between the axion and light quarks can be obtained using the soft axion theorem,

$$\begin{aligned}
<\beta, a|\mathcal{L}|\alpha> &= -\frac{i}{F_a} <\beta|[Q_5^a, \mathcal{L}]|\alpha> \\
&= \frac{i}{F_a} <\beta| \int d^3x' [\frac{1}{2(1+Z)}(u^\dagger \gamma_5 u + Z d^\dagger \gamma_5 d), m_u \bar{u}u + m_d \bar{d}d]|\alpha> \\
&= \frac{im_u}{F_a(1+Z)} <\beta|\bar{u}\gamma_5 u + \bar{d}\gamma_5 d|\alpha>
\end{aligned} \qquad (112)$$

Thus we know that the heavy quark axion couples to the light quarks with the same strength.

The heavy quark axion does not couple to leptons at order m_ℓ/F_a.

5.3. The Peccei-Quinn-Weinberg-Wilczek Axion

The PQWW axion arises at the electroweak scale and is little bit more complicated than the heavy quark axion, as we will see below. For the axion degree, one introduces an additional Higgs doublet. The key point is the same as before: one introduces the global axial symmetry (PQ symmetry). One considers the following complex fields, coupling to the quarks,

$$Y = \begin{pmatrix} u \\ d \end{pmatrix}_L \quad u_R \quad d_R \quad \phi_1 \quad \phi_2 \qquad (113)$$
$$ \tfrac{1}{6} \quad\quad \tfrac{2}{3} \quad -\tfrac{1}{3} \quad \tfrac{1}{2} \quad -\tfrac{1}{2}$$

where the left-handed quark doublet is called q_L. The kinetic energy terms of these fields will have 5 global $U(1)$ symmetries. However, consideration of all possible Yukawa couplings will leave only 1 global $U(1)$ which is the baryon number B. B is not axial and is not a candidate for the PQ symmetry. So one must remove some couplings to have the PQ symmetry. For this, we apply the Glashow-Weinberg criterion for the removal of the flavor changing processes through neutral Higgs fields: couple one Higgs field to the u-type quarks and the other to the d-type quarks,

$$\mathcal{L}_y^{PQ} = -f_u \bar{q}_L u_R \phi_2 - f_d \bar{q}_L d_R \phi_1 + \text{h.c.} \qquad (114)$$

If we denote the complex phases of the fields as α_{field}, the PQ phase is

$$\frac{\alpha_1 + \alpha_2}{2} = \alpha_q - \frac{\alpha_u + \alpha_d}{2}$$

where α's are related by the Yukawa coupling $\alpha_1 = \alpha_q - \alpha_d$ and $\alpha_2 = \alpha_q - \alpha_u$. So two Higgs doublets with the coupling (114) can have the desired symmetry, but we must preserve it in the Higgs potential too. We should not have

$$\sum_{i,j} c_{ij}\phi_i\phi_j\phi_i\phi_j + \text{h.c.}$$

Because of the PQ symmetry the above choice of the Yukawa coupling and the Higgs potential is consistent. Below the electroweak symmetry breaking scale, the axion a will appear. It must be orthogonal to Z_l, the longitudinal component of Z. So concentrate on the neutral Higgs couplings

$$-f_u \bar{u}_L u_R \phi_2^0 - f_d \bar{d}_L d_R \phi_1^0 + \text{h.c.}$$

Let the PQ transformation be

$$U(1)_{PQ} : \phi_1 \to e^{i\alpha\Gamma_1}\phi_1, \; d_L \to e^{i\alpha\Gamma_1/2}d_L, \; d_R \to e^{-i\alpha\Gamma_1/2}d_R$$
$$\phi_2 \to e^{i\alpha\Gamma_2}\phi_2, \; u_L \to e^{i\alpha\Gamma_2/2}u_L, \; u_R \to e^{-i\alpha\Gamma_2/2}u_R \quad (115)$$

where $\Gamma_i (i = 1,2)$ will be determined. Representing ϕ_1 and ϕ_2 as

$$\phi_1^0 = \frac{1}{\sqrt{2}}(v_1 + \rho_1)e^{iP_1/v_1}$$
$$\phi_2^0 = \frac{1}{\sqrt{2}}(v_2 + \rho_2)e^{iP_2/v_2} \quad (116)$$

Representing P_1 and P_2 in terms of a and Z_l

$$P_1 = \cos\theta\, a - \sin\theta Z_l$$
$$P_2 = \sin\theta\, a + \cos\theta\, Z_l \quad (117)$$

one determines θ in terms of Γ_i from the PQ symmetry ($a \to a + \epsilon$, $Z_l \to Z_l$),

$$\cos\theta = \frac{v_1\Gamma_1}{\sqrt{v_1^2\Gamma_1^2 + v_2^2\Gamma_2^2}} \quad \sin\theta = \frac{v_2\Gamma_2}{\sqrt{v_1^2\Gamma_1^2 + v_2^2\Gamma_2^2}}$$

or
$$a = \frac{v_1\Gamma_1 P_1 + v_2\Gamma_2 P_2}{\sqrt{v_1^2\Gamma_1^2 + v_2^2\Gamma_2^2}} \tag{118}$$

A similar treatment of Z_l gives

$$Z_l = \frac{v_1 Q_Z(\phi_1^0)P_1 + v_2 Q_Z(\phi_2^0)P_2}{\sqrt{v_1^2 Q_Z(\phi_1^0)^2 + v_2^2 Q_Z(\phi_2^0)^2}}$$

where the Z-boson charge is $Q_Z = I_3 - Y$. Since $Q_Z(\phi_1^0) = -1$ and $Q_Z(\phi_2^0) = +1$, we have

$$Z_l = \frac{-v_1 P_1 + v_2 P_2}{\sqrt{v_1^2 + v_2^2}} \tag{119}$$

Thus the orthogonality of a and Z_l determines the PQ charges

$$\Gamma_1 : \Gamma_2 = x : \frac{1}{x} \tag{120}$$

where

$$x = \frac{v_2}{v_1} \tag{121}$$

Thus the axion component is found as

$$a = \frac{v_2 P_1 + v_1 P_2}{\sqrt{v_1^2 + v_2^2}} \tag{122}$$

Now it is a simple exercise to write the PQ current,

$$J_\mu^{PQ} = v\partial_\mu a - \frac{1}{2x}\sum_i \bar{u}_i \gamma_\mu \gamma_5 u_i - \frac{x}{2}\sum_i \bar{d}_i \gamma_\mu \gamma_5 d_i + \text{leptonic terms} \tag{123}$$

where N_g is the number of families. The divergence is

$$\partial^\mu J_\mu^{PQ} = -(\frac{1}{x} + x) N_g \frac{1}{32\pi^2} F_{\mu\nu}^a \tilde{F}^{a\mu\nu} \tag{124}$$

To avoid the a–π^0 mixing, we construct the the anomaly-free current as in the previous section,

$$J_\mu^a = v\partial_\mu a - \frac{1}{2x}\sum i\bar{u}_i\gamma_\mu\gamma_5 u_i - \frac{x}{2}\sum i\bar{d}_i\gamma_\mu\gamma_5 d_i$$
$$+ \frac{N_g}{2}(\frac{1}{x}+x)(\frac{1}{1+Z}\bar{u}\gamma_\mu\gamma_5 u + \frac{Z}{1+Z}\bar{d}\gamma_\mu\gamma_5 d) \qquad (125)$$

from which we calculate

$$m_a = N_g(\frac{1}{x}+x)\frac{\sqrt{Z}}{1+Z}\frac{f_\pi m_{\pi^0}}{v} \qquad (126)$$

In contrast to the heavy quark axion, the PQWW axion can have the leptonic couplings which depends on models.[21]

5.4. THE DFSZ MODEL

The PQWW axion was not phenomenologically viable, which led to the original heavy quark invisible axion. In this case one has to introduce another scale ($\neq$ the electroweak scale). It must be that the scale is generated by an $SU(2) \times U(1)$ singlet not to interfere with the successful standard model. We have seen that it is provided by a complex scalar σ.[41] In the heavy quark model, we introduced the singlet quark Q to have the color anomaly.

If one tries to get the anomaly through the light quarks, the singlet scalar must couple to Higgs doublets which in turn couples to light quarks. But as we have seen in the PQWW axion model, one Higgs doublet will not lead to an axial $U(1)$ global symmetry. One needs at least two Higgs doublets. So in this scheme one must introduce an additional Higgs doublet instead of the heavy quark. Indeed this idea works as shown by Dine, Fischler and Srednicki, and independently by Zhitniskii.[42] So the DFSZ model introduces the following complex scalars,

$$\phi_1, \ \phi_2, \ \sigma$$

The Yukawa couplings of the light quarks with ϕ_1 and ϕ_2 are the same as those of the PQWW model. In addition, one has to relate the phase of σ with the PQ phase of the previous section. So we may couple

$$V = c\phi_1^T \phi_2 \sigma\sigma + \text{h.c.} + \text{trivial terms} \tag{127}$$

where we have not written the rest which preserves the needed symmetry. The PQ symmetry is

$$U(1)_{PQ} : \sigma \to e^{i\alpha \Gamma_\sigma}\sigma$$
$$\phi_1 \to e^{-i\alpha \Gamma_1}\phi_1 \tag{128}$$
$$\phi_2 \to e^{-i\alpha \Gamma_2}\phi_2$$

where $\Gamma_1 : \Gamma_2 = x : (1/x)$ as given in the previous section. The potential V determines the PQ quantum number of σ as

$$\Gamma_\sigma : \Gamma_1 : \Gamma_2 = x + \frac{1}{x} : 2x : \frac{2}{x} \tag{129}$$

Thus the PQ current is

$$J_\mu^{PQ} = \tilde{v}\partial_\mu P_0 - \frac{2x}{x+x^{-1}}v_1\partial_\mu P_1 - \frac{2x^{-1}}{x+x^{-1}}v_2\partial_\mu P_2$$
$$+ \frac{x^{-1}}{x+x^{-1}}\sum_i \bar{u}_i\gamma_\mu\gamma_5 u_i + \frac{x}{x+x^{-1}}\sum_i \bar{d}_i\gamma_\mu\gamma_5 d_i \tag{130}$$

where $\tilde{v} = \sqrt{2}<\sigma>$, etc, and P_0, P_1 and P_2 are the phases of the singlet and doublet scalar fields. Separating the axion field from the Z_l component gives

$$J_\mu^{PQ} = \sqrt{\tilde{v}^2 + (\frac{2x}{x+x^{-1}})^2 v_1^2 + (\frac{2x^{-1}}{x+x^{-1}})^2 v_2^2}\,\partial_\mu a + \text{quark currents} \simeq \tilde{v}\partial_\mu a + \cdots \tag{131}$$

where in the last relation we used $\tilde{v} \gg v_1, v_2$. The divergence of the PQ current is exactly the anomaly term,

$$\partial^\mu J_\mu^{PQ} = \frac{N_g}{32\pi^2}\left(\frac{2x^{-1}}{x+x^{-1}} + \frac{2x}{x+x^{-1}}\right)F_{\mu\nu}^a \tilde{F}^{a\mu\nu} = \frac{2\cdot N_g}{32\pi^2}F_{\mu\nu}^a \tilde{F}^{a\mu\nu} \tag{132}$$

which must be the case for the axion. As in the previous sections, we can estimate the axion mass from Eqs. (131) and (132),

$$m_a \simeq \frac{f_\pi m_{\pi^0}}{\tilde{v}} \frac{\sqrt{Z}}{1+Z} \cdot 2 \cdot N_g \tag{133}$$

As in the PQWW axion model, the DFSZ axion can couple to leptons.[21]

5.5. OTHER AXIONS

Another obvious possibility for the axion is that it can be made a composite of fermions. Indeed the invisible axion of this type has been constructed by introducing additional confining force.[43]

Another more interesting axions are the superstring axions. Here also, one derives a 4 dimensional effective interaction of the pseudoscalar a after compactification of the extra dimensions.[44] There can be model-dependent axions. But here let us briefly discuss the model independent (MI) axion, because the appearance of the MI axion is different from the axions we have discussed so far. Superstring models have the second rank antisymmetric tensor field $B_{\mu\nu}$. The gauge invariant coupling of this comes through the field strength $H_{\mu\nu\rho}$. For example the fermion coupling is

$$\sim H_{\mu\nu\rho}\bar{\psi}\gamma^\mu\gamma^\nu\gamma^\rho\psi$$

Under the duality transformation of $H_{\mu\nu\rho}$, $H_{\mu\nu\rho} = \epsilon_{\mu\nu\rho\sigma}\partial^\sigma a$, the above coupling is proportional to

$$\sim \partial^\sigma a \bar{\psi}\gamma_\sigma\gamma_5\psi \propto a\partial_\mu\bar{\psi}\gamma^\mu\gamma_5\psi \propto F^a_{\mu\nu}\tilde{F}^{a\mu\nu}$$

Thus the equation of motion of a (The kinetic energy term comes from $(H_{\mu\nu\rho})^2$.) can be written as

$$\partial^2 a = \frac{1}{32\pi^2}\frac{a}{M}F^a_{\mu\nu}\tilde{F}^{a\mu\nu} \tag{134}$$

where M is a scale determined by the compactification. Thus we can interpret $H_{\mu\nu\rho}$ as an axion. The MI axion coupled with gravity has been shown recently to have the cosmologically interesting wormhole solution.[50] But the MI axion has the decay constant larger than 10^{15} GeV and has the cosmological energy density problem.[51]

5.6. AXION COUPLINGS

Below the PQ symmetry breaking scale but above the QCD chiral symmetry breaking scale, the axion couplings with light fermions are given by the PQ charges of the complex fields.[52] The simplest of these is the axion-lepton coupling which is given in Table 3. Here we focus on the axion-photon and axion-nucleon couplings which are more interesting theoretically[53] and also in the axion search experiments[54–56] and in the stellar evolutions.[57–59]

Above 1 GeV, the axion has the anomalous coupling,

$$-\frac{a}{F_a}\left(\frac{g_3^2}{32\pi^2}F^a_{\mu\nu}\tilde{F}^{a\mu\nu} + \bar{c}_{a\gamma\gamma}\frac{e^2}{32\pi^2}F^{em}_{\mu\nu}\tilde{F}^{\mu\nu}_{em}\right) \tag{135}$$

where $\bar{c}_{a\gamma\gamma}$ is determined by the electromagnetic charges of the fermions. It appears from $U(1)_{PQ}$-photon-photon anomaly.

The heavy quark axion with the quark charge $Q_{em} = ae$ gives

$$\bar{c}_{a\gamma\gamma} = a^2 \tag{136}$$

The PQWW axion has three possibilities: (I) ϕ_1 couples to leptons, (II) ϕ_2 couples to leptons, and (III) another (PQ neutral) Higgs doublet ϕ_3 couples to leptons. For these cases,

$$\begin{aligned}(I)\ \ &\bar{c}_{a\gamma\gamma} = \frac{4}{3}(x+\frac{1}{x})\\ (II)\ \ &\bar{c}_{a\gamma\gamma} = \frac{1}{3}(x+\frac{1}{x})\\ (III)\ \ &\bar{c}_{a\gamma\gamma} = \frac{1}{3}x+\frac{4}{3x}\end{aligned} \tag{137}$$

The DFSZ axion gives

$$\bar{c}_{a\gamma\gamma} = N_g\{(-1)^2 + 3\cdot(\frac{2}{3})^2 + 3\cdot(-\frac{1}{3})^2\}\cdot\frac{2}{2N_g} = \frac{8}{3} \quad (138)$$

which is often called the unification relation, and corresponds to (I) of the PQWW axion. In this case ϕ_1 couples to both d quark and electron. As in the $SU(5)$ case, d quark and electron obtain masses from the same Higgs field and can be easily unified in a multiplet. Cases (II) and (III) of the PQWW axion can be considered here too.[21]

Below 1 GeV, a is mixed with pion and our master formula can be used to derive the axion-photon coupling, treating a as a background field. In Eq.(135), let us remove the axion-gluon-gluon term, choosing $\alpha = -a/F_a^*$ in Eq.(74). The axion-photon-photon coupling is then changed,

$$c_{a\gamma\gamma} \equiv \bar{c}_{a\gamma\gamma} + 6\sum_{i=u,d}\alpha_i(Q_i^{em})^2$$

Let us consider u and d quarks only. As before we determine

$$\begin{aligned}\alpha_u &= \frac{m_d}{m_u+m_d}(-\frac{a}{F_a})\\ \alpha_d &= \frac{m_u}{m_u+m_d}(-\frac{a}{F_a})\end{aligned} \quad (139)$$

For any invisible axion model, this term is the same. So we have the following invisible axion-photon-photon coupling coefficient

$$c_{a\gamma\gamma} = \bar{c}_{a\gamma\gamma} - \frac{2}{3}\frac{4+Z}{1+Z} \simeq \bar{c}_{a\gamma\gamma} - 1.92 \quad (140)$$

where we used $Z = 0.56$ [60] in the last equation. The relevant invisible axion couplings are shown in Table 3.

---

* We have already shifted a such that θ parameter disappears.

For the PQWW axion model, PQ charges have the x dependence and we obtain†

$$c_{a\gamma\gamma} \simeq \bar{c}_{a\gamma\gamma} - 1.92 \frac{x + x^{-1}}{2} \quad (141)$$

where $\bar{c}_{a\gamma\gamma}$ is given in Eq.(137).

Table 3. Coupling constants in hadronic (KSVZ) and DFSZ invisible axion models. Positive $X_u(= \frac{2x^{-1}}{x+x^{-1}})$ and $X_d(= \frac{2x}{x+x^{-1}})$ are the PQ charges. Penomenological values of $SU(3)_{flavor}$ are $D = 0.81$, $F = 0.44$ and $0.1 \leq S \leq 2.2$.

Couplings	Hadronic axion	DFSZ axion
$c_{a\gamma\gamma}$	$\bar{c}_{a\gamma\gamma} - 1.92$	0.75
g_{aee}	~ 0	$1.4 \times 10^{-11} X_d \frac{3}{N_g} (m_a/[\text{eV}])$
g_{aNN}	$7.8 \times 10^{-8} (m_a/[\text{eV}]) \cdot \{[\frac{1-Z}{1+Z}(D+F)\tau_3 + \frac{3F-D+2S}{3} \mathbf{1}\}$	$2.6 \times 10^{-8} \frac{3}{N_g} (m_a/[\text{eV}]) \cdot \{\frac{D+F}{2}(X_u - X_d - N_g \frac{1-Z}{1+Z})\tau_3 + [\frac{1}{6}(3F-D)(X_u + X_d - N_g) + \frac{1}{3}S(X_u + 2X_d - N_g)\mathbf{1}]\}$

For the case of the axion-nucleon-nucleon coupling, one has to be careful how to use the couplings. Table 3 is good for the lowest order amplitude. But in the core of the neutron star[61] or SN1987A,[62] the relevant process involves virtual nucleon and pion. Here we follow the discussion of Choi et al.[63]‡

Both the axion and pions are treated as Goldstone bosons for which there exists an excellent book.[64] Let N be the nucleon doublet, and $\Sigma = \exp(2i\pi/f_\pi)$

---

† In Eq.(6.25) of Ref. [21], the x dependence is hidden after $x = 1$ is inserted.
‡ This part is added after the talk.

where $\pi = \vec{\pi}\cdot\vec{\tau}/2$. Consider the following CP conserving effective Lagrangian,

$$\mathcal{L} = \bar{N}i\gamma^\mu\partial_\mu N - m(\bar{N}_L\Sigma N_R + h.c.) + \lambda(\bar{N}_L\Sigma i\gamma^\mu\partial_\mu\Sigma^+ N_L$$
$$+ \bar{N}_R\Sigma^+i\gamma^\mu\partial_\mu\Sigma N_R) - (F_a)^{-1}\bar{N}(\gamma^\mu\partial_\mu a)\gamma_5 XN + (f_\pi^2/4)Tr\partial_\mu\Sigma\partial^\mu\Sigma^+ \quad (141)$$
$$+ (1/2)\partial_\mu a\partial^\mu a + \cdots$$

where X denotes the real constant 2×2 diagonal matrix of the PQ charges, λ is a real constant and m is the nucleon mass. Here the isospin–symmetry breaking terms which are of order m_q are neglected.

The choice of the nucleon basis is up to us depending on the convenience. Consider the $a-$ and π–dependent phase transformation of the nucleon,

$$N'_L = U^\dagger N_L, \quad N'_R = UN_R \quad (142.a)$$

with

$$U = \exp i(2h\pi/f_\pi + aY/F_a) \quad (142.b)$$

where h is a constant and Y a constant 2×2 diagonal matrix. Then it is well-known[65] that both the exact and tree–level on–mass–shell amplitudes calculated from Eq.(141) and from the new Lagrangian defined by

$$\mathcal{L}' = \mathcal{L}(N_L \to UN_L, \ N_R \to U^\dagger N_R) \quad (143)$$

are identical independently of any specific choices of h and Y in Eq.(142.b). It is then clear that there are many equivalent coupling schemes possible. In particular let us consider the interaction Lagrangian $\mathcal{L}'_{int}$ relevant to the nucleon bremsstrahlung of axions in the one–pion–exchange approximation to the order

$(1/F_a f_\pi)$,

$$\begin{aligned}\mathcal{L}'_{int} = &-(2m/f_\pi)(1-2h)\bar{N}i\gamma_5\pi N + (2/f_\pi)(h-\lambda)\bar{N}\gamma^\mu\partial_\mu\pi\gamma_5 N \\&+ (2m/F_a)\bar{N}i\gamma_5 aYN + (1/F_a)\bar{N}\gamma^\mu\partial_\mu a\gamma_5(Y-X)N \\&- (2m/F_a f_\pi)(1-2h)\bar{N}a\{\pi,Y\}N \\&+ (1/F_a f_\pi)(h-2\lambda)\bar{N}ia[Y,\gamma^\mu\partial_\mu\pi]N \\&+ (h/F_a f_\pi)\bar{N}i\gamma^\mu\partial_\mu a[\pi,Y-2X]N\end{aligned}$$ (144)

Note for the hadronic axions with the PQ-charge matrix $X = g\mathbf{1}$, which is the case considered effectively in Refs.[61,59,62], that the interaction Lagrangian $\mathcal{L}'_{int}$ reduces to only the derivative-coupling terms for the convenient choice $h = 1/2$ and $Y = 0$,

$$\mathcal{L}'_{int} = (1/f_\pi)(1-2\lambda)\bar{N}\gamma^\mu\partial_\mu\pi\gamma_5 N - (g/F_a)\bar{N}\gamma^\mu\partial_\mu a\gamma_5 N \qquad (145.a)$$

while we get the non-derivative pseudoscalar couplings only for another convenient choice $h = \lambda$ and $Y = X$,

$$\begin{aligned}\mathcal{L}'_{int} = &-(2m/f_\pi)(1-2\lambda)\bar{N}i\gamma_5\pi N + (2mg/F_a)\bar{N}i\gamma_5 a N \\&- (4mg/F_a f_\pi)(1-2\lambda)\bar{N}a\pi N\end{aligned}$$ (145.b)

Also the case of the derivative pion coupling and pseudoscalar axion coupling can be obtained when $h = 1/2$ and $Y = X$,

$$\mathcal{L}'_{int} = (1/f_\pi)(1-2\lambda)\bar{N}\gamma^\mu\partial_\mu\pi\gamma_5 N + (2mg/F_a)\bar{N}i\gamma_5 aN \qquad (145.c)$$

Similarly, the case of the derivative axion coupling and pseudoscalar pion coupling follows for the choice $h = \lambda$ and $Y = 0$. Since the S-matrix is independent of U, any one of the possible coupling schemes from Eq.(144) is correct and in particular the three cases in Eqs.(145) are all equivalent with the usual coupling constants $(1-2\lambda)/f_\pi = f/m_\pi$ and $g_{an} = 2mg/F_a$.

Hence the last row of Table 3 is good at order $1/F_a$, but is not enough for calculation of amplitudes of order $1/f_\pi F_a$ which is important for obtaining a bound on F_a from SN1987A.[62]

6. Axions and Cosmology

6.1. AXIONIC STRINGS

Axion models provide strings, domain walls, and coherent axion oscillations which can be cosmologically significant. Strings appear at the earliest epoch followed by the domain wall formation and coherent oscillation.

Axion strings[66] appear at the PQ symmetry breaking scale $\sim 10^{11}$ GeV. It is because one dimensional topological defects appear when a $U(1)$ symmetry (global or gauge) is spontaneously broken. The $U(1)$ symmetry here is $U(1)_{PQ}$. This situation is shown in Fig. 5.

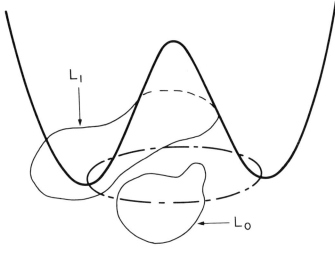

Fig. 5

In the figure, two typical loops on the surface of the Mexican hat shape potential is drawn. The arrows represent increasing φ (e.g. the argument of the complex scalar in the PQWW, KSVZ, and DFSZ axion models). The ground state is degenerate at this scale and is shown as a dash-dot circle. Imagine a closed loop l_0 in space. Along this loop, let us assign VEV's of a (effective) scalar field which can be the loops shown in the figure, i.e. the loop L_0. When the classical scalar field settles at the vacuum, the loop L_0 becomes a unclosed segment on the dash-dot circle. We can shrink this segment to a point by a gauge or global transformation. Another nearby loop l'_0 to l_0 has the same behavior. The transformation moving the segment corresponding to L_0 to that of L'_0 does not cause energy. In space, we can continuously shrink l_0 to l'_0 without causing energy. In fact, l_0 can be shrunk to a point, implying that there is no topological defect detected inside the loop l_0. Next consider the loop L_1 in the figure whose VEV's are assigned on the loop l_1 in space. When the classical scalar field settles at vacuum, L_1 become the closed dash-dot circle in the figure which cannot be shrunk to a point. To shrink it to a point, one has to supply energy so that the loop L_1 goes over the top of the Mexican hat potential. In space, one detects a topological defect inside the loop l_1 which must be a one-dimensional object and is called a string. In the figure, one can consider more loops going around the Mexican hat n times. Then different n's define different strings.

Explicitly, the string solutions can be obtained by solving the classical field equations. It has been shown that the gauge string has a finite line energy density.[67] For the global string present in the infinite volume space, the line energy density is not finite; but it can appear as defects in cosmology. For the global string, one solves

$$\frac{1}{r}\frac{\partial}{\partial r} r \frac{\partial}{\partial r} \rho + \frac{1}{r^2}\frac{\partial^2}{\partial \varphi^2} \rho + (v+\rho)[\lambda(v+\rho)^2 - \mu^2] = 0$$
$$\frac{1}{r}\frac{\partial}{\partial r} r \frac{\partial}{\partial r} a + \frac{1}{r^2}\frac{\partial^2}{\partial \varphi^2} a = 0 \qquad (146)$$

where $v = \sqrt{2} <\sigma>$ as usual, and ρ and a are the real and phase fields of σ. We are interested in the solution of a which is

$$a = n\,\theta\,v \qquad n = \text{integer} \qquad (147)$$

Hence any axion models have cosmological strings.

6.2. DOMAIN WALLS

For a brief period (forever in some axion models) after the QCD chiral symmetry breaking, the domain walls are formed.[68] As noted by Zel'dovich et al,[69] the spontaneous breaking of global symmetries lead to domain walls. The axion is defined by

$$\mathcal{L} = \frac{1}{2}(\partial_\mu a)^2 + \frac{1}{32\pi^2}\frac{a}{F_a} F^a_{\mu\nu}\tilde{F}^{a\mu\nu} + \cdots \qquad (148)$$

where F_a as defined above is the axion decay constant. Thus the QCD θ at vacuum is a/F_a. Since the vacuum of the axion potential shown in Fig.6 has a discrete symmetry, we expect domain walls. Note that the axion potential obtained from Eq.(148) appears below the QCD chiral symmetry breaking scale, and the form was obtained from the θ periodicity and nonzero axion mass.

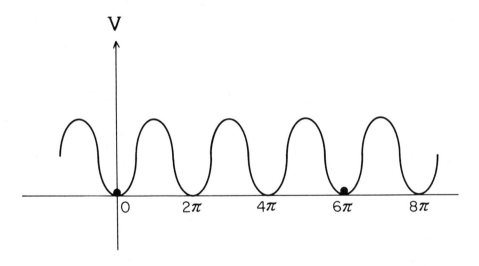

Fig. 6

However, the field vaiable a may not come to its original value after a shift of $2\pi F_a$. Imagine a loop L_3 in Fig. 5 and the corresponding one l_3 in space. One circumnavigation of l_3 corresponds to three circumnavigation around the potential of Fig.5. The phase field a is identified after one circumnavigation around the space loop l_3. The $2\pi F_a$ shift of a which is one circumnavigation around the space loop l_3. The $2\pi F_a$ shift of a which is one circumnavigation around the potential of Fig.5, implies $2\pi/3$ rotation around l_3. Thus the potential $V[a]$ in terms of a is Fig.6 with $6\pi F_a$ identified with 0. Note our angle variables, $\theta = a/F_a$ and φ = phase of σ.

Let us discuss the DFSZ model with the axion coupling $(2N_g a/\tilde{v})$ $(1/32\pi^2)$ $F^a_{\mu\nu}\tilde{F}^{a\mu\nu}$. We can identify

$$F_a = \frac{\tilde{v}}{2N_g} \qquad \sigma = \frac{\tilde{v}+\rho}{\sqrt{2}} e^{(ia/\tilde{v})} \quad (\varphi = \frac{a}{\tilde{v}}) \tag{149}$$

Namely $\varphi = 2\pi$ rotation corresponds to $a \to a + 2N_g 2\pi F_a$, and we have $2N_g$ degenerate vacua, leading to the domain wall problem.[68]

Also, the PQWW axion model has the domain walls as can be seen from its interaction to the gluon anomaly, but the axion is short lived compared to the age of the universe and these walls would not have survived until today.

The heavy quark axion model with one heavy quark does not have the domain wall problem as we show below.

Let us proceed to discuss the other interesting solutions of the domain wall problem. The original scheme embedding the discrete group of the PQ symmetry (disctre vacuum points) in nonabelian gauge groups is due to Lazarides and Shafi.[70] Here we start with the scheme[71] due to Barr, Gao, Reiss, Choi, and Kim, which is easier to understand. Let us consider two would-be axions a_1 and a_2 which have the following couplings,

$$\frac{a_1}{F_1} F^a_{\mu\nu}\tilde{F}^{a\mu\nu} + \frac{a_2}{F_2} F^a_{\mu\nu}\tilde{F}^{a\mu\nu} \tag{150}$$

As discussed above we can identify

$$\begin{aligned}< a_1 > \equiv 2\pi N_1 F_1 + < a_1 > \\ < a_2 > \equiv 2\pi N_2 F_2 + < a_2 >\end{aligned} \tag{151}$$

where N_1 and N_2 are given from a model. An example with $N_1 = 2$ and $N_2 = 5$ is shown in Fig.7. Looking at Eq.(150), one would interpret $(a_1/F_1 + a_2/F_2)$ as the axion and the orthogonal combination to it as a Goldstone boson. A constant axion direction corresponds to shifting of the Goldstone field, which is shown as solid lines in Fig.7. Because of the identification given in Eq. (151), all the seemingly different vacua are connected and gives

$$\mathcal{L}_{int} = \frac{1}{32\pi^2} \frac{a}{F_a} F^a_{\mu\nu} \tilde{F}^{a\mu\nu} \quad \text{with} \quad N_{DW} = 1 \;,\; F_a = \frac{F_1 F_2}{\sqrt{F_1^2 + F_2^2}} \tag{152}$$

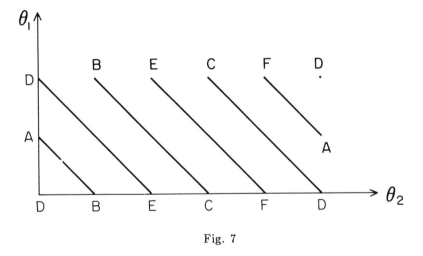

Fig. 7

The case $N_{DW} = 1$ is obtained because N_1 and N_2 are relatively prime. The domain wall number is the greatest common divisor of N_1 and N_2. This simple example is the case with spontaneously broken two global $U(1)$'s. If one $U(1)$ is broken but the other is not, then we arrive at the same conclusion also. So far we have considered a continuous symmetry to connect the vacua, keeping the classical axion value constant.

This leads to the possibility of discrete shift, keeping the axion constant.

Consider the $SU(N)$ group has group elements

$$g = \begin{pmatrix} e^{i\alpha/N} & 0 & \cdots & 0 \\ 0 & e^{i\alpha/N} & \cdots & 0 \\ \cdots & \cdots & \cdots & \cdots \\ 0 & 0 & \cdots & e^{i\alpha/N} \end{pmatrix} \qquad (153.a)$$

where

$$\alpha = 2\pi k \qquad (k = 0, 1, 2, \cdots, N) \qquad (153.b)$$

If $\alpha \neq 2\pi k$, g does not belong to $SU(N)$ ($Det \neq 1$). Consider a symmetry $SU(N)_{gauge} \otimes U(1)_{PQ}$. The global rotation gives an anomaly, since PQ symmetry is chiral. The gauge rotation does not give an anomaly, since it is supposed to be renormalizable. But the anomaly term vanishes when $\theta = 2n\pi$ where n is an integer as discussed in Sec.3.2. Thus by making N of $SU(N)$ an integer multiple of the vacuum degeneracy of the axion model, the bundle of the gauge rotation can be made to meet at $\theta = a/F_a = 2\pi \cdot$(integer) of the PQ rotation. Otherwise, starting from $< a >= 0$, gauge rotations do not cross the next degenerate vacuum point of $< a >$. In practice, N is chosen as the vacuum degeneracy to have $N_{DW} = 1$. When the vacuum degeneracy is divisible by N,

$$N_{DW} = \frac{\text{vacuum degeneracy}}{N}. \qquad (154)$$

The disctre group elements, g, form the center of $SU(N)$. In other nonabelian gauge groups also, the above solution of the domain wall problem can be worked out if the center matches the vacuum degeneracy. This method of solving the domain wall problem is called the Lazarides- Shafi mechanism.[70] Of course, this type of solution works also for the global nonabelian groups.[72] Thus for the application to the domain wall problem, we list the centers of nonabelian groups,

$$SU(N) , N \geq 2 : Z_N$$
$$SO(2\ell + 1) , \ell \geq 2 : Z_2$$
$$SO(4k) , k \geq 2 : Z_2 \otimes Z_2$$
$$SO(4k + 2) : Z_4$$
$$Sp(2n) , n \geq 3 : Z_2 \qquad (155)$$
$$E_6 : Z_3$$
$$E_7 : Z_2$$
$$E_8 , F_4 , G_2 : \text{trivial}$$

The heavy quark axion model can have 1 or 3 heavy quarks to have $N_{DW} = 1$. In the latter case, 3 of the color-$SU(3)$ is equal to the vacuum degeneracy.

For the DFSZ model $N_{DW} = 6$ is not reduced by the color-$SU(3)$ or by the weak-$SU(2)$, since u_L and d_L quarks are fundamental representations of $SU(3)$ and $SU(2)$.

Thus we see that the domain wall problem is in some sence related to the flavor problem. A three generation model with $N_{DW} = 1$ has been found in $SU(9)$.[73]* This model introduces fermions, $\psi^{abcd}(35)_{-5}$, $\psi^{abc}(21)_3$, two $\psi_{ab}(7)_7$, and four $\psi_a(1)_{-1}$, where $a, b, \cdots$ are the $SU(9)$ indices, taking the antisymmetric combinations as usual, and the number of quarks in the representation are shown in the bracket. Two types of the Higgs fields are introduced: $(H^{ab})_2$ and $(H^{abc})_{-6}$. All possible Yukawa couplings leaves one global symmetry whose quantum numbers are shown as subscripts. In fact, it is the PQ symmetry, since the divergence of this current is

$$\partial^\mu J_\mu = \frac{1}{32\pi^2} F^a_{\mu\nu} \tilde{F}^{a\mu\nu} \cdot (-5 \cdot 35 + 3 \cdot 21 + 7 \cdot 2 \cdot 7 - 1 \cdot 4) = \frac{1}{32\pi^2} F \tilde{F} \cdot (-18)$$

Since the PQ quantum numbers of the Higgs fields are multiples of 2, the current contains $2\tilde{v}\partial_\mu a + \cdots$, which implies that the coefficient of the anomalous axion

---

* For a four generation $SU(9)$ model, see Ref.[74].

interaction is -18/2$\tilde{v}$. Quark condensations coming from two ψ's also have even PQ quantum numbers. Thus the vacuum degeneracy is 9. However, it is embedded in the center of $SU(9)$, and we obtain $N_{DW} = 1$. It can be easily shown that this model has three families.

The evolution of universe in the presence of axionic strings and domain walls is discussed in Ref.[75].

6.3. Cosmological Energy Density

Because there exist excellent introductory discussions on this subject, we will be brief here. The axion potential is extremely flat, and hence the vacuum will not start to oscillate until the Hubble parameter (cosmic expansion rate) becomes smaller than the vacuum oscillation rate (axion mass) around the minimum of the axion potential. This oscillation lauching time t_1 is important to estimate the present oscillation amplitude, since at t_1 we expect the amplitude is of order F_a. Since the invisible axions interact very weakly, the present coherent vacuum oscillation of axions does not dissipate the energy enough for the axion oscillation not to be important. Between t_1 and now ($\equiv t_f$), these coherent axions lose energy only through the Hubble expansion. As a result, the vacuum axions can have enough oscillating energy now exceeding the critical energy if $F_a \geq 10^{12}$ GeV. The equation describing the classical axion field in an expanding universe is

$$\frac{d^2}{dt^2} <a> + 3H(t)\frac{d}{dt} <a> + m_a(t)^2 \simeq 0 \qquad (156)$$

which can be solved by substituting

$$<a> = A(t)\cos(m_a t)$$

where $A(t)$ is the amplitude of oscillation. Eq. (156) is solved with $H = \dot{R}/R$

and gives
$$\frac{(m_a A^2)_f}{(m_a A^2)_1} = \left(\frac{R_1}{R_f}\right)^3 \tag{157}$$

T_1 (temperature at t_1) is estimated from

$$m_a(T_1) \simeq 3 H(T_1) \tag{158}$$

which can be solved for T_1 from the knowledge of the high temperature axion mass[77] and the temperature dependence of the Hubble parameter to give[76]

$$T_1 \simeq 1 \text{ GeV} \tag{159}$$

Since the coherent oscillation energy density is $m_a^2 A(t)^2/2$, the axion number density is $n_a = m_a A(t)^2/2$. Now we can express the present axion energy in terms of known quantities,

$$\begin{aligned}\rho_a(T_\gamma) &= m_a(T_\gamma) n_a(T_\gamma) \\ &= m_a(T_\gamma) \frac{1}{2} m_a(T_1) A(T_1)^2 \left(\frac{T_\gamma}{T_1}\right)^3 \left(\frac{g_{*f}}{g_{*1}}\right)\end{aligned} \tag{160}$$

which can be compared to give

$$\Omega_a = \frac{\rho_a(T_\gamma)}{\rho_c} \simeq 2 \times 10^7 h_{1/2}^{-2} \left(\frac{\text{GeV}}{T_1}\right) \left(\frac{T_\gamma}{2.7 \text{ K}}\right)^3 \left(\frac{A(T_1)}{F_a}\right)^2 \left(\frac{F_a}{M_p}\right) \tag{161}$$

where $h_{1/2}$ is the present Hubble parameter in units of 50 km/sec/Mpc. Since $A(T_1)$ is of order F_a, $\Omega_a < 1$ gives

$$F_a < 0.5 \times 10^{12} h_{1/2}^2 \text{ GeV} \tag{162}$$

Still the coherent axions oscillate. In other words, the vacuum angle θ still oscillates and gives an oscillating neutron electric dipole moment. This oscillating dipole moment will radiate photons. But a gram of nucleus put inside a cavity near zero temperature gives a signal off from a detectable rate by a factor of 10^{-10}.

6.4. Cosmic Axion Detection

As we have seen in the previous section, the universe can be closed by the coherent cold axions if $F_a \sim 10^{11}$ GeV. Sikivie[54] has shown that these axions can be detected by low energy cavity experiments, probing the axion-photon-photon coupling

$$\mathcal{L}_{a\gamma\gamma} = -\frac{c_{a\gamma\gamma}}{2}\frac{\alpha_{em}}{\pi}\frac{a}{F_a}\mathbf{E}\cdot\mathbf{B} \qquad (163)$$

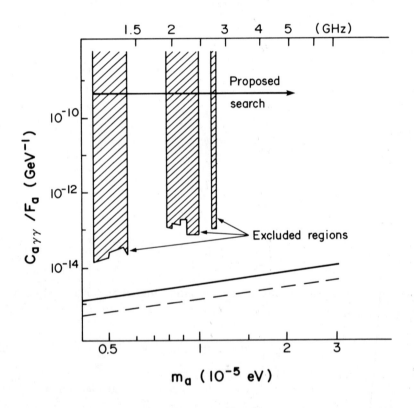

Fig. 8

Because of the small velocity $\beta \simeq 10^{-3}$ of typical galactic objects, the galactic axions will have a small spread of kinetic energy $(1/2)\beta^2 \simeq O(10^{-6})$. If all the missing mass of the universe is inside the galactic halos, the galactic halo axion energy density will be of order 10^{-24} g/cm^3. These axions can be converted to microwave photons which can be collected inside the cavity detector. It is possible to collect these photons because of the small (10^{-6}) spread of the photon energy. The first result of these cavity experiment has been reported.[56] The Rochester–BNL–Fermilab experimental data as of Feb., 1988 is shown in Fig. 8. The shaded region is excluded. The solid line is the hadronic axion prediction with $c_{a\gamma\gamma} = -1.92$, the dashed line is the DFSZ model prediction. We note that the sensitivity of the experiment is, however, a factor ~ 30 off from the invisible axion detection.

7. Conclusion

In this lecture, we have discussed that a good extension of the standard model must explain the *parameters* of the standard model. One excellent example is GUTs. Another good extension is introduction of axions which solves the θ-parameter problem. We focused on the theoretical aspects of the axion solution. The invisible axion is expected to be discovered in a near future, or the remaining window of the axion scale $10^9 \sim 10^{12}$ is washed out. In the latter case, the possibility $m_u = 0$ is the remaining attractive solution for the part of the strong CP problem but as a parameter problem it is not understood and belongs to the unsolved flavor problem.

Acknowledgements:

I thank the Organizing Committee of TASI–88 for the kind hospitality extended to me during the summer school. This work has been supported in part by U.S. Department of Energy under the Contract No. DE-AC02-76ER03130.A026 and the Korean Science and Engineering Foundation.

References

1. S. L. Glashow, Nucl. Phys. **22**, 579 (1961); S. Weinberg, Phys. Rev. Lett. **19**, 1264 (1967); A. Salam, in Elementary Particle Theory, ed. N. Svartholm (Almqvist and Wiksells, Stockholm, 1969) p. 367.

2. S. Kobayashi and K. Maskawa, Prog. Theor. Phys. **49**, 652 (1973).

3. R. D. Peccei, this proceedings.

4. J. E. Kim et al, Rev. Mod. Phys. **53**, 211 (1981); U. Amaldi et al, Phys. Rev. **D26**, 1385 (1987).

5. P. Langacker, Phys. Rep. **72**, 185 (1981); H. Georgi, J. de Phys. C3 Suppl. No. **12**, C3-705 (1982); A. Zee, *Unity of Forces in the Universe* (World Scientific Publishing Co., Singapore, 1982); For a recent excellent introduction, see, M. Srednicki, TASI-86 lecture, in *From the Planck Scale to the Weak Scale : toward a Theory of the Universe*, ed. H. E. Haber (World Scientific Publishing Co., Singapore, 1987).

6. J. Wess and J. Bagger, *Supersymmetry and Supergravity* (Princeton University Press, Princeton, 1983).

7. H. Georgi and S. L. Glashow, Phys. Rev. Lett. **32**, 438 (1974).

8. J. C. Pati and A. Salam, Phys. Rev. **D10**, 275 (1974).

9. Howard Georgi, Nucl. Phys. **B156**, 126 (1979).

10. R. N. Cahn, Phys. Lett. **104B**, 282 (1982).

11. H. Georgi, H. R. Quinn and S. Weinberg, Phys. Rev. Lett. **33**, 451 (1974).

12. U. Amaldi et al, Phys. Rev. **D26**, 1385 (1987).

13. G. 't Hooft, Phys. Rev. Lett. **37**, 8 (1976).

14. G. 't Hooft, Nucl. Phys. **B35**, 167 (1972).

15. For an excellent review, see, P. Langacker, Phys. Rep. **72**, (1981).

16. P. Langacker, in *Inner Space/Outer Space*, eds. E. W. Kolb et al (Univ. of Chicago Press, Chicago, 1986).

17. G. Blewitt et al, Phys. Rev. Lett. **55**, 2114 (1985).

18. A. A. Belavin, A. Polyakov, A. Schwartz, and V. Tyupkin, Phys. Lett. **59B**, 85 (1974).

19. C. G. Callan, R. F. Dashen and D. J. Gross, Phys. Lett. **63B**, 334 (1976); R. Jackiw and C. Rebbi, Phys. Rev. Lett. **37**, 172 (1976).

20. G. 't Hooft, in *Recent Development in Gauge Theories*, ed. G. 't Hooft et al (Plenum Press, New York and London, 1979) p.135.

21. J. E. Kim, Phys. Rep. **150**, 1 (1987).

22. M. A. B. Beg and H.-S. Tsao, Phys. Rev. Lett. **41**, 278 (1978); R. N. Mohapatra and G. Senjanovic, Phys. Lett. **79B**, 283 (1978); H. Georgi, Hadronic J. **1**, 155 (1979); G. Segre and A. Weldon, Phys. Rev. Lett. **42**, 1191 (1979); S. M. Barr and P. Langacker, Phys. Rev. Lett. **42**, 1654 (1979). Spontaneous weak CP violation but with the axion has been discussed by many people. See, for example, M. Shin, Phys. Lett. **160B**, 411 (1985).

23. A. Nelson, Phys. Lett. **136B**, 387 (1984); S. M. Barr, Phys. Rev. Lett. **53**, 329 (1984).

24. H.-Y. Cheng, Phys. Rep. **158**, 1 (1988).

25. V. Baluni, Phys. Rev. **D19**, 2227 (1979); R. Crewther, P. DiVecchia, G. Veneziano, and E. Witten, Phys. Lett. **89B**, 123 (1979).

26. S. Adler, Phys. Rev. **177**, 2426 (1969); J. S. Bell and R. Jackiw, Nuovo Cim. **60A**, 47 (1969).

27. K. Fujikawa, Phys. Rev. **D21**, 2848 (1980).

28. R. F. Dashen, Phys. Rev. **D3**, 1879 (1971).

29. V. Baluni, Phys. Rev. **D19**, 2227 (1979).

30. I. S. Altarev et al, Pis'ma Zh. Eksp. Teor. Fiz. **29**, 794 (1979) [JETP Lett. **29**, 730 (1979)].

31. T. D. Lee, Phys. Rev. **D8**, 1226 (1973).

32. J. Gasser and H. Leutwyler, Phys. Rep. **87**, 78 (1982).

33. H. M. Georgi, Hadronic J. **1**, 155 (1979).

34. P. Langacker and H. Pagels, Phys. Rev. **D19**, 2070 (1979).

35. D. B. Kaplan and H. V. Manohar, Phys. Rev. Lett. **56**, 2004 (1986).

36. G. 't Hooft, Phys. Rev. **D14**, 3432 (1976).

37. S. Weinberg, Phys. Rev. **D14**, 3583 (1975).

38. K. Choi, C. W. Kim and W. K. Sze, Phys. Rev. Lett. **61**, 794 (1988);
For an earlier discussion, see, H. M. Georgi and I. N. McArthur, Harvard preprint HUTP–81/A011 (1981).

39. R. D. Peccei and H. R. Quinn, Phys. Rev. Lett. **38**, 1440 (1977); Phys. Rev. **D16**, 1791 (1977).

40. S. Weinberg, Phys. Rev. Lett. **40**, 223 (1978);
F. Wilczek, Phys. Rev. Lett. **40**, 279 (1978).

41. J. E. Kim, Phys. Rev. Lett. **43**, 103 (1979); M. A. Shifman, V. I. Vainstein and V. I. Zakharov, Nucl. Phys. **B166**, 493 (1980).

42. M. Dine, W. Fischler and M. Srednicki, Phys. Lett. **104B**, 199 (1981);
A. P. Zhitniskii, Sov. J. Nucl. Phys. **31**, 260 (1980).

43. J. E. Kim, Phys. Rev. **D31**, 1733 (1985); K. Choi and J. E. Kim, Phys. Rev. **D32**, 1828 (1985).

44. E. Witten, Phys. Lett. **149B**, 351 (1984); Phys. Lett. **153B**, 243 (1985); Phys. Lett. **155B**, 151 (1985);
K. Choi and J. E. Kim, Phys. Lett. **154B**, 393 (1985); Phys. Lett. **165 B**, 71 (1985).

45. C. Vafa and E. Witten, Phys. Rev. Lett. **53**, 535 (1984).

46. H. Georgi, in *Grand Unified Theories and Related Topics*, eds. M. Konuma and T. Maskawa (World Scientific Publishing Co., Singapore, 1981) p.209.

47. H. Georgi and L. Randall, Nucl. Phys. **B276**, 241 (1986).

48. R. D. Peccei, T. T. Wu and T. Yanagida, Phys. Lett. **172B**, 435 (1986); L. M. Krauss and F. Wilczek, Phys. Lett. **173B**, 139 (1986).

49. W. A. Bardeen and S.-H. H. Tye, Phys. Lett. **74B**, 229 (1978).

50. S. B. Giddings and A. Strominger, Nucl. Phys. **B306**, 890 (1988); A. Strominger, this Proceedings.

51. K. Choi and J. E. Kim, Phys. Lett. **154B**, 393 (1985).

52. See, for example, H. Georgi, D. B. Kaplan and L. Randall, Phys. Lett. **169B**, 73 (1986).

53. D. B. Kaplan, Nucl. Phys. **B260**, 215 (1985); M. Srednicki, Nucl. Phys. **B260**, 689 (1985).

54. P. Sikivie, Phys. Rev. Lett. **51**, 1415 (1983).

55. P. Sikivie, M. Sullivan and D. Tanner, "A Search for the Invisible Axion: Experimental Proposal", 1984; A. C. Melissinos et al, "A Search for Galactic Axions", Dec., 1984; D. E. Morris, LBL Report No. LBL-17915 (1984); A. C. Melissinos et al, "Search for the Coherent Production of Light Scalar and Pseudoscalar Particles", 1988.

56. P. DePanfilis et al, Phys. Rev. Lett. **59**, 839 (1987).

57. D. A. Dicus et al, Phys. Rev. **D22**, 839 (1980).

58. M. Fukugita, S. Watamura and M. Yoshimura, Phys. Rev. **D26**, 1840 (1982).

59. A. Pantziris and K. Kang, Phys. Rev. **D33**, 3509 (1986).

60. S. Weinberg, in *Festschrift for I. I. Rabi*, Trans. of New York Academy of Sciences, Vol. **38**, 185 (1977).

61. N. Iwamoto, Phys. Rev. Lett., **53**, 1198 (1984).

62. G. Raffelt and D. Seckel, Phys. Rev. Lett. **60**, 1793 (1988); M. S. Turner, Phys. Rev. Lett., **60**, 1797 (1988); R. Mayle et al, Phys. Lett. **203B**, 188 (1988); T. Hatsuda and M. Yoshimura, Phys. Lett. **203B**, 469 (1988).

63. K. Choi, K. Kang and J. E. Kim, Brown-HET-671 (1988).

64. H. Georgi, *Weak Interactions and Modern Particle Theory* (Benjamin-Cummings, 1984).

65. S. Coleman, J. Wess and B. Zumino, Phys. Rev. **177**, 2239 (1969).

66. For a review, see, A. Vilenkin, Phys. Rep. **121**, 263 (1985).

67. H. B. Nielson and P. Olesen, Nucl. Phys. **B61**, 45 (1973).

68. P. Sikivie, Phys. Rev. Lett. **48**, 1156 (1982).

69. Ya. B. Zel'dovich, I. Y. Kobzarev and L. B. Okun, Zh. Eksp. Teor. Fiz. **67**, 3 (1974) [JETP **40**, 1 (1975)].

70. G. Lazarides and Q. Shafi, Phys. Lett. **115B**, 21 (1982).

71. S. M. Barr, X. C. Gao and D. B. Reiss, Phys. Rev. **D26**, 2637 (1982); K. Choi and J. E. Kim, Phys. Rev. Lett. **55**, 2637 (1985).

72. S. M. Barr, D. B. Reiss and A. Zee, Phys. Lett. **116B**, 227 (1982).

73. S. Dimopoulos et al, Phys. Lett. **117B**, 185 (1982).

74. K. Kang et al, Phys. Lett. **133B**, 79 (1983).

75. S. M. Barr, K. Choi and J. E. Kim, Nucl. Phys. **B283**, 591 (1987).

76. J. Preskill, M. B. Wise and F. Wilczek, Phys. Lett. **120B**, 127 (1983); L. F. Abbot and P. Sikivie, Phys. Lett. **120B**, 133 (1983); M. Dine and W. Fischler, Phys. Lett. **120B**, 137 (1983).

77. D. Gross, R. D. Pisarski and L. G. Yaffe, Rev. Mod. Phys. **53**, 43 (1981).

Decoupling Anomalies

R. D. Ball
Institute for Theoretical Physics,
State University of New York at Stony Brook
Stony Brook, NY 11794-3840

Seminar given at TASI-88, Brown University, Providence, RI, June 1988

Abstract

We examine the effective low energy field theories resulting from the decoupling of Higgs bosons and Yukawa coupled chiral fermions, in an attempt to shed light on the possible consistency of anomalous chiral gauge theories in four dimensions.

Introduction

It is usually taken for granted that the effect of heavy particles on low energy dynamics will be unobservable since it may be absorbed into the renormalised parameters of the low energy theory. Indeed, this may be demonstrated explicitly term by term in the perturbation series of a renormalisable theory[1]. However, when the heavy particles gain their large masses only in some strong coupling limit, as in the case with a Higgs particle in a spontaneously broken theory, or a Yukawa coupled heavy fermion, the decoupling theorem breaks down; extra terms (not present at the tree level) are introduced into the low energy effective Lagrangian by closed loops of the heavy particles. It is these quantum effects (or "decoupling anomalies") that we will be concerned with here.

To be more specific, consider a renormalisable theory of coupled scalar, fermion and vector fields with Lagrangian (in the notation of Ref [2,3])

$$\mathcal{L} = \frac{1}{4g_A^2} tr F_{\mu\nu} F_{\mu\nu} + \bar{\psi}(\slashed{D} + g_\chi M)\psi + \tfrac{1}{2} tr D_\mu \hat{M} D_\mu M + \tfrac{1}{4!}\lambda tr(\hat{M}M - v^2)^2. \quad (1)$$

We work throughout in four dimensional Euclidean space (for convenience), with chiral group $\mathcal{G} = G_L \times G_R$ broken spontaneously down to a subgroup $\mathcal{H}$. The fermions ψ (which are assumed throughout to be in some anomaly free representation r_ψ of $\mathcal{G}$) may then be naturally separated into two species λ and χ; the fermions λ transform in a representation r_λ of $\mathcal{H}$, and thus remain massless, whereas the fermions χ transform in a representation r_χ

of $\mathcal{G}$, and become massive. Separating out the scalars and pseudoscalars (which in the spontaneously broken phase are Higgs bosons and Goldstone bosons respectively) by diagonalising $M = \hat{u}\{\sigma_i\}u^\dagger, u^\dagger u = 1$, we have

$$\begin{aligned}\mathcal{L} &= \frac{1}{4g_A^2}trF_{\mu\nu}F_{\mu\nu} + \bar{\lambda}\not{D}\lambda + \bar{\chi}(\not{D} + g_\chi \hat{u}\{\sigma_i\}u^\dagger)\chi \\ &+ \sum_{ij}\sigma_i\sigma_j(\hat{u}^\dagger D_\mu \hat{u})_{ij}(u^\dagger D_\mu u)_{ji} + \tfrac{1}{2}\sum_i \sigma_i^2 (D_\mu u^\dagger D_\mu u)_{ii} \\ &+ \tfrac{1}{2}\sum_i \partial_\mu \sigma_i \partial_\mu \sigma_i + \tfrac{1}{4!}\lambda \sum_i (\sigma_i^2 - v^2)^2. \end{aligned} \quad (2)$$

We may now attempt to decouple the fermions χ by sending the Yukawa coupling g_χ to infinity with the fields σ_i taking on their vacuum expectation value v, which is kept fixed[4]. To do this successfully we must also send the scalar self coupling λ to infinity to enforce the constraint $\sigma_i = v$; if λ is held fixed, sending g_χ to infinity will force σ_i to zero. So the Higgs bosons will also become infinitely massive ($m_H^2 = \tfrac{1}{3}\lambda v^2$) and decouple. The fermions and Higgs bosons should thus be integrated out together, to give the low energy effective action

$$S_{eff}[A, \lambda, u] = S_A[A] + S_\lambda[A, \lambda] + \Gamma_\sigma[A, u] + \Gamma_\chi^+[A, u] + \Gamma_\chi^-[A, u]. \quad (3)$$

The first two terms of this expression are just the kinetic terms for the vectors and the fermions λ. The third term is the effective action obtained from the last three terms of (2) by integrating out the Higgs bosons σ_i in the limit $\lambda \to \infty$; classically this yields the nonlinear σ-model. The fourth and fifth terms are the normal and abnormal parity pieces of the effective action for the massive fermion χ[2,3,4]; the latter is the Wess-Zumino action which contains the anomalies of χ.

The low energy effective action (3) raises several interesting questions. Firstly, is it possible to treat it as a quantum field theory in its own right, perhaps involving a cutoff Λ? Is this theory renormalisable, in the sense that low energy amplitudes are independent of the precise form of the cutoff, but depend only on a finite set of renormalised parameters? Is the S-matrix on the Hilbert space of physical states (with Euclidean momenta less than Λ) unitary? And what is the physical spectrum of the theory? Since (3) is not only nonpolynomial, but also contains terms with more than two time derivatives of the scalar fields, these questions have no easy answers; in the first part of this talk we will attempt to show how they may be decided.

One objection to the above decoupling programme should be mentioned immediately. It is now generally believed that scalar field theories are free: as the cutoff $\Lambda \to \infty$, the renormalised coupling $\lambda \to 0$[5]. This means that the mass of a Higgs boson, $m_H^2 \sim \frac{1}{3}\lambda v^2$, is necessarily bounded above, since as m_H is increased, Λ must be reduced until they collide, at which point the theory stops making sense. In the standard model, for example, such bounds on m_H are now believed to be of the order of 800GeV[6]. Thus the Higgs bosons (or indeed the Yukawa coupled fermions χ) cannot be decoupled. On the other hand, the nonlinear action (3) contains no Higgs boson fields. We will show below how this paradox may be resolved; the Higgs boson reappears as a chiral soliton in the nonlinear model.

Related to this point, it is interesting to note that a scalar field theory is unnatural (in the sense of 't Hooft[7]) unless the cutoff Λ is finite, and the coupling λ small - in fact naturalness requires that $\lambda \sim m^2/\Lambda^2$. Radiative corrections to m^2 ($\sim \lambda\Lambda^2/32\pi^2$) will necessarily imply fine tuning of the bare values of m^2 and λ unless the naturalness requirement is met. The absence of fine tuning is thus a sufficient (but not necessary) condition for an upper bound on the Higgs mass. Naturalness will be an important ingredient of our analysis of low energy effective Lagrangians below.

One of the main motivations for this work is its application to the study of anomalous chiral gauge theories and it is to this that the second part of the talk will be devoted. It is well known that the regulation of an anomalous chiral gauge theory violates gauge invariance (unless extra fields are introduced), and it has been generally believed that this leads to the loss of renormalisability and unitarity[8]. However it has recently been found that in two dimensions (where renormalisability is not an issue) an anomalous abelian or nonabelian theory may be unitary, subject to certain restrictions on the renormalised parameters[9]. This may be understood rather simply by considering a nonanomalous theory in which some of the fermions are decoupled, leaving behind a unitary low energy theory with an anomalous fermion content, and it is not unreasonable to hope that a similar procedure may also lead to consistent anomalous theories in four dimensions[4,10].

More specifically, given a representation r_λ of $\mathcal{H}$ we choose r_ψ to be such that $r_\psi = r_\lambda \oplus r_\chi$ is anomaly free (for example the fermions χ may have opposite chirality to the fermions $\lambda; r_\chi = \hat{r}_\lambda$). The effective action for the chiral gauge theory of the anomalous fermion λ is then of the form (3), and so on understanding of the physical spectrum of this action will tell us whether a consistent quantisation has been achieved. The consequences of this approach may be summarised as follows:

1) Vector bosons which have anomalous couplings to a particular species of chiral fermions will always become massive in the present approach, with masses $m_A^2 \lesssim v^2$. The possibility of massless vector bosons is a separate issue [11] which cannot be decided by fine tuning in the massive phase, since the limit is singular.

2) a) When $\mathcal{H}$ is Abelian, the anomalous low energy theory is consistent. In particular axial QED is expected to be consistent, as suggested independently by Rajaraman and Rajeev [12].

b) When $\mathcal{H}$ is non-Abelian, and r_λ irreducible (so that there is only one species of massless fermions λ), the quantised low energy theory will always contain in addition fermionic solitons in the representation r_χ, with masses $m_* \lesssim v$, so that the total fermionic content $r_\lambda \oplus r_\chi$ is necessarily anomaly free. This is a result not of the gauge anomalies, but of geometry anomalies hidden in the $SU(2) \subset \mathcal{H}$, which are required to cancel in order that the theory be reparameterisation invariant. Global and gauge anomalies are then cancelled automatically.

c) When $\mathcal{H}$ is non-Abelian, but r_λ is reducible, we have situations a) or b) above depending on whether the multiplicities of the irreducible components of r_λ are even or odd, respectively.

These results have interesting physical consequences. If axial QED is consistent we have a solution of the strong CP problem; axions may be eaten by very massive axial gauge bosons. If the number of fermion generations is odd, we still require that the numbers of quarks and leptons be equal, and further that the weak gauge bosons must be massive (without having to invoke the Higgs mechanism). One may even speculate that the generations may be radial excitations of chiral solitons!

1) Non-linear σ-models

We will begin our study of the effective action (3) by considering firstly the non-linear σ-model $\Gamma_\sigma[U]$, obtained formally by integrating out the Higgs boson field σ_i in the infinite coupling limit of the linear model. The Lagrangian $\mathcal{L}_\sigma(U)$ will be assumed to be local, and (at least at low Euclidean momenta, $p^2 \ll \Lambda^2$) expandable in a power series in derivatives:

$$\mathcal{L}_\sigma(U) = \epsilon\Lambda^2 \mathcal{L}_\sigma^{(2)}(U) + \epsilon \mathcal{L}_\sigma^{(4)}(U) + \frac{\epsilon}{\Lambda^2}\mathcal{L}_\sigma^{(6)}(U) + \ldots \quad (4)$$

where
$$\mathcal{L}_\sigma^{(2)}(U) = \tfrac{1}{4} tr(\partial_\mu \hat{U} \partial_\mu U)$$
$$\mathcal{L}_\sigma^{(4)}(U) = \tfrac{1}{4} tr(a_0 \partial^2 \hat{U} \partial^2 U + a_1 [\partial_\mu \hat{U}, \partial_\nu U]^2 + a_2 (\partial_\mu \hat{U} \partial_\mu U)^2)$$

The parameters in this Lagrangian will be natural provided that the coefficients $a_i, i = 0, 1, \ldots$ are all of order unity. When (4) is 'derived' in the loop expansion from the linear model, it is found that $\epsilon \sim \ln(\Lambda^2/m^2) + \ldots$ to leading order[13], terms with higher numbers of derivatives appearing at higher and higher orders. Goldstone boson loops, which diverge as powers of $ln\Lambda^2/v^2$[14] may thus be renormalised by including the effects of the Higgs boson loops. The limit $\Lambda^2 \to \infty$, $m^2 \to \infty$ must be taken in such a way that $v^2 = \epsilon\Lambda^2$ is held fixed. Thus

$$m^2 \sim \Lambda^2 e^{-v^2/\Lambda^2} < \Lambda^2, \quad \epsilon < 1$$

as required.

Now the Lagrangian (4) is nonrenormalisable if we define the propagator in the usual way by means of $\mathcal{L}_\sigma^{(2)}$ alone, since at large Euclidean momenta it behaves as $1/p^2$, individual diagrams diverge in arbitrarily large powers of Λ^2/v^2 and an infinite number of counterterms are required. However (4) is renormalisable if we also include $\mathcal{L}_\sigma^{(4)}(U)$ in the definition of the propagator, since at large Euclidean momenta it now behaves as $1/p^4$, the divergences are at worst quadratic, and moreover may all be absorbed into the four renormalised parameters v^2, a_0, a_1 and a_2[15]. The irrelevant operators $\mathcal{L}_\sigma^{(6)}, \ldots$ may also be included, but with no further advantage; we will ignore them in what follows.

The problem with setting up the perturbation theory in this way is that it is now more difficult to demonstrate perturbative unitary: in momentum space the propagator is

$$G(p^2) = (p^2 + a_o p^4/\Lambda^2)^{-1}.$$

When $a_0 < 0$ there is a tachyon at $p^2 = \Lambda^2/|a_o|$, so the S-matrix is not well defined since there are no asymptotic states. When $a_0 > 0$ we have a massive ghost at $p^2 = -\Lambda^2/a_0$; for $a_0 > 1$ this ghost pole results in a loss of unitarity since it lies below the cutoff, whereas for $0 < a_0 < 1$ we have unitarity on the subspace of states with Euclidean momenta below the cutoff, since the ghost threshold lies above it. Thus the S-matrix derived from the effective Lagrangian (4) has "effective unitarity" provided $0 < a_0 < 1$.

The limit $\Lambda \to \infty$ is still problematic, since the perturbation theory described above has a coupling $\sim 1/\epsilon$, and thus breaks down as $\epsilon \to 0$. It is possible to repair this by summing "cocoon", diagrams, to obtain a new perturbation theory in which all diagrams behave as positive powers of ϵ in the limit; this is a common technique in nonpolynomial theories[16]. Perhaps using this technique one can demonstrate the unitarity of the non-linear σ-model even as $\Lambda \to \infty$.

This is an interesting point, since not only the non-linear σ-model, but also massive gauge theories and possibly some anomalous chiral gauge theories may, as we shall see below, be formulated in such a way that they are renormalisable and effectively unitary. The same remarks apply to higher derivative quantum gravity (with Lagrangian $\sim \Lambda^2 R + R^2$), which has the further advantage over traditional Euclidean quantum gravity that the action is positive definite.

We next turn to an examination of the spectrum of the Lagrangian (4). Besides the Goldstone bosons, there will be chiral solitons when $\mathcal{H}$ is non-Abelian, since the map $U(\underline{x}) : S^3 \to \mathcal{G}/\mathcal{H}$ may be nontrivial because $\pi_3(\mathcal{G}/\mathcal{H}) \neq 0$. The soliton then has a conserved winding number, and is topologically stable. Solitons may be constructed explicitly by embedding the $SU(2)$ soliton $U(\underline{x}) \sim \exp\left(i\underline{\tau} \cdot \underline{x} f(|\underline{x}|)\right)$ in $\mathcal{G}/\mathcal{H}$. The energy (or mass) of the soliton may be determined as a function of its size R from the

Lagrangian (4);

$$E(R) = \tfrac{1}{2}\epsilon(\lambda\Lambda^2 R + S(\frac{1}{\Lambda^2 R^2}))\tag{5}$$

where $\quad S(\frac{1}{\Lambda^2 R^2}) = \sum_{i=0}^{2} a_i \lambda_i + 0(\frac{1}{\Lambda^2 R^2})$

for $R \gg \Lambda^{-1}$, and λ_i are constants of order unity, depending on the details of the soliton configuration. If we retain only the first term of (5), the soliton will collapse[17]. However, when we include the second term as well, the soliton may stabilise, if the a_i are chosen such that $S(0) > 0$ (see Fig. 1).

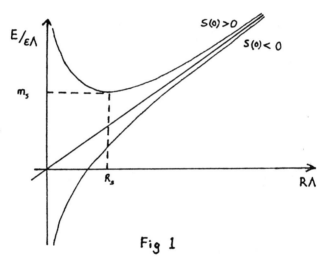

Fig 1

The soliton radius and mass are given by

$$R_s = (S(0)/\lambda)^{\frac{1}{2}}\Lambda^{-1} \sim \sqrt{\epsilon}/v,$$
$$m_s = (\lambda S(0))^{\frac{1}{2}}\epsilon\Lambda \sim \sqrt{\epsilon}v,$$

so that since $\epsilon < 1$,

$$m_s \lesssim v < \Lambda \sim 1/R_s.$$

Furthermore as $\epsilon \to 0$ (i.e. $\Lambda \to \infty$), but the radius and mass decrease as $\sqrt{\epsilon}$; the soliton becomes light and pointlike. The chiral soliton is thus the 'shadow' of the 'decoupled' Higgs boson in the strongly coupled linear model. The situation is summarised in Fig. 2., a plot of the Higgs mass vs the cutoff.

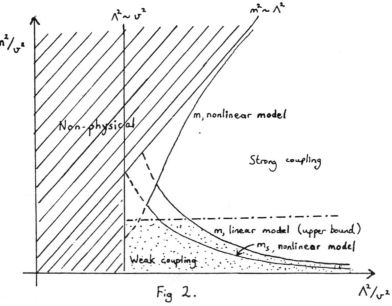

Fig 2.

The non-linear model is thus reconciled with triviality in the linear model; the 'decoupling' of the Higgs boson in the former is an illusion. To make this more concrete would require a better understanding of the relativistic second quantisation of solitons.

Adding weakly coupled vector fields to the above system is now comparatively simple (though strongly coupled vector fields may change the qualitative nature of the spectrum dramatically, of course). The terms in the Lagrangian (4) with two and four derivatives are now

$$\mathcal{L}_\sigma^{(2)}(A,U) = \tfrac{1}{4}tr(D_\mu \hat{U} D_\mu U + \frac{a_3}{\Lambda^2} F_{\mu\nu} F_{\mu\nu}),$$

$$\mathcal{L}_\sigma^{(4)}(A,U) = \tfrac{1}{4}tr(a_0 D^2 \hat{U} D^2 U + a_1 [D_\mu \hat{U}, D_\nu U]^2 + a_2 (D_\mu \hat{U} D_\mu U)^2$$

$$+ a_4 D_\mu \hat{U} F_{\mu\nu} D_\nu U + a_5 (F_{\mu\nu} \hat{U} \hat{F}_{\mu\nu} U - F_{\mu\nu} F_{\mu\nu})),$$

where $D_\mu U \equiv \partial_\mu U + \hat{A}_\mu U - U A_\mu$. The chiral gauge fields become massive by eating Goldstone bosons; $m_A = \sqrt{2} g_A v$. The only way to decouple the Higgs bosons (chiral solitons in this Stuckelberg formulation) is to let $v \to \infty$, but then the chiral gauge bosons also decouple! So, in the absence of heavy fermions, we must have Higgs bosons, or new physics, at scales of order v.

Note that in the 'unitary' gauge $U = 1$, the propagator for the gauge fields is

$$G_{\mu\nu}(p^2) = \frac{1}{p^2 + m_A^2}\{(\delta_{\mu\nu} + \frac{p_\mu p_\nu}{m_A^2}) - \frac{p_\mu p_\nu}{m_A^2} \frac{(1-\alpha)p^2}{((1-\alpha)p^2 + m_A^2)}\},$$

where $\alpha = 1 - 2\epsilon a_0 g_A^2$, so we still have a massive ghost pole at $p^2 = -\Lambda^2/a_0$, and a unitary S-matrix only for states below this threshold.

In an Abelian theory, we may let $\epsilon \to 0$ with impunity, since $\partial_\mu \hat{U} \partial_\mu U = (\partial_\mu \theta)^2$ is renormalisable without the need for terms with four derivatives.

The solitons then unwind, since $\pi_3(\mathcal{G}/\mathcal{H}) = 0$, and $R_s, m_s \to 0$. So there is no need for the Higgs bosons; massive Abelian gauge theory is consistent without them.

Finally, we add the light and heavy fermions χ and λ respectively, and consider the full effective action (3). Since $\Gamma_\chi^+[A,U]$ is of the same form as $\Gamma_\sigma[A,U]$ (but with the coefficients fixed) it may be absorbed, and $\Gamma_\chi^-[A,U]$ is (up to irrelevant operators) the Wess-Zumino term $\Gamma_{WZ}[A,U] = W[{}^uA] - W[A]$, so

$$S_{eff}[S,\lambda,U] = S_A[A] + S_\lambda[A,\lambda] + \Gamma_\sigma[A,U] + \Gamma_{WZ}[A,U] \qquad (6)$$

The coefficient of Γ_{WZ} is fixed topologically, and is thus not renormalised when graphs with closed loops of vector fields are considered[18]. On the other hand, Γ_{WZ} may spoil the naturalness of the resummed perturbation theory for Γ_σ as $\epsilon \to 0$, though it is tempting to speculate that it does not.

It is important to realise that the inclusion of chiral fermions makes it necessary to couple the vector fields to the Goldstone boson fields U, through $\Gamma_{WZ}[A,U]$. This in turn implies that $\Gamma_\sigma[A,U]$ will generally be non-zero (it is generated by radiative corrections), and thus that chiral gauge bosons will be massive. This dynamical mass generation is closely analogous to that in the chiral Schwinger model (in two dimensions); for this reason we wil refer to it as the 'Schwinger mechanism'.

Though the Wess-Zumino term in the effective action (6) will have no effect on the static properties (and in particular the mass and radius) of the chiral solitons, since it contains precisely one time derivative, it may have important consequences for their spin and statistics. When $\mathcal{G}$ is non-Abelian, it necessarily contains $SU(2)$ as a subgroup, and the global geometry anomaly of this embedded $SU(2)$[19] means that the soliton must be quantised as a boson if the number of species of fermion χ (i.e. the multiplicities of the irreducible components of r_χ) is even, but as a fermion if it is odd[20]. The spin of the soliton is then integer or half-integer respectively, consistent with the spin-statistics theorem. Furthermore, the winding number of the soliton can be shown to correspond to the fermion number of the fermions χ, and in fact it carries all their other quantum numbers as well.

Thus when we attempt to decouple fermions, there is a conspiracy between them and the Higgs bosons. For non-Abelian theories, an even number of fermions χ may be decoupled with impunity, but a single fermion (in an irreducible representation r_χ) may not be decoupled; a Higgs boson may only be decoupled when such a fermion is present (with which it can form a bound state). However, in Abelian theories both fermions and Higgs bosons may be readily decoupled. Finally, whenever either is decoupled, the corresponding gauge bosons will become massive by the Schwinger mechanism.

It is interesting to speculate on the implications of these results for the dynamical breaking of supersymmetric theories; though at high energies the spectrum may be supersymmetric at low energies there may be little trace of supersymmetery left, since some of the bosons may be 'eaten' by the fermions to form low energy fermionic bound states, presumably in several 'generations' (radial excitations of the soliton). This possibility deserves further consideration.

2) Consistent Anomalous Gauge Theories?

Consider an anomalous representation of light chiral fermions r_λ (of the chiral group $\mathcal{H}$). Using Pauli-Villars regulators φ, coupled as $\bar\varphi(\not{D} + \Lambda 1)\varphi$, to cancel off the integrability obstruction to the construction of the fermion determinant [21,2,3] the effective action for the fermions λ takes the form

$$\tilde\Gamma_{\lambda-\varphi}[A] = \Gamma_\lambda^+[A] - \Gamma_\Lambda^+[A] + \Gamma_\lambda^-[A] - W[A], \tag{7}$$

where $\Gamma_\lambda^\pm[A]$ are gauge invariant but nonlocal, $\Gamma_\Lambda^+[A]$ is gauge variant but local (and could thus be modified or removed by the addition of local counterterms), and $W[A] = \int_{\mathcal{M}^6} \Omega_5^0(A)$ is gauge variant but nonlocal, giving rise to the anomalies of λ. To attempt a consistent quantisation of the gauge fields, we will now proceed by a process of 'perpetual optimism', tackling the issues of gauge invariance, renormalisability, and effective unitarity successively, so that the question of consistency eventually becomes one of dynamics alone.

The Pauli-Villars regulators φ pick out a particular section in the space of gauge potentials, so to ensure gauge invariance we must take care to integrate over the whole space of gauge potentials [22,11]. Using the Faddeev-Popov trick, the partition function is given by

$$\begin{aligned} Z &= \int \mathcal{D}A\mathcal{D}\lambda e^{-(S_A[A]+S_\lambda[A,\lambda])} \\ &= \int \mathcal{D}A e^{-(S_A[A]+\tilde{\Gamma}_{\lambda-\varphi}[A])} \\ &= \int \mathcal{D}A\mathcal{D}U \delta(f(^{u^{-1}}A))\Delta_f[A] e^{-(S_A[A]+\tilde{\Gamma}_{\lambda-\varphi}[A])} \\ &= \int (\mathcal{D}A \delta(f(A))\Delta_f[A])\mathcal{D}U e^{-(S_A[A]+\tilde{\Gamma}_{\lambda-\varphi}[^uA])}, \end{aligned} \quad (8)$$

where the last line follows by a change of variables $A \to^u A \equiv \hat{U}DU$. So if we insist on gauge invariance, the anomaly means that the gauge field must be coupled to the field U. Indeed, we could have obtained (8) by using the usual gauge field quantisation, but coupling the regulators as $\bar{\varphi}(\displaystyle{\not}D + \Lambda U)\varphi$ (i.e. gauge invariantly); U is the 'phase' of the regulating fields φ against which the chiral phase of the fermions λ is determined. The gauge invariant effective action for the fermions λ is now

$$\tilde{\Gamma}_{\lambda-\varphi}[A,U] \equiv \tilde{\Gamma}_{\lambda-\varphi}[^uA] = \Gamma_\lambda^+[A] - \Gamma_\Lambda^+[^uA] + \Gamma_\lambda^-[A] - W[^uA]. \quad (9)$$

Another way of obtaining a gauge invariant quantisation of the theory is to add massive anomaly cancelling fermions χ (in a representation $r_\chi = \hat{r}_\lambda$), coupled as $\bar{\chi}(\hat{\displaystyle{\not}D} + \Lambda \hat{U})\chi$. Since they have masses of the order of the cutoff, these fermions may be regarded as a part of the regularisation of the theory, just as the fermions φ were above [10]. By definition $\Gamma^\pm[\hat{A}] = \pm\Gamma^\pm[A]$, so the effective action for the fermions λ is now

$$\tilde{\Gamma}_{\lambda+\chi}[A,U] = \Gamma_\lambda^+[A] + \Gamma_\Lambda^+[^uA] + \Gamma_\lambda^-[A] - W[^uA]. \quad (10)$$

The only difference between this and the previous formulation (9) is in the sign of the second term.

In two dimensions $\Gamma_\lambda^+[{}^u A] \sim (\hat{U} D_\mu U)^2$, no renormalisation of the actions (9) and (10) is necessary, and it is not difficult to see that while the former leads to a nonunitary S-matrix (because of the minus sign in the second term, implying massless ghosts), the latter is unitary[9]. This is precisely as one would expect; the (Pauli-Villars) regulators φ have negative metric, whereas the (anomaly cancelling) regulars χ have positive metric, and thus do not spoil unitarity[10]. One may interpolate between the two formulations by modifying the regularisation by the addition of a counterterm $\Gamma_\sigma[A, U]$, with some arbitrary coefficient; $\tilde{\Gamma}_\lambda = \tilde{\Gamma}_{\lambda+\chi} + (a-2)\Gamma_\sigma$, with unitarity for $a > 1$.

In four dimensions, the situation is rather more involved, since it is here necessary to add the counterterms $\Gamma_\sigma[A, U]$ (which now involves terms with both two and four derivatives) to ensure renormalisability. Absorbing Γ_λ^+, we then have

$$\tilde{\Gamma}_\lambda[A, U] = \Gamma_\lambda^+[A] + \Gamma_\sigma[A, U] + \Gamma_\lambda^-[A] - W[{}^u A], \tag{11}$$

so that in place of (8)

$$Z = \int (\mathcal{D} A \delta(f(A)) \Delta_f[A]) \mathcal{D} U e^{-(S_A[A] + \tilde{\Gamma}_\lambda[A, U])}. \tag{12}$$

To summarise, the integrability obstruction to the definition of the effective action for λ leads to a modification of the fermionic measure typified by the introduction of φ or χ, which in turn requires a coupling to the field U to ensure gauge invariance. This is essential, because without it we would lose reparameterisation invariance (even in the absence of dynamical gauge fields); the Wess-Zumino term in (11) (note that $\Gamma_\lambda^-[A] - W[{}^u A] = \Gamma_\lambda^-[A] - W[A] - \Gamma_{WZ}[A,U]$) is required to cancel reparameterisation anomalies in the non-linear σ-model $S_\lambda + \Gamma_\sigma$[23]. Indeed, it is possible to cancel these anomalies by adding such a Wess-Zumino term only if the model which results can be obtained by the dynamical breakdown of a theory containing anomaly cancelling fermions χ[24]. So the construction we have used is not only sufficient but also necessary to obtain a reparameterisation invariant theory.

The formulation (12) gives us a gauge invariant, gauge independent (on shell), renormalisable, and, for $0 < a_0 < 1$, effectively unitary formulation of the chiral gauge theory of the light fermions λ, which is essentially unique up to the choice of representation r_χ. The chiral gauge bosons are necessarily massive, by the Schwinger mechanism. The fermion content of the physical spectrum must, however, take into account the possibility of the existence of the light fermionic solitons described previously; this leads directly to the results for the alternatives a) b) and c) advertised in the introduction.

A more substantial account of this work may be found in [25].

Acknowledgements

I would like to thank I. Aitchison, J. Bagger, S. Coleman, C. Fraser and H. Georgi for stimulating discussions which helped to shape this work, the organisers of TASI-88 for giving me the opportunity to present it here, and R. Brandenberger and J. McCarthy for useful suggestions on the form of the manuscript.

REFERENCES

1. K. Symanzik, *Comm. Math. Phys.* **34** (1973) 7.
   T. Appelquist and J. Carrazone, *Phys. Rev.* **D11** (1975) 2856.
2. R. Ball and H. Osborn, *Phys. Lett.* **165B** (1985) 410;
   Nucl. Phys. **B263** (1986) 245.
3. R. Ball, in 'Superstrings, Anomalies and Unification' eds. M. Martinis and I. Andric (Dubrovnik, 1986) (World Scientific, 1987).
4. E. D'Hoker and E. Farhi, *Nucl. Phys.* **B248** (1984) 59, 77.
5. K. Wilson, *Phys. Rev.* **B4** (1971) 3184.
   K. Wilson and J. Kogut, *Phys. Rep* **12C** (1974) 75.
6. P. Hasenfrantz and J. Nager, Bern preprint BUTP-87/18.
   J. Kuti, L. Liu and Y. Shen, UCSD—PTH 87-19.
7. G. 't Hooft in 'Recent Developments in Gauge Theories', eds. G. 't Hooft– (Cargese, 1979) (Plenum, NY, 1980).
8. D. Gross and R. Jackiw, *Phys. Rev.* **D6** (1972) 477.
9. R. Jackiw and R. Rajaraman, *Phys. Rev. Lett.* **54** (1985) 1219, 2060(E); **55** (1985) 2224 (C).
   R. Jackiw in TASI Proceedings (Yale Univ., 1985);
   APS Proceedings (Eugene, OR, 1985)
   R. Rajaraman, *Phys. Lett.* **154B** (1985) 305; **162B** (1985) 148.
   J. Lott and R. Rajaraman, *Phys. Lett.* **165B** (1985) 321.
10. R. Ball, *Phys. Lett.* **B183** (1987) 315.
11. L. Faddeev, *Phys. Lett.* **B145** (1984) 81;
    Theor. Math. Phys. **60** (1984) 770.
    Nuffield Lectures, 1985 (unpublished).
    L. Faddeev and S. Shatashvili, *Phys. Lett.* **B167** (1986) 225.
12. R. Rajaraman, *Phys. Lett.* **B184**(1987) 369.
    S. Rajeev, MIT preprint CTP1405 (1986, unpublished).
13. M. Veltman, *Acta Phys. Pol.* **B8** (1978) 475.
    T. Appelquist and C. Bernard, *Phys. Rev.* **D22** (1980) 200.
    R. Akhoury and Y.-P Yao, *Phys. Rev.* **D25** (1982) 3361.
14. S. Weinberg, *Physica* **96A** (1979) 327.
15. A. Slavnov, *Nucl. Phys.* **B31** (1971) 301.
16. A. Salam, C. Isham and J. Strathdee, *Phys. Rev.* **D3** (1971) 1805;**D5** (1972) 2548.
    A. Salam, Proc. of Amsterdam Int. Conf. EC Part., 1971;
    Proc. of Rochester Meeting of APS/DDPF, 1971.
    J. Honerkamp and K. Meetz, *Phys. Rev.* **D3** (1971) 1976.

G. Ecker and J. Honerkamp, *Nucl. Phys.* B36 (1972) 130.

J. Honerkamp, F. Krause and M. Scheunert, *Nucl. Phys.* B69 (1973) 168.

17. G. Derrick, *Jour. Math. Phys.* 5 (1964) 1253.
18. S. Adler and W. Bardeen, *Phys. Rev.* 182 (1969) 1517.
19. E. Witten *Phys. Lett.* 117B (1982) 324.
20. E. Witten *Nucl. Phys.* B223 (1983) 433.
21. H. Leutwyler, *Phys. Lett.* 152B (1985) 78.
22. D. Foerster, H. Nielsen and M. Ninomiya, *Phys. Lett.* B94 (1980) 135.

    D. Babelon, F. Shaposnik and C. Viallet, *Phys. Lett.* B177 (1986) 385.

    K. Harada and I. Tsutsui, *Phys. Lett.* B183 (1987) 30.
23. A. Manohar and G. Moore, *Nucl. Phys.* B243 (1984) 55.

    G. Moore and P. Nelson, *Phys. Rev. Lett.* 53 (1984) 1519.
24. L. Alvarez-Gaumé and P. Ginsparg, *Nucl. Phys.* B262 (1985) 439.

    A. Manohar, G. Moore and P. Nelson, *Phys. Lett.* 152B 91985) 68.

    J. Bagger, D. Nemeschansky and S. Yankielowicz in 'Anomalies, Geometry and Topology', ed. A. White (World Scientific, Singapore, 1987).
25. R. Ball, *Phys. Rep.* (to appear).

A SHORT TOUR OF THE UNIVERSE: AN INTRODUCTION TO SUPERNOVAE, THE SOLAR NEUTRINO PUZZLE AND THE MISSING MASS PROBLEM

D.N. Spergel
Princeton University Observatory
Princeton, NJ 08544
USA

ABSTRACT

The purpose of these lectures is to introduce particle physics students to a few of the exciting problems in astrophysics: In the first lecture, I will describe the physics of supernova explosions, focusing on the recent detection of neutrinos from SN 1987a. In the second lecture, I will discuss the solar neutrino problem and some of its suggested solutions. In the third lecture, I will review the astrophysical evidence for missing mass and discuss experimental efforts to search for proposed candidates. The goal of these lectures is not to review the vast literature in these fields, but rather to provide a pedagogical introduction.

Like many rock stars, massive stars live fast and die young in a brilliant flash. Rock stars are only a small fraction of the human population, similarly, massive stars that end their lives as supernova are a small but very noticeable portion of the stellar population.

This lecture will focus on the short but brilliant life of Sandulak 69-202, a blue supergiant in a nearby galaxy, the Lesser Magellenic Cloud (LMC), that ended its life in a brilliant flash visible to terrestial observers as SN-1987a. I will begin by reviewing supernova theory pre-1987a and then compare the recent supernova observations with theory (I know that this is an old fashioned practice, but astrophysicists are often a little behind the times.)

In the 1930s, Fritz Zwicky classified supernovae into two types: type I and type II. The two different types have different light curves (brightness as a function of time). Baade and Zwicky (1934) suggested that if only a small fraction of the gravitational binding energy released when the core of a red giant star collaped to a neutron star was transfered to the envelope of a red giant star, then core collapse might power supernova. (This is our current picture of type II supernova, the focus of this lecture.) In the 1960s, Fowler and Hoyle (1960) emphasized the importance of thermonuclear explosive energy, the source of energy for type I supernova. Colgate & White (1966) constructed the first "neutrino transport" model. Over the past 20 years, the discovery of weak neutral currents, improved treatments of hydrodynamics and neutrino transport, and better models of pre-supernovae stars have contributed to an improved although not yet complete picture of supernova dynamics (see Woolsey & Weaver 1986 for review).

Prior to the observation of neutrinos of SN 1987a, our picture of supernova evolution was based on observations of photons from the expanding supernova shell. These photons carried off only 0.01% of the supernova's energy— 99% of the supernova's energy is released in the form of neutrinos and most of the remaining energy is in the form of kinetic energy of the expanding gas. The measurement of the energies and arrival times of the neutrinos from SN1987a tested our basic understanding of supernova and confirmed the basic picture of core collapse.

THE BRIEF LIFE OF A MASSIVE STAR

Like our Sun, a massive star relies on pressure gradients to provide support against gravitational collapse. These pressure gradients are mostly due to large temperature gradients. These massive star are so luminous (luminosity scales roughly as mass to the fourth power) that they rapidly exhaust their fuel supplies. While the lifetime of our Sun is measured in billions of years, the lifetime of a 10 solar mass star is measured in millions of years.

A massive star takes only 7 million years to exhaust its supply of hydrogen. Its core then contracts, heats up, and begins burning helium to carbon, while an outer shell still burns hydrogen to helium. Helium burning supports the star for another 500,00 years, then the pace of evolution quickens as each reaction yields less energy and the ever denser stellar core radiates a larger fraction of its energy in neutrinos. The star burns its way through the periodic table, building up an onion-like structure: carbon burning lasts for 600 years, neon burning for a year, oxygen burning for 6 months and silicon burning lasts for a day. After this final day, the iron core (left over from silicon burning) collapses.

Prior to 1987, based on previous observational experience, we expected this explosion to take place while the star was a red supergiant. The Sandulak star, however, was a blue supergiant prior to its dramatic death. Several hypothesis have been suggested to explain this discrepancy: (a) the supernova explosion occurred in the red giant phase in models that assumed solar metal abundances; when the lower LMC abundances were used, the supernova went off in the blue supergiant phase (b) mass loss from the star can alter its evolution. Woolsey (1988) reviews these possible solutions and discusses their relative merits.

BEYOND THE CHANDRASEKHAR LIMIT, CORE COLLAPSE

Iron has the highest binding energy per nucleon— thus once a star's core is mainly iron— it has nothing to burn and it can not maintain a thermal gradient. The star must rely on the degeneracy pressure of its electrons for support against the gravitational collapse of its core.

A star can tolerate only so much degeneracy in its core. When the mass supported by degeneracy pressure exceeds a critical mass called the Chandrasekhar mass, gravity overcomes degeneracy pressure.

We can quickly derive this fundamental mass scale, which sets the size of the supernova explosion. Consider N_e electrons supporting a core of radius, R. The number density of electrons is roughly,

$$n_e \sim \frac{N_e}{R^3}$$

Thus, each electron has a Fermi energy,

$$E_F \sim p_F c \sim \hbar n_e^{1/3} c \sim \frac{\hbar N^{1/3} c}{R}$$

This Fermi energy must exceed the gravitational binding energy per electron:

$$E_G \sim \frac{GNm_B^2}{R} \frac{1}{Y_e}$$

m_B is the mass of a baryon in the star. Y_e is the number of electrons per baryon ($\sim 1/2$). Requiring that $E_F > E_G$ places an upper limit on the number of electrons in a sphere supported by degeneracy pressure, which in turn contrains the mass of the sphere to by less than a critical mass, $M_{Ch.}$:

$$M_{Ch} = \frac{Nm_B}{Y_e} = \frac{Y_e^{1/2}\hbar c}{Gm_B^2} = \frac{Y_e^{1/2}M_{Pl}^3}{m_B^2}$$

M_{Pl} is of course the Planck mass.

Once the mass of the iron core exceeds the Chandrasekhar mass, collapse begins. During collapse, the core temperature rises, previouly this stopped contraction; however, this does not save the ill fated star— energy now goes into the photodisociation of iron, which reduces the pressure. Electron pressure eventually becomes high enough for neutronization,

$$e^- + p \rightarrow n + \nu_e,$$

to be energetically favorable. This process softens the electrons equation of state (decreases $\frac{d\log p}{d\log \rho}$) and speeds the pace of collapse. The inner iron core collapse homologously ($v \propto r$), thus it retains its profile as it collaspes towards nuclear density (Goldreich and Weber 1980). The outer core collapses supersonically ($v \propto r^{-1/2}$), accreting onto the inner core.

When the central density reaches nuclear densities ($\sim 10^{14}$ g/cm^3), the nuclear forces become strong enough to stop collapse. The core "bounces", producing a shock wave that propagates outward carrying off some of the gravitational energy released in collapse. There is an intense debate in the supernova community about how this shock couples its energy to the loosely bound envelope of the star. The shock must climb through the infalling material from the outer core. As the shock moves through the shells of iron and silicon, energy losses through neutrino emission and dissociation of iron deplete the shock's energy and slows shock propagation. Baron et al. (1985) using a soft equation of state find that the shock does have sufficent energy to climb outwards and produce a "prompt" explosion. Wilson et al. (1985) argue that neutrino energy transport is needed to reenergize the shock and generate a "delayed" explosion. Unfortunately, SN1987a did not provide enough infromation to discriminate between these two scenarios— further theoretical work and perhaps another supernova will be needed before this question is answered.

COOLING A HOT NEUTRON STAR— THE NEUTRINO SIGNAL

Regardless of how the shock propogates outward, there is a consensus in the supernova community that the core collapse of a massive star leaves behind a hot neutron star. The temperature in the center of this star reaches 60 MeV— the bulk of the core gravitational binding energy is converted into thermal energy.

The densities in the neutron star and the surrounding material are so large that cooling by "normal" means (conduction or photon radiation) is exteremely inefficent. The neutron star cools by emitting neutrinos. Even these neutrinos have difficulties escaping: the neutrino mean free path is a small fraction of the neutron star radius. The neutrino effectively radiate from the outer surface of the neutron star— a neutrinosphere, akin to the photosphere of the Sun.

Simple estimates can yield approximate values for the neutrino luminosity and temperature. These values are consistent with numbers obtained using sophisticated numerical codes that include far more physics than I will cover in this lecture. For pedagogical purposes, I will follow Dar (1988) in an approximate derivation of neutrino properties.

Measurements of the binary pulsar suggest that neutron stars often form with masses near the Chandrasekhar mass, ~ 1.6 solar masses. Observations of X-ray bursting neutron stars suggest that these stars have radii, $R \sim 15 \pm 5$ km. Thus, the energy emitted in neutrino ought to be on order the binding energy:

$$E_\nu \simeq E_{grav.} \sim \frac{GM_*^2}{R_*}$$

The temperature of the neutrinos depends on the location of the neutrinosphere. The density of the hot neutron star falls off exponentially— neutrinos will only escape when their mean free path reaches a substational fraction of the neutron star radius:

$$\lambda_\nu \sim \alpha R$$

The dominant source of opacity for the $\bar{\nu}_e$'s detected by the terrestial detectors is absorbtion by protons,

$$\bar{\nu}_e + p \to e^+ + n$$

The cross-section for this reaction is temperature-dependent,

$$\sigma = 9 \times 10^{-43} (T/\text{MeV})^2 \text{ cm}^2,$$

thus, the neutrino mean free path depends on temperature and density:

$$\lambda_{\nu_e} \simeq (\rho N_A Y_e \sigma),$$

where N_A is Avogadro's number and Y_e is the proton fraction. The iron core starts with $Y_e \sim 0.45$, the cold neutron star that remains after core collapse has $Y_e \sim 0.04$. The equation of state for the collapsing neutron star, $p \propto \rho^{4/3}$ yields a density profile that is strongly correlated with local temperature:

$$\rho \sim 1.13 \times 10^{10} (T/\text{MeV})^3 \text{ gm/cm}^3$$

Combining these equations:

$$T \sim (m_p/12 Y_p \rho_0 \sigma_0 \alpha R)^{1/5}$$

Using $R \sim 20$ km, $0.1 < \alpha < 1$ and $0.04 < Y_p < 0.45$, suggests that the electron anti-neutrino temperatuer temperature ought to be $\sim 4\,\text{MeV}$. Detailed numerical simulations (see e.g. Burrows and Lattimore 1987, Mayle, Wilson and Schramm 1987) yield similar values and suggest that the ν_μ's and ν_τ's with their lower cross-sections ought to be emitted at a higher temperature, $T \sim 8$ MeV. These models suggest that each neutrino species carries off roughly equal shares of the neutron star's binding energy— cooling the core in a few seconds.

The detection of neutrinos from SN1987a tested this standard picture of core collapse. Initially, both Kamiokande and IMB reported neutrino detections with their water Cherenkov detectors. Both reactors are sensitive primarily to $\bar{\nu}_e$'s through $\bar{\nu}_e + p \to n + e^+$. The recoiling electrons are expected to be isotropically distributed. These detectors can also detect neutrino scattering, $\nu + e^- \to \nu + e^-$. The scattering events, which are expected to be primarily due to electron scattering, all point directly away from the supernova. The Kamiokande experiment reported the detection of 11 events. The larger IMB experiment, which is sensitive high energy neutrinos because of its high threshold, saw 8 events. After reanalyzing their experiment, the Baksan experiment concluded that it saw 5 events in its small proton-rich plastic scintillator experiment.

My collaborators and I (Spergel et al. 1988), as well as many other scientists, analyzed the neutrino signal from SN1987a. Most groups reached similar conclusions about the neutrino signal. The anti-neutrino temperature, $T \sim 4$ MeV, and nearly thermal spectrum was consistent with pre-supernova expectations. The energy carried in electron anti-neutrinos, $E_{\bar{\nu}_e} \sim 6 \times 10^{52}$ ergs represented about 1/6th the binding energy of a neutron star. The signal lasted several seconds, consistent with the detailed numerical models. While the supernova did not provide enough signal to discriminate between specific models, it did confirm the broad-brush picture of core collapse.

Much has been said about the angular distribution of events in the IMB and Kamiokande detectors. The two events in Kamiokande were forward scattering events. This suggests that one and perhaps both of the events were ν_e events from the neutronization burst, rather than $\bar{\nu}_e$'s from the thermal cooling of the neutron star. (The neutron star also cooled by emitting ν_e's however, their cross-section is much smaller than the $\bar{\nu}_e$'s.) If the MSW effect operates in the Sun (see lecture 2) and converts solar ν_e's to ν_μ's, then it ought to operate in the envelope of Sandulak 69-202, whose density profile is more gradual then the Sun. If either of those events are part of a neutronization pulse, then we can rule out the MSW effect as a solution to the solar neutrino problem.

In the IMB detector, a large fraction of the events are pointed 30° and 90° away from the LMC. These events are not close enough to 0° to be scattering events, however, their distribution does not appear to be the isotropic distribution expected of $\bar{\nu}_e$ absorbtion. Given the small number of events, none of the angular effects are so rare that we can securely make profound conclusions. For the moment, I believe that the excitement about the neutrino angular distribution may be much ado about nothing.

The Monte-Blanc scintillator detector also reported seeing events in the day before the first reported optical detection of SN1987a. However, the Monte-Blanc detection was not at the same time as the other 3 detectors and any signal seen by the small Monte-Blanc detector ought to have been seen by the Kamiokande detector. The Monte-Blanc detection would require that the supernova emit more than the rest-mass energy of its core in electron neutrinos or the invention of some new physics. I am hesitant to base whole new theories on an experimental result that can not be repeated and is contradicted by another experiment. The Monte-

Blanc events may well be noise and bad luck.

PARTICLE PHYSICS IMPLICATIONS OF SN 1987a

The detection of neutrinos from SN1987a probed not only the convential picture of core collapse, but also tested both the standard model of electroweak interactions and some of its proposed extensions.

The arrival of the neutrinos suggest that neutrinos are long-lived $\gamma\tau > 5 \times 10^{12}$ seconds. Although even this remark needs to be tempered with an asterisk (see Frieman et al. 1988).

If the neutrino had a mass, then its arrival time would depend upon its mass. Neutrino would experience a time of flight delay,

$$t_\nu \simeq 2.7 \left(\frac{m_\nu}{10eV}\right)^2 \left(\frac{10\,\text{MeV}}{E_\nu}\right)^2,$$

that would have produced a correlation between event energy and arrival time. The lack of correlation in the events from SN1987a implies that the electron neutrino mass is less than 16 eV. The observation of ν_μ's and ν_τ's from a future supernova with a neutral current detector (e.g. Sudbury) could set similar limits on their masses, an improvement of six orders of magnitude over laboratory limits.

Similarly, the lack of correlation between event energy and arrival time constrains the charge of the neutrino. An electric charged neutrino would follow a path from the LMC to the earth, whose length depended on its energy. The electron neutrino's charge can not exceed $2 \times 10^{-17} e$ (Barbiellini and Cocconi 1987)

The normality of the neutrino signal also constrains extensions of the standard model. Majorons and axions, proposed light particles, could have carried off the bulk of the supernova's binding energy— they didn't. This limits the strengths of their couplings to ordinary matter (see e.g. Kolb and Turner 1987). The normality of the neutrino signal suggests that nothing is radically amiss with standard electroweak theory when applied in the high temperature (MeV) and density (10^{11} gm/cm^3) neutron star environment.)

EXPLOSION!

Even if we do not understand how the supernova shock climbs through the infalling silicon and iron shells, Sandulak 69-202 knew what had to be done. The shock reached the oxygen layer and propogated to the surface carrying $\sim 10^{51}$ ergs. The shock reached the outer layers of the star a few hours after the neutrino burst: the optical brightening was consistent with a supernova exploding in a blue supergiant star.

The supernova ejected the outer layers of the Sandulak star, which cooled— after a few days, the envelopes effective termperature settled down to about 5000 K, a little colder than the surface of the Sun. On-going observations of the spectrum and luminosity of the star has revealed new information about the onion skin structure of Sandulak 69-202. (See Woolsey 1988).

While kinetic energy in the shock powered the luminosity of SN1987a for its first few weaks, radioactive decay is now the most important energy source. In the neutron rich environment of the supernova explosion, ^{56}Ni is the energtically favored isotope. Some of this nickel is carried out in the supernova ejecta, where it decays to ^{56}Co. The cobalt decays to iron with a 114 day half-life. Most of the gamma rays are absorbed by the supernova shell— this energy is quickly reemited as thermal radiation. A fraction of the gammas escape at are detected directly as cobalt decay lines. The observations suggest that SN1987a produced 0.07 solar masses of cobalt and that this material mixed with other layers in the supernova.

We suspect that the supernova explosion left behind a neutron star. However, we have not yet observed this explosion remnant. If the neutron star is rotating and has a magnetic field, it will eventually be observed as a pulsar. The pulsar emits low frequency radiation which also heats the supernova ejecta. This source has not yet made an important contribution to the supernova's total luminosity. However, as the abundance of cobalt exponentially falls, the neutron star may become visible. There is still much to be learned from SN 1987a!

Annotated Bibliography

Bethe, H.A. and Brown, G., *Scientific American*,**221**, 60 (1985). This beautiful written article is an excellent introduction to the theory of supernova.

Woolsey, S.E. & Weaver, T.A., *Ann. Rev. Astron. Astrophys.*, **24**, 205 (1986). A more detailed review of supernova explosions for those who want to delve deeper into the topic.

References

Baade, W. & Zwicky, F., *Phys. Rev.*, **45**, 138 (1934).

Baron, E., Cooperstein, J. & Kahana, S., *Nucl Phys. A*, **A440**, 744 (1985).

Barbiellini, G. and Cocconi, G., *Nature*, **329**, 21 (1987).

Burrows,A. and Lattimore, J.M. *Ap.J. Lett.*, **318**, L631 (1987).

Colgate, S.A. & White, R.H., *Ap.J.*, **143**, 626 (1966).

Frieman, J.A., Haber, H.E. and Freese, K., *Phys. Lett.*, **200B**, 115 (1988).

Goldreich, P. and Weber, A., *Ap. J.*, **238**, 991 (1980).

Hoyle, F. & Fowler, W.A., *Ap. J.*, **132**, 565 (1960).

Kolb, E.W. and Turner, M.S., FERMILAB-PUB-97/223 (1987).

Mayle, R., Wilson,J.R. & Schramm, D.N., *Ap.J.*, **318**, 288 (1987).

Spergel, D.N., Piran, T., Loeb, A., Goodman, J., and Bahcall, J.N., *Science*, **237**, 1471 (1987).

Wilson, J.R., Mayle, R., Woosley, S.E., Weaver, T.A., *Proc Texas Symp. Relative.*

Astrophys. 12th Ann. NY Acad. Sci, (1985).

Woosley, S.E., "SN 1987a: After the Peak", Lawerence Livermore preprint UCRL-98001.

LECTURE II. SOLAR NEUTRINO PROBLEM

In this lecture, I will introduce the solar neutrino problem. I will begin by reviewing stellar structure, focusing on the nuclear burning reactions that produce neutrinos. I will then describe the existing experiments, Ray Davis' chlorine detector and the Kamiokande water Chernkov detector. Both of these detectors measure far fewer solar neutrinos than predicted by detailed theoretical models. A variety of hypothesis have been promoted as solutions to the solar neutrino problem. I will review these hypothesis and then describe how future experiments will test these models and help resolve the solar neutrino puzzle.

I. STELLAR STRUCTURE AND NEUTRINO GENERATION

A star's life is a constant struggle against gravitational collapse. In order to support itself against its own gravity, a star must maintain a pressure gradient in order to remain in hydrostatic equilibrium. This pressure gradient is mostly due to the large thermal gradient between the Sun's core ($T \sim 1$ keV) and its surface ($T \sim 0.5$ eV). The Sun must constantly convert hydrogen to helium to maintain this thermal gradient.

The details of stellar structure are determined by four equations: (1) the equation of hydrostatic equilibrium; (2) energy transport and generation; (3) conservation of mass and (4) the equation of state. Atomic and ionic opacities and the efficency of convection determine the rate of energy transport. Energy generation rates depend on nuclear cross-sections that are measured in the laboratory.

Our Sun, like Gaul, can be divided into three regions: an inner radiative zone, an intermediate convective zone, and outer radiative zone. We will focus on the inner radiative zone, where nuclear burning transpires and neutrinos are produced.

The Sun generates its energy by transforming hydrogen to helium through one of several nuclear reactions pathways. Table 1, taken from Bahcall 1988, lists the avaliable channels, their frequency of use and the energy of the neutrinos produced in each reaction.

Most of the hydrogen is burned in the PPI chain, in which two ^{3}He nuclei are combined to make one α particle and two protons. The neutrinos produced in this channel, the "pp" neutrinos, are not energetic enough to be detected by existing experiments. A fraction ($\sim$ 15%) of the ^{3}He interacts with ^{4}He to form ^{7}Be. This Beryllium nucleus can capture an electron and emit a neutrino that is barely energetic enough to be detected in the Davis' experiment. Current experiments are sensitive to the far more energetic neutrinos produced in the PPIII chain. These neutrinos, produced by the decay of ^{8}B, represent only a small fraction of the Sun's nuclear burning. Since roughly two neutrinos are produced for every helium nucleus, we can estimate by neutrino flux by assigning one neutrino to every 12.5 MeV released in nuclear reaction. Thus, the Earth recieves in addition to solar photon flux of 1.4×10^6 erg/cm^2/s, approximately 6×10^{10} neutrinos/cm^2/s. Most of this flux is in low energy pp and ^{7}Be neutrinos, the ^{8}B flux is expected to be only 5.8 $\times 10^6$.

The predicted ^{8}B flux is a sensitive function of the Sun's central temperature. Bahcall (1988) estimates that the expected flux of these high energy neutrinos scales at T_{core}^{18}! The pp flux, on the other hand, is weakly anti-correlated with core temperature: $\phi_{pp} \propto T^{-1.2}$. Thus, experiments that measure the Boron flux are sensitive solar thermometers.

Table 1. The pp chain in the Sun. The average number of pp neutrinos produced per termination in the Sun is 1.85. For all other neutrinos sources, the average number of neutrinos produced per termination is equal to (the termination percentage/100). [This table is taken from Bahcall 1988]

Reaction	Number	Termination[†] (%)	ν Energy (MeV)
$p + p \to {}^2H + e^+ + \nu_e$	1a	99	≤ 0.420
or			
$p + e^- + p \to {}^2H + \nu_e$	1b (pep)	0.4	1.442
${}^2H + p \to {}^3He + \gamma$	2	100	
${}^3He + {}^3He \to \alpha + 2\,p$	3	85	
or			
${}^3He + {}^4He \to {}^7Be + \gamma$	4	15	
${}^7Be + e^- \to {}^7Li + \nu_e$	5	15	(90%) 0.861
			(10%) 0.383
${}^7Li + p \to 2\,\alpha$	6	15	
or			
${}^7Be + p \to {}^8B + \gamma$	7	0.02	
${}^8B \to {}^8Be^* + e^+ + \nu_e$	8	0.02	< 15
${}^8Be^* \to 2\,\alpha$	9	0.02	
or			
${}^3He + p \to {}^4He + e^+ + \nu_e$	10 (hep)	0.00002	≤ 18.77

[†]The termination percentage is the fraction of terminations of the pp chain, $4p \to \alpha + 2e^+ + 2\nu_e$, in which each reaction occurs. The results are averaged over the entire model of the current Sun. Since in essentially all terminations at least one pp neutrino is produced and in a few terminations one pp and one pep neutrino are created, the total of pp and pep terminations exceeds 100%

II. EXISTING EXPERIMENTS

The cornerstone of neutrino astronomy is Ray Davis' Chlorine experiment, a 10^5 gallon tank of C_2Cl_4 sitting in the Homestake mine in North Dakota. This experiment is sensitive to high energy neutrinos that transform Chlorine to Argon:

$$\nu_e + {}^{37}Ar \rightarrow e^- + {}^{37}Ar$$

Argon, a noble gas, is chemically seperated from the tank by bubbling though Helium. Davis is able to count individual Argon atoms by monitoring the decays of radioactive ^{37}Ar.

Davis detects 0.46±0.06 atoms per day. This corresponds to 2±0.3 SNUs(1 σ error). A SNU (Solar Neutrino Unit) is 10^{-36} events per target per second. Theory predicts 7.9±0.9 SNUs(1 σ error). This large discrepancy is the solar neutrino problem.

Recently, a new detector has broadened neutrino astronomy. The Kamiokande detector, a large tank of pure water surrounded by photomultiplier tubes in a deep mine in Japan, was originally built to look for proton decay predicted in SU(5) GUT theory, but has recently been reconfigured to search for solar neutrinos. Kamiokande detects the Cherkhov radiation from scattered Kamiokande is sensitive to Cherenkhov radiation from scattered electrons:

$$\nu + e^- \rightarrow \nu + e^-$$

Since the electron neutrino scattering cross-section is much larger than the muon or tau neutrino-electron cross-section, Kamiokande is much more sensitive to electron neutrino flux.

Because of radioactive background, Kamiokande is sensitive only to neutrinos with energies above 9 MeV: boron neutrinos. Kamiokande has not yet reported a detection. However, the current Kamiokande upper limit on the solar Boron neutrino flux ($3.2 \times 10^6/\text{cm}^2/\text{s}$) is already 40% below the predicted solar rate.

III. SOLUTIONS TO THE SOLAR NEUTRINO PROBLEM

A. Blame the Neutrino: the MSW solution

The most popular resolution of the solar neutrino puzzle is the Mikeyev-Smirnov-Wolfenstein solution: blame the neutrino. This solution suggests that the Sun emits the number of electron neutrinos predicted in the standard solar model, but these neutrinos are resonantly converted to another neutrino species. The Davis' experiment is not sensitive to μ or τ neutrinos and the Kamiokande experiment does not place any significant constraints on the μ or τ neutrino flux.

I will follow Bethe's (1986) treatment of the MSW effect in the adiabatic regime (the neutrino oscillation wavelength is longer than the density scale length.) The electron and muon neutrinos are assumed to be linear combinations of 2 neutrino mass eigenstates:

$$|\nu_e\rangle = |\nu_1\rangle \cos\theta + |\nu_2\rangle \sin\theta$$

$$|\nu_\mu\rangle = -|\nu_1\rangle \sin\theta + |\nu_2\rangle \cos\theta$$

Thus, in vacuum, the square of the mass matrix represented in the interaction eigenstate is:

$$\mathcal{M} = \frac{1}{2}(m_1^2 + m_2^2)I + \frac{\Delta}{2}\begin{pmatrix} -\cos 2\theta & \sin 2\theta \\ \sin 2\theta & \cos 2\theta \end{pmatrix}$$

where I is the identity matrix and $\Delta = m_2^2 - m_1^2$.

In a plasma (e.g. the Sun), the electron neutrino propogates differently from the muon neutrino. The electron neutrino interacts through both the neutral and charged weak vector boson exchange, while the muon neutrino interacts only through the neutral weak boson exchange. Thus a ν_e propogating through the Sun interacts with a potential that depends on the electron number density, N_e:

$$V = 2\sqrt{2} G_F N_e$$

where G_F is the Fermi constant.

This potential affects the propogation of the neutrino by adding an effective mass term to the dispersion relation:

$$k^2 + m^2 = (E - V)^2 \simeq E^2 - 2EV$$

This effective mass, $m_{eff}^2 = 2EV$, is akin to the effective plasma mass of a photon moving through an electron gas. The addition of these terms to the mass matrix,

$$\mathcal{M} = \frac{1}{2}(m_1^2 + m_2^2 + m_{eff}^2)I + \frac{1}{2}\begin{pmatrix} m_{eff}^2 - \Delta\cos 2\theta & \Delta\sin 2\theta \\ \Delta\sin 2\theta & -m_{eff}^2 + \Delta\cos 2\theta \end{pmatrix}$$

alters the neutrino propogation properties. The mass eigenstates now depend sensitively on the plasma density through the effective mass:

$$m_\nu^2 = \frac{1}{2}(m_1^2 + m_2^2 + m_{eff}^2) \pm \frac{1}{2}\sqrt{(\Delta\cos 2\theta - m_{eff}^2)^2 + m_{eff}^4 \sin^2(2\theta)}$$

A level crossing occurs when the square root term approaches zero.

Figure 1 shows that the mass eigenstate that is predominately $|\nu_e>$ in the high density solar center is adiabatically converted to an eigenstate that is predominately $|\nu_\mu>$ in vacuum.

The Davis and Kamiokande detectors are sensitive to neutrinos with energies around 8 MeV. This implies that an adiabatic crossing from electron to muon neutrino can occur when

$$(m_2^2 - m_1^2)\cos 2\theta \simeq 6 \times 10^{-5} \text{eV}$$

This is the beauty of the MSW effect: only a small mixing angle is required to produce a large effect.

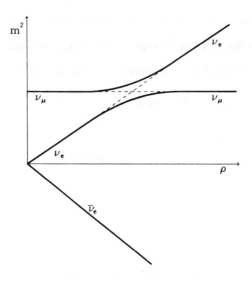

Figure 1. (from Bahcall 1988)

If the density gradient is steeper than the neutrino oscillation frequency, then the adiabatic assumption is no longer valid. Rosen and Gelb (1986) provide a pedagogical discussion of this "non-adiabatic solution". This solution reduces the predicted SNU rate to the observed value when $(m_2^2 - m_1^2)\sin 2\theta \simeq 10^{-7.4}$. The MSW effect can also reduce the flux of pp neutrinos. However, since the pp neutrinos are emitted from a large region of the Sun then the ^{8}B neutrinos, the reduction factor need not be the same. In fact, almost any rate measured by the Gallium experiment is consistent with the MSW effect. Testing this effect will require several experiments which will probe the solar neutrino energy spectrum.

Okun et al. (1986) suggested another possible neutrino "fault" that could prevent it from being detected. They proposed that the neutrino has a significant magnetic moment $(10^{-10}\mu_B)$— in the Sun's magnetic field, the neutrino rotates from a left-handed neutrino to a sterile right-handed neutrino. Both the chlorine and Kamiokande detectors are insensitive to the right-handed neutrino flux. Since the strength of the solar magnetic field determine the fraction of neutrinos rotated to a sterile state, this explanation explains the reported (but constested) anti-correlation of detected neutrino flux with sunspot number. However, the required neutrino magnetic moment is many orders of magnitude larger than predicted in the standard model of electroweak interactions.

B. Blame the Sun: the WIMPy solution

The problem may lie not with the neutrino but with our solar models. The solar models are based on standard physics: the atomic and ionic cross-sections are either measured in the laboratory or calculated from well-established; the nuclear reaction rates are based on experimental rates. However, new physical assumption could alter the Sun's thermal profile and change the predicted SNU flux.

My collaborators and I suggested that energy transport is altered in the stellar interior. We posited the existence of a new particle that carried much of the energy in the core of the Sun. Bill Press and I suggested that a weakly interacting particles could be extremely efficient at energy transport (Spergel and Press 1985). After publication, we learned that John Faulkner and Ron Gilliland had considered this possibility several years earlier, but were dissuaded from publishing their results. [Most of their conclusions were summarized in Steigman et al. (1978). Their full paper finally appeared seven years later (Faulkner and Gilliland 1985)].

Particles with cross-sections of order 10^{-36} cm² are ideal for transporting energy in the Sun. In the conductive (large cross-section) regime, energy transport scales as the mean free path. As the cross-section decreases, the cosmion travels through a larger temperature gradient between collisions and is thus more effective at transporting energy. In the small cross-section regime, collisions are so rare that the energy transport scales as the collision rate. The cross-over between these two regimes occurs at the optimal cross-section for energy transport: when $\sigma \approx 10^{-36}$ cm^2 and the cosmion's mean free path is its orbital radius. The cosmion can deposit a large fraction of its kinetic energy at aphelion and can increase its kinetic energy at perihelion.

The cosmions are extremely efficient at energy transport. The timescale for the cosmion to transfer energy from the center of the Sun to a scale height, the free fall time (≈ 100 s), is much shorter than the timescale for photons to diffuse the same distance, the Kelvin-Helmholtz time ($\approx 10^6$ years). In fact, only 10^{-11} cosmions per baryon is sufficient to significantly alter energy transport in the solar core and lower the predicted SNU flux to the observed value.

The net effect of the cosmion on the temperature distribution in the Sun is to cool the central core of the Sun while heating the region near the aphelion of the typical orbit (Spergel and Press 1985, Nauenberg 1986 and Gould 1987a). The scale height of the cosmion distribution can be estimated by equating the cosmion's thermal energy with its potential energy,

$$r_x = 0.13 \left(\frac{m_p}{m_x}\right)^{1/2} R_\odot$$

Most of the B^8 neutrinos are produced in the inner 0.05 $R_\odot$, while most of the Sun's luminosity is produced in the inner 0.2 $R_\odot$. Thus a cosmion with mass between 2 and 10 GeV will reduce the B^8 neutrino production rate without reducing the solar luminosity or affecting the production rate of pp neutrinos. Hence the predicted count rate from a solar model cum cosmions for the pp-neutrino sensitive ^{71}Ga experiment does not differ significantly from a standard model. Cosmions more massive than 10 GeV will be too centrally concentrated to affect the thermal structure in most of the ^{8}B neutrino producing region (Gilliland, Faulkner, Press and Spergel 1986).

The Sun will capture weakly interacting particles from the galactic halo. The escape velocity from the Sun's surface is 617 km/s, while the escape velocity from the core is over 1000 km/s. A halo cosmion with typical velocity 30 km/s will fall into the Sun where it can be captured through a single collision as long as its mass is less than ~ 50 proton masses. Thus the solar capture rate is approximately the cross-sectional area of the Sun, $\pi R_\odot^2$, divided by the typical cosmion halo velocity, times the escape velocity squared, since gravitational focusing is always important in any elastic capture processes. Press and Spergel (1985) discuss these effects and find that the capture rate is sufficient for the Sun to accumulate a significant number of cosmions in the solar lifetime. If we multiply the capture rate by the lifetime, we find that we can achieve a significant concentration of cosmions relative to baryons,

$$\frac{n_x}{n_b} \simeq 3 \times 10^{-10} \left(\frac{\rho_x}{1 M_\odot/pc^3}\right)\left(\frac{v_{esc}}{\bar{v}}\right)\left(\frac{\sigma}{\sigma_{crit}}\right)\left(\frac{m_p}{m_x}\right)$$

Recall that a concentration of 10^{-11} of cosmions with cross-section of 4×10^{-36} cm^2 will resolve the solar neutrino problem. If the cosmions compose the halo $(\rho_{HALO} \approx 10^{-2} M_\odot/pc^3, v_{HALO} \approx 300 km/s)$, then their cross-section must be within a factor of 2 of σ_{crit}. If cosmions compose the disc $(\rho_{DISC} \approx 10^{-1} M_\odot/pc^3, v_{DISC} \approx 50 km/s)$, then they can resolve the solar neutrino problem, if their baryon scattering cross-section is between 10^{-37} and $10^{-34} cm^2$.

Halo particles will also be captured by the Earth and the other planets (Freese 1986, Krauss, Wilczek & Srednicki 1986, Gould 1987b and Bouquet and Salati 1987). For a particle to be captured, its velocity at infinity must be less than the escape velocity from the planet's core. Thus for the *capturable* particles, gravitational focusing is always important, even if it is not important for the average particle in the distribution. Note that for stars, for whom most of the flux is capturable, $F \propto v_{esc}^2 R^2 \propto MR$, while for planets, which can capture only a fraction of the flux, $F \propto v_{esc}^4 R^2 \propto M^2$.

Cosmions do not significantly alter main sequence stellar evolution. Since they are centrally concentrated within the luminosity producing region, they do not alter the mass of available fuel (hydrogen). Many other proposed solutions to the solar neutrino problem that reduce the thermal gradient in the inner core also reduce the gradient throughout the luminosity producing region, thus increasing its volume. With more available hydrogen, the main sequence lifetime of a star is longer. Such solutions aggravate the discrepancy between the Hubble time and the inferred age of globular clusters. Cosmions have little effect in the evolution of massive stars. Since their number density in the star grows linearly with time, few cosmions can accumulate in the short-lived massive stars.

Renzini (1986) argues that cosmions can significantly alter the evolution of horizontal branch stars. Spergel and Faulkner (1988), however, point out that Renzini severely overestimated the cosmion energy transport. Cosmions are only effective at energy transport when their mean free path is comparable to an orbital length.

Since cosmions alter the solar thermal structure, they affect the seismology of the Sun. Solar seismology, which measures the sound speed as a function of radius, might detect the variations in density and temperature induced by cosmion energy transport. Dappen et al. (1986) and Faulkner et al. (1986) suggest that cosmions can eliminate the discrepancy between the observed p-wave spectrum and the standard solar model. Bahcall and Ulrich (1988) argue that this discrepancy may not be not significant. Cosmions would have more dramatic effects on the still unobserved g-wave spectrum, which is more sensitive to the core conditions.

The good news is a halo population of cosmions of the correct cross-section ($\approx 10^{-36}$ cm^2) and mass (5–10 GeV) will be captured by the Sun in sufficient number to resolve the SNU problem, and yet will not alter other aspects of stellar evolution. We now must turn to the bad news: Once captured by the Sun, cosmions can be lost either through annihilation or through evaporation.

Most of the cosmions in the Sun are tightly bound: their typical velocities, $\sqrt{3kT/2m_x} \approx 300$ km/s, is much less than the escape velocity from the core $v_{esc}^2 = 1400$ km/s, so scatterings that produce $v \geq v_{esc}$ are rare. In the conclusion of Spergel and Press (1985), the evaporation rate is estimated as the fraction of cosmion distribution with energy sufficient to escape divided by the time to repopulate the tail. This estimate suggests that evaporation is negligible for cosmions with m $> 4m_p$.

The cosmion distribution function may differ significantly from Maxwellian in its high energy tail. Nauenberg (1986), Greist and Seckel (1987) and Gould (1987a) have considered this effect and conclude that this lowers the evaporation limit to 3.5 m_p.

Annihilation can also reduce the number of cosmions in the core of the Sun. If the cosmion is a Majorana particle, it is its own anti-particle and will self-annihilate. If the cosmion is a Dirac particle and the Sun contains both it and its anti-particle in equal numbers, annihilation will also reduce its solar abundance. The cosmion annihilation timescale in the Sun can be estimated,

$$t_{ann} = (n_x \sigma_{ann} v)^{-1} = \left(\frac{n_p}{n_x}\right)\left(\frac{\sigma_{ann}}{\sigma_{bx}}\right) t_{coll} ,$$

where σ_{ann} is the cosmion annihilation cross-section and σ_{bx} is the cosmion-baryon scattering cross-section. If the cosmion is to resolve the Solar neutrino problem, $t_{coll} \approx t_{dynamical} \approx 100$ seconds and $n_p/n_x \approx 10^{11}$. Most of the attractive particle physics cosmion candidates, photinos, scalar neutrinos, massive and Dirac neutrinos, have scattering cross-sections less than or on the order of their annihilation cross-sections; this implies $t_{ann} < 10^{13}$ seconds, much shorter than the age of the Sun (Krauss, Freese, Spergel and Press 1986). Cosmions are more centrally concentrated than baryons; this enhances their annihilation rate and exacerbates the problem.

This annihilation constraint makes it difficult to construct viable particle physics models that contain solar cosmions. Gelmini et al. (1986) suggested two models that avoid the annihilation problem. One model uses a cosmic assymmetry between particle and anti-particles (just like the assymmetry between baryon and anti-baryons). Another model suppress cosmion annihilation cross-section. Raby and West postulate a fourth generation neutrino with a large magnetic moment due to a Higgs particle with similar mass. Neither of these models were motivated by other particle physics consideration.

There are other ways of modifying the Sun and solving the solar neutrino problem. Boyd and his collaborators suggested that there might be a heavy quark that stabalizes a 5 nucleon Lithium nucleus (Boyd et al. 1983). This changes solar nuclear chemistry— a new reaction chain in which "heavy" ^{4}He catalyzes hydrogen burning is possible. This new channel speeds fusion and would allow the Sun to have a lower central temperature. This lower temperature implies a lower neutrino detection rate in both the Davis and Kamiokande experiments and also lowers the solar oscillation frequency.

IV. CONCLUSION

The solar neutrino problem suggests that something is amiss in our understanding of basic physics. Either we need to modify the standard electroweak model or we need to understand some new collective effect that modifies the structure of the Sun. The new physics solutions may reveal new the physics beyond the standard model: neutrino masses, neutrino magnetic moments or exotic particles. I have described only a few of the solutions proposed to the solar neutrino problem.

There are several new experiments planned that will measure the neutrino flux at lower energies. These experiments will discriminate between different solutions and may provide an answer to the solar neutrino problem.

BIBLIOGRAPHY

Bahcall, J.N. and Ulrich, R.K. *Rev. Mod. Phys.*, **60**, 297 (1988). This paper uses a detailed standard model to compute the expected neutrino spectrum. It discusses the effects of several modifications of the standard model. It is a **must** for students who want to delve deeper in this field.

Bahcall, J.N. *Neutrino Astrophysics*, in preperation. This book which should appear soon is likely to become the standard text for anyone interested in the solar neutrino problem.

REFERENCES

Bethe, H.A., *Phys. Rev. Lett.*, **56**, 1305 (1986).

Bouquet, A. and Salati, P., "Life and death of cosmions in stars", LAPP-TH-192/87 (1987).

Boyd, R.N., Turner, R.E., Wiescher, M., and Rybarcyk, N., *Phys. Rev. Lett.*, **51**, 609 (1983).

Dappen, W. Gilliland,R.L. and Christensen-Dalsgaard, J. *Nature*, **321**, 229 (1986).

Faulkner, J. and Gilliland, R.L. *Ap. J.*, **299**, 994 (1985).

Faulkner, J. Gough, D.O. and Vahia, M.N. *Nature*, **321**, 226 (1986).

Gelmini, G.B. Hall, L.J. and Lin, M.J., *Nucl.Phys.B* **281** , 726 (1986).

Gilliland, R., Faulkner, J., Press, W.H. and Spergel, D.N., *Ap.J.*, **306**, 703 (1986).

Gould, Andrew *Ap. J.*, **321**, 560 (1987a).

Gould, Andrew *Ap. J.*, **321**, 571 (1987b).

Griest, K. and Seckel, D., *Nucl.Phys.B* **283**, 681 (1987).

Krauss, L.M., Freese, K., Spergel, D.N. and Press, W.H., *Ap. J.*, **299**, 1001 (1985).

Krauss, L.M., Srednicki, M., and Wilczek, F., *Phys. Rev.*, **D33**, 3079 (1988).

Nauenberg, Michael, *Phys. Rev.*, **D36**, 1080 (1987).

Okun, L.B., Voloshin, M.B., and Vysotskii, M.I., *Sov. J. Nucl. Phys.*, **44**, 546 (1986).

Press, W.H. and Spergel, D.N., *Ap. J.*, **296**, 663 (1985).

Raby, S. and West, B.G., *Phys. Lett.*, **202B**, 47 (1988).

Renzini, Alvio, *Astronomy and Astrophysics*, **171**, 121 (1986).

Rosen, S.P. and Gelb, J.M., *Phys. Rev.*, **D34**, 969 (1986).

Spergel, D.N. and Faulkner,J., *Ap. J. Lett.*, **331**, L21 (1988).

Spergel, D.N. and Press, W.H., *Ap.J.*, **294**, 679 (1985).

Srednicki, M., Olive, K.A. and Silk, J., *Nucl.Phys.B* **279**, 804 (1987).

Steigman, G., Sarazin, C.L., Quintana, H. and Faulkner, J. *A.J.*, **83**, 1050 (1978).

LECTURE III. DARK MATTER

The third lecture reviews the evidence for this "missing mass" and discusses several of the proposed candidates: baryons, neutrinos, axions and WIMPs. Baryonic dark matter, the simplest hypothesis, is not consistent with a flat universe and standard big bang light element nucleosynthesis. The non-detection of time of flight delay in the arrival time of the neutrinos from SN1987a rules out the electron neutrino has a galactic dark matter candidate. Recent observations of a nearby dwarf galaxy may rule out the possibility that any neutrino species is the dark matter ubiquitiously seen in all galaxies. Axions and WIMPs may be detected or ruled out by current and future experiments.

I. ASTRONOMICAL EVIDENCE FOR GALACTIC DARK MATTER

I will begin the lecture by reviewing the astronomical data that suggests that the dark matter problem is ubiquitous: all galaxies, whether spirals or elliptical, dwarf and normal, seem to have halos of dark matter.

The galactic missing mass problem is over 50 years old. Jan Oort's studies of stellar motions in the galactic neighborhood suggested that mass in visible stars was sufficient to account for the observed velocities. In 1933, Zwicky applied the virial theorem to rich clusters of galaxies and concluded that 10 - 100 times the observed mass in luminous galaxies was required to keep the clusters bound.

In the 1970s, the astronomical community began to recognize the importance of this problem. Ostriker, Peebles and Yahil (1974) and Einasto, Kraasik and Saar (1974) suggested that all galaxies have massive halos. The astronomical evidence collected in the intervening 15 years have confirmed their conclusions. Observations of the rotation curves in spiral galaxies, measurements of the X-ray emitting gas in elliptical galaxies and the high velocities of galaxies in binaries, groups and clusters suggest that roughly 90% of the mass of the universe is in non-luminous form.

Infering the existence of dark matter from astronomical observations relies on Newton's laws, the low energy limit of general relativity. Milgrom and Bekenstein (1986) and others have suggested that perhaps there is no missing mass, but rather that gravity in the limit of very weak accelerations deviates from Newtonian physics.

A. Spirals

Most galaxies fall into one of two catagories: spirals and ellipticals. Spiral galaxies contain cold gas and are the site of on-going star formation. These young stars lie in spiral arms in the disk of the galaxy. Elliptical galaxy are as the name suggests elliptical shaped in projection and consists only of older stars. We live in a spiral galaxy.

Spiral galaxies have three distinct components:

(a) The Spheroid: This spheroidally shaped component consists of older stars and is centrally concentrated. In our own galaxy, it contains roughly 3.4×10^{10} $M_\odot$ solar masses ($M_\odot$) of stars (Ostriker and Caldwell 1983). Most of this mass is concentrated in the inner 2 Kpc of our galaxy. (A parsec (pc) is 3×10^{18}

centimeters and is the standard unit of distance in astronomy. Its origin dates to the first measurements of the stellar parallax: the nearest stars have a *par*allax of 1 *arcsec*ond and are located 1 parsec from the Sun.)

(b) The Disk: The disk of our galaxy consists of both young and old stars, gas and dust. Our Sun is located in the disk of our galaxy. Stars and gas in the disk move on nearly circular orbits around the center of our galaxy. The surface density of our disk falls of exponentially with distance from the galactic center, R:

$$\Sigma(disk) \simeq 75 M_\odot / \text{pc}^2 \exp\left[\frac{(R - 8\,\text{Kpc})}{4\,\text{Kpc}}\right]$$

Thus, most of the mass of our disk is located within the orbit of the Sun (8 Kpc).

Pictures of distant spiral galaxies reveal their beautiful spiral arms. Spiral density waves, often stimulated by nearby galaxies or rotating bars in the center of the galaxy, concentrate and shock the gas in the disk stimulating star formation. The massive stars live short brilliant lives enhancing the luminosity of these structures. The spiral arms usually contain only a small fraction of the mass of a galactic disk.

(c) The Halo: the existence of this spheroidally shaped component is infered only through its gravitational effects. We do not know the composition of this component (Hence, this lecture.)

As promised, we will now use Newton's laws to infer the existence of the halo. Since the gas in stellar disks move on circular orbits, we can relate the centripetal acceleration to the gravitational acceleration and infer the mass interior to the rotating gas:

$$\frac{v_{circ.}^2}{r} = \frac{GM(r)}{r^2}$$

or

$$M(r) = \frac{v_{circ.}^2 \, r}{G}$$

The planets in our own Solar System move on nearly circular orbits. Most of the solar system's mass is in the Sun: $M(r) = M_\odot$. The circular velocities of the planets fall off with radius, $v_{circ.} \propto \sqrt{r}$, Kepler's law. Since most of the luminosity of our galaxy is concentrated within the sun's orbit, we would expect stellar velocities to follow Kepler's law outside 8 Kpc. This is not observed.

The observation of other spiral galaxies provide corroborating evidence: the rotation curves of *all* observed spiral galaxies are either flat or rising (e.g. Ruben et al. 1985). This implies that $M(r) \propto r$. The light in galaxies, however, is centrally concentrated.

Most of the light in galaxies like our own comes from the inner 8 Kpc. Most of the mass, however, lies outside this radius. In spiral galaxies, the galactic mass inferred from rotation curves exceeds the mass in observed stars by about one order of magnitude!

B. Ellipticals

Elliptical galaxies consist of old stars. The interstellar medium in these galxies is filled with hot gas. This hot gas represents only a small fraction of the mass of the system, however, it is an excellent tracer of the overall mass distribution.

Since the free-fall time in an elliptical galaxy is much shorter than the age of the universe, we can assume that the gas obeys the equation of hydrostatic equilibrium:

$$\frac{dp}{dr} = -\frac{GM(r)\rho}{r^2}$$

ρ is the gas density. The equation of state of the 10^6 K gas is well described by the ideal gas law,

$$p = nk_B T$$

Combining these equations, we can relate the mass of the system to the observed gas density and temperature:

$$M(r) = \frac{k_B T(r)}{G \mu m_p} \left[-\frac{d \log \rho}{d \log r} - \frac{d \log T}{d \log R} \right]$$

μ is the mean molecular weight of the gas.

Forman et al. (1985) model the X-ray observations of several elliptical galaxies and conclude that all of them show evidence for massive halos. Uncertainties in estimating the temperature profile in the galaxy with existing data is the major source of ambiguity in these estimates. AXAF, the next generation X-ray satellite, will obtain detailed temperature profiles and remove this source of ambiguity. (Sarazin (1986) reviews the X-ray observations of elliptical galaxies and the associated theoretical uncertainties.)

C. Clusters

Galaxies seem to love companionship. Many galaxies are found in binary systems and in small groups. A smaller fraction are found in rich clusters of thousands of galaxies.

Since the free-fall time of these systems is much shorter than the age of the universe, we can assume that they are in virial equilibrium — their total kinetic energy is one-half their total potential energy:

$$G \sum_{i,j, i \neq j} \frac{M_i M_j}{r_{ij}} = \sum M_i v_i^2$$

M_i and v_i are the masses and velocities of individual galaxies in the system and r_{ij} is the seperation between the i-th and j-th galaxy.

Observations of binary galaxies, groups and clusters all suggest that the dark matter composes $\sim$ 90% of the mass of the system. Each of these measurements are beset with observational and theoretical uncertainties (see the IAU Symposium on Dark Matter), however, the weight of the evidence points towards the existence of large amounts of dark matter in all galactic systems.

D. Cosmology

The evidence for cosmological dark matter is purely theoretical and aesthetic. Cosmologists describe the density of the universe in terms of a dimensionless parameter, Ω, the ratio of the mean density of the universe, ρ, to the critical density, $\rho_{crit.} = 2 \times 10^{-30}$ gm/cm^3:

$$\Omega = \frac{\rho}{\rho_{crit.}}$$

If $\Omega > 1$, the universe is closed and will eventually collapse back on itself. If $\Omega < 1$, the universe is open and will expand forever.

Many theorists believe that Ω ought to be very near 1. $\Omega = 1$ is an unstable fixed point in the theory and we are living in a very special epoch if the observations that suggest $\Omega \simeq 0.1$ are correct. If $\Omega = 0.1$ today, then $\Omega = 0.9999$ at the epoch of recombination and $\Omega = 1 - 10^{-10}$ at the epoch of nucleosynthesis. Many theorists believe that this is fine tuning.

The theory of inflation, which successfully resolves problems of isotropy and prevents monopoles from overclosing the universe, also predicts $\Omega = 1$. The beauty of inflationary theory has strenghtened theorists' belief that there must be missing mass on the cosmological scale.

Ultimately, we might hope to measure the curvature of the universe and determine the mass density of the universe. Brave attempts (e.g. Loh and Spillar 1985) to find standard candles for these observations are plagued by systematic errors and uncertainties due to galactic evolution.

II. DETECTING GALACTIC DARK MATTER

In this half of the lecture, I will describe several particles and objects from the zoo of candidates for this "missing mass" and discuss experimental and astronomical techniques employed in the search for dark matter. These candidates range from 10^{-6} eV axions to 10^{39} gram black holes. This range of 78 orders of magnitude in mass reveals our state of complete ignorance of the true nature of the dark matter.

How can we begin to search this enormous mass range? The concept of maturalness is a promising guide which will hopefully suggest fruitful places to search for this "missing mass". The most "natural" candidate is baryonic dark matter. Baryons exist. We know that at least 10% of our galaxy is made of baryons, why not the rest? We will first consider this possibility. Neutrinos are probably the next best choice. We know neutrinos exist and are produced in the standard big bang model in roughly the same number as the microwave background photons. After discussing neutrinos, I will describe candidates for which there is no direct evidence but only theoretical desire: WIMPs and axions.

A. Baryons

Baryons are the obvious candidate for dark matter. We know that baryons compose most of the mass of the Earth, the Sun and other stars in our galaxy. Why not the halo?

Baryonic dark matter does not satisfy the theorists, who long for a flat $\Omega = 1$ universe. The abundance of light elements, deuterium, ^{3}He, ^{4}He and Lithium constrain the baryonic contribution to closure density: $\Omega_{baryon} < 0.1$ (Yang et al. 1984). [This limit, derived from the standard big bang model, can be evaded by either modifying light element big bang nucleosynthesis (e.g. Hogan, Applegate and Scherrer 1987, Dimopoulus et al. 1988) or ignored by those willing to live in an open universe with $\Omega_{tot} = 0.1$.]

If the halo is baryonic, the hydrogen can be ineither in hot nor cold gas. X-ray observations limit the mass of hot gas in the halo to be a tiny fraction of the masss needed to explain rotation curves. Radio observations constrain the mass in cold gas. Asteroids and comets are composed mostly of heavier elements— it would be difficult to imagine how the halo could be made of heavy elements, while the rest of the galaxy is made primarily of hydrogen and helium.

Baryonic dark matter must be clumped. These clumps must be bigger than 10^16 grams, otherwise these hydrogen snowballs would evaporate (Hegyi and Olive 1986). These clumps must be smaller than 0.08 solar masses- otherwise, they would be visible as stars. The present limits rule out the possibility of even low mass stars comprise the halo. The remaining possibility is either Jupiter-like objects or massive black holes. Objects of either type could evade contemporary detection efforts.

B. Neutrinos

Neutrinos, known non-baryonic matter, are a natural dark matter candidate. In the hot big bang, neutrinos are produced in roughly equal numbers as photons. These weakly interacting particles decouple from the thermal bath of electrons, positrons, muons and photons while there are still relativistic. Thus, the abundance of light neutrinos is independent of their mass. Hence, the neutrino contribution to closure density depends scales with mass:

$$\Omega_\nu \simeq \left(\frac{m_\nu}{30\,\text{eV}}\right) \left(\frac{H_0}{50\,\text{km}\,\text{s}^{-1}\,\text{Mpc}^{-1}}\right)$$

ITEP, a Soviet β-decay experiment reported measuring an electron neutrino mass lies between 17 and 40 eV(S. Boris et al. 1987) — just what was needed to close the universe. Subsequent experiments did not confirm this value: the Zurich tritium endpoint experiment puts a 18eV upper limit on the electron neutrino mass (M. Fritschi et al. 1986) and the Los Alamos experiment rules out electron neutrinos more massive than 27 eV (J.F. Wilkerson et al. 1987). The lack of correlation between the arrival times and the energies of the electron anti-neutrinos detected from SN 1987a place a 16 eV upper limit on the electron neutrino mass. It is important to emphasize that these experiments impose no limits on the mass of either the μ or τ neutrinos.

Experimental limits on the τ and μ neutrino mass are not cosmologically interesting. The only hope for measuring τ and μ neutrino masses in the 30 - 100 eV range is the detection of time of flight delay in the arrival of low energy neutrinos from a galactic supernova (see Dar and Dado 1987). The construction of the Sudbury heavy water detector, which is sensitive to neutral current reactions, and a galactic supernova could improve experimental limits by many order of magnitude and eliminate the possibility of cosmologically interesting neutrinos.

Neutrinos have fallen from favor as astrophysically popular dark matter candidates. In N-body simulation of galaxy formation in a neutrino dominated universe, clusters form before galaxies. Observations suggest that galaxies formed before clusters. Neutrino-dominated universes have structures like massive X-ray emitting gas haloes that are not observed in our universe (See Hut & White (1987) for review of neutrino pancake models.)

Cosmic strings have revived interest in neutrino-dominated universes. Cosmic strings provide seeds for the formation of structure. In a cosmic string-seeded universe, galaxies form before clusters. This scenario is promising and is worthy of further exploration (see Bertschinger and Watts 1988, Brandenberger et al. 1988).

Recent observations (Carrignan and Freeman 1988) of a nearby dwarf galaxy, DDO 154, may pose a serious problem for neutrino dark matter advocates. Tremaine and Gunn (1979) showed that neutrinos as fermions have a maximum phase space density. This observation of a halo with very high phase space density constrains the neutrino mass to exceed 96 eV, if the neutrino is to be the missing mass of DDO 154. This implies that the age of the universe is only 6.7 Gyr. A variety of astrophysical arguments suggest that our universe is at least 10 Gyr— a neutrino-dominated halo for DDO 154 is incompatible with standard big bang cosmology (Spergel et al. 1988). Neutrino proponents must contend that neutrinos make up the dark matter is some but not all halos. An unaesthetic proposition.

C. Axions One of the outstanding problems in the standard model is understanding the lack of CP violation in the strong interaction. An elegant solution to this problem is the exsitence of a new U(1) symmetry (the Peccei-Quinn symmetry) that is spontaneously broken. The pseudo-goldstone boson associated with this broken symmetry is called the axion.

The axion acquires a mass through non-perturbative QCD effects— in its original incarnation (Weinberg 1978, Wilczek 1978), the axion mass was $\sim$MeV. This particle was ruled out by several laboratory experiments. Theorists revived the axion in a new form: the "invisible axion", an extremely weakly interacting low mass particle. If the scale of the Peccei-Quinn symmetry breaking is $\sim 10^{12}$ GeV, then this particle can make up the missing mass (Dine, Fischler and Srednicki 1981). However, even the invisible axion can be detected: Axion couple to photons through the axial anomaly; stars can exploit this $a - \gamma - \gamma$ coupling as a alternative cooling channel. Red giants do not cool use this cooling mechanism— this constrains the axion mass to be less than 10^{-2} eV (Dearborn and Raffelt

1987). The best limits on the Peccei-Quinn symmetry breaking scale comes from SN1987A: since we did see neutrinos, we know that axions did not carry off the lion's share of the thermal energy of the hot neutron star. This constrain the P-Q scale to be greater than $\sim 10^{10}$ GeV (Turner 1988)— not yet ruling out cosmologically interesting axions.

Halo axions are potentially detectable in the laboratory through their photon coupling (Sikivie 1985). Ongoing experiments have not yet reached the sensitivities needed for axion detection (De Panfilis et al. 1987); however, future experiments may rule out axions as dark matter (or detect them!)

D. WIMPs

A variety of proposed new particles hae earned the epithet WIMP. (Weakly Interacting Massive Particle). The first candidate to fall into this catagory was the massive fourth generation neutrino. Other WIMP candidates include the solar cosmion (see the last lecture) and the technobaryon. The most popular WIMP are predicted by supersymmetry. In supersymmetry, there is an additional symmetry associated with the correspondence between fermions and bosons, R parity. The lowest mass particle with a given R parity is stable. In many supersymmetric theories, this mass eigenstate is a linear combination of photino, higgsino and zino eigenstates. The photino is the fermionic supersymmetric partner of the photon. The higgsino is the partner of the not-yet-detected Higgs and the zino is the supersymmetric partner of the Z-boson.

All of these WIMPs are possible relics from the early universe. At high temperatures, the WIMPs and in equilibrium with the thermal bath of photons, electrons, positrons, etc.:

$$\gamma + \gamma \to X + \bar{X}$$

X denotes WIMP. Eventually, the temperature of the universe drops below the WIMP mass. Photon-photon collisions are no longer energetic enought to produce WIMPs. WIMPs, however, can continue to annihilate:

$$X + \bar{X} \to \gamma + \gamma, e^+ + e + -, \text{etc.}$$

This annihilation reduces the WIMP abundance until it is so low that WIMP annihilation mean free path exceeds the size of the universe. The final WIMP abundance depends on teh the WIMP annihilation cross-section, $\sigma_{X\bar{X}}$:

$$\Omega_X \simeq \left(\frac{<\sigma_{X\bar{X}}\beta>}{4 \times 10^{-37} \text{cm}^2} \right),$$

β is the WIMP velocity.

In many theories, WIMPs have the right cross-section (SUPRISE!) to close the universe and be the galactic missing mass. If the halo of our galaxy is composed of WIMPs, then millions of these particles are streaming through a square centimeter every second. Goodman and Witten(1985) and Wasserman (1986) realized that this flux could be experimentally detectable. There are, however, two difficulties involved with detecting WIMPs:

(1) THEY DON'T DO MUCH: All of the proposed WIMPs have some conserved quantum number (R-parity for SUSY particles and fourth generation lepton number for massive neutrinos); hence, the end-product of an interaction with a nucleus is at best the deposition of a few keV of energy.

(2) THEY DON'T DO IT OFTEN: The lowest mass supersymmetric particle (which is stable in many theories) is usually some linear combination of photino and higgsino interaction eigenstates. This Majorana particle has only axial couplings with quarks and thus has a typical elastic nuclear cross-section $\sim 10^{-37}$ cm^2 for WIMPs through its spin-dependent interactions with nuclei. If these sparticles were the galactic missing mass, they would produce $\sim 10^{-1} - 1$ counts in a

kilogram of detector. Scalar neutrinos and massive Dirac neutrinos have vector couplings with quarks, and thus the neutrons in the nucleus constructively interfere to yield a much larger cross-section $\sim 10^{-34}$ cm^2 and a higher count rate $\sim 10^3$ counts/kg/day. In either case, the search for WIMPs requires a detector with very good background rejection and a low energy threshold.

We already have limits on weakly interacting halo dark matter from germanium spectrometers. These limits, however, only exclude particles more massive than 12 GeV and it does not seem likely that these limits could be extended below 8 GeV (Ahlen et al. 1987, Caldwell et al. 1988). Silicon may be sensitive to several GeV particles (Sadoulet et al. 1988). However, searching for 2 GeV particles (the Lee-Weinberg limit for neutrinos) may require new technologies. Particles with spin-dependent interactions (e.g. photinos, higgsinos, and Majorana neutrinos) also evade detection since most of the abundant isotopes of silicon and germanium have zero spin. (They have an even number of both protons and neutrons.)

This motivates the use of new types of detectors in the search for dark matter. There are several schemes for direct detection of dark matter. One scheme relies on the principle of "simple is beautiful": the use of a single kilogram crystal of silicon to detect phonons produced by WIMP-nucleon scattering (Cabrerra et al. 1985, Sadoulet 1987, Marthoff 1987). There are several groups actively exploring the possibities of phonon detectors: Cabrerra and Sadoulet and their collaborators at Stanford and Berkeley; Smith and coworkers in Oxford; Moseley and collaborators at NASA.

Other groups guided by the philosophy of "small is beautiful" are focusing on superconducting grains. This lecture will focus on these grains which detect the change in state due to the nuclear recoil of a WIMP. (Drukier and Vallette 1972, Drukier et al. 1975, Drukier and Stodolsky 1984 Gonzalez-Mestres and

Perret-Gallix 1985, Drukier 1987). Several groups are now actively working on developing grains and grain detectors: A. Drukier (Applied Research Corp.) and his collaborators at UBC, Vancouver; G. Waysand and his collaborators in Orsay, Saclay and Annecy; L. Stodolsky, K. Pretzel and collaborators at MPI für Physik und Astrophysik.

In a detector, these grains would be kept in a superheated superconducting state. The deposition of a few keV of energy in an elastic scatter of a WIMP off of a nucleus in the grain heats the grain and flips it from the superconducting to the normal state. The background magnetic field can now permeate the grain. This produces a change in the magnetic flux through a loop surrounding the grains equivalent to the addition of a dipole whose strength is proportional to the product of the grain's cross-sectional area and the background magnetic field. SQUIDs (superconducting quantum interference devices) would be ideal for detecting this small change in flux. The grain could be composed of any type I superconductor. These include gallium and aluminium, both of which are mostly composed of isotopes with an odd number of neutrons, and thus have large cross-sections for particles with spin-dependent couplings. The grains would probably be coated with a dielectric composed of low Z material. This dielectric separates the grains and suppresses diamagnetic interactions with neighboring grains. The energy threshold of a grain, the minimum amount of energy needed to flip the grain, is set by its composition, size, and temperature. Micron size grains are needed to detect photinos of a few GeV.

The major source of background in the detector is the radioactive decay of trace contaminants in the grains and the surrounding dielectric. Most of these decays produce MeV electrons. Since the dialectric is composed of low Z material, these electrons, which lose energy through Coulomb interactions, will deposit most of their energy in the grains. Thus radioactive decays will flip multiple grains, which will produce a large change in flux in the SQUID loop. The scatter of a WIMP off of a grain will flip only a single grain. Uniform grains, which have similar energy thresholds, are needed for this background suppression mechanism to work. Most of the grains must be in the superheated state so that there is little dead material into which the β particle can deposit its energy.

Here, at Brown, Lanou, Maris and Seidel are exploring another attractive technology. Superfluid helium has excitation called "rotons". A recoiling Helium nucleus excites these rotons, which propogate to the surface, where they evaporate Helium. The number of evaporated Helium atoms is a measure of the energy of the energy of the recoiling nucleus. The purity of Helium makes this technology very attractive.

Seperating signal from background is a challenge for all of these technologies. Since both radioactive decays and the elastic scatter of a halo WIMP produce the same signal in the detectors, we must find a characteristic of the signal that will allow differentiation from the background. Failing this, any "detection" could be written off as a misunderstanding of the background. Fortunately, the earth's motion around the Sun provides a significant modulation in the background rate. (Drukier, Freese and Spergel 1985).

The Sun is moving around the Galactic Center at a velocity of about 250 km/s. The non-dissipative dark matter in the galactic halo, on the other hand, never collapsed into a disc; thus, it is not rapidly rotating. As a result of the Sun's motion, the detector is moving relative to the galactic halo. Since the grains have an energy threshold, the anisotropy in the velocity distribution alters the predicted count rate.

The earth's motion around the Sun modulates the velocity of the detector relative to the halo. The earth moves around the Sun with a velocity of 30 km/s. Since the ecliptic is inclined at 62° relative to the galactic plane, only a fraction of this velocity is added to the Sun's motion. In January, when the Sun is in Saggitarius (the location of the Galactic Center), the earth is moving at $\sim$ 235 km/s relative to the halo. (The earth's motion around the Sun is counter-clockwise, while the Sun's motion around the Galactic Center is clockwise.) In July, more energetic particles will stream through the detector when the earth is moving at $\sim$ 265 km/s relative to the galactic halo. Since the detector has an energy threshold, the flux of more energetic particles produces a higher count rate. Drukier, Freese and Spergel estimate a modulation in the signal of $\sim$ 12% in a detector sensitive to 20% of the incident flux. This calculation assumes that the galactic halo has an isothermal distribution of velocities.

The existence of this modulation effect does not depend upon details of models of the galactic halo (Spergel & Richstone 1988). It only requires that the earth's velocity relative to the rest frame of the halo changes with time. The assumption that the halo is composed of WIMPs implies that it is non-dissipative; hence, its rest frame should differ from that of the Sun which is composed of baryons which collected in the disc through dissipation. The only other astronomical requirement for the modulation effect is that the earth moves around the Sun.

WIMPs may also be detected through their annihilation products. Joe Silk and his collaborators have been actively exploring this possible way of searching for WIMPs. Since WIMPs annihilate in the early universe, they also annihilate today. Annihilation products include electrons, positrons, protons, anti-protons and gamma -rays, all potentially detectable. (See Silk, Olive and Srednicki 1985, Rudaz, Stecker and Walsh 1985, Silk and Srednicki 1986). All of these particles are potentially observable— this alternative route in the WIMP search that may

may prove fruitful.

CONCLUSIONS

We do not know what makes up 90% of our galaxy. The list of candidates are limited by experimental and astrophysical evidence and our imagination. Searching for this missing mass requires contributions from astrophysicists, particle physicists, solid state physicists, chemists, experimentalists and theorists.

BIBLIOGRAPHY

Trimble, Virginia, *Ann. Rev. Astron. & Astrophys.*, This article, complete with numerous references reviews the astrophysical evidence for dark matter and outlines several possible candidates.

Primack, J.R., Seckel, D. and Sadoulet, B., *Ann. Rev. of Nuc. & Particle Science*, in press (1988). This article reviews particle physics dark matter candidates and discusses several proposed detection schemes.

REFERENCES

Ahlen, S., Avignone III, F.T., Brodzinsky, R., Drukier, A. K. Gelmini, G. and D.N. Spergel, *Phys. Lett.*, **B195**, 603 (1987).

Applegate, J.H., Hogan, C.J. and Scherrer, R.J., *Phys. Rev.*, **D35**, 1151 (1987).

Bertschinger, E. and Watts, P.N., *Ap. J*, **329**, 23 (1988). ϒ

Boris, S., et al., *Phys. Rev. Lett.*, **58**, 2019 (1987).

Brandenberger, R., Kaiser, N., Schramm, D.N., and Turok, N., *Phys. Rev. Lett.*, **59**, 2371 (1987).

Carrignan, C. and Freeman, K.C., *Ap.J. Lett.*, **332**, L33 (1988).

Dar, A. and Dado, S., Technion preprint TECH-87-55 (1987).

Dearborn, D.S.P. and Raffelt, G., *Phys. Rev.*, **D 36**, 2211 (1987).

DePanfilis, S., et al., *Phys. Rev. Lett*, **59**, 839 (1987).

Dimopoulos, S., Esmailzadeh, R., Hall, L.J., and Starkman, G.D., *Ap.J.*, **330**, 545 (1988).

Dine, M., Fischler, W. and Srednicki, M., *Phys. Lett*, **104B**, 199 (1981).

Drukier, A.K., Freese, K. and Spergel, D.N., *Phys. Rev.* **D**, **33**, 3495 (1985).

Drukier, A.K. and Valette, C. *NIM* **105**, 285 (1972).

Drukier, A.K. and Stodolsky, L. *Phys. Rev.* **D 30**, 2295 (1984).

Einasto, J., Krassik, A., and Saar, J., *Nature*, **250**, 309 (1974).

Forman, W., Jones, C. and Tucker, W., *Ap. J. Lett.*, **293**, 102 (1985).

Fritschi, M. et al., *Phys. Lett.* **B 173**, 485 (1986).

Gaisser, T.K., Steigman, G. and Tilav, S., *Phys.Rev.D* **34**, 2206 (1986).

Goodman, M. and Witten, E., *Phys. Rev.* **D 31**, 3059 (1985).

Hegyi, D. and Olive, K.A.,*Ap. J.*, **303**, 56 (1987).

Hut,P. and White, S.D.M., *Nature*, **310**, 637 (1987).

Krauss, L.M., Freese, K., Spergel, D.N. and Press, W.H., *Ap.J.*, **299**, 1001 (1985).

Krauss, L.M., Srednicki, M. and Wilczek, K., *Phys. Rev.* **D 33**, 2079 (1986).

Lanou, R.E., Maris, H.J. and Seidel, G.M. *Phys.Rev.Lett* **58**, 2498 (1987).

Loh, E.O. and Spillar, E.J., *Ap. J. Lett.*, **301**, L1 (1986).

Marthoff, C.J., *Science*, **237**, 507 (1987).

Milgrom, M. and Bekenstein, J., in *Dark Matter in the Universe*, ed. Kormendy, J. and Knapp, G.R., Reidel Pub. Co., 1987.

Ostriker J.P., Peebles, P.J.E. and Yahil, A., *Ap. J. Lett.*, **193**, L1 (1974).

Ostriker, J.P. and Caldwell, N., *Ap. J.*, **251**, 61 (1983).

Press, W.H. and Spergel, D.N., *Ap.J.*, **296**, 663 (1985).

Ruben, V.C., Burstein, D., Ford, W.K., Jr., and Thonnard, N., *Ap.J.*, **289**, 81 (1987).

Sadoulet, Bernard, "Prospects for detecting dark matter particles by elastic scattering" LBL preprint LBL-23098, Presented at 13th Texas Symp. on Relativistic Astrophysics, Chicago, IL, Dec 14-19, 1986.

Sadoulet, B., Rich, J., Spiro, M. & Caldwell, D.O., LBL-23779 (1987).

Sarazin, C., *Rev. Mod. Phys.*, **58**,1 (1986).

Sikivie, P., *Phys. Rev.* , **D32**, 2988 (1985).

Silk, J., Olive, K.A. and Srednicki, M., *Phys. Rev. Lett*, **55**, 257.

Silk, J. and Srednicki, M., *Phys. Rev. Lett*, **53**, 624.

Spergel, D.N. and Bahcall, J.N., *Phys. Lett.*, **B200**, 366 (1988).

Spergel, D.N., Weinberg, D.H., and Gott, J.R., III, *Phys. Rev.*, **D38**, xxxx (1988).

Srednicki, M., Olive, K.A. and Silk, J., *Nucl.Phys.B* **279** , 804.

Stecker, F.W., Rudaz, S. and Walsh, T.F. *Phys. Rev. Lett*, **55**, 2622 (1985).

Tremaine, S. and Gunn, J.E., *Phys. Rev. Lett.*, **42**, 407 (1979).

Turner, M.S., *Phys. Rev. Lett.*, **60**, 1797 (1988).

Wasserman, Ira, *Phys. Rev. D*, **33**, 2071 (1986).

Weinberg, S. *Phys. Rev. Lett.*, **40**, 223 (1978).

Wilczek, F. *Phys. Rev. Lett*, **40**, 279 (1978).

Wilkerson, J.F., et al., *Phys. Rev. Lett.*, **58**, 2023 (1987).

Yang, J., Turner, M.S., Steigman, G., Schramm D.N. and Olive, K., *Ap. J.*, **281**, 493 (1984).

COSMOLOGY AND PARTICLE PHYSICS*

Michael S. Turner
NASA/Fermilab Astrophysics Center
Fermi National Accelerator Laboratory
Batavia, Illinois 60510
and
Departments of Physics and Astronomy and Astrophysics
Enrico Fermi Institute
The University of Chicago
Chicago, Illinois 60637

INTRODUCTION

In the past five years or so progress in both elementary particle physics and in cosmology has become increasingly dependent upon the interplay between the two disciplines. On the particle physics side, the $SU(3)_C \times SU(2)_L \times U(1)_Y$ model seems to very accurately describe the interactions of quarks and leptons at energies below, say, 10^3 GeV. At the very least, the so-called standard model is a satisfactory, effective low energy theory. The frontiers of particle physics now involve energies of much greater than 10^3 GeV—energies which are not now available in terrestrial accelerators, nor are ever likely to be available in terrestrial accelerators. For this reason particle physicists have turned both to the early Universe with its essentially unlimited energy budget (up to 10^{19} GeV) and high particle fluxes (up to $10^{107} cm^{-2} s^{-1}$), and to various unique, contemporary astrophysical environments (centers of main sequence stars where temperatures reach 10^8 K, neutron stars where densities reach $10^{14} - 10^{15} g cm^{-3}$, our galaxy whose magnetic field can impart 10^{11} GeV to a Dirac magnetic charge, etc.) as non-traditional laboratories for studying physics at very high energies and very short distances.

On the cosmological side, the hot big bang model, the so called standard model of cosmology, seems to provide an accurate accounting of the history of the Universe from about 10^{-2} s after 'the bang' when the temperature was about 10 MeV, until today, some 10-20 billion years after 'the bang' and a temperature of about 3 K ($\simeq 3 \times 10^{-13}$ GeV). Extending our understanding to earlier times and higher temperatures, requires knowledge about the fundamental particles (presumably quarks and leptons) and their interactions at very high energies. For this reason, progress in cosmology has become linked to progress in elementary particle physics.

In these lectures I will try to illustrate the two-way nature of the interplay between these fields by focusing on a few selected topics. In Lecture 1 I will review the standard cosmology, especially concentrating on primordial nucleosynthesis, and discuss how the standard cosmology has been used to place constraints on the properties of various particles. Grand Unification makes two striking predictions: (1)

---
* Reprinted from *The Early Universe*, (Riedel, Holland) pp 19–114.

baryon number nonconservation; (2) the existence of stable, superheavy magnetic monopoles. Both have had great cosmological impact. In Lecture 2 I will discuss baryogenesis, the very attractive scenario in which the B, C, CP violating interactions in GUTs provide a dynamical explanation for the predominance of matter over antimatter, and the present baryon-to-photon ratio. Baryogenesis is so cosmologically attractive, that in the absence of observed proton decay it has been called 'the best evidence for some kind of unification'. Monopoles on the other hand started out as a cosmological disaster; however, efforts to solve the problem of monopole overproduction in the standard cosmology led to one of the most exciting payoffs of the Inner Space/Outer Space connection: the inflationary Universe scenario. In the third lecture I will discuss how the very early ($t \lesssim 10^{-34}$sec) dynamical evolution of a very weakly coupled scalar field which is initially displaced from the minimum of its potential has the potential to explain a handful of very fundamental cosmological facts—facts which can be accommodated by the standard cosmology, but which are not 'explained' by it.

By selecting just a few topics I have left out some other very important and interesting topics—monopoles and astrophysics, axions, galaxy formation, the deconfinement/chiral transition of QCD, supersymmetry/supergravity and cosmology, superstrings, and cosmology in extra dimensions—to mention just a few. I refer the interested reader to other reviews of the early Universe (refs. 1) and papers on some of these topics (refs. 2 and 3).

LECTURE 1:
THE STANDARD COSMOLOGY AND ITS SUCCESSES

The hot, big bang cosmology—the so-called standard cosmology, neatly accounts for the (Hubble) expansion of the Universe, the 2.7 K microwave background radiation (see Figs. 1.1, 1.2), and, through primordial nucleosynthesis, the cosmic abundances of the light elements D and ^4He (and in all likelihood ^3He and ^7Li as well; see Fig. 1.3). The most distant galaxies and QSO's observed to date have redshifts in excess of 3—the current record holders are: for galaxies $z = 3.2$ (ref. 4) and for QSO's $z = 4.0$ (ref. 5). The light we observe from an object with redshift $z = 3$ left that object only 1–2 Byr after the bang. Observations of even the most distant galaxies and QSO's are consistent with the standard cosmology, thereby testing it back to times as early as 1 Byr (see, e.g., ref. 6). The surface of last scattering for the microwave background is the Universe at an age of a few $\times 10^5$ yrs and temperature of about 3000 K. Measurements at wavelengths from 0.05 cm to 80 cm indicate that it is consistent with being radiation from a blackbody of temperature 2.75 K ± 0.05 K (see Fig. 1.1 and ref. 7). Measurements of the isotropy indicate that the temperature is uniform to a part in 1000 on angular scales ranging from $10''$ to $180°$—to a part in 10^4 after the dipole component is removed (see Fig. 1.2 and ref. 8). The observations of the microwave background test the standard cosmology back to times as early as 100,000 yrs. According to the standard cosmology, when the Universe was 0.01 sec–300 sec old, corresponding to temperatures of $\simeq$ 10 MeV– 0.1 MeV, conditions were right for 'the synthesis of a number of light nuclei. The predicted abundances of D, ^3He, ^4He, and ^7Li are consistent with their observed abundances provided that the baryon-to-photon ratio is

$$\eta \equiv n_b/n_\gamma \simeq (4-7) \times 10^{-10} \tag{1.1}$$

The baryon-to-photon ratio and the fraction of critical density contributed by baryons are related by: $\Omega_b h^2/T_{2.7}^3 \simeq 3.53 \times 10^7 \eta$ where $T_{2.7}$ is the microwave temperature in units of 2.7K and h is the present value of the Hubble constant in units of 100 km s^{-1} Mpc^{-1}. The allowed range for η corresponds to: $0.014 \lesssim \Omega_b h^2/T_{2.7}^3 \lesssim 0.025$, implying that baryons alone cannot provide the closure density. The concordance of theory and observation for D and ^{4}He is particularly compelling evidence in support of the standard cosmology as there are no known contemporary astrophysical sites which can simultaneously account for the primordial abundances of both these isotopes (see Fig. 1.3; see ref. 9 for further discussion of primordial nucleosynthesis). In sum, all the available evidence indicates that the standard cosmology provides an accurate accounting of the evolution of the Universe from 0.01 sec after the bang until today, some 15 or so Byr later—quite a remarkable achievement!

I will now briefly review the standard cosmology (more complete discussions of the standard cosmology are given in ref. 6). Throughout I will use high energy physics units, where $\hbar = k = c = 1$. The following conversion factors may be useful.

$$1\text{GeV}^{-1} = 0.197 \times 10^{-13}\text{cm}$$
$$1\text{GeV}^{-1} = 0.658 \times 10^{-24}\text{sec}$$
$$1\text{GeV} = 1.160 \times 10^{13}K$$
$$1\text{GeV}^4 = 2.32 \times 10^{17}\text{gcm}^{-3}$$
$$1M_\odot = 1.99 \times 10^{33}g \simeq 1.2 \times 10^{57}\text{baryons}$$
$$1\text{pc} = 3.26\text{ light} - \text{year} \simeq 3.09 \times 10^{18}\text{cm}$$
$$1\text{Mpc} = 3.09 \times 10^{24}\text{cm}$$
$$G_N = 6.673 \times 10^{-8}\text{cm}^3 g^{-1}\text{sec}^{-2} \equiv m_{pl}^{-2}$$
$$(m_{pl} = 1.22 \times 10^{19}\text{GeV})$$

On large scales ($\gg$ 100Mpc) the Universe is isotropic and homogeneous, as evidenced by the uniformity of the 2.7 K background radiation, the x-ray background, and counts of galaxies and radio sources, and so the standard cosmology is based on the maximally-symmetric Robertson–Walker line element

$$ds^2 = -dt^2 + R^2(t)[dr^2/(1-kr^2) + r^2 d\theta^2 + r^2 \sin^2\theta d\phi^2] \tag{1.2}$$

where ds^2 is the square of the proper separation between two space-time events, k is the curvature signature (and can, by a suitable rescaling of R, be set equal to -1, 0, or

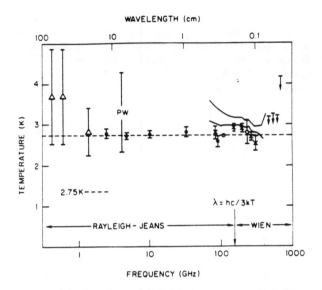

Figure 1.1: Summary of microwave background temperature measurements from $\lambda \simeq 0.05$ to 80 cm (see refs. 7). Measurements indicate that the background radiation is well described as a 2.75 ± 0.05K blackbody. PW denotes the discovery measurement of Penzias and Wilson.

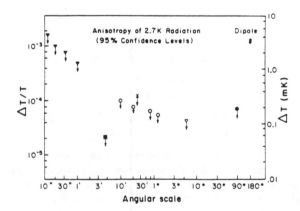

Figure 1.2: Summary of microwave background anisotropy measurements on angular scales from $10''$ to $180°$ (see ref. 8). With the exception of the dipole measurements, the rest are 95% confidence upper limits to the anisotropy.

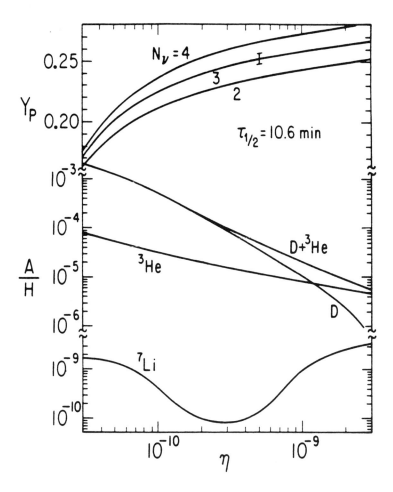

Figure 1.3: Big bang nucleosynthesis predictions for the primordial abundances of D, ^{3}He, ^{4}He, and ^{7}Li. Y_p = mass fraction of ^{4}He, shown for N_ν = 2, 3, 4 light neutrino species. Present observational data suggest: $0.23 \leq Y_p \leq 0.25$, $(D/H)_p \geq 1 \times 10^{-5}$, $[(D + ^3He)/H]_p \leq 10^{-4}$, and $(^7Li/H)_p \simeq (1.1 \pm 0.4) \times 10^{-10}$. Concordance requires $\eta \simeq (4-7) \times 10^{-10}$. For further discussion see ref. 9.

+1), and $R(t)$ is the cosmic scale factor. The expansion of the Universe is embodied in $R(t)$—as $R(t)$ increases all proper (i.e., physical—as measured by meter sticks) distances scale with $R(t)$. The coordinates r, θ, and φ are comoving coordinates: test particles initially at rest will have constant comoving coordinates, and the velocity of NR test particles moving with respect to the comoving coordinates decrease ($\propto R(t)^{-1}$). The distance between two objects comoving with the expansion, e.g., two galaxies, simply scales up with $R(t)$. The momentum of any freely-propagating particle decreases as $1/R(t)$. In particular, the wavelength of a photon $\lambda \propto R(t)$, i.e., is redshifted by the expansion of the Universe

The coordinate distance at which curvature effects become noticeable is $|k|^{-1/2}$, which corresponds to the physical (or proper) distance

$$R_{curv} \simeq R(t)|k|^{-1/2} \qquad (1.3)$$

—which one might call the curvature radius of the Universe. Note that R_{curv} also just scales with the cosmic scale factor $R(t)$.

The evolution of the cosmic scale factor and of the stress energy in the Universe are governed by the Friedmann equations:

$$H^2 \equiv (\dot{R}/R)^2 = 8\pi G\rho/3 - k/R^2 \qquad (1.4)$$

$$d(\rho R^3) = -p d(R^3) \qquad (1.5)$$

where ρ is the total energy density and p is the isotropic pressure. [The assumption of isotropy and homogeneity require that the stress-energy tensor take on the perfect fluid form: $T^\mu_\nu = diagonal(-\rho, p, p, p)$.] Because $\rho \propto R^{-n}$ ($n = 3$ for matter, $n = 4$ for radiation) it follows from Eqn(4) that model Universes with $k < 0$ expand forever, while those with $k > 0$ must necessarily recollapse.

The expansion rate H (also known as the Hubble parameter) sets the characteristic timescale for the growth of $R(t)$: H^{-1} is the e-folding time for R. The present value of H is

$$H_0 = 100h \text{ km sec}^{-1}\text{Mpc};$$
$$\simeq h(10^{10}\text{yr})^{-1}$$

where the observational data strongly suggest that $0.4 \leq h \leq 1$ (ref. 10).

The sign of the spatial curvature k—and the ultimate fate of the Universe can be determined from measurements of ρ and H:

$$k/H^2 R^2 = \rho/(3H^2/8\pi G) - 1$$
$$\equiv \Omega - 1 \qquad (1.6)$$

where $\Omega = \rho/\rho_{crit}$ and $\rho_{crit} = 1.88h^2 \times 10^{-29}\text{gcm}^{-3} = 1.05 \times 10^4 h^2 \text{eVcm}^{-3}$. The curvature radius, R_{curv}, is related to Ω by

$$(R_{curv}/H^{-1})^2 = 1/|\Omega - 1| \qquad (1.7)$$

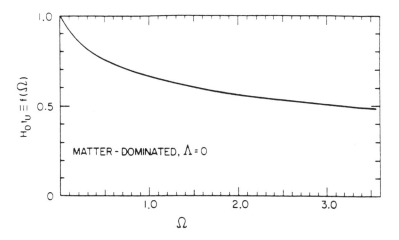

Figure 1.4a: The age of a matter-dominated, $\Lambda = 0$ model universe in Hubble units, $f(\Omega) = H_0 t_u$, as a function of Ω.

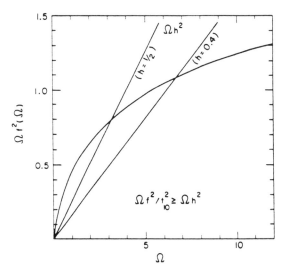

Figure 1.4b: The functions Ωf^2 and Ωh^2 ($h = 0.4$ and 0.5). The function $\Omega f^2/t_{10}^2$ bounds Ωh^2 from above. For $t_{10} \geq 1$ and $h \geq 0.4(0.5)$, this implies $\Omega h^2 \leq 1.1(0.8)$. The age of the Universe $t_u = t_{10} 10$ Gyr.

A reliable and definitive determination of Ω has thus far eluded cosmologists. Based upon the luminous matter in the Universe (which is relatively easy to keep track of) we can set a lower bound to Ω

$$\Omega \geq \Omega_{LUM} \simeq 0.01$$

Based on dynamical techniques—which all basically involve Kepler's third law in one guise or another, the observational data seem to indicate that the material that clusters with visible galaxies on scales $\leq$ 10-30 Mpc accounts for

$$\Omega_{GAL} \simeq 0.1 - 0.3$$

Although Ω can, in principle, be determined by measurements of the deceleration parameter q_0

$$\begin{aligned} q_0 &\equiv -(\ddot{R}/R)/H^2, \\ &= \Omega(1 + 3p/\rho)/2, \end{aligned} \quad (1.8)$$

because of the difficulty of reliably determining q_0, the observations probably only restrict Ω to be less than a few[10]. [For a more thorough discussion of the amount of matter in the Universe see ref. 11.]

The best upper limit to Ω comes from the age of the Universe. The age of the Universe is related to the Hubble time H^{-1} by

$$t_u = f(\Omega) H_0^{-1} \quad (1.9)$$

where $f(\Omega)$ is a monotonically decreasing function of Ω; $f(0) = 1$ and $f(1) = 2/3$ for a matter-dominated Universe and $1/2$ for a radiation-dominated Universe. The dating of the oldest stars and the elements strongly suggest that the Universe is at least 10 Byr old—the best estimate being around 15 Byr old[12]. From Eqn(9) and $t_u \geq t_{10} 10$ Byr it follows that $\Omega f^2/t_{10}^2 \geq \Omega h^2$. The function Ωf^2 is monotonically increasing and bounded above by $\pi^2/4$, implying that independent of h, $\Omega h^2 \leq 2.5/t_{10}^2$. Requiring $h \geq 0.4$ and $t_{10} \geq 1$, it follows that $\Omega h^2 \leq 1.1$ (see Fig. 1.4).

The energy density of the Universe quite naturally splits up into that contributed by relativistic particles—today the microwave photons and cosmic neutrino backgrounds, and that contributed by non-relativistic particles—baryons and whatever else! The energy density contributed by non-relativistic particles decreases as $R(t)^{-3}$—just due to the increase in the proper volume of the Universe, while that of relativistic particles varies as $R(t)^{-4}$—the additional factor of R being due to the fact that the momenta of relativistic particles are redshifted by the expansion. [Both of these results follow directly from Eqn(1.5).]

The energy density contributed by relativistic particles at temperature T is

$$\rho_R = g_*(T) \frac{\pi^2}{30} T^4 \quad (1.10)$$

where $g_*(T)$ counts the effective number of degrees of freedom (weighted by their temperature) of all the relativistic particle species (those with $m \ll T$):

$$g_*(T) = \sum_{Bose} g_B(T_i/T)^4 + 7/8 \sum_{Fermi} g_F(T_i/T)^4, \quad (1.11)$$

here T_i is the temperature of the species i, and T is the photon temperature.

Today the energy density contributed by relativistic particles (photons and three neutrino species) is very small ($g_* = 3.36$)

$$\Omega_{\gamma\,3\nu}h^2 \simeq 4 \times 10^{-5} T_{2.7}^4$$

However, because $\rho_R \propto R^{-4}$, while $\rho_{NR} \propto R^{-3}$, at early times the energy density contributed by relativistic particles dominated that of non-relativistic particles. To be specific, the Universe was radiation-dominated for

$$t \leq t_{EQ} \simeq 4 \times 10^{10} \text{sec} (\Omega h^2)^{-2} T_{2.7}^6,$$
$$R \leq R_{EQ} \simeq 4 \times 10^{-5} R_{today} (\Omega h^2)^{-1} T_{2.7}^4,$$
$$T \geq T_{EQ} \simeq 5.8 \text{eV} (\Omega h^2) T_{2.7}^3.$$

Therefore, at very early times Eqn(1.4) simplifies to

$$H = (\dot{R}/R) = (4\pi^3 g_*/45)^{1/2} T^2/m_{pl},$$
$$= 1.66 g_*^{1/2} T^2/m_{pl} \qquad (1.12)$$

[Note since the curvature term varies as $R(t)^{-2}$ it too is negligible compared to the energy density in relativistic particles.] For reference, $g_*(\text{few MeV}) = 10.75$ (γ, $e^{\pm}$, $3\nu\bar{\nu}$); $g_*(100\text{GeV}) \simeq 110$ (γ, $8G$, $W^{\pm}Z$, 3 families of quarks and leptons, and 1 Higgs doublet).

So long as thermal equilibrium is maintained, the second Friedmann equation, Eqn(1.5), implies that the entropy per comoving volume, $S \propto sR^3$, remains constant. Here s is the entropy density which is dominated by the contribution from relativistic particles, and is

$$s = (\rho + p)/T \simeq (2\pi^2/45) g_* T^3. \qquad (1.13)$$

The entropy density is just proportional to the number density of relativistic particles. Today the entropy density is just 7.04 times the number density of photons. The constancy of S means that $s \propto R^{-3}$, or that the ratio of any number density to s is just proportional to the number of that species per comoving volume ($N \propto nR^3 \propto n/s$). The baryon number-to-entropy ratio is

$$n_B/s \simeq (1/7)\eta \simeq (6-10) \times 10^{-11}$$

and since today the number density of baryons is much greater than that of antibaryons, this ratio is also the net baryon number per comoving volume—which is conserved so long as the rate of baryon-number non-conserving reactions is small.

The constancy of S implies that

$$T \propto g_*(T)^{-1/3} R(t)^{-1}. \qquad (1.14)$$

Whenever g_* is constant, this means that $T \propto R(t)^{-1}$. Together with Eqn(1.12) this gives

$$R(t) = R(t_0)(t/t_0)^{1/2},$$
$$t \simeq 1/2H^{-1} \simeq 0.3 g_*^{-1/2} m_{pl}/T^2, \qquad (1.15)$$
$$\simeq 2.4 \times 10^{-6} \text{sec } g_*^{-1/2} (T/\text{GeV})^{-2}.$$

Finally, let me mention one more important feature of the standard cosmology, the existence of particle horizons. In the standard cosmology the distance a photon could have traveled since the bang is finite, meaning that at a given epoch the Universe is comprised of many causally-distinct domains. Photons travel on paths characterized by $ds^2 = 0$; for simplicity and without loss of generality consider a trajectory with $d\varphi = d\theta = 0$. The coordinate distance traversed by a photon since 'the bang' is

$$\int_0^t dt'/R(t')$$

which corresponds to the physical distance (measured at time t)

$$d_H(t) = R(t) \int_0^t dt'/R(t'). \qquad (1.16)$$

If $R(t) \propto t^n$ and $n < 1$, then the horizon distance $d_H(t)$ is finite and $d_H(t) = t/(1-n) = nH^{-1}/(1-n) \simeq t$.

Note that even if $d_H(t)$ diverges (e.g., if $R(t) \propto t^n$ with $n > 1$), the Hubble radius H^{-1} still sets the scale of the 'Physics Horizon'. All physical distances scale with $R(t)$. Thus microphysical processes operating on a timescale $\gtrsim H^{-1}$ will have their effects distorted by the expansion, strongly suggesting that a coherent microphysical process can only operate over a time interval of order H^{-1}. Then, causally-coherent microphysical processes can only operate on distances $\leq$ the Hubble radius, H^{-1}. The intuitive notion that the Hubble radius acts as the 'Physics Horizon' is borne out quantitatively time and time again, and so it is useful to think of H^{-1} as the maximum scale for microphysical processes.

During the radiation-dominated era $n = 1/2$ and $d_H(t) = 2t$; the entropy and baryon number within the horizon at a given time are easily computed:

$$S_{HOR} = (4\pi/3)t^3 s,$$
$$\simeq 0.05 g_*^{-1/2} (m_{pl}/T)^3,$$
$$N_{B-HOR} = (n_B/s) S_{HOR},$$
$$\simeq 10^{-12} (m_{pl}/T)^3,$$
$$\simeq 10^{-2} M_\odot (T/\text{MeV})^{-3}.$$

We can compare these numbers to the entropy and baryon number contained within the present horizon volume:

$$S_U \simeq 10^{88},$$
$$N_{BU} \simeq 10^{78}.$$

Evidently, in the standard cosmology the comoving volume which corresponds to the part of the Universe which is presently observable contained many, many horizon volumes at early times. This is an important point to which we shall return shortly.

Although our verifiable knowledge of the early history of the Universe only takes us back to $t \simeq 10^{-2}$ s and $T \simeq 10$ MeV (the epoch of primordial nucleosynthesis), nothing in our present understanding of the laws of physics suggests that it is unreasonable to extrapolate back to times as early as $\simeq 10^{-43}$ s and temperatures as high as $\simeq 10^{19}$ GeV. At high energies the interactions of quarks and leptons are asymptotically free (and/or weak), justifying the dilute gas approximation made in Eqn(1.10). At energies below 10^{19} GeV, quantum corrections to general relativity are expected to be small. I hardly need to remind the reader that 'reasonable' does not necessarily mean 'correct'. Making this extrapolation, I have summarized 'The Complete History of the Universe' in Fig. 1.5.

1.1. PRIMORDIAL NUCLEOSYNTHESIS

At present the most stringent test of the standard cosmology is big bang nucleosynthesis. Here I will briefly review primordial nucleosynthesis, discuss the concordance of the predictions with the observations, and mention one example of how primordial nucleosynthesis has been used as a probe of particle physics—constraining the number of light neutrino species.

Two fundamental assumptions underlie big bang nucleosynthesis: General Relativity is valid and the Universe was once hotter than a few MeV. An additional assumption (which, however, is not necessary) is that the lepton number of the Universe, $n_L/n_\gamma = (n_{e^-} - n_{e^+})/n_\gamma + (n_\nu - n_{\bar{\nu}})/n_\gamma \simeq \eta + (n_\nu - n_{\bar{\nu}})/n_\gamma$, like the baryon number ($\eta \simeq 4 - 7 \times 10^{-10}$), is small ($\lesssim 1$). Having swallowed these assumptions, the rest follows like 1-2-3.

Frame 1: $t \simeq 10^{-2}$sec, $T \simeq 10$MeV. The energy density of the Universe is dominated by relativistic species: $\gamma, e^+e^-, \nu_i\bar{\nu}_i$ ($i = e, \mu, \tau, \ldots$); $g_* \simeq 10.75$ (assuming 3 neutrino species). Thermal equilibrium is maintained by weak interactions ($e^+ + e^- \leftrightarrow \nu_i + \bar{\nu}_i, e^+ + n \leftrightarrow p + \bar{\nu}_e, e^- + p \leftrightarrow n + \nu_e$) as well as electromagnetic interactions ($e^+ + e^- \leftrightarrow \gamma + \gamma, \gamma + p \leftrightarrow \gamma + p$, etc.), both of which are occurring rapidly compared to the expansion rate $H = \dot{R}/R$. Thermal equilibrium implies that $T_\nu = T_\gamma$ and that $n/p = \exp(-\Delta m/T)$; here n/p is the neutron to proton ratio and $\Delta m = m_n - m_p$. At high temperatures $\gtrsim 0.3$ MeV) all the light isotopes are in nuclear statistical equilibrium with very small abundances, due to the very high photon to nucleon ratio, $\eta^{-1} \simeq 10^{10}$:

$$n_D/n_B \simeq \eta(T/m_N)^{3/2}\exp(2.2\,\text{MeV}/T),$$
$$\simeq 10^{-13}$$
$$n_{He}/n_B \simeq 0.2\eta^3(T/m_N)^{9/2}\exp(28\,\text{MeV}/T),$$
$$\simeq 4 \times 10^{-39},$$

where the abundances were evaluated for $T \simeq 10$ MeV, and 2.2 MeV and 28 MeV

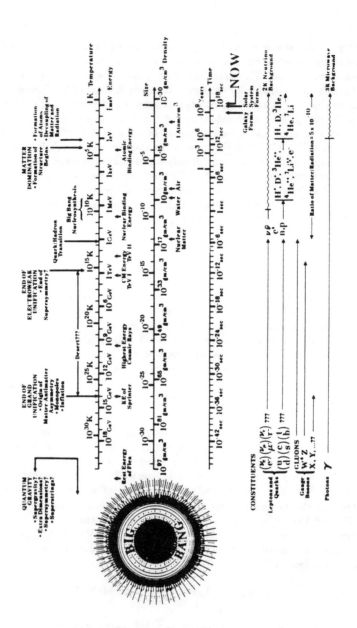

Figure 1.5: 'The Complete History of the Universe' according to the standard cosmology and currently fashionable ideas in elementary particle physics.

are the binding energies of D and ^4He respectively. The fact that very little nucleosynthesis has taken place when $T \simeq 1$ MeV is clearly traceable to the large value of η^{-1}. Of course, had the binding energies of the light nuclei been more like 30-100 MeV rather than 3-30 MeV nucleosynthesis would already be occurring when $T \simeq 10$ MeV. Since D is the first stepping stone in the path of nucleosynthesis, the absence of substantial nucleosynthesis until much lower temperatures is usually blamed on its small equilibrium abundance and binding energy, and this phenomenon is often referred to as 'the deuterium bottleneck'.

Frame 2: $t \simeq 1$ sec, $T \simeq 1$ MeV. At about this temperature the weak interaction rates become slower than the expansion rate and thus weak interactions effectively cease occurring. The neutrinos decouple and thereafter expand adiabatically ($T_\nu \propto R^{-1}$). This epoch is the surface of last scattering for the neutrinos; detection of the cosmic neutrino seas would allow us to directly view the Universe as it was 1 sec after 'the bang'. From this time forward the neutron to proton ratio no longer 'tracks' its equilibrium value, but instead 'freezes out' a value $\simeq 1/6$, decreasing very slowly due to occasional free neutron decays and other β-reactions. A little bit later ($T \simeq m_e/3$), the $e^\pm$ pairs annihilate and transfer their entropy to the photons, heating the photons relative to the neutrinos, so that from this point on $T_\nu \simeq (4/11)^{1/3} T_\gamma$. The so-called 'deuterium bottleneck' continues to operate, preventing nucleosynthesis.

Frame 3: $t \simeq 200$ sec, $T \simeq 0.1$ MeV. If ^4He were to track its equilibrium abundance, then n_{He}/n_B would reach order unity at a temperature of about 0.3 MeV. However, the equilibrium abundances of ^3H, ^3He, and D are too small at this temperature to allow ^4He to be produced rapidly enough to achieve its equilibrium value. At a temperature of about 0.1 MeV, there is sufficient ^3H, ^3He, and D to produce ^4He at a rate comparable to the expansion rate, and nucleosynthesis begins in earnest. Essentially all the neutrons present are quickly incorporated into ^4He nuclei. As the D and ^3He are depleted, the rates at which they are burned into ^4He fall ($\Gamma \propto n(^3He)$ or $n(D)$); eventually they drop below the expansion rate and so trace amounts of D and ^3He remain unburned. Substantial nucleosynthesis beyond ^4He is prevented by the lack of stable isotopes with A = 5 and 8, and by coulomb barriers. A small amount of ^7Li is synthesized by $^4He(t,\gamma)^7Li$ (for $\eta \lesssim 3 \times 10^{-10}$) and by $^4He(^3He,\gamma)^7Be$ followed by the eventual β-decay of ^7Be (via electron capture) to ^7Li (for $\eta \gtrsim 3 \times 10^{-10}$).

The nucleosynthetic yields depend upon η, N_ν (which I will use to parameterize the number of light ($\lesssim 1$ MeV) species present, other than γ and $e^\pm$), and in principle all the nuclear reaction rates which go into the reaction network. In practice, most of the rates are known to sufficient precision that the yields only depend upon a few rates. ^4He production depends only upon η, N_ν, and $\tau_{1/2}$, the neutron half-life, which determines the rates for all the weak processes which interconvert neutrons and protons. The mass fraction Y_p of ^4He produced increases monotonically with increasing values of η, N_ν, and $\tau_{1/2}$—a fact which is simple to understand. Larger η means that the 'deuterium bottleneck' breaks earlier, when the value of

n/p is larger. More light species (i.e., larger value of N_ν) increases the expansion rate (since $H \propto (G\rho)^{1/2}$), while a larger value of $\tau_{1/2}$ means slower weak interaction rates ($\propto \tau_{1/2}^{-1}$)—both effects cause the weak interactions to freeze out earlier, when n/p is larger. The yield of ^4He is determined by the n/p ratio when nucleosynthesis commences, $Y_p \simeq 2(n/p)/(1+n/p)$, so that a higher n/p ratio means more ^4He is synthesized. At present the value of the neutron half-life is only known to an accuracy of about 2%: $\tau_{1/2} = 10.6$ min $\pm$ 0.2 min. Since ν_e and ν_μ are known (from laboratory measurements) to be light, $N_\nu \geq 2$. Based upon the luminous matter in galaxies, η is known to be $\gtrsim 0.3 \times 10^{-10}$. If all the mass in binary galaxies and small groups of galaxies (as inferred by dynamical measurements) is baryonic, then η must be $\gtrsim 2 \times 10^{-10}$.

To an accuracy of about 10%, the yields of D and ^3He only depend upon η, and decrease rapidly with increasing η. Larger η corresponds to a higher nucleon density and earlier nucleosynthesis, which in turn results in less D and ^3He remaining unprocessed. Because of large uncertainties in the rates of some of the reactions which create and destroy ^7Li, the predicted primordial abundance of ^7Li is only accurate to within about a factor of 2.

In 1946 Gamow[13] suggested the idea of primordial nucleosynthesis. In 1953, Alpher, Follin, and Herman[14] all but wrote a code to determine the primordial production of ^4He. Peebles (in 1966) and Wagoner, Fowler, and Hoyle (in 1967) wrote codes to calculate the primordial abundances[15]. Yahil and Beaudet[16] (in 1976) independently developed a nucleosynthesis code and also extensively explored the effect of large lepton number $(n_\nu - n_{\bar\nu} \simeq 0(n_\gamma))$ on primordial nucleosynthesis. Wagoner's 1973 code[17] has become the 'standard code' for the standard model. In 1981 the reaction rates were updated by Olive et al.[18], the only significant change which resulted was an increase in the predicted ^7Li abundance by a factor of 0(3). In 1982 Dicus et al.[19] corrected the weak rates in Wagoner's 1973 code for finite temperature effects and radiative/coulomb corrections, which led to a systematic decrease in Y_p of about 0.003. Figs. 1.3, 1.6 show the predicted abundances of D, ^3He, ^4He, and ^7Li, as calculated by the most up to date version of Wagoner's 1973 code[20]. The numerical accuracy of the predicted abundances is about 1%. Now let me discuss how the predicted abundances compare with the observational data. [This discussion is a summary of the collaborative work in ref. 20.]

The abundance of D has been determined in solar system studies and in UV absorption studies of the local ($\lesssim 200$ pc) interstellar medium (ISM). The solar system determinations are based upon measuring the abundances of deuterated molecules in the atmosphere of Jupiter and inferring the pre-solar (i.e., at the time of the formation of the solar system) D/H ratio from meteoritic and solar data on the abundance of ^3He. These determinations are consistent with a pre-solar value of $(D/H) \simeq (2 \pm 0.5) \times 10^{-5}$. An average ISM value for $(D/H) \simeq 2 \times 10^{-5}$ has been derived from UV absorption studies of the local ISM with individual measurements spanning the range $(1-4) \times 10^{-5}$. Note that these measurements are consistent with the solar system determinations of D/H.

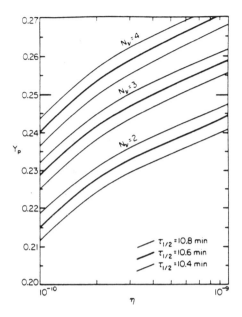

Figure 1.6: The predicted primordial abundance of ^4He. Note that Y_p increases with increasing values of $\tau_{1/2}, \eta$, and N_ν. Hence lower bounds to η and $\tau_{1/2}$ and an upper bound to Y_p imply an upper bound to N_ν. Taking $\tau_{1/2} \geq 10.4$ min, $\eta \geq 4 \times 10^{-10}$ (based on $D + {}^3He$ production), and $Y_p \lesssim 0.25$, it follows that N_ν must be ≤ 4.

The deuteron being very weakly-bound is easily destroyed and hard to produce, and to date, it has been difficult to find an astrophysical site where D can be produced in its observed abundance[21]. Thus, it is generally accepted that the presently-observed deuterium abundance provides a *lower* bound to the primordial abundance. Using $(D/H)_p \gtrsim 1 \times 10^{-5}$ it follows that η must be less than about 10^{-9} in order for the predictions of primordial nucleosynthesis to be concordant with the observed abundance of D. [Note: because of the rapid variation of $(D/H)_p$ with η, this upper bound to η is rather insensitive to the precise lower bound to $(D/H)_p$ used.] This implies an upper bound to $\Omega_b : \Omega_b \lesssim 0.035h^{-2}T_{2.7}^3 K \leq 0.19$—baryons alone cannot close the Universe. One would like to also exploit the sensitive dependence of $(D/H)_p$ upon η to derive a *lower* bound to η for concordance; this is not possible because D is so easily destroyed. However, as we shall soon see, this end can be accomplished instead by using the abundances of both D and ^3He.

The abundance of ^3He has been measured in solar system studies and by observations of the $^3He^+$ hyperfine line in galactic HII regions (the analog of the 21 cm line of H). The abundance of ^3He in the solar wind has been determined by analyzing gas-rich meteorites, lunar soil, and the foil placed upon the surface of the moon by the Apollo astronauts. Since D is burned to ^3He during the sun's approach to the main sequence, these measurements represent the pre-solar sum of D and ^3He. These determinations of $D + {}^3He$ are all consistent with a pre-solar $[(D+{}^3He)/H] \simeq (4.0\pm 0.3) \times 10^{-5}$. Earlier measurements of the $^3He^+$ hyperfine line in galactic HII regions and very recent measurements lead to derived present abundances of $^3He : {}^3He/H \simeq (3-20) \times 10^{-5}$. The fact that these values are higher than the pre-solar abundance is consistent with the idea that the abundance of ^3He should increase with time due to the stellar production of ^3He by low mass stars.

^3He is much more difficult to destroy than D. It is very hard to efficiently dispose of ^3He without also producing heavy elements or large amounts of ^4He (environments hot enough to burn ^3He are usually hot enough to burn protons to ^4He). In ref. 20 we have argued that in the absence of a Pop III generation of very exotic stars which process essentially all the material in the Universe and in so doing destroy most of the ^3He without overproducing ^4He or heavy elements, ^3He can have been astrated (i.e. reduced by stellar burning) by a factor of no more than $f_a \simeq 2$. [The youngest stars, e.g., our sun, are called Pop I; the oldest observed stars are called Pop II. Pop III refers to a yet to be discovered, hypothetical first generation of stars.] Using this argument and the inequality

$$[(D + {}^3He)/H]_p \leq (D/H)_{ps} + f_a({}^3He/H)_{ps}$$
$$\leq (1-f_a)(D/H)_{ps} + f_a(D + {}^3He/H)_{ps};$$

the pre-solar abundances (noted by 'ps') of D and $D + {}^3He$ can be used to derive an upper bound to the primordial abundance of $D + {}^3He : [(D + {}^3He)/H]_p \lesssim 8 \times 10^{-5}$. [For a very conservative astration factor, $f_a \simeq 4$, the upper limit becomes 13×10^{-5}.] Using 8×10^{-5} as an upper bound on the primordial $D + {}^3He$ production implies that for concordance, η must be greater than 4×10^{-10} (for the upper bound of 13×10^{-5}, η must be greater than 3×10^{-10}). To summarize, consistency between the predicted big bang abundances of D and ^3He, and the derived abundances observed today requires η to lie in the range $\simeq (4-10) \times 10^{-10}$.

Until very recently, our knowledge of the ^7Li abundance was limited to observations of meteorites, the local ISM, and Pop I stars, with a derived present abundance of $^7Li/H \simeq 10^{-9}$ (to within a factor of 2). Given that ^7Li is produced by cosmic ray spallation and some stellar processes, and is easily destroyed (in environments where $T \gtrsim 2 \times 10^6 K$), there is not the slightest reason to suspect (or even hope!) that this value accurately reflects the primordial abundance. Recently, Spite and Spite[22] have observed ^7Li lines in the atmospheres of 13 unevolved halo and old disk stars (Pop II) with very low metal abundances ($Z \simeq Z_\odot/12 - Z_\odot/250$), whose masses span the range of $\simeq (0.6-1.1)M_\odot$. [Note that $Z \equiv$ mass fraction of metals, a metal being any isotope with $A \geq 4$. $Z_\odot \simeq 0.02$.] Stars less massive than about $0.7M_\odot$ are expected to astrate (by factors $\geq O(10)$) their ^7Li abundance during their approach to the MS, while stars more massive than about $1M_\odot$ do not seem to significantly astrate ^7Li in their outer layers. Indeed, they see this trend in their data, and deduce a primordial ^7Li abundance of: $^7Li/H \simeq (1.12 \pm 0.38) \times 10^{-10}$. Remarkably, this is the predicted big bang production for η in the range $(2-5) \times 10^{-10}$. If we take this to be the primordial ^7Li abundance, and allow for a possible factor of 2 uncertainty in the predicted abundance of Li (due to estimated uncertainties in the reaction rates which affect ^7Li), then concordance for ^7Li restricts η to the range $(1-7) \times 10^{-10}$. Note, of course, that their derived ^7Li abundance is the pre-Pop II abundance, and may not necessarily reflect the true primordial abundance (e.g., if a Pop III generation of stars processed significant amounts of material).

In sum, the concordance of big bang nucleosynthesis predictions with the derived abundances of D and ^3He requires $\eta \simeq (4-10) \times 10^{-10}$; moreover, concordance for D, ^3He, and ^7Li further restricts η: $\eta \simeq (4-7) \times 10^{-10}$.

In the past few years the quality and quantity of ^4He observations has increased markedly. In Fig. 1.7 all the ^4He abundance determinations derived from observations of recombination lines in HII regions (galactic and extragalactic) are shown as a function of metalicity Z (more precisely, 2.2 times the mass fraction of ^{16}O). [Astronomers refer to ionized hydrogen as 'HII'; an HII region then is a region of ionized hydrogen; typical sizes are 10's of pc and typical temperatures $\gg 10,000$ K.]

Since ^4He is also synthesized in stars, some of the observed ^4He is not primordial. Since stars also produce metals, one would expect some correlation between Y and Z, or at least a trend: lower Y where Z is lower. Such a trend is apparent in Fig. 1.7. From Fig. 1.7 it is also clear that there is a large primordial component to ^4He: $Y_p \simeq 0.22-0.26$. Is it possible to pin down the value of Y_p more precisely?

There are many steps in going from the line strengths (what the observer actually measures), to a mass fraction of ^4He (e.g., corrections for neutral ^4He, reddening, etc.). In galactic HII regions, where abundances can be determined for various positions within a given HII region, variations are seen within a given HII region. Observations of extragalactic HII regions are actually observations of a superposition of several HII regions. Although observers have quoted statistical uncertainties of $\Delta Y \simeq \pm 0.01$ (or lower), from the scatter in Fig. 1.7 it is clear that the systematic uncertainties must be larger. For example, different observers have derived ^4He abundances of between 0.22 and 0.25 for I Zw18, an extremely metal-poor dwarf emission line galaxy.

Perhaps the safest way to estimate Y_p is to concentrate on the ^4He determinations for metal-poor objects. From Fig. 1.7 $Y_p \simeq 0.23-0.25$ appears to be consistent with all the data (although Y_p as low as 0.22 or high as 0.26 could not be ruled out). Recently Kunth and Sargent[23] have studied 13 metal-poor $(Z \lesssim Z_\odot/5)$ Blue Compact galaxies. From a weighted average for their sample they derive a primordial abundance $Y_p \simeq 0.245 \pm 0.003$; allowing for a 3σ variation this suggests

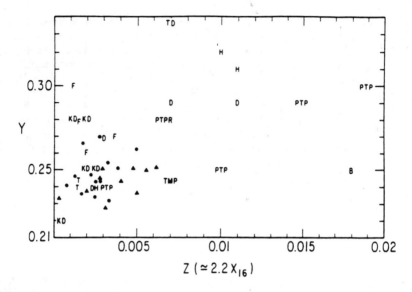

Figure 1.7: Summary of ^4He abundance determinations (galactic and extragalactic) from recombination lines in HII regions vs. mass fraction of heavy elements Z ($\simeq$ 2.2 mass fraction of ^{16}O). Note, observers do not usually quote errors for individual objects—scatter is probably indicative of the uncertainties. The triangles and filled circles represent two data sets of note: circles $\simeq$ 13 very metal poor emission line galaxies (Knuth and Sargent[23]); triangles - 9 metal poor, compact galaxies (Lequeux et al.[23]).

$0.236 \leq Y_p \leq 0.254$.

For the concordance range deduced from D, ^3He, and ^7Li ($\eta \geq 4 \times 10^{-10}$) and $\tau_{1/2} \geq 10.4$ min, the predicted ^4He abundance is

$$Y_p \geq \begin{cases} 0.230 & N_\nu = 2. \\ 0.244 & N_\nu = 3. \\ 0.256 & N_\nu = 4. \end{cases}$$

[Note, that $N_\nu = 2$ is permitted only if the τ-neutrino is heavy ($\gtrsim$ few MeV) and unstable; the present experimental upper limit to its mass is 160 MeV.] Thus, since $Y_p \simeq 0.23 - 0.25$ (0.22 - 0.26?) there are values of η, N_ν, and $\tau_{1/2}$ for which there is agreement between the abundances predicted by big bang nucleosynthesis and the primordial abundances of D, ^3He, ^4He, and ^7Li derived from observational data.

To summarize, the only isotopes which are predicted to be produced in significant amounts during the epoch of primordial nucleosynthesis are: D, ^3He, ^4He, and ^7Li. At present there is concordance between the predicted primordial abundances of all four of these elements and their observed abundances for values of $N_\nu, \tau_{1/2}$, and η in the following intervals: $2 \leq N_\nu \leq 4$; $10.4 \text{ min} \leq \tau_{1/2} \leq 10.8 \text{ min}$; and $4 \times 10^{-10} \leq \eta \leq 7 \times 10^{-10}$ (or 10×10^{-10} if the ^7Li abundance is not used). This is a truly remarkable achievement—note that the predicted abundances span some nine orders of magnitude—and strong evidence that the standard model is valid back as early as 10^{-2} sec after 'the bang'.

The standard model will be in serious straits if the primordial mass fraction of ^4He is unambiguously determined to be less than 0.22. What alternatives exist if $Y_p \leq 0.22$? If a generation of Pop III stars that efficiently destroyed ^3He and ^7Li existed, then the lower bound to η based upon D, ^3He, (and ^7Li) no longer exists. The only solid lower bound to η would then be that based upon the amount of luminous matter in galaxies (i.e., the matter inside the Holmberg radius): $\eta \geq 0.3 \times 10^{-10}$. In this case the predicted Y_p could be as low as 0.15 or 0.16. Although small amounts of anisotropy increase[24] the primordial production of ^4He, recent work[25] suggests that larger amounts could decrease the primordial production of ^4He. Another possibility is neutrino degeneracy; a large lepton number ($n_\nu - n_{\bar{\nu}} \simeq O(n_\gamma)$) drastically modifies the predictions of big bang nucleosynthesis[26]. Finally, one might be forced to discard the standard cosmology altogether. [For the most up-to-date review of primordial nucleosynthesis see ref. 27.]

1.2. PRIMORDIAL NUCLEOSYNTHESIS AS A PROBE

If, based upon its apparent success, we accept the validity of the standard model, we can use primordial nucleosynthesis as a very powerful probe of cosmology and particle physics. For example, concordance requires: $4 \times 10^{-10} \lesssim \eta \lesssim 7 \times 10^{-10}$ and $N_\nu \leq 4$. This is the most precise determination we have of η and implies that

$$0.014 h^{-2} T_{2.7}^3 \lesssim \Omega_b \lesssim 0.024 h^{-2} T_{2.7}^3 \tag{1.17}$$

$$0.014 \lesssim \Omega_b \lesssim 0.14 \tag{1.18}$$

$$n_B/s \simeq n/7 \simeq (6-10) \times 10^{-11}.$$

If, as some dynamical studies suggest, $\Omega > 0.14$, then some other non-baryonic form of matter must account for the difference between Ω and Ω_b. [For a recent review of the dynamical measurements of Ω, see refs. 28, 29.] Numerous candidates have been proposed for the dark matter, including primordial black holes, axions, quark nuggets, photinos, gravitinos, relativistic debris, massive neutrinos, sneutrinos, monopoles, pyrgons, maximons, etc. [A discussion of some of these candidates is given in ref. 30.]

With regard to the limit on N_ν, Schvartsman[31] first emphasized the dependence of the yield of ^4He on the expansion rate of the Universe during nucleosynthesis, which in turn is determined by g_*, the effective number of massless degrees of freedom. As mentioned above the crucial temperature for ^4He synthesis is $\simeq 1$ MeV— the freeze out temperature for the n/p ratio. At this epoch the massless degrees of freedom include: $\gamma, \nu\bar{\nu}, e^\pm$ pairs, and any other light particles present, and so

$$g_* = g_\gamma + 7/8(g_{e^\pm} + N_\nu g_{\nu\bar{\nu}}) + \underset{Bose}{\sum g_i(T_i/T)^4} + 7/8 \underset{Fermi}{\sum g_i(T_i/T)^4}$$

$$= 5.5 + 1.75 N_\nu + \sum_{Bose} g_i(T_i/T)^4 + 7/8 \sum_{Fermi} g_i(T_i/T)^4. \qquad (1.19)$$

Here T_i is the temperature of species i, T is the photon temperature, and the total energy density of relativistic species is: $\rho = g_* \pi^2 T^4/30$. The limit $N_\nu \leq 4$ is obtained by assuming that the only species present are: γ, $e^\pm$, and N_ν neutrinos species, and follows because for $\eta \geq 4 \times 10^{-10}$, $\tau_{1/2} \geq 10.4 min$, and $N_\nu \geq 4$, the mass fraction of ^4He produced is ≥ 0.254 (which is greater than the observed abundance). More precisely, $N_\nu \leq 4$ implies

$$g_* \lesssim 12.5 \qquad (1.20)$$

or

$$1.75 \gtrsim 1.75(N_\nu - 3) + \sum_{Bose} g_i(T_i/T)^4 + 7/8 \sum_{Fermi} g_i(T_i/T)^4. \qquad (1.21)$$

At most one additional light ($\lesssim$ MeV) neutrino species can be tolerated; many more additional species can be tolerated if their temperatures T_i are $< T$. [Big bang nucleosynthesis limits on the number of light ($\lesssim$ MeV) species have been derived and/or discussed in refs. 32.]

The number of neutrino species can also be determined by measuring the width of the $Z°$ boson: each neutrino flavor less massive than $0(M_Z/2)$ contributes $\simeq$ 190 MeV to the width of the $Z°$. Preliminary results on the width of the $Z°$ imply that $N_\nu \lesssim 0(10)$[33] and e^+e^- collider experiments looking for $e^+e^- \to X + \gamma$ (X = undetected) also imply $N_\nu \lesssim 0(10)$. Note that while big bang nucleosynthesis and the width of the $Z°$ both provide information about the number of neutrino flavors, they 'measure' slightly different quantities. Big bang nucleosynthesis is sensitive to the number of light ($\lesssim$ MeV) neutrino species, and all other light degrees of freedom, while the width of the $Z°$ is determined by the number of particles less massive than about 50 GeV which couple to the $Z°$ (neutrinos among them). This issue has been recently reviewed in ref. 34.

Given the important role occupied by big bang nucleosynthesis, it is clear that continued scrutiny is in order. The importance of new observational data cannot be overemphasized: extragalactic D abundance determinations (Is the D abundance universal? What is its value?); more measurements of the ^3He abundance (What is its primordial value? Is it true that ^3He has not been significantly astrated?); continued improvement in the accuracy of ^4He abundances in very metal poor objects (Recall, the difference between $Y_p = 0.22$ and $Y_p = 0.23$ is crucial); and further study of the ^7Li abundance in very old stellar populations (Has the primordial abundance of ^7Li already been measured?). Data from particle physics will prove useful too: a high precision determination of $\tau_{1/2}$ (i.e., $\Delta \tau_{1/2} \leq \pm 0.05$ min) will all but eliminate the uncertainty in the predicted ^4He primordial abundance (in the standard cosmology); an accurate measurement of the width of the recently-found $Z°$ vector boson will determine the total number of neutrino species (less massive than about 50 GeV) and thereby bound the total number of light neutrino species.

All these data will not only make primordial nucleosynthesis a more stringent test of the standard cosmology, but they will also make primordial nucleosynthesis a more powerful probe of the early Universe.

1.3. DEPARTURE FROM THERMAL EQUILIBRIUM, 'FREEZE-OUT' AND THE MAKING OF A RELIC SPECIES

Thus far I have tacitly assumed that the Universe is always in thermal equilibrium. Of course if that were always the case, the Universe would be a very boring place (Fe nuclei at 3K today!). In the strictest sense the Universe can never be in thermal equilibrium because of its expansion. However, if particle reaction rates (Γ) are rapid compared to the expansion rate ($\Gamma \gg H$), then the Universe will pass through a succession of nearly equilibrium states characterized by $T \propto g_*^{-1/3} R^{-1}$. Interestingly, a massless (or very relativistic, $T \gg m$), non-interacting species which initially has an equilibrium phase space distribution will continue to do so with a temperature $T \propto R^{-1}$. The same is true for a very massive (i.e., $m \gg T$), non-interacting species, except that $T \propto R^{-2}$. [Both of these facts are straightforward to verify and follow from $p \propto R^{-1}$ and $V \propto R^{-3}$.]

Photons and ionized matter remain in thermal equilibrium until $T \simeq \frac{1}{3}\text{eV}, R \simeq 10^{-3} R_{today}$, and $t \simeq 6.5 \times 10^{12} (\Omega h^2)^{-1/2}$ sec, when it becomes energetically favorable for the ions and electrons present to form neutral atoms (at which time the scattering cross section for photons drops precipitously from the Thomson cross section $\sigma_T \simeq \alpha^2/m_e^2 \simeq 0.66 \times 10^{-24} \text{cm}^2$). This is the so-called decoupling or recombination epoch. After decoupling, the photons freely expand, and the expansion preserves their thermal distribution with a temperature $T \propto R^{-1}$.

A given particle species can only remain in 'good thermal contact' with the photons if the reactions that are important for keeping it in thermal equilibrium are occurring rapidly compared to the rate at which T is decreasing (which is set by the expansion rate $-\dot{T}/T \simeq \dot{R}/R = H$). Roughly-speaking the criterion is

$$\Gamma \gtrsim H, \qquad (1.22)$$

where $\Gamma = n\langle \sigma v \rangle$ is the interaction rate per particle, n is the number density of target particles and $\langle \sigma v \rangle$ is the thermally-averaged interaction cross section. When Γ drops below H, that reaction is said to 'freeze-out' or 'decouple'. The temperature T_f (or T_d) at which $H = \Gamma$ is called the freeze-out or decoupling temperature. [Note that if $\Gamma = aT^n (n > 2)$ and the Universe is radiation-dominated so that $H = (2t)^{-1} \simeq 1.67 g_*^{1/2} T^2/m_{pl}$, then the total number of interactions which occur for $T \lesssim T_f$ is just: $\int_{T_f}^0 \Gamma dt \simeq (\Gamma/H)|_{T_f}/(n-2) \simeq (n-2)^{-1}$]. If the species in question is very relativistic ($T_f \gg m_i$) (or very non-relativistic ($T_f \ll m$)) when it decouples, then its phase space distribution (in momentum space) remains thermal (i.e., Bose-Einstein or Fermi-Dirac) with a temperature $T_i \propto R^{-1}$ (or $T_i \propto R^{-2}$).

[It is interesting to note that based upon just the known interactions, one would not expect the Universe to be in thermal equilibrium during its earliest epochs. At high temperatures, the cross section for renormalizable interactions which transfer significant momentum ($\simeq$ few T) and which are mediated by light gauge bosons scale as: $\sigma \simeq \alpha^2/T^2$ (α = appropriate gauge coupling constant), and the number density of particles $n \simeq g_* T^3$. Taking the expansion rate to be: $H \simeq g_*^{1/2} T^2/m_{pl}$, it follows that $(\Gamma \simeq n\langle \sigma v \rangle) \geq H$, only for $T \lesssim g_*^{1/2} \alpha^2 m_{pl} \simeq 10^{16}$GeV (for $g_*^{1/2} \alpha^2 \simeq 10^{-3}$).]

Now consider the evolution of the temperature of a decoupled, relativistic species relative to that of the photons. For the decoupled species $T_i \propto R^{-1}$. However, due to the entropy release when various massive species annihilate (e.g., $e^\pm$ pairs when $T \simeq 0.1$ MeV), the photon temperature does not always decrease as R^{-1}. [More precisely, when various massive species transfer their entropy to the EM plamsa.] Entropy conservation ($S \propto g_* T^3$ = constant) of course, can be used to calculate its evolution; if g_* is decreasing, then T will decrease less rapidly than R^{-1}. As an example consider neutrino freeze-out. The cross section for processes like $e^+ e^- \leftrightarrow \nu \bar\nu$ is: $\langle \sigma v \rangle \simeq 0.2 G_F^2 T^2$ (here $G_F \simeq 1.1 \times 10^{-5}$ GeV^{-2} is the Fermi coupling constant), and the number density of targets $n \simeq T^3$, so that $\Gamma \simeq 0.2 G_F^2 T^5$. Equating this to H it follows that

$$T_f \simeq (30 m_{pl}^{-1} G_F^{-2})^{1/3} \qquad (1.23)$$
$$\simeq \text{few MeV}, \qquad (1.24)$$

i.e., neutrinos freeze out before $e^\pm$ annihilations and do not share in subsequent entropy transfer. For $T \lesssim$ few MeV, neutrinos are decoupled and $T_\nu \propto R^{-1}$, while the entropy density in $e^\pm$ pairs and γ's $\propto R^{-3}$. Using the fact that before $e^\pm$ annihilation the entropy density of the $e^\pm$ pairs and γ's is: $s \propto (7g_{e^\pm}/8 + g_\gamma)T^3 = 5.5T^3$ and that after $e^\pm$ annihilation $s \propto g_\gamma T^3 = 2T^3$, it follows that after the $e^\pm$ annihilations

$$T_\nu/T = [g_\gamma/(g_\gamma + 7g_{e^\pm}/8)]^{1/3}$$
$$= (4/11)^{1/3}. \qquad (1.25)$$

Similarly, the temperature at the time of primordial nucleosynthesis T_i of a species which decouples at an arbitrary temperature T_d can be calculated:

$$T_i/T = [(g_\gamma + \frac{7}{8}(g_{e^\pm} + N_\nu g_{\nu\bar\nu}))/g_{*d}]^{1.3}$$
$$\simeq (10.75/g_{*d})^{1/3} \qquad \text{(for N}_\nu = 3\text{)}. \qquad (1.26)$$

Here $g_{*d} = g_*(T_d)$ is the number of species in equilibrium when the species in question decouples. Species which decouple at a temperature 30 MeV $\simeq m_\mu/3 \lesssim T \lesssim$ few 100 MeV do not share in the entropy release from $\mu^\pm$ annihilations, and $T_i/T \simeq 0.91$; the important factor for limits based upon primordial nucleosynthesis $(T_i/T)^4 \simeq 0.69$. Species which decouple at temperatures $T_d \gtrsim \Lambda_{QCD}$, the temperature of the quark/hadron transition $\simeq$ few 100 MeV, do not share in the entropy transfer when the quark-gluon plasma $[g_* \simeq g_\gamma + g_{Gluon} + \frac{7}{8}(g_{e^\pm} + g_{\mu^\pm} + g_{\nu\bar\nu} + g_{u\bar u} + g_{d\bar d} + g_{s\bar s} + \ldots) \gtrsim 62]$ hadronizes, and $T_i/T \simeq 0.56$; $(T_i/T)^4 \simeq 0.10$.

'Hot' relics

Consider a stable particle species X which decouples at a temperature $T_f \gg m_x$. For $T < T_f$ the number density of X's, n_x, just decreases as R^{-3} as the Universe expands. In the absence of entropy production the entropy density s also decreases as R^{-3}, and hence the ratio n_x/s ($\propto$ number of X's per comoving volume) remains constant. At freeze-out

$$n_x/s = \frac{(g_{xeff}\varsigma(3)/\pi^2)}{(2\pi^2 g_{*d}/45)},$$
$$\simeq \frac{0.278 g_{xeff}}{g_{*d}}, \qquad (1.27)$$

where $g_{xeff} = g_x$ for a boson or $3g_x/4$ for a fermion, $g_{*d} = g_*(T_d)$, and $\varsigma(3) = 1.20206\ldots$. Today $s \simeq 7.04 n_\gamma$, so that the number density and mass density of X's are

$$n_x \simeq (2g_{xeff}/g_{*d})n_\gamma, \qquad (1.28)$$
$$\Omega_x = \rho_x/\rho_c \simeq 7.6(m_x/100eV)(g_{xeff}/g_{*d})h^{-2}T_{2.7}^3. \qquad (1.29)$$

Note, that if the entropy per comoving volume s has increased since the X decoupled, e.g., due to entropy production in a phase transition, then these values are decreased by the same factor that the entropy increased. As discussed earlier, Ωh^2 must be $\lesssim O(1)$, implying that for a *stable* particle species

$$m_x/100eV \lesssim 0.13 g_{*d}/g_{xeff}; \qquad (1.30)$$

for a neutrino species: $T_d \simeq$ few MeV, $g_{*d} \simeq 10.75$, $g_{xeff} = 2 \times (3/4)$, so that $n_{\nu\bar\nu}/n_\gamma \simeq 3/11$ and m_ν must be $\lesssim 96 eV$. Note that for a species which decouples very early (say $g_{*d} = 200$), the mass limit (1.7 keV for $g_{xeff} = 1.5$) which $\propto g_{*d}$ is much less stringent.

Constraint (1.24) obviously does not apply to an unstable particle with $\tau < 10 - 15$ billion yrs. However, any species which decays radiatively is subject to other very stringent constraints, as the photons from its decays can have various unpleasant astrophysical consequences, e.g., dissociating D, distorting the microwave background, 'polluting' various diffuse photon backgrounds, etc. The astrophysical/cosmological constraints on the mass/lifetime of an unstable neutrino species and the photon spectrum of the Universe are shown in Figs. 1.8, 1.9.

'Cold' relics

Consider a stable particle species which is still coupled to the primordial plasma ($\Gamma > H$) when $T \simeq m_x$. As the temperature falls below m_x, its equilibrium abundance is given by

$$n_x/n_\gamma \simeq \left(\frac{g_x}{2\varsigma(3)}\right)\left(\frac{\pi}{8}\right)^{1/2}\left(\frac{m_x}{T}\right)^{3/2}\exp\left(-\frac{m_x}{T}\right), \qquad (1.31)$$

$$n_x/s \simeq 0.14\left(\frac{g_x}{g_*}\right)\left(\frac{m_x}{T}\right)^{3/2}\exp\left(-\frac{m_x}{T}\right), \qquad (1.32)$$

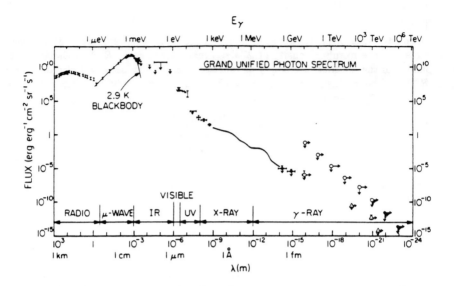

Figure 1.8: The diffuse photon spectrum of the Universe from $\lambda = 1$ km to 10^{-24} m. Vertical arrows indicate upper limits; horizontal arrows indicate integrated flux, i.e., Flux $(> E)$.

and in order to maintain an equilibrium abundance X's must diminish in number (by annihilations since by assumption the X is stable). So long as $\Gamma_{ann} \simeq n_x \langle \sigma v \rangle_{ann} \gtrsim H$ the equilibrium abundance of X's is maintained. When $\Gamma_{ann} \simeq H$, when $T = T_f$, the X's 'freeze-out' and their number density n_x decreases only due to the volume increase of the Universe, so that for $T \lesssim T_f$:

$$n_x/s \simeq (n_x/s)|_{T_f}. \tag{1.33}$$

The equation for freeze-out $(\Gamma_{ann} \simeq H)$ can be solved approximately, giving

$$\frac{m_x}{T_f} \simeq \ln[0.04(\sigma v)_o m_x m_{pl} g_x g_*^{-1/2}]$$
$$+ (\frac{1}{2} - n) \ln\{[0.04(\sigma v)_o m_x m_{pl} g_x g_*^{-1/2}]\},$$
$$\simeq 39 + \ln[(\sigma v)_o m_x] + (\frac{1}{2} - n) \ln[39 + \ln[(\sigma v)_o m_x], \tag{1.34}$$

$$\frac{n_x}{s} \simeq \frac{5\{\ln[0.04(\sigma v)_o m_x m_{pl} g_x g_*^{-1/2}]\}^{1+n}}{[(\sigma v)_o m_x m_{pl} g_*^{1/2}]},$$
$$\simeq \frac{4 \times 10^{-19}\{39 + \ln[(\sigma v)_o m_x]\}^{1+n}}{[(\sigma v)_o m_x g_*^{1/2}]} \tag{1.35}$$

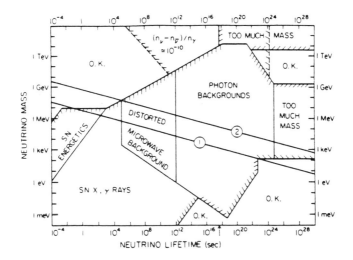

Figure 1.9: Summary of astrophysical/cosmological constraints on neutrino masses/lifetimes[38]. Lines 1 and 2 represent mass/lifetime relationships: $\tau = a \times 10^{-6} \sec(m_\mu/m_\nu)^5$, for $a = 1, 10^{12}$.

where $\langle\sigma v\rangle_{ann}$ is taken to be $(\sigma v)_o (T/m_x)^n$ ($n = 0$ corresponds to s-wave annihilation, $n = 1$ to p-wave, etc.). In the second form of each equation $g_x = 2, g_* \simeq 100$, and all dimensional quantities are to be measured in GeV units.

[The 'correct way' to solve for n_x/s is to integrate the Boltzmann equation which governs the X abundance, $d/dt(n_x/s) = -\langle\sigma v\rangle s[(n_x/s)^2 - (n_{xeq}/s)^2]$. This has been done carefully in ref. 35 (also see ref. 36), and the 'freeze-out' approximation used in Eqns(1.33-1.35) is found to be reasonably good. As discussed in ref. 35 a more accurate analytic approximation gives

$$\frac{m_x}{T_f} \simeq \ln[0.04(n+1)(\sigma v)_o m_x m_{pl} g_x g_*^{-1/2}]$$
$$- (n + \frac{1}{2}) \ln\{\ln[0.04(n+1)(\sigma v)_o m_x m_{pl} g_x g_*^{-1/2}]\},$$
$$\frac{n_x}{s} \simeq \frac{3.79(n+1)(m_x/T_f)^{n+1}}{[(\sigma v)_o m_x m_{pl} g_*^{1/2}]}.$$

This analytic approximation for n_x/s agrees with the numerical results for n_x/s to better than 5%.]

As an example, consider a 'heavy' (Dirac) neutrino species ($m_x \gg$ MeV), for which $\langle\sigma v\rangle \simeq 0(1) m_x^2 G_F^2$. In the absence of annihilations this species would decouple at $T \simeq$ few MeV which is $\ll m_x$, and so the X will become a 'cold relic'. Using

Eqns(1.34, 1.35), we find that today:

$$n_x/s \simeq 5 \times 10^{-9}(m_x/\text{GeV})^{-3}, \qquad (1.36)$$
$$\Omega_x h^2 \simeq 2(m_x/\text{GeV})^{-2}, \qquad (1.37)$$

implying that a stable, heavy neutrino species must be more massive than a few GeV. [This calculation was first done by Lee and Weinberg[37], and independently by Kolb[37].] Note that $\rho_x \propto n_x m_x \propto (\sigma v)_o^{-1}$—implying that the more weakly-interacting a particle is, the more 'dangerous' it is cosmologically. If a particle species is to saturate the mass density bound and provide most of the mass density today $(\Omega_x h^2 \simeq 1)$ then its mass and annihilation cross section must satisfy the relation:

$$(\sigma v)_o g_*^{1/2} \simeq 10^{-10}\{39 + \ln[m_x(\sigma v)_o]\}^{1+n} \qquad (1.38)$$

where as usual all dimensional quantities are in GeV units. Note also that the relic abundance of a species $(\Omega_x h^2)$ 'determines' its annihilation cross section!

1.4. PARTICLES WHICH DECAY OUT-OF-EQUILIBRIUM

Consider a particle species 'X', whose abundance per comoving volume, n_X/s, has frozen out at a value $n_X/s = r$, due to the ineffectiveness of its annihilations. If the species X is not stable, then eventually it will decay when the age of the Universe is about equal to its lifetime, $\tau \equiv \Gamma^{-1}$. In the process of decaying X particles will produce entropy, thereby increasing the entropy per comoving volume, S, of the Universe.

Suppose the X particles become NR before they decay. Until they decay their energy density relative to that of relativisitc particles steadily increases, $\rho_X/\rho_{rad} \propto R(t)$. At the temperature of order rm_X the energy density in X's begins to exceed that in radiation. If this occurs before they decay, then when they decay, X decays will significantly increase the entropy of the Universe.

The equations which govern this process are

$$\dot\rho = -3H\rho_X - \Gamma\rho_X, \qquad (1.39)$$
$$(\Rightarrow \rho_X = \rho_{X0}(R/R_0)^{-3}\exp[-\Gamma(t-t_0)])$$
$$\dot S = (2\pi^2 g_*/45)^{1/3}\Gamma R^4 \rho_X/S^{1/3}, \qquad (1.40)$$
$$H^2 = 8\pi G(\rho_{rad} + \rho_X)/3. \qquad (1.41)$$

Eqn(1.39) is the usual exponential decay law for particle decays (where I am assuming that the X's are NR). Eqn(1.40) follows from the first law of thermodynamics, $dS = dQ/T$, where dQ, the heat released due to decays per comoving volume, is $R^3\rho_X dt$, and I have assumed that the decay products are relativistic and rapidly thermalize so that $S = 4\rho_{rad}/3T = (2\pi^2 g_*/45)T^3$. [If g_* is constant, then Eqn(1.40) can also be written as: $\dot\rho_{rad} = -4H\rho_{rad} + \Gamma\rho_X$.] Eqn(1.41) is the usual Friedmann equation and I have assumed that only X's and R particles contribute significantly to the energy density.

This set of equations is straightforward to solve. Let me discuss the case where the entropy release is significant (final entropy/initial entropy $\equiv S_f/S_0 \gg 1$). In this case the temperature of the Universe decreases as $g_*^{-1/3}R^{-1}$ until the energy released by decays per expansion time becomes comparable to $\rho_{rad}(\Gamma t \rho_X \simeq \rho_{rad})$. After this time the energy density in particles produced by X decays exceeds that in the original radiation component and $\rho_{rad} \propto t^{-1}$, $T \propto R^{-3/8}$. The temperature at the decay epoch ($t \simeq \Gamma^{-1}$; for $t \geq \Gamma^{-1}$, the X's rapidly disappear) is

$$T_{decay} \simeq 0.6 g_*^{-1/4}(\Gamma m_{pl})^{1/2},$$

which is a factor of order $g_*^{1/12}(rm_x)^{1/3}/(\Gamma m_{pl})^{1/6}$ greater than what it would have been in the absence of X decays. Contrary to intuition, however, the Universe is not suddenly 'heated up' at $t \simeq \tau = \Gamma^{-1}$; rather, due to the exponential decay law, the entropy produced by X decays is steadily released, resulting in the Universe cooling more slowly, $T \propto R^{-3/8}$ (compared to the usual R^{-1}). The increase in entropy per comoving volume is given by

$$S_f/S_0 = 1.83 g_*^{1/4}(rm_x)/(\Gamma m_{pl})^{1/2}, \quad (1.42)$$

where $g_* \simeq$ the value of g_* when $t \simeq \Gamma^{-1}$.

The equations for particles which decay out of equilibrium will arise in many interesting situations, including the reheating of the new inflationary Universe. A more detailed discussion of decaying particles is given in ref. 102 and in Lecture 3.

LECTURE 2: BARYOGENESIS

I'll begin by briefly summarizing the evidence for the baryon asymmetry of the Universe and the seemingly insurmountable problems that render baryon symmetric cosmologies untenable. For a more detailed discussion of these I refer the reader to Steigman's review of the subject[39]. For a review of recent attempts to reconcile a symmetric Universe with both baryogenesis and the observational constraints, I refer the reader to Stecker[40].

2.1 EVIDENCE FOR A BARYON ASYMMETRY

Within the solar system we can be very confident that there are no concentrations of antimatter (e.g., antiplanets). If there were, solar wind particles striking such objects would be the strongest γ-ray sources in the sky. Also, NASA has yet to lose a space probe because it annihilated with antimatter in the solar system.

Cosmic rays more energetic than 0(0.1 GeV) are generally believed to be of "extrasolar" origin, and thereby provide us with samples of material from throughout the galaxy (and possibly beyond). The ratio of antiprotons to protons in the cosmic rays is about 3×10^{-4}, and the ratio of anti-^4He to ^4He is less than 10^{-5} (ref. 41). Antiprotons are expected to be produced as cosmic-ray secondaries (e.g. $p + p \to 3p + \bar{p}$) at about the 10^{-4} level. At present both the spectrum and total flux of cosmic-ray antiprotons are at variance with the simplest model of their production as secondaries. A number of alternative scenarios for their origin have been proposed including the possibility that the detected $\bar{p}$'s are cosmic rays from distant antimatter galaxies. Although the origin of these $\bar{p}$'s remains to be resolved, it is clear that they do not provide evidence for an appreciable quantity of antimatter in our galaxy. [For a review of antimatter in the cosmic rays we refer the reader to ref. 41.]

The existence of both matter and antimatter galaxies in a cluster of galaxies containing intracluster gas would lead to a significant γ-ray flux from decays of π°'s produced by nucleon-antinucleon annihilations. Using the observed γ-ray background flux as a constraint, Steigman[39] argues that clusters like Virgo, which is at a distance ≈ 20 Mpc ($\approx 10^{26}$ cm) and contains several hundred galaxies, must not contain both matter and antimatter galaxies.

Based upon the above-mentioned arguments, we can say that if there exist equal quantities of matter and antimatter in the Universe, then we can be absolutely certain they are separated on mass scales greater than 1 $M_\odot$, and reasonably certain they are separated on scales greater than (1-100) $M_{galaxy} = 10^{12} - 10^{14} M_\odot$. As I will discuss below, this fact is virtually impossible to reconcile with a symmetric cosmology.

It has often been pointed out that we derive most of our direct knowledge of the large-scale Universe from photons, and since the photon is a self-conjugate particle we obtain no clue as to whether the source is made of matter or antimatter. Neutrinos, on the other hand, are not self-conjugate, and can in principle reveal information about the matter-antimatter composition of their source. Large neutrino detectors may someday provide direct information about the matter-antimatter composition of the Universe on the largest scales.

Baryons account for only a tiny fraction of the particles in the Universe, the 3K-microwave photons being the most abundant species (yet detected). The number density of 3K photons is $n_\gamma = 399 T_{2.7}^3$ cm^{-3}. The baryon density is not nearly as well determined. Luminous matter (baryons in stars) contribute at least 0.01 of closure density ($\Omega_{lum} > 0.01$), and as discussed in Lecture 1 the age of the Universe requires that Ω_{tot} (and Ω_b) must be $< 0(2)$. These direct determinations place the baryon-to-photon ratio $\eta \equiv n_b/n_\gamma$ in the range 3×10^{-11} to 6×10^{-8}. As I also discussed in Lecture 1 the yields of big-bang nucleosynthesis depend directly on η, and the production of amounts of D, ^3He, ^4He, and ^7Li that are consistent with their present measured abundances restricts η to the narrow range (4-7) $\times 10^{-10}$.

Since today it appears that $n_b \gg n_{\bar{b}}$, η is also the ratio of net baryon number to photons. The number of photons in the Universe has not remained constant, but has increased at various epochs when particle species have annihilated (e.g. $e^\pm$ pairs at T ≈ 0.5 MeV). Assuming the expansion has been isentropic (i.e., no significant entropy production), the entropy per comoving volume ($\equiv S \propto sR^3$) has remained constant. The "known entropy" is presently about equally divided between the 3K photons and the three cosmic neutrino backgrounds (e, μ, τ). Taking this to be the present entropy, the ratio of baryon number to entropy is

$$n_B/s \simeq \eta/7 \simeq (6-10) \times 10^{-11}, \qquad (2.1)$$

where $n_B \equiv n_b - n_{\bar{b}}$ and η is taken to be in the range (4-7) $\times 10^{-10}$. So long as the expansion is isentropic and baryon number is at least effectively conserved this ratio remains constant and is what I will refer to as the baryon number of the Universe. As discussed earlier the net baryon number per comoving volume is $\propto n_B/s$.

Although the matter-antimatter asymmetry appears to be "large" today (in the sense that $n_B \approx n_b \gg n_{\bar{b}}$), the fact that $n_B/s \simeq 10^{-10}$ implies that at very early times the asymmetry was "tiny" $(n_B \ll n_b)$. To see this, let us assume for simplicity that nucleons are the fundamental baryons. Earlier than 10^{-6} s after the bang the temperature was greater than the mass of a nucleon. Thus nucleons and antinucleons should have been about as abundant as photons, $n_N \approx n_{\bar{N}} \approx n_\gamma$. The entropy density s is $\approx g_* n_\gamma \approx g_* n_N \approx 0(10^2) n_N$. The constancy of $n_B/s \approx 0(10^{-10})$ requires that for $t < 10^{-6}s$, $(n_N - n_{\bar{N}})/n_N (\approx 10^2 n_B/s) \approx 0(10^{-8})$. During its earliest epoch, the Universe was nearly (but not quite) baryon symmetric.

2.2 THE TRAGEDY OF A SYMMETRIC COSMOLOGY

Suppose that the Universe was initially locally baryon symmetric. Earlier than 10^{-6} s after the bang nucleons and antinucleons were about as abundant as photons. For $T < 1$ GeV the equilibrium abundance of nucleons and antinucleons is $(n_N/n_\gamma)_{EQ} \approx (m_N/T)^{3/2} \exp(-m_N/T)$, and as the Universe cooled the number of nucleons and antinucleons would decrease tracking the equilibrium abundance as long as the annihilation rate $\Gamma_{ann} \approx n_N \langle \sigma v \rangle_{ann} \approx n_N m_\pi^{-2}$ was greater than the expansion rate H. At a temperature T_f annihilations freeze out $((\Gamma_{ann}/H)|_{T_f} \simeq 1)$, nucleons and antinucleons being so rare they can no longer find each other to annihilate. Using Eqn(1.34) we can compute T_f: $T_f \simeq 0(20\,\text{MeV})$. Because of the incompleteness of the annihilations, residual nucleon and antinucleon to photon ratios (given by Eqn(1.35)) $n_{\bar{N}}/n_\gamma = n_N/n_\gamma \simeq 10^{-18}$ are "frozen in". Even if the matter and antimatter could subsequently be separated, n_N/n_γ is a factor of 10^8 too small. To avoid 'the annihilation catastrophe', matter and antimatter must be separated on large scales before $t \approx 3 \times 10^{-3}s (T \approx 20\,\text{MeV})$. I'll consider two possible mechanisms.

Statistical fluctuations:

One possible mechanism for doing this is statistical (Poisson) fluctuations. The comoving volume that encompasses our galaxy today contains $\simeq 10^{12} M_\odot \approx 10^{69}$ baryons and $\simeq 10^{79}$ photons. Earlier than 10^{-6} s after the bang this same comoving volume contained $\simeq 10^{79}$ photons and $\simeq 10^{79}$ baryons and antibaryons. In order to avoid the annihilation catastrophe, this volume would need an excess of baryons over antibaryons of $\simeq 10^{69}$, but from statistical fluctuations one would expect $N_b - N_{\bar{b}} \approx 0(N_b^{1/2}) \simeq 3 \times 10^{39}$—a mere 29.5 orders of magnitude too small!

A Causal 'Mystery Interaction'

Clearly, statistical fluctuations are of no help, so consider a hypothetical interaction that separates matter and antimatter. In the standard cosmology the distance over which light signals (and hence causal effects) could have propagated since the bang (the horizon distance) is finite and $\simeq 2t$. When $T \approx 20\,\text{MeV} (t \simeq 10^{-3}\,\text{s})$ causally-coherent regions contained only about $10^{-5} M_\odot$. Thus, in the standard cosmology causal processes could have only separated matter and antimatter into lumps of mass $\lesssim 10^{-5} M_\odot \ll M_{galaxy} \approx 10^{12} M_\odot$. [In the next lecture I will discuss inflationary scenarios; in these scenarios it is possible that the Universe is globally symmetric, while asymmetric locally (within our observable region of the Universe). This is possible because inflation removes the causality constraint.]

It should be clear that the two observations, $n_b \gg n_{\bar{b}}$ on scales at least as large as $10^{12} M_\odot$ and $n_b/n_\gamma \approx (4-7) \times 10^{-10}$, effectively render all baryon-symmetric cosmologies untenable. A viable pre-GUT cosmology needed to have as an initial condition a tiny baryon number, $n_B/s \simeq (6-10) \times 10^{-11}$—a very curious initial condition at that!

2.3. THE INGREDIENTS NECESSARY FOR BARYOGENESIS

More than a decade ago Sakharov[42] suggested that an initially baryon-symmetric Universe might dynamically evolve a baryon excess of $0(10^{-10})$, after which baryon-antibaryon annihilations would destroy essentially all of the antibaryons, leaving the one baryon per 10^{10} photons that we observe today. In his 1967 paper Sakharov outlined the three ingredients necessary for baryogenesis: (a) B-nonconserving interactions; (b) a violation of both C and CP; (c) a departure from thermal equilibrium.

It is clear that B(baryon number) must be violated if the Universe begins baryon symmetric and then evolves a net B. In 1967 there was no motivation for B nonconservation. After all, the proton lifetime is at least 35 orders of magnitude longer than that of any unstable elementary particle—pretty good evidence for B conservation. Of course, grand unification provides just such motivation, and proton decay experiments are likely to detect B nonconservation in the next decade if the proton lifetime is $\lesssim 10^{34}$ years.

Under C (charge conjugation) and CP (charge conjugation combined with parity), the baryon number of a state (B) changes sign. Thus a state that is either C or CP invariant must have $B = 0$. If the Universe begins with equal amounts of matter and antimatter, and without a preferred direction (as in the standard cosmology), then its initial state is both C and CP invariant. Unless both C and CP are violated, the Universe will remain C and CP invariant as it evolves, and thus cannot develop a net baryon number even if B is not conserved. Both C and CP violations are needed to provide an arrow to specify that an excess of matter be produced. C is maximally violated in the weak interactions, and both C and CP are violated in the K° - $\bar{K}^\circ$ system. Although a fundamental understanding of CP violation is still lacking at present, GUTs can (and must) accommodate CP violation. It would be very surprising if CP violation only occurred in the K° -

$\bar{K}^\circ$ system and not elsewhere in the theory also (including the B-nonconserving sector). In fact, without miraculous cancellations the CP violation in the neutral kaon system will give rise to CP violation in the B-nonconserving sector at some level.

The necessity of a departure from thermal equilibrium is a bit more subtle. It is a straightforward exercise to show that CPT and unitary alone are sufficient to guarantee that equilibrium particle phase space distributions are given by: $f(p) = [\exp(\mu/T + E/T) \pm 1]^{-1}$. In equilibrium, processes like $\gamma + \gamma \leftrightarrow b + \bar{b}$ imply that $\mu_b = -\mu_{\bar{b}}$, while processes like (but not literally) $\gamma + \gamma \leftrightarrow b + b$ require that $\mu_b = 0$. Since $E^2 = p^2 + m^2$ and $m_b = m_{\bar{b}}$ by CPT, it follows that in thermal equilibrium, $n_b \equiv n_{\bar{b}}$. [Note, the number density $n = \int d^3p f(p)/(2\pi)^3$.]

Because the temperature of the Universe is changing on a characteristic timescale H^{-1}, thermal equilibrium can only be maintained if the rates for reactions that drive the Universe to equilibrium are much greater than H. Departures from equilibrium have occurred often during the history of the Universe. For example, because the rate for $\gamma + \text{matter} \to \gamma' + \text{matter}'$ is $\ll H$ today, matter and radiation are not in equilibrium, and nucleons do not all reside in ^{56}Fe nuclei (thank God!).

2.4. THE STANDARD SCENARIO: OUT-OF-EQUILIBRIUM DECAY

The basic idea of baryogenesis has been discussed by many authors[43-48]. The model that incorporates the three ingredients discussed above and that has become the "standard scenario" is the so-called out-of-equilibrium decay scenario. I now describe the scenario in some detail.

Denote by "X" a superheavy ($\gtrsim 10^{14}$ GeV) boson whose interactions violate B conservation. X might be a gauge or a Higgs boson (e.g., the XY gauge bosons in SU(5), or the color triplet component of the 5 dimensional Higgs). [Scenarios in which the X particle is a superheavy fermion have also been suggested.] Let its coupling strength to fermions be $\alpha^{1/2}$, and its mass be M. From dimensional considerations its decay rate $\Gamma_D = \tau^{-1}$ should be

$$\Gamma_D \approx \alpha M. \qquad (2.2)$$

At the Planck time ($\simeq 10^{-43}$ s) let us assume that the Universe is baryon symmetric ($n_B/s = 0$), with all fundamental particle species (fermions, gauge and Higgs bosons) present with equilibrium distributions. At this epoch $T \simeq g_*^{-1/4} m_{pl} \simeq 3 \times 10^{18}$ GeV $\gg M$. (Here I have taken $g_* \simeq O(100)$; in minimal SU(5) $g_* \approx 160$.) At the Planck time $X, \bar{X}$ bosons should be very relativistic (as $T \gg M$) and up to statistical factors about as abundant as photons: $n_x = n_{\bar{x}} \approx n_\gamma$. Nothing of importance occurs until $T \approx M$.

For $T \lesssim M$ the equilibrium abundance of $X, \bar{X}$ bosons relative to photons is

$$X_{EQ} \simeq (M/T)^{3/2} \exp(-M/T), \qquad (2.3)$$

where $X \equiv n_x/n_\gamma$ is just the number of $X, \bar{X}$ bosons per comoving volume. In order for $X, \bar{X}$ bosons to maintain an equilibrium abundance as T falls below M, they must be able to diminish in number rapidly compared to $H = |\dot{T}/T|$. The most important process in this regard is decay; other processes (e.g., annihilation) are higher order in α and self-limiting. If $\Gamma_D \gg H$ for $T = M$, then X, $\bar{X}$ bosons can adjust their abundance (by decay) rapidly enough so that X "tracks" the equilibrium value. In this case thermal equilibrium is maintained and no asymmetry is expected to evolve.

More interesting is the case where $\Gamma_D < H \approx 1.66 g_*^{1/2} T^2/m_{pl}$ when $T = M$, or equivalently $M > g_*^{-1/2} \alpha 10^{19} \text{GeV}$. In this case, X, $\bar{X}$ bosons are not decaying on the expansion timescale ($\tau > t$) and so remain as abundant as photons ($X \simeq 1$) for $T \lesssim M$; hence they are overabundant relative to their equilibrium number. This overabundance (indicated with an arrow in Fig. 2.1) is the departure from thermal equilibrium. Much later, when $T \ll M$, $\Gamma_D \approx H$ (i.e. $t \approx \tau$), and X, $\bar{X}$ bosons begin to decrease in number as a result of decays. To a good approximation they decay freely since the fraction of fermion pairs with sufficient center-of-mass energy to produce an X or $\bar{X}$ is $\simeq \exp(-M/T) \ll 1$, which greatly suppresses inverse decay processes ($\Gamma_{ID} \approx \exp(-M/T)\Gamma_D \ll H$). Fig. 2.1 summarizes the time evolution of X; Fig. 2.2 shows the relationship of the various rates (Γ_D, Γ_{ID}, and H) as a function of $M/T (\propto t^{1/2})$.

Now consider the decay of X and $\bar{X}$ bosons: suppose X decays to channels 1 and 2 with baryon numbers B_1 and B_2, and branching ratios r and (1-r). Denote the corresponding quantities for $\bar{X}$ by $-B_1, -B_2, \bar{r}$, and $(1-\bar{r})$ [e.g., $1 = (\bar{q}\bar{q}), 2 = (q\ell), B_1 = -2/3$, and $B_2 = 1/3$]. The mean net baryon number of the decay products of the X and $\bar{X}$ are, respectively, $B_x = rB_1 + (1-r)B_2$ and $B_{\bar{x}} = -\bar{r}B_1 - (1-\bar{r})B_2$. Hence the decay of an X, $\bar{X}$ pair on average produces a baryon number ε,

$$\varepsilon \equiv B_x + B_{\bar{x}} = (r - \bar{r})(B_1 - B_2). \tag{2.4}$$

If $B_1 = B_2$, or $r = \bar{r}$, $\varepsilon = 0$. If $B_1 = B_2$ X could have been assigned a baryon number B_1, and B would not be violated by X, $\bar{X}$ bosons.

It is simple to show that $r = \bar{r}$ unless both C and CP are violated. Let $\bar{X}$ = the charge conjugate of X, and $r_\uparrow, r_\downarrow, \bar{r}_\uparrow, \bar{r}_\downarrow$ denote the respective branching ratios in the upward and downward directions. [For simplicity, I have reduced the angular degree of freedom to up and down.] The quantities r and $\bar{r}$ are branching ratios averaged over angle: $r = (r_\uparrow + r_\downarrow)/2, \bar{r} = (\bar{r}_\uparrow + \bar{r}_\downarrow)/2$ and $\varepsilon = (r_\uparrow - \bar{r}_\uparrow + r_\downarrow - \bar{r}_\downarrow)/2$. If C is conserved, $r_\uparrow = \bar{r}_\uparrow$ and $r_\downarrow = \bar{r}_\downarrow$, and $\varepsilon = 0$. If CP is conserved $r_\uparrow = \bar{r}_\downarrow$ and $r_\downarrow = \bar{r}_\uparrow$, and once again $\varepsilon = 0$.

When the X, $\bar{X}$ bosons decay ($T \ll M, t \approx \tau$), $n_x = n_{\bar{x}} \approx n_\gamma$. Therefore, the net baryon number density produced is $n_B \approx \varepsilon n_\gamma$. The entropy density $s \approx g_* n_\gamma$, and so the baryon asymmetry produced is $n_B/s \approx \varepsilon/g_* \approx 10^{-2}\varepsilon$.

Recall that the condition for a departure from equilibrium to occur is: $(\Gamma_D/H)|_{T=M} < 1$ or $M > g_*^{-1/2} \alpha m_{pl}$. If X is a gauge boson then $\alpha \approx 1/45$, and so M must be $\gtrsim 10^{16} \text{GeV}$. If X is a Higgs boson, then α is essentially arbitrary, although $\alpha \approx (m_f/M_W)^2 \alpha_{gauge} \approx 10^{-3} - 10^{-6}$ if the X is in the same representation as the light Higgs bosons responsible for giving mass to the fermions

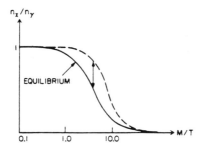

Figure 2.1: The abundance of X bosons relative to photons. The broken curve shows the actual abundance, while the solid curve shows the equilibrium abundance.

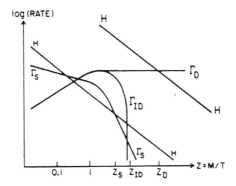

Figure 2.2: Important rates as a function of $z = M/T$. H is the expansion rate, Γ_D the decay rate, Γ_{ID} the inverse decay rate, and Γ_s the $2 \leftrightarrow 2$ $\not{B}$ scattering rate. Upper line marked H corresponds to case where $K \ll 1$; lower line the case where $K > 1$. For $K \ll 1$, X's decay when $z = z_D$; for $K > 1$, freeze out of IDs and S occur at $z = z_{ID}$ and z_s.

(here m_f = fermion mass, M_W = mass of the W boson ≈ 83 GeV). It is apparently easier for Higgs bosons to satisfy this mass condition than it is for gauge bosons. If $M > g_*^{-1/2} \alpha m_{pl}$, then only a modest C, CP-violation ($\varepsilon \approx 10^{-8}$) is necessary to explain $n_B/s \approx (6-10) \times 10^{-11}$. As I will discuss below ε is expected to be larger for a Higgs boson than for a gauge boson. For both these reasons a Higgs boson is the more likely candidate for producing the baryon asymmetry.

2.5 NUMERICAL RESULTS

Boltzmann equations for the evolution of n_B/s have been derived and solved numerically in refs. 49, 50. They basically confirm the correctness of the qualitative picture discussed above, albeit, with some important differences. The results can best be discussed in terms of

$$K \equiv \frac{\Gamma_D}{2H(M)} \simeq \alpha m_{pl}/3g_*^{1/2}M, \tag{2.5}$$

$$\simeq 3 \times 10^{17} \alpha \text{GeV}/M. \tag{2.6}$$

K measures the effectiveness of decays, i.e., rate relative to the expansion rate. K also measures the effectiveness of B-nonconserving processes in general because the decay rate characterizes the rates in general for B nonconserving processes, for $T \lesssim M$ (when all the action happens):

$$\Gamma_{ID} \simeq (M/T)^{3/2} \exp(-M/T) \, \Gamma_D, \tag{2.7}$$

$$\Gamma_s \simeq A\alpha(T/M)^5 \, \Gamma_D, \tag{2.8}$$

where Γ_{ID} is the rate for inverse decays (ID), and Γ_S is the rate for $2 \leftrightarrow 2$ B nonconserving scatterings (S) mediated by X. [A is a numerical factor which depends upon the number of scattering channels, etc, and is typically 0(100-1000).]

[It is simple to see why $\Gamma_s \propto \alpha(T/M)^5\Gamma_D \propto \alpha^2 T^5/M^4$. $\Gamma_s \simeq n\langle\sigma v\rangle$; $n \simeq T^3$ and for $T < M$, $\langle\sigma v\rangle \propto \alpha T^2/M^4$. Note, in some supersymmetric GUTs, there exist fermionic partners of superheavy Higgs which mediate B-nonconservation (and also lead to dim-5 $\cancel{B}$ operators). In this case $\langle\sigma v\rangle \propto \alpha^2/M^2$ and $\Gamma_s \simeq A\alpha(T/M)^3\Gamma_D$, and $2 \leftrightarrow 2$ $\cancel{B}$ scatterings are much more important.]

The time evolution of the baryon asymmetry $(n_B/s$ vs $z = M/T \propto t^{1/2})$ and the final value of the asymmetry which evolves are shown in Figs. 2.3 and 2.4 respectively. For $K < 1$ all B nonconserving processes are ineffective (rate $<$ H) and the asymmetry which evolves is just ε/g_* (as predicted in the qualitative picture). For $K_c > K > 1$, where K_c is determined by

$$K_c(\ln K_c)^{-2.4} \simeq 300/A\alpha, \tag{2.9}$$

$2 \leftrightarrow 2$ B nonconserving scatterings 'freeze-out' before ID's do and can be ignored. Equilibrium is maintained to some degree (by Ds and IDs), however a sizeable asymmetry still evolves

$$n_B/s \simeq (\varepsilon/g_*)0.3K^{-1}(\ln K)^{-0.6}. \tag{2.10}$$

This is the surprising result: for $K_c > K \gg 1$, equilibrium is not well maintained and a significant n_B/s evolves, whereas the qualitative picture would suggest that for $K \gg 1$ no asymmetry should evolve. For $K > K_c$, S are very important, and the n_B/s which evolves becomes exponentially small:

$$n_B/s \simeq (\varepsilon/g_*)(AK\alpha)^{1/2} \exp[-4(AK\alpha)^{1/4}/3]. \quad (2.11)$$

[In supersymmetric models which have dim-5 $\not{B}$ operators, $K_c(\ln K_c)^{-1.2} \simeq 18/A\alpha$ and the analog of Eqn(2.9) for $K > K_c$ is: $n_B/s \simeq (\varepsilon/g_*)A\alpha K \exp[-2(A\alpha K)^{1/2}]$.]

For the XY gauge bosons of SU(5) $\alpha \simeq 1/45$, $A \simeq$ few $\times 10^3$, and $M \simeq$ few $\times 10^{14} GeV$, so that $K_{XY} \simeq 0(30)$ and $K_c \simeq 100$. The asymmetry which could evolve due to these bosons is $\simeq 10^{-2}(\varepsilon_{XY}/g_*)$. For a color triplet Higgs $\alpha_H \simeq 10^{-3}$ (for a top quark mass of 40 GeV) and $A \simeq$ few $\times 10^3$, leading to $K_H \simeq 3 \times 10^{14}$ GeV/M_H and $K_c \simeq$ few $\times 10^3$. For $M_H \lesssim 3 \times 10^{14}$ GeV, $K_H < 1$ and the asymmetry which could evolve is $\simeq \varepsilon_H/g_*$.

2.6. VERY OUT-OF-EQUILIBRIUM DECAY

If the X boson decays very late, when $M \gg T$ and $\rho_x > \rho_{rad}$, the additional entropy released in its decays must be taken into account. This is very easy to do. Before the Xs decay, $\rho = \rho_x + \rho_{rad} \simeq \rho_x = Mn_x$. After they decay $\rho_x \simeq \rho_{rad} = \pi^2 g_* T_{RH}^4/30 \simeq 0.75 s T_{RH} (s, T_{RH}$ are the entropy density and temperature after the X decays). As usual assume that on average each decay produces a mean net baryon number ε. Then the result in n_B/s produced is

$$\frac{n_B}{s} = \frac{\varepsilon n_x}{s}, \quad (2.12)$$

$$\simeq 0.75\varepsilon(T_{RH}/M). \quad (2.13)$$

[Note, I have assumed that when the X's decay $\rho_x \gg \rho_{rad}$ so that the initial entropy can be ignored compared to entropy produced by the decays; this assumption guarantees that $T_{RH} \lesssim M$. I have also assumed that $T \ll M$ so that IDs and S processes can be ignored. Finally, note that how the X's produce a baryon number of ε per X is irrelevant; it could be by $X \to$ q's l's, or equally well by $X \to \phi's \to$ q's l's (ϕ = any other particle species).]

Note that the asymmetry produced depends upon the ratio T_{RH}/M and not T_{RH} itself—this is of some interest in inflationary scenarios in which the Universe does not reheat to a high enough temperature for baryogenesis to proceed in the standard way (out-of-equilibrium decays). For reference T_{RH} can be calculated in terms of $\tau_x \simeq \Gamma^{-1}$; when the Xs decay ($t \simeq \tau_x, H \simeq t^{-1} \simeq \Gamma$): $\Gamma^2 = H^2 = 8\pi\rho_x/3m_{pl}^2$. Using the fact that $\rho_x \simeq g_*(\pi^2/30)T_{RH}^4$ it follows that

$$T_{RH} \simeq g_*^{-1/4}(\Gamma m_{pl})^{1/2} \quad (2.14)$$

2.7. THE C, CP VIOLATION ε

The crucial quantity for determining n_B/s is ε—the C, CP violation in the superheavy boson system. Lacking 'The GUT', ε cannot be calculated precisely, and hence n_B/s cannot be predicted, as, for example, the ^4He abundance can be.

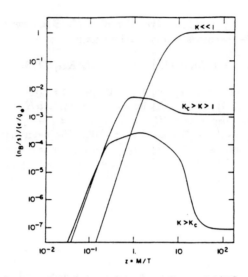

Figure 2.3: Evolution of n_B/s as a function of $z = M/T (\sim t^{1/2})$. For $K \ll 1$, n_B/s is produced when the X, $\bar{X}$ bosons decay out-of-equilibrium ($z \gg 1$). For $K_c > K > 1$, $n_B/s \propto z^{-1}$ (due to IDs) until the IDs freeze out ($z \simeq 10$). For $K > K_c$ $2 \leftrightarrow 2$ $\not{B}$ scatterings are important, and n_B/s decreases very rapidly until they freeze out ($z \simeq z_S$).

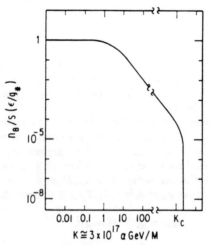

Figure 2.4: The final baryon asymmetry (in units of ε/g_*) as a function of $K \simeq 3 \times 10^{17} \alpha$ GeV/M. For $K \lesssim 1$, n_B/s is independent of K and $\simeq \varepsilon/g_*$. For $K_c > K > 1$, n_B/s decreases slowly, $\propto 1/(K(\ln K)^{0.6})$. For $K > K_c$ (when $2 \leftrightarrow 2$ $\not{B}$ scatterings are important), n_B/s decreases exponentially with $K^{1/4}$.

The quantity $\varepsilon \propto (r - \bar{r})$; at the tree graph (i.e., Born approximation) level $r - \bar{r}$ must vanish. Non-zero contributions to $(r - \bar{r})$ arise from higher order loop corrections due to Higgs couplings which are complex[47,51-55]. For these reasons, it is generally true that:

$$\varepsilon_{Higgs} \lesssim 0(\alpha^N) \sin \delta \qquad (2.15)$$
$$\varepsilon_{gauge} \lesssim 0(\alpha^{N+1}) \sin \delta, \qquad (2.16)$$

where α is the coupling of the particle exchanged in loop (i.e., $\alpha = g^2/4\pi$), $N \geq 1$ is the number of loops in the diagram which make the lowest order, non-zero contributions to $(r - \bar{r})$, and δ is the phase of some complex coupling. The C, CP violation in the gauge boson system occurs at 1 loop higher order than in the Higgs because gauge couplings are necessarily real. Since $\alpha \lesssim \alpha_{gauge}$, ε is at most $0(10^{-2})$—which is plenty large enough to explain $n_B/s \simeq 10^{-10}$. Because K for a Higgs is likely to be smaller, and because C, CP violation occurs at lower order in the Higgs boson system, the out-of-equilibrium decay of a Higgs is the more likely mechanism for producing n_B/s. [No additional cancellations occur when calculating $(r-\bar{r})$ in supersymmetric theories, so these generalities also hold for supersymmetric GUTs.]

In minimal SU(5)—one $\underline{5}$ and one $\underline{24}$ of Higgs, and three families of fermions, $N = 3$. This together with the smallness of the relevant Higgs couplings implies that $\varepsilon_H \lesssim 10^{-15}$ which is not nearly enough[47,51,52]. With 4 families the relevant couplings can be large enough to obtain $\varepsilon_H \simeq 10^{-8}$ - if the top quark and fourth generation quark/lepton masses are $0(M_W)$ (ref. 53). By enlarging the Higgs sector (e.g., by adding a second $\underline{5}$ or a $\underline{45}$), $(r - \bar{r})$ can be made non-zero at the 1-loop level, making $\varepsilon_H \simeq 10^{-8}$ easy to achieve.

In more complicated theories, e.g., E6, SO(10), etc., $\varepsilon \simeq 10^{-8}$ can also easily be achieved. However, to do so restricts the possible symmetry breaking patterns. Both E6 and SO(10) are C-symmetric, and of course C-symmetry must be broken before ε can be non-zero. In general, in these models ε is suppressed by powers of M_C/M_G where $M_C(M_G)$ is the scale of C(GUT) symmetry breaking, and so M_C cannot be significantly smaller than M_G.

It seems very unlikely that ε can be related to the parameters of the $K^\circ - \bar{K}^\circ$ system, the difficulty being that not enough C, CP violation can be 'fed up' to the superheavy boson system. It has been suggested that ε could be related to the electric dipole moment of the neutron[54].

Although baryogenesis is nowhere near being on the same firm footing as primordial nucleosynthesis, we now at least have for the first time a very attractive framework for understanding the origin of $n_B/s \simeq 10^{-10}$. A framework which is so attractive, that in the absence of observed proton decay, the baryon asymmetry of the Universe is probably the best evidence for some kind of quark/lepton unification.

There are several very interesting aspects of baryogenesis which I have not mentioned. Affleck and Dine have discussed a very interesting mechanism for generating a baryon asymmetry at low temperatures in supersymmetric models[56]. Very recently, Kuzmin, Rubakov, and Shaposnikov have pointed out that the rate for B nonconservation through the electroweak anomaly (B+L is anomalous in the standard SU(2) × U(1) theory) can be large ($\Gamma \gg H$) at temperatures $\simeq 100 - 200$ GeV, and can strongly damp any baryon and lepton asymmetry with $B - L = 0$. (In SU(5) any baryon/lepton asymmetry produced must necessarily have $B - L = 0$ since $B - L$ is conserved in SU(5); for larger gauge groups, e.g., SO(10), E6, etc., $B - L$ is not necessarily conserved, and $B - L \simeq 0(B, L)$ also typically evolves.) B nonconservation via the electroweak anomaly and its cosmological implications are discussed in detail in refs. 57, 58. [In writing up this lecture I have borrowed freely and heavily from the review on baryogenesis written by myself and E. W. Kolb (ref. 55), and refer the interested reader there for a more thorough discussion of the details of baryogenesis.]

LECTURE 3:
TOWARD THE INFLATIONARY PARADIGM

3.1. OVERVIEW

Guth's inflationary Universe scenario has revolutionized our thinking about the very early Universe. The inflationary scenario offers the possibility of explaining a handful of very fundamental cosmological facts—the homogeneity, isotropy, and flatness of the Universe, the origin of density inhomogeneities and the origin of the baryon asymmetry, while at the same time avoiding the monopole problem. It is based upon microphysical events which occurred early ($t \leq 10^{-34}$ sec) in the history of the Universe, but well after the planck epoch ($t \geq 10^{-43}$ sec). While Guth's original model was fundamentally flawed, the variant based on the slow-rollover transition proposed by Linde, and Albrecht and Steinhardt (dubbed 'new inflation') appears viable. Although old inflation and the earliest models of new inflation were based upon first order phase transitions associated with spontaneous-symmetry breaking (SSB), it now appears that the inflationary transition is a much more generic phenomenon, being associated with the evolution of a weakly-coupled scalar field which for some reason or other was initially displaced from the minimum of its potential. Models now exist which are based on a wide variety of microphysics: SSB, SUSY/SUGR, compactification of extra dimensions, R^2 gravity, induced gravity, and some random, weakly-coupled scalar field. While there are several models which successfully implement the inflation, none is particularly compelling and all seem somewhat ad hoc. The common distasteful feature of all the successful models is the necessity of a small dimensionless number in the model—usually in the form of a dimensionless coupling of order 10^{-15}. And of course, all inflationary scenarios rely upon the assumption that vacuum energy (or equivalently a cosmological term) was once dynamically very significant, whereas today there exists every evidence that it is not (although we have no understanding why it is not). For these reasons

I have entitled this lecture *Toward the Inflationary Paradigm*. I have divided my lecture into the following sections: Shortcomings of the standard cosmology; New inflation—the slow-rollover transition; Scalar field dynamics; Origin of density inhomogeneities; Specific models, I. Interesting failures; Lessons learned—prescription for successful inflation; Two models that work; The Inflationary paradigm; Loose ends; and Inflation confronts observation.

3.2. SHORTCOMINGS OF THE STANDARD COSMOLOGY

The standard cosmology is very successful—it provides us with a reliable framework for describing the history of the Universe as early as 10^{-2} sec after the bang (when the temperature was about 10 MeV) and perhaps as early as 10^{-43} sec after the bang (see Fig. 1.5). In sum, the standard cosmology is a great achievement. [There is nothing in our present understanding of physics that would indicate that it is incorrect to extrapolate the standard cosmology back to times as early as 10^{-43} sec—the fundamental constituents of matter, quarks and leptons, are point-like particles and their known interactions should remain 'weak' up to energies as high as 10^{19} GeV—justifying the dilute gas approximation made in writing $\rho_r \propto T^4$. (This fact was first pointed out by Collins and Perry[59]). However, at times earlier than 10^{-43} sec, corresponding to temperatures greater than 10^{19} GeV, quantum corrections to general relativity—a classical theory, should become very significant.]

However, it is not without its shortcomings. There are a handful of very important and fundamental cosmological facts which, while it can accommodate, it in no way elucidates. I will briefly review these puzzling facts.

(1-2) Large-scale Isotropy and Homogeneity

The observable Universe (size $\simeq H^{-1} \simeq 10^{28}$cm $\simeq 3000$ h^{-1} Mpc) is to a high degree of precision isotropic and homogeneous on the largest scales, say > 100Mpc. [Of course, our knowledge of the Universe outside our past light cone is very limited; see ref. 60.] The best evidence for the isotropy and homogeneity is provided by the uniformity of the cosmic background temperature (see Fig. 1.2): $(\delta T/T) < 10^{-3}$ (10^{-4} if the dipole anisotropy is interpreted as being due to our motion relative to the cosmic rest frame). Large-scale density inhomogeneities or anisotropic expansion would result in temperature fluctuations of comparable magnitude (see refs. 61, 62). The smoothness of the observed Universe is puzzling if one wishes to understand it as being due to causal, microphysical processes which operated during the early history of the Universe. Our Hubble volume today contains an entropy of about 10^{88}. At decoupling ($t \simeq 6 \times 10^{12}(\Omega h^2)^{-1/2}$ sec, $T \simeq 1/3$eV), the last epoch when matter and radiation were known to be interacting vigorously and particle interactions might have been able to smooth the radiation, the entropy within the horizon was only about 8×10^{82}; that is, the comoving volume which contains the presently-observable Universe, then was comprised of about 2×10^5 causally-distinct regions. How is it that they came to be homogeneous? Put another way, the particle horizon at decoupling only subtends an angle of about 1/2° on the sky today—how is it that the cosmic background temperature is so uniform on angular scales much greater than this?

The standard cosmology can accommodate these facts—after all the FRW cosmology is exactly isotropic and homogeneous, but at the expense of very special initial data. In 1973 Collins and Hawking[63] showed that the set of initial data which evolve to a Universe which globally is as smooth as ours has measure zero (provided that the strong and dominant energy conditions are always satisfied).

(3) Small-scale Inhomogeneity

As any real astronomer will gladly testify, the Universe is very lumpy—stars, galaxies, clusters of galaxies, superclusters, etc. Today, the density contrast on the scale of galaxies is: $\delta\rho/\rho \simeq 10^5$. The fact that the microwave background radiation is very uniform even on very small angular scales ($\ll 1°$) indicates that the Universe was smooth even on the scale of galaxies at decoupling. [The relationship between the angle on the sky and mass contained within the corresponding length scale at decoupling is: $\theta \simeq 1'(M/10^{12}M_\odot)^{1/3}\Omega^{-1/3}h^{1/3}$.] On small angular scales: $\delta T/T \simeq c(\delta\rho/\rho)_{dec}$, where the numerical constant $c \simeq 10^{-1} - 10^{-2}$ (see ref. 62 for further details). Whence came the structure which today is so conspicuous?

Once matter decouples from the radiation and is free of the pressure support provided by the radiation, any density inhomogeneities present will grow via the Jeans (or gravitational instability)—in the linear regime, $\delta\rho/\rho \propto R(t)$. [If the mass density of the Universe is dominated by a collisionless particle species, e.g., a light, relic neutrino species or relic axions, density perturbations in these particles can begin to grow as soon as the Universe becomes matter-dominated.] In order to account for the present structure, density perturbations of amplitude $\sim 10^{-3}$ or so at decoupling are necessary on the scale of galaxies. The standard cosmology sheds no light as to the origin or nature (spectrum and type—adiabatic or isothermal) of the primordial density perturbations so crucial for understanding the structure observed in the Universe today. [For a review of the formation of structure in the Universe according to the gravitational instability picture, see ref. 64.]

(4) Flatness (or Oldness) of the Universe

The observational data suggest that

$$0.01 \leq \Omega \leq \text{few}.$$

Ω is related to both the expansion rate of the Universe and the curvature radius of the Universe:

$$\Omega = 8\pi G\rho/3H^2 \equiv H_{crit}^2/H^2, \tag{3.1}$$

$$|\Omega - 1| = (H^{-1}/R_{curv})^2, \tag{3.2}$$

The fact that Ω is not too different from unity today implies that the present expansion rate is close to the critical expansion rate and that the curvature radius of the Universe is comparable to or larger than the Hubble radius. As the Universe expands Ω does not remain constant, but evolves away from 1

$$\Omega = 1/(1 - x(t)), \tag{3.3}$$

$$x(t) = (k/R^2)/(8\pi G\rho/3), \tag{3.4}$$

$$x(t) \propto \begin{cases} R(t)^2 & \text{radiation - dominated} \\ R(t) & \text{matter - dominated} \end{cases}$$

That Ω is still of order unity means that at early times it was equal to 1 to a very high degree of precision:

$$|\Omega(10^{-43} \text{ sec}) - 1| \simeq O(10^{-60}),$$
$$|\Omega(1 \text{ sec}) - 1| \simeq O(10^{-16}).$$

This in turn implies that at early times the expansion rate was equal to the critical rate to a high degree of precision and that the curvature of the Universe was much, much greater than the Hubble radius. If it were not, i.e., suppose that $|(k/R^2)/(8\pi G\rho/3)| \simeq O(1)$ at $t \simeq 10^{-43}$ sec, then the Universe would have collapsed after a few Planck times $(k > 0)$ or would have quickly become curvature-dominated, $(k < 0)$, in which case $R(t) \propto t$ and $t(T = 3K) \leq 10^{-11}$ sec! Why was this so?

The so-called flatness problem has sometimes been obscured by the fact that it is conventional to rescale $R(t)$ so that $k = -1$, 0, or $+1$, making it seem as though there are but three FRW models. However, that clearly is not the case; there are an infinity of models, specified by the curvature radius $R_{curv} = R(t)|k|^{-1/2}$ at some given epoch, say the planck epoch. Our model corresponds to one with a curvature radius that exceeds its initial Hubble radius by 30 orders-of-magnitude. Again, this fact can be accommodated by FRW models, but the extreme flatness of our Universe is in no way explained by the standard cosmology. [The flatness problem and the naturalness of the $k = 0$ model have been emphasized by Dicke and Peebles[65].]

(5) Baryon Number of the Universe

There is ample evidence (see ref. 66) for the dearth of antimatter in the observable Universe. That fact together with the baryon-to-photon ratio ($\eta \simeq 4 - 7 \times 10^{-10}$) means that our Universe is endowed with a net baryon number, quantified by the baryon number-to-entropy ratio

$$n_B/s \simeq (6 - 10) \times 10^{-11},$$

which in the absence of baryon number non-conserving interactions or significant entropy production is proportional to the constant net baryon number per comoving volume which the Universe has always possessed. Until five or so years ago this very fundamental number was without explanation. Of course it is now known that in the presence of interactions that violate B, C, and CP a net baryon asymmetry will evolve dynamically. Such interactions are predicted by Grand Unified Theories (or GUTs) and 'baryogenesis' is one of the great triumphs of the marriage of grand unification and cosmology. [See ref. 67 for a review of grand unification.] If the baryogenesis idea is correct, then the baryon asymmetry of the Universe is subject to calculation just as the primordial Helium abundance is. Although the idea is very attractive and certainly appears to be qualitatively correct, a precise calculation of the baryon number-to-entropy ratio cannot be performed until The Grand Unified Theory is known. [Baryogenesis is discussed in Lecture 2 and is reviewed in ref. 68.]

(6) The Monopole Problem

If the great success of the marriage of GUTs and cosmology is baryogenesis, then the great disappointment is 'the monopole problem'. 't Hooft-Polyakov monopoles[69] are a generic prediction of GUTs. In the standard cosmology (and for the simplest GUTs) monopoles are grossly overproduced during the GUT symmetry-breaking transition, so much so that the Universe would reach its present temperature of 3K at the very tender age of 30,000 yrs! [For a detailed discussion of the monopole problem, see refs. 70, 71.] Although the monopole problem initially seemed to be a severe blow to the Inner Space/Outer Space connection, as it has turned out it provided us with a valuable piece of information about physics at energies of order 10^{14} GeV and the Universe at times as early as 10^{-34} sec—the standard cosmology and the simplest GUTs are definitely incompatible! As it turned out, it was the search for a solution to the monopole problem which in the end led Guth to come upon the inflationary Universe scenario[72,73].

(7) The Smallness of the Cosmological Constant

With the possible exception of supersymmetry/supergravity and superstring theories, the absolute scale of the scalar potential $V(\phi)$ is not specified (here ϕ represents the scalar fields in the theory, be they fundamental or composite). A constant term in the scalar potential is equivalent to a cosmological term (the scalar potential contributes a term $V g_{\mu\nu}$ to the stress energy of the Universe[74]). At low temperatures (say temperatures below any scale of spontaneous symmetry-breaking) the constant term in the potential receives contributions from all the stages of SSB—chiral symmetry breaking, electroweak SSB, GUT SSB, etc. The observed expansion rate of the Universe ($H = 100h$ km sec^{-1}Mpc^{-1}) limits the total energy density of the Universe to be

$$\rho_{TOT} \leq O(10^{-46}\,\text{GeV}^4).$$

Making the seemingly very reasonable assumption that all stress energy self-gravitates (which is dictated by the equivalence principle) it follows that the vacuum energy of our $SU(3) \times U(1)$ vacuum must be less than 10^{-46}GeV^4. Compare this to the scale of the various contributions to the scalar potential: $0(M^4)$ for physics associated with a symmetry breaking scale of M

$$V_{today}/M^4 \leq \rho_{TOT}/M^4 \leq \begin{cases} 10^{-122} & M = m_{pl} \\ 10^{-102} & M = 10^{14}\text{GeV} \\ 10^{-56} & M = 300\text{GeV} \\ 10^{-46} & M = 1\text{GeV} \end{cases}$$

At present there is no explanation for the vanishingly small value of the energy density of our very unsymmetrical vacuum. It is easy to speculate that a fundamental

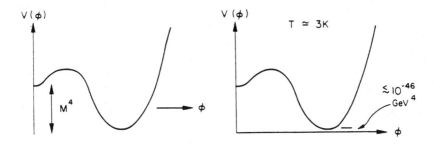

Figure 3.1: In gauge theories the vacuum energy is a function of one or more scalar fields (here denoted collectively as ϕ); however, the absolute energy scale is not set. Vacuum energy behaves like a cosmological term; the present expansion rate of the Universe constrains the value of the vacuum energy today to be $\leq 10^{-46}$ GeV^4.

understanding of the smallness of the cosmological constant will likely involve an intimate link between gravity and quantum field theory.

Today we can be certain the vacuum energy is small and plays a minor role in the dynamics of the expansion of the Universe (compared to the potential role that it could play). If we accept this as an empirical determination of the absolute scale of the scalar potential $V(\phi)$, it then follows that the energy density associated with an expectation value of ϕ near zero is enormous—of order M^4 (see Fig. 3.1) and therefore could have played an important role in the dynamics of the very early Universe. Accepting this empirical determination of the zero of vacuum energy—which is a very great leap of faith, is the starting point for inflation. In fact, the rest of the journey is downhill.

All of these cosmological facts can be accommodated by the standard model, but seemingly at the expense of highly special initial data (the possible exception being the monopole problem). Over the years there have been a number of attempts to try to understand and/or explain this apparent dilemma of initial data. Inflation is the most recent attempt and I believe shows great promise. Let me begin by briefly mentioning the earlier attempts:

* ⋆ Mixmaster Paradigm–Starting with a solution with a singularity which exhibits the features of the most general singular solutions known (the so-called mixmaster model) Misner and his coworkers hoped that they could show that particle viscosity would smooth out the geometry. In part because horizons still effectively exist in the mixmaster solution 'the chaotic cosmology program' has

proven unsuccessful (for further discussion see refs. 75).

* Nature of the Initial Singularity–Penrose[76] explored the possibility of explaining the observed smoothness of the Universe by restricting the kinds of initial singularities which are permitted in Nature (those with vanishing Weyl curvature). In a sense his approach is to postulate a law of physics governing allowed initial data.

* Quantum Gravity Effects–The first two solutions involve appealing to classical gravitational effects. A number of authors have suggested that quantum gravity effects might be responsible for smoothing out the space-time geometry (deWitt[77]; Parker[78]; Zel'dovich[79]; Starobinskii[80]; Anderson[81]; Hartle and Hu[82]; Fischetti et al.[83]). The basic idea being that anisotropy and/or inhomogeneity would drive gravitational particle creation, which due to back reaction effects would eliminate particle horizons and smooth out the geometry. Recently, Hawking and Hartle[84] have advocated the Quantum Cosmology approach to actually compute the initial state. All of these approaches necessarily involve events at times $\lesssim 10^{-43}$ sec and energy densities $\gtrsim m_{pl}^4$.

* Anthropic Principle–Some (see, e.g., refs. 85) have suggested (or in some cases even advocated) 'explaining' many of the puzzling features of the Universe around us (and in some cases, even the laws of physics!) by arguing that unless they were as they are intelligent life would not have been able to develop and observe them! Hopefully we will not have to resort to such an explanation.

The approach of inflation is somewhat different from previous approaches. Inflation (at least from my point-of-view) is based upon well-defined and reasonably well-understood microphysics (albeit, some of it very speculative). That microphysics is:

* Classical Gravity (general relativity), at least as an effective, low-energy theory of gravitation.

* 'Modern Particle Physics'–grand unification, supersymmetry/supergravity, field theory limit of superstring theories, etc. at energy scales $\lesssim m_{pl}$

As I will emphasize, in all viable models of inflation the inflationary period (at least the portion of interest to us) takes place well after the planck epoch, with the energy densities involved being far less than m_{pl}^4 (although semi-classical quantum gravity effects might have to be included as non-renormalizable terms in the effective Lagrangian). Of course, it could be that a resolution to the cosmological puzzles discussed above involves both 'modern particle physics' and quantum gravitational effects in their full glory (as in a fully ten dimensional quantum theory of strings).

I will not take the time or the space here to review the historical development of our present view of inflation; I refer the interested reader to the interesting paper on this subject by Lindley[86]. It suffices to say that Guth's very influential paper of 1981 was the 'shot heard 'round the world' which initiated the inflation revolution,[73] and that Guth's doomed original model (see Guth and Weinberg[87]; Hawking et al[87]) was revived by Linde's[88] and Albrecht and Steinhardt's[89] variant, 'new inflation'. I will focus all of my attention on the present status of the 'slow-rollover' model of Linde[88] and Albrecht and Steinhardt[89].

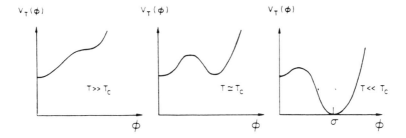

Figure 3.2: The finite temperature effective potential as a function of T (schematic). The Universe is usually assumed to start out in the high temperature, symmetric minimum ($\phi = 0$) of the potential and must eventually evolve to the low temperature, asymmetric minimum ($\phi = \sigma$). The evolution of ϕ from $\phi = 0$ to $\phi = \sigma$ can prove to be very interesting—as in the case of an inflationary transition.

3.3. NEW INFLATION—THE SLOW-ROLLOVER TRANSITION

The basic idea of the inflationary Universe scenario is that there was an epoch when the vacuum energy density dominated the energy density of the Universe. During this epoch $\rho \simeq V \simeq$ constant, and thus $R(t)$ grew exponentially ($\propto \exp(Ht)$), allowing a small, causally-coherent region (initial size $\leq H^{-1}$) to grow to a size which encompasses the region which eventually becomes our presently-observable Universe. In Guth's original scenario[73], this epoch occurred while the Universe was trapped in the false ($\phi = 0$) vacuum during a strongly, first-order phase transition. In new inflation, the vacuum-dominated, inflationary epoch occurs while the region of the Universe in question is slowly, but inevitably, evolving toward the true, SSB vacuum. Rather than considering specific models in this section, I will try to discuss new inflation in the most general context. For the moment I will however assume that the epoch of inflation is associated with a first-order, SSB phase transition, and that the Universe is in thermal equilibrium before the transition. As we shall see later new inflation is more general than these assumptions. But for definiteness (and for historical reasons), let me begin by making these assumptions.

Consider a SSB phase transition characterized by an energy scale M. For $T \geq T_c \simeq 0(M)$ the symmetric ($\phi = 0$) vacuum is favored, i.e., $\phi = 0$ is the global minimum of the finite temperature effective potential $V_T(\phi)$ (=free energy density). As T approaches T_c a second minimum develops at $\phi = 0$, and at $T = T_c$, the two minima are degenerate. At temperatures below T_c the SSB ($\phi = \sigma$) minimum is the global minimum of $V_T(\phi)$ (see Fig. 3.2). However, the Universe does not instantly make the transition from $\phi = 0$ to $\phi = \sigma$; the details and time required are a question of dynamics. [The scalar field ϕ is the order parameter for the SSB transition under discussion; in the spirit of generality ϕ might be a gauge singlet field or might have nontrivial transformation properties under the gauge group, possibly even responsible for the SSB of the GUT.] Once the temperature of the

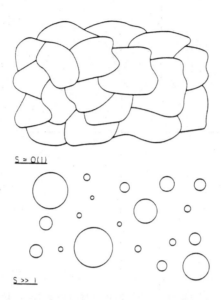

Figure 3.3: If the tunneling action is large ($S \gg 1$), barrier penetration will proceed via bubble nucleation, while in the case that it becomes small ($S \simeq O(1)$), the Universe will fragment into irregularly-shaped fluctuation regions. The very large scale (scale $\gg$ bubble or fluctuation region) structure of the Universe is determined by whether $S \simeq O(1)$—in which case the Universe is comprised of irregularly-shaped domains, or $S \gg O(1)$—in which case the Universe is comprised of isolated bubbles.

Universe drops below $T_c \simeq O(M)$, the potential energy associated with ϕ being far from the minimum of its potential, $V \simeq V(0) \simeq M^4$, dominates the energy density in radiation ($\rho_r < T_c^4$), and causes the Universe to expand exponentially. During this exponential expansion (known as a deSitter phase) the temperature of the Universe decreases exponentially causing the Universe to supercool. The exponential expansion continues so long as ϕ is far from its SSB value. Now let's focus on the evolution of ϕ.

Assuming a barrier exists between the false and true vacua, thermal fluctuations and/or quantum tunneling must take ϕ across the barrier. The dynamics of this process determine when and how the process occurs (bubble formation, spinodal decomposition, etc.) and the value of ϕ after the barrier is penetrated. If the action for bubble nucleation remains large, $S_b \gg 1$, then the barrier will be overcome by the nucleation of Coleman-deLuccia bubbles;[90] on the other hand if the action for bubble nucleation becomes of order unity, then the Universe will undergo spinodal decomposition, and irregularly-shaped fluctuation regions will form (see Fig. 3.3; for a more detailed discussion of the barrier penetration process see refs. 89–91). For definiteness suppose that the barrier is overcome when the temperature is T_{MS} and that after the barrier is penetrated the value of ϕ is ϕ_0. From this point the journey to the true vacuum is downhill (literally). For the moment let us assume that the evolution of ϕ is adequately described by semi-classical equations of motion:

$$\ddot{\phi} + 3H\dot{\phi} + \Gamma\dot{\phi} + V' = 0, \qquad (3.5)$$

where ϕ has been normalized so that its kinetic term in the Lagrangian is $1/2\partial_\mu\phi\partial^\mu\phi$, and prime indicates a derivative with respect to ϕ. The subscript T on V has been dropped; for $T \ll T_c$ the temperature dependence of V_T can be neglected and the zero temperature potential ($\equiv V$) can be used. The $3H\dot{\phi}$ term acts like a frictional force, and arises because the expansion of the Universe 'redshifts away' the kinetic energy of $\phi(\propto R^{-3})$. The $\Gamma\dot{\phi}$ term accounts for particle creation due to the time-variation of ϕ[refs. 92, 93]. The quantity Γ is determined by the particles which couple to ϕ and the strength with which they couple ($\Gamma^{-1} \simeq$ lifetime of a ϕ particle). As usual, the expansion rate H is determined by the energy density of the Universe:

$$H^2 = 8\pi G\rho/3, \qquad (3.6)$$

$$\rho \simeq 1/2\dot{\phi}^2 + V(\phi) + \rho_r, \qquad (3.7)$$

where ρ_r represents the energy density in radiation produced by the time variation of ϕ. [For $T_{MS} \ll T_c$ the original thermal component makes a negligible contribution to ρ.] The evolution of ρ_r is given by

$$\dot{\rho}_r + 4H\rho_r = \Gamma\dot{\phi}^2, \qquad (3.8)$$

where the $\Gamma\dot{\phi}^2$ term accounts for particle creation by ϕ.

In writing Eqns(3.5–3.8) I have implicitly assumed that ϕ is spatially homogeneous. In some small region (inside a bubble or a fluctuation region) this will be a good approximation. The size of this smooth region will turn out to be unimportant; take it to be of order the 'Physics Horizon', H^{-1}—certainly, it is not likely to be larger. Now follow the evolution of ϕ within the small, smooth patch of size H^{-1}.

If $V(\phi)$ is sufficiently flat somewhere between $\phi = \phi_0$ and $\phi = \sigma$, then ϕ will evolve very slowly in that region, and the motion of ϕ will be 'friction-dominated' so that $3H\dot{\phi} \simeq -V'$ (in the slow growth phase particle creation is not important[94]). If V is sufficiently flat, then the time required for φ to transverse the flat region can be long compared to the expansion timescale H^{-1}; for definiteness say, $\tau_\phi = 100H^{-1}$. During this slow growth phase $\rho \simeq V(\phi) \simeq V(\phi = 0)$; both ρ_r and $1/2\dot{\phi}^2$ are $\ll V(\phi)$. The expansion rate H is then just

$$\begin{aligned} H &\simeq (8\pi V(0)/3m_{pl}^2)^{1/2} \\ &\simeq O(M^2/m_{pl}), \end{aligned} \qquad (3.9)$$

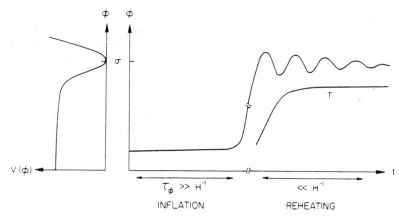

Figure 3.4: Evolution of ϕ and the temperature inside the bubble or fluctuation region (schematic). Early on ϕ evolves slowly (relative to the expansion timescale), then as the potential steepens ϕ evolves rapidly (on the expansion timescale). The oscillations of ϕ are damped by particle creation, which leads to the reheating of the bubble or fluctuation region.

where $V(0)$ is assumed to be of order M^4. While $H \simeq$ constant, R grows exponentially: $R \propto \exp(Ht)$; for $\tau_\phi = 100 H^{-1}$, R expands by a factor of e^{100} during the slow rolling period, and the physical size of the smooth region increases to $e^{100} H^{-1}$.

As the potential steepens, the evolution of ϕ quickens. Near $\phi = \sigma$, ϕ oscillates around the SSB minimum with frequency m_ϕ : $m_\phi^2 \simeq V''(\sigma) \simeq O(M^2) \gg H^2 \simeq M^4/m_{pl}^2$. As ϕ oscillates about $\phi = \sigma$ its motion is damped both by particle creation and the expansion of the Universe. If $\Gamma^{-1} \ll H^{-1}$, then coherent field energy density $(V + 1/2\dot\phi^2)$ is converted into radiation in less than an expansion time $(\Delta t_{RH} \simeq \Gamma^{-1})$, and the patch is reheated to a temperature $T \simeq O(M)$—the vacuum energy is efficiently converted into radiation ('good reheating'). On the other hand, if $\Gamma^{-1} \gg H^{-1}$, then ϕ continues to oscillate and the coherent field energy redshifts away with the expansion: $(V + 1/2\dot\phi^2) \propto R^{-3}$—the coherent energy behaves like non-relativistic matter. Eventually, when $t \simeq \Gamma^{-1}$ the energy in radiation begins to dominate that in coherent field oscillations, and the patch is reheated to a temperature $T \simeq (\Gamma/H)^{1/2} M \simeq (\Gamma m_{pl})^{1/2} \ll M$ ('poor reheating'). The evolution of ϕ is summarized schematically in Fig. 3.4. In the next section I will discuss the all-important scalar field dynamics in great detail.

For the following discussion let us assume 'good reheating' ($\Gamma \gg H$). After reheating the patch has a physical size $e^{100} H^{-1}$ ($\simeq 10^{17}$cm for $M \simeq 10^{14}$GeV), is at a temperature of order M, and in the approximation that ϕ was initially constant throughout the patch, the patch is exactly smooth. From this point forward the region evolves like a radiation-dominated FRW model. How have the cosmological conundrums been 'explained'?

First, *the homogeneity and isotropy*; our observable Universe today ($\simeq 10^{28}$cm) had a physical size of about 10cm ($= 10^{28}$cm $\times$ $3K/10^{14}$GeV) when T was 10^{14}GeV—thus it lies well within one of the smooth regions produced by the inflationary epoch. Put another way, inflation has resulted in a smooth patch which contains an entropy of order $(10^{17}$cm$)^3 \times (10^{14}$GeV$)^3 \simeq 10^{134}$, which is much, much greater than that within the presently-observed Universe ($\simeq 10^{88}$). Before inflation that same volume contained only a very small amount of entropy, about $(10^{-23}$cm$)^3 (10^{14}$GeV$)^3 \simeq 10^{14}$. The key to inflation then is the highly nonadiabatic event of reheating (see Fig. 3.5). The very large-scale cosmography depends upon the state of the Universe before inflation and how inflation was initiated (bubble nucleation or spinodal decomposition); see ref. 96 for further discussion.

Since we have assumed that ϕ is spatially constant within the bubble or fluctuation region, after reheating the patch in question is precisely uniform, and at this stage *the inhomogeneity puzzle* has not been solved. Inflation has produced a smooth manifold on which small fluctuations can be impressed. Due to deSitter space produced quantum fluctuations in ϕ, ϕ is not exactly uniform even in a small patch. Later, I will discuss the density inhomogeneities that result from the quantum fluctuations in ϕ.

The flatness puzzle involves the smallness of the ratio of the curvature term to the energy density term. This ratio is exponentially smaller after inflation: $x_{\text{after}} \simeq e^{-200} x_{\text{before}}$ since the energy density before and after inflation is $0(M^4)$, while k/R^2 has exponentially decreased (by a factor of e^{200}). Since the ratio x is reset to an exponentially small value, the inflationary scenario predicts that today Ω should be $1 \pm 0(10^{-BIG\#})$.

If the Universe is reheated to a temperature of order M, a baryon asymmetry can evolve in the usual way, although the quantitative details may be different[55,94]. If the Universe is not efficiently reheated ($T_{RH} \ll M$), it may be possible for n_B/s to be produced directly in the decay of the coherent field oscillations[92-95] (which behave just like NR ϕ particles); this possibility will be discussed later. In any case, it is absolutely necessary to have baryogenesis occur after reheating since any baryon number (or any other quantum number) present before inflation is diluted by a factor of $(M/T_{MS})^3 \exp(3H\tau_\phi)$—the factor by which the total entropy increases. Note that if C, CP are violated spontaneously, then ϵ (and n_B/s) could have a different sign in different patches—leading to a Universe which on the very largest scales ($\gg e^{100} H^{-1}$) is baryon symmetric.

Since the patch that our observable Universe lies within was once (at the beginning of inflation) causally-coherent, the Higgs field could have been aligned throughout the patch (indeed, this is the lowest energy configuration), and thus there is likely to be ≤ 1 monopole within the entire patch which was produced as a topological defect. *The glut of monopoles* which occurs in the standard cosmology does not occur. [The production of other topological defects (such as domain walls, etc.) is avoided for similar reasons.] Some monopoles will be produced after reheating

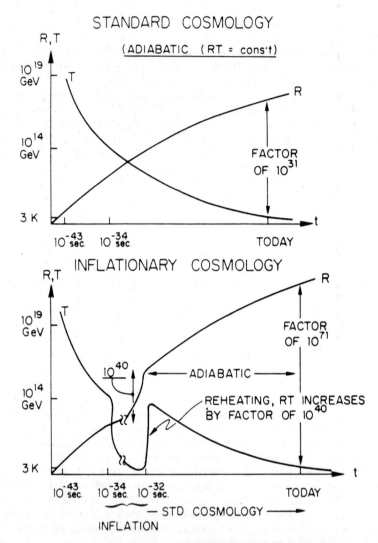

Figure 3.5: Evolution of the scale factor R and temperature T of the Universe in the standard cosmology and in the inflationary cosmology. The standard cosmology is always adiabatic ($RT \simeq$ const), while the inflationary cosmology undergoes a highly non-adiabatic event (reheating) after which it is adiabatic.

in rare, very energetic particle collisions[97]. The number produced is both exponentially small and exponentially uncertain. [In discussing the resolution of the monopole problem I am tacitly assuming that the SSB of the GUT is occurring during the SSB transition in question, or that it has already occurred in an earlier SSB transition; if not then one has to worry about the monopoles produced in the subsequent GUT transition.] Although monopole production is intrinsically small in inflationary models, the uncertainties in the number of monopoles produced are exponential and of course, it is also possible that monopoles might be produced as topological defects in a subsequent phase transition[98] (although it may be difficult to arrange that they not be overproduced).

Finally, the inflationary scenario sheds no light upon *the cosmological constant puzzle*. Although it can potentially successfully resolve all of the other puzzles in my list, inflation is, in some sense, a house of cards built upon the cosmological constant puzzle.

3.4. SCALAR FIELD DYNAMICS

The evolution of the scalar field ϕ is key to understanding new inflation. In this section I will focus on the semi-classical dynamics of ϕ. Later, I will return to the question of the validity of the semi-classical approach. Much of what I will discuss here is covered in more detail in ref. 99.

Stated in the most general terms, the current view of inflation is that it involves the dynamical evolution of a very weakly-coupled scalar field (hereafter referred to as ϕ) which is, for one reason or another, initially displaced from the minimum of its potential (see Fig. 3.6). While it is displaced from its minimum, and is slowly-evolving toward that minimum, its potential energy density drives the rapid (exponential) expansion of the Universe, now known as inflation.

The usual assumptions which are made (often implicitly) in order to analyze the scalar field dynamics inflation are:

⋆ A FRW spacetime with scale factor $R(t)$ and expansion rate

$$H^2 \equiv (\dot{R}/R)^2 = 8\pi\rho/3m_{pl}^2 - k/R^2 \qquad (3.10)$$

where the energy density is assumed to be dominated by the stress energy associated with the scalar field (in any case, other forms of stress energy rapidly redshift away during inflation and become irrelevant).

⋆ The scalar field ϕ is spatially constant (at least on a scale $\gtrsim H^{-1}$) with initial value $\phi_i \neq \sigma$, where $V(\sigma) = V'(\sigma) = 0$.

⋆ The semi-classical equation of motion for ϕ provides an accurate description of its evolution; more precisely,

$$\phi(t) = \phi_{cl}(t) + \Delta\phi_{QM}$$

where the quantum fluctuations (characterized by size $\Delta\phi_{QM} \simeq H/2\pi$) are assumed to be a small perturbation to the classical trajectory $\phi_{cl}(t)$. From

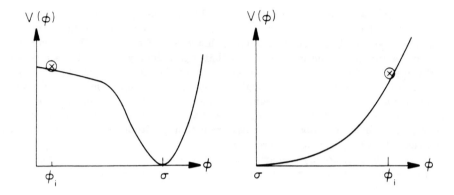

Figure 3.6: Stated in the most general terms, inflation involves the dynamical evolution of a scalar field which was initially displaced from the minimum of its potential, be that minimum at $\sigma = 0$ or $\sigma \neq 0$.

this point forward I will drop the subscript 'cl'. I will return later to these assumptions to discuss how they have been or can be relaxed and/or justified.

Consider a classical scalar field (minimally coupled) with lagrangian density given by

$$\mathcal{L} = -\frac{1}{2}\partial_\mu \phi \partial^\mu \phi - V(\phi). \tag{3.11}$$

For now I will ignore the interactions that ϕ must necessarily have with other fields in the theory. As it will turn out they must be weak for inflation to work, so that this assumption is a reasonable one. The stress-energy tensor for this field is then

$$T_{\mu\nu} = -\partial_\mu \phi \partial_\nu \phi - \mathcal{L} g_{\mu\nu} \tag{3.12}$$

Assuming that in the region of interest ϕ is spatially-constant, $T_{\mu\nu}$ takes on the perfect fluid form with energy density and pressure given by

$$\rho = \frac{1}{2}\dot{\phi}^2 + V(\phi)(+(\nabla\phi)^2/2R^2), \tag{3.13a}$$

$$p = \frac{1}{2}\dot{\phi}^2 - V(\phi)(-(\nabla\phi)^2/6R^2), \tag{3.13b}$$

where I have included the spatial gradient terms for future reference. [Note, once inflation begins the spatial gradient terms decrease rapidly, $(\nabla\phi)^2/R^2 \propto R^{-2}$, for

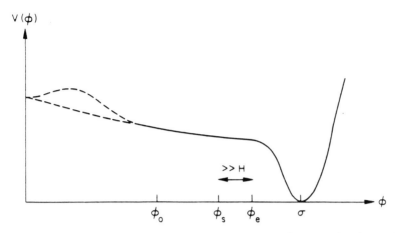

Figure 3.7: Schematic plot of the potential required for inflation. The shape of the potential for $\phi \ll \sigma$ determines how the barrier between $\phi = 0$ and $\phi = \sigma$ (if one exists) is penetrated. The value of ϕ after barrier penetration is taken to be ϕ_0; the flat region of the potential is the interval $[\phi_s, \phi_e]$.

wavelengths $\gtrsim H^{-1}$, and quickly become negligible.] That the spatial gradient term in ϕ be unimportant is crucial to inflation; if it were to dominate the pressure and energy density, then $R(t)$ would grow as t (since $p = -\frac{1}{3}\rho$) and not exponentially.

The equations of motion for ϕ can be obtained either by varying the action or by using $T^{\mu\nu}_{;\nu} = 0$. In either case the resulting equation is:

$$\ddot{\phi} + 3H\dot{\phi}(+\Gamma\dot{\phi}) + V'(\phi) = 0. \qquad (3.14)$$

I have explicitly included the $\Gamma\dot{\phi}$ term which arises due to particle creation. The $3H\dot{\phi}$ friction term arises due to the expansion of the Universe; as the scalar field gains momentum, that momentum is redshifted away by the expansion.

This equation, which is analogous to that for a ball rolling with friction down a hill with a valley at the bottom, has two qualitatively different regimes, each of which has a simple, approximate, analytic solution. Fig. 3.7 shows schematically the potential $V(\phi)$.

(1) The slow-rolling regime

In this regime the field rolls at terminal velocity and the $\ddot{\phi}$ term is negligible. This occurs in the interval where the potential is very flat, the conditions for sufficient flatness being[95]:

$$|V''| \leq 9H^2, \qquad (3.15a)$$
$$|V'm_{pl}/V| \leq (48\pi)^{1/2}, \qquad (3.15b)$$

Condition (3.15a) usually subsumes condition (3.15b), so that condition (3.15a) generally suffices. During the slow-rolling regime the equation of motion for ϕ reduces to

$$\dot{\phi} \simeq -V'/3H. \qquad (3.16)$$

During the slow-rolling regime particle creation is exponentially suppressed[94] because the timescale for the evolution of ϕ (which sets the energy/momentum scale of the particles created) is much greater than the Hubble time (which sets the physics horizon), i.e., any particles radiated would have to have wavelengths much larger than the physics horizon, which results in the exponential suppression of particle creation during this epoch. Thus, the $\Gamma\dot{\phi}$ term can be neglected during the 'slow roll'.

Suppose the interval where conditions (3.15a,b) are satisfied is $[\phi_s, \phi_e]$, then the number of e-folds of expansion which during the time ϕ is evolving from $\phi = \phi_s$ to $\phi = \phi_e$ ($\equiv N$) is

$$N \equiv \int H dt \simeq -3 \int_{\phi_s}^{\phi_e} H^2 d\phi/V'(\phi) \simeq -(8\pi/m_{pl}^2) \int V(\phi) d\phi/V'(\phi). \qquad (3.17)$$

[Note that $R_e/R_s \equiv \exp(N)$ since $\dot{R}/R \equiv H$.] Taking H^2/V' to be roughly constant over this interval and approximating V' as $\simeq \phi V''$ (which is approximately true for polynominal potentials) it follows that

$$N \approx 3H^2/V'' \geq 3.$$

If there is a region of the potential where the evolution is friction-dominated, then N will necessarily be greater than 1 (by condition (3.15a)).
(2) Coherent field oscillations
   In this regime

$$|V''| \gg 9H^2,$$

and ϕ evolves rapidly, on a timescale $\ll$ the Hubble time H^{-1}. Once ϕ reaches the bottom of its potential, it will oscillate with an angular frequency equal to $m_\phi \equiv V''(\sigma)^{1/2}$. In this regime it proves useful to rewrite Eqn(3.14) for the evolution of ϕ as

$$\dot{\rho}_\phi = -3H\dot{\phi}^2 - \Gamma\dot{\phi}^2. \qquad (3.18)$$

where

$$\rho_\phi \equiv 1/2\dot{\phi}^2 + V(\phi).$$

Once ϕ is oscillating about $\phi = \sigma$, $\dot{\phi}^2$ can be replaced by its average over a cycle

$$<\dot{\phi}^2>_{cycle} = \rho_\phi,$$

and Eqn(3.18) becomes

$$\dot{\rho}_\phi = -3H\rho_\phi - \Gamma\rho_\phi \qquad (3.19)$$

which is nothing else but the equation for the evolution of the energy density of zero momentum, massive particles with a decay width Γ.

Referring back to Eqn(3.13) we can see that the cycle average of the pressure (i.e., space-space components of $T_{\mu\nu}$) is zero—as one would expect for NR particles. The coherent ϕ oscillations are in every way equivalent to a very cold condensate of ϕ particles. The decay of these oscillations due to quantum particle creation is equivalent to the decay of zero-momentum ϕ particles.

The complete set of semi-classical equations for the reheating of the Universe is

$$\dot\rho_\phi = -3H\rho_\phi - \Gamma\rho_\phi, \qquad (3.20a)$$
$$\dot\rho_r = -4H\rho_r + \Gamma\rho_\phi, \qquad (3.20b)$$
$$H^2 = 8\pi G(\rho_r + \rho_\phi)/3, \qquad (3.20c)$$

where $\rho_r = (\pi^2/30)g_* T^4$ is the energy density in the relativistic particles produced by the decay of the coherent field oscillations. [I have tacitly assumed that the decay products of ϕ rapidly thermalize; Eqn(3.20b) is correct whether or not the decay products thermalize, so long as they are relativistic.] The evolution for the energy density in the scalar is easy to obtain

$$\rho_\phi = M^4(R/R_e)^{-3}\exp[-\Gamma(t - t_e)], \qquad (3.21)$$

where I have set the initial energy equal to M^4, the initial epoch being when the scalar field begins to evolve rapidly (at $R = R_e$, $\phi = \phi_e$, and $t = t_e$).

From $t = t_e$ until $t \simeq \Gamma^{-1}$, the energy density of the Universe is dominated by the coherent sloshings of the scalar field ϕ, set into motion by the initial vacuum energy associated with $\phi \ll \sigma$. During this phase

$$R(t) \propto t^{2/3}$$

that is, the Universe behaves as if it were dominated by NR particles—which it is!

Interestingly enough it follows from Eqn(3.20) that during this time the energy density in radiation is actually decreasing ($\rho_r \propto R^{-3/2}$—see Fig. 3.8). [During the first Hubble time after the end of inflation ρ_r does increase.] However, the all important entropy per comoving volume is increasing

$$S \propto R^{15/8}.$$

When $t \simeq \Gamma^{-1}$, the coherent oscillations begin to decay exponentially, and the entropy per comoving volume levels off—indicating the end of the reheating epoch. The temperature of the Universe at this time is,

$$T_{RH} \simeq g_*^{-1/4}(\Gamma m_{pl})^{1/2} \qquad (3.22)$$

If Γ^{-1} is less than H^{-1}, so that the Universe reheats in less than an expansion time, then all of the vacuum is converted into radiation and the' Universe is reheated to a temperature

$$T_{RH} \simeq g_*^{-1/4} M \quad (\text{if } \Gamma \geq H) \qquad (3.22')$$

the so-called case of good reheating.

To summarize the evolution of the scalar field ϕ: early on ϕ evolves very slowly, on a timescale $\gg$ the Hubble time H^{-1}; then as the potential steepens (and $|V''|$ becomes $> 9H^2$) ϕ begins to evolve rapidly, on a timescale $\ll$ the Hubble time H^{-1}. As ϕ oscillates about the minimum of its potential the energy density in these oscillations dominates the energy density of the Universe and behaves like NR matter ($\rho_\phi \propto R^{-3}$); eventually when $t \simeq \Gamma^{-1}$, these oscillations decay exponentially, 'reheating' the Universe to a temperature of $T_{RH} \simeq g_*^{-1/4}(\Gamma m_{pl})^{1/2}$ (if $\Gamma > H$, so that the Universe does not e-fold in the time it takes the oscillations to decay, then $T_{RH} \simeq g_*^{-1/4} M$). Saying that the Universe reheats when $t \simeq \Gamma^{-1}$ is a bit paradoxical as the temperature has actually been *decreasing* since shortly after the ϕ oscillations began. However, the fact that the temperature of the Universe was actually once greater than T_{RH} for $t < \Gamma^{-1}$ is probably of no practical use since the entropy per comoving volume increases until $t \simeq \Gamma^{-1}$—by a factor of $(M^2/\Gamma m_{pl})^{5/4}$, and any interesting objects that might be produced (e.g., net baryon number, monopoles, etc.) will be diluted away by the subsequent entropy production. By any reasonable measure, T_{RH} is the reheat temperature of the Universe. The evolution of ρ_ϕ, ρ_r, and S are summarized in Fig. 3.8.

Armed with our detailed knowledge of the evolution of ϕ we are ready to calculate the precise number of e-folds of inflation necessary to solve the horizon and flatness problems and to discuss direct baryon number production. First consider the requisite number of e-folds required for sufficient inflation. To solve the homogeneity problem we need to insure that a smooth patch containing an entropy of at least 10^{88} results from inflation. Suppose the initial bubble or fluctuation region has a size $H^{-1} \simeq m_{pl}/M^2$—certainly it is not likely to be significantly larger than this. During inflation it grows by a factor of $\exp(N)$. Next, while the Universe is dominated by coherent field oscillations it grows by a factor of

$$(R_{RH}/R_e) \simeq (M^4/T_{RH}^4)^{1/3},$$

where T_{RH} is the reheat temperature. Cubing the size of the patch at reheating (to obtain its volume) and multiplying its volume by the entropy density ($s \approx T_{RH}^3$), we obtain

$$S_{patch} \simeq e^{3N} m_{pl}^3/(M^2 T_{RH}).$$

Insisting that S_{patch} be greater than 10^{88}, it follows that

$$N \geq 53 + \frac{2}{3}\ln(M/10^{14}\text{GeV}) + \frac{1}{3}\ln(T_{RH}/10^{10}\text{GeV}). \qquad (3.23)$$

Varying M from 10^{19}GeV to 10^8GeV and T_{RH} from 1GeV to 10^{19}GeV this lower bound on N only varies from 36 to 68.

The flatness problem involves the smallness of the ratio

$$x = (k/R^2)/(8\pi G\rho/3)$$

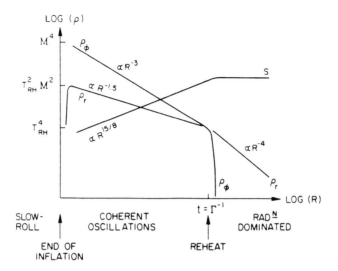

Figure 3.8: The evolution of ρ_ϕ, ρ_r, and S during the epoch when the Universe is dominated by coherent ϕ-oscillations. The reheat temperature $T_{RH} \simeq g_*^{-1/4}(\Gamma m_{pl})^{1/2}$. The maximum temperature achieved after inflation is actually greater, $T_{max} \simeq (T_{RH} M)^{1/2}$.

required at early times. Taking the pre-inflationary value of x to be x_i and remembering that

$$x(t) \propto \begin{cases} R^{-2} & \rho = \text{cons't} \\ R & \rho \propto R^{-3} \\ R^2 & \rho \propto R^{-4} \end{cases}$$

it follows that the value of x today is

$$x_{today} = x_i e^{-2N} (M/T_{RH})^{4/3} (T_{RH}/10eV)^2 (10eV/3K).$$

Insisting that x_{today} be at most of order unity implies that

$$N \geq 53 + \ln(x_i) + \frac{2}{3}\ln(M/10^{14}\text{GeV}) + \frac{1}{3}\ln(T_{RH}/10^{10}\text{GeV})$$

—up to the term $\ln(x_i)$, precisely the same bound as we obtained to solve the homogeneity problem. Solving the isotropy problem depends upon the initial anisotropy present; during inflation isotropy decreases exponentially (see refs. 100).

Finally, let's calculate the baryon asymmetry that can be directly produced by the decay of the ϕ particles themselves. Suppose that the decay of each ϕ particle results in the production of net baryon number ϵ. This net baryon number might be produced directly by the decay of a ϕ particle (into quarks and leptons) or indirectly through an intermediate state ($\phi \to X\bar{X}$; X, $\bar{X} \to$ quarks and leptons; e.g., X might be a superheavy, color triplet Higgs[101]). The baryon asymmetry produced per volume is then

$$n_B \simeq \epsilon n_\phi.$$

On the other hand we have

$$(g_*\pi^2/30)T_{RH}^4 \simeq n_\phi m_\phi.$$

Taken together it follows that[93,102]

$$n_B/s \simeq (3/4)\epsilon T_{RH}/m_\phi. \tag{3.24}$$

This then is the baryon number per entropy produced by the decay of the ϕ particles directly. If the reheat temperature is not very high, baryon number non-conserving interactions will not subsequently reduce the asymmetry significantly. Note that the baryon asymmetry produced only depends upon the *ratio* of the reheat temperature to the ϕ particle mass. This is important, as it means that a very low reheat temperature can be tolerated, so long as the ratio of it to the ϕ particle mass is not too small.

3.5. ORIGIN OF DENSITY INHOMOGENEITIES

To this point I have assumed that ϕ is precisely uniform within a given bubble or fluctuation region. As a result, each bubble or fluctuation region resembles a perfectly isotropic and homogeneous Universe after reheating. However, because of deSitter space produced quantum fluctuations, ϕ cannot be exactly uniform, even within a small region of space. It is a well-known result that a massless and non-interacting scalar field in deSitter space has a spectrum of fluctuations given by (see, e.g., ref. 103)

$$(\Delta\phi)^2 \equiv (2\pi)^{-3}k^3|\delta\phi_k|^2 = H^2/16\pi^3, \tag{3.25}$$

where

$$\delta\phi = (2\pi)^{-3}\int d^3k \delta\phi_k e^{-ikx}, \tag{3.26}$$

and $\vec{x}$ and $\vec{k}$ are comoving quantities. This result is applicable to inflationary scenarios as the scalar field responsible for inflation must be very weakly-coupled and nearly massless. [That Universe is not precisely deSitter during inflation, i.e., $\rho + p = \dot{\phi}^2 \neq 0$, does not affect this result significantly; this point is addressed in ref. 104.] These deSitter space produced quantum fluctuations result in a calculable spectrum of adiabatic density perturbations. These density perturbations were first calculated by the authors of refs. 105-108; they have also been calculated by the authors of refs. 109 who addressed some of the technical issues in more detail. All the calculations done to date arrive at the same result. I will briefly describe the calculation in ref. 108; my emphasis here will be to motivate the result. I refer the reader interested in more details to the aforementioned references.

It is conventional to expand density inhomogeneities in a Fourier expansion

$$\delta\rho/\rho = (2\pi)^{-3} \int \delta_k e^{-ikx} d^3k. \quad (3.27)$$

The physical wavelength and wavenumber are related to comoving wavelength and wave number, λ and k, by

$$\lambda_{ph} = (2\pi/k)R(t) \equiv \lambda R(t),$$
$$k_{ph} = k/R(t).$$

The quantity most people refer to as $\delta\rho/\rho$ on a given scale is more precisely the RMS mass fluctuation on that scale

$$(\delta\rho/\rho)_k \equiv <(\delta M/M)^2>_k \simeq \Delta_k^2 \equiv (2\pi)^{-3} k^3 |\delta_k|^2, \quad (3.28)$$

which is just related to the Fourier component δ_k on that scale.

The cosmic scale factor is often normalized so that $R_{today} = 1$; this means that given Fourier components are characterized by the physical size that they have today (neglecting the fact that once a given scale goes non-linear ($\delta\rho/\rho \gtrsim 1$) objects of that size form bound 'lumps' that no longer participate in the universal expansion and remain roughly constant in size). The mass (in NR matter) contained within a sphere of radius $\lambda/2$ is

$$M(\lambda) \simeq 1.5 \times 10^{11} M_\odot (\lambda/\text{Mpc})^3 \Omega h^2.$$

Although physics depends on physical quantities (k_{ph}, λ_{ph}, etc.), the comoving labels k, M, and λ are the most useful way to label a given component as they have the affect of the expansions already scaled out.

I want to emphasize at the onset that the quantity $\delta\rho/\rho$ is not gauge invariant (under general coordinate transformations). This fact makes life very difficult when discussing Fourier components with wavelengths larger than the physics horizon (i.e., $\lambda_{ph} \gtrsim H^{-1}$). The gauge non-invariance of $\delta\rho/\rho$ is not a problem when $\lambda_{ph} \lesssim H^{-1}$, as the analysis is essentially Newtonian. The usual approach is to pick a convenient gauge (e.g., the synchronous gauge where $g_{oo} = -1$, $g_{oi} = 0$) and work very carefully (see refs. 110, 111). The more elegant approach is to focus on gauge-invariant quantities; see ref. 112. I will gloss over the subtleties of gauge invariance in my discussion, which is aimed at motivating the correct answer.

The evolution of a given Fourier component (in the linear regime—$\delta\rho/\rho \ll 1$) separates into two qualitatively different regimes, depending upon whether or not the perturbation is inside or outside the physics horizon ($\simeq H^{-1}$). When

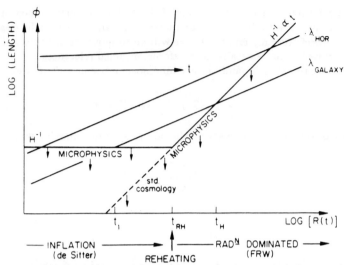

Figure 3.9: The evolution of the physical size of galactic- and (present) horizon-sized perturbations ($\lambda_{ph} \propto R$) and the size of the physics horizon H^{-1}. Causally-coherent microphysics operates only on scales $\leq H^{-1}$. In the standard cosmology a perturbation crosses the horizon but once as $H^{-1} \propto R^n (n > 1)$, making it impossible for microphysics to create density perturbations at early times. In the inflationary cosmology a perturbation crosses the horizon twice (since $H^{-1} \simeq$ const during inflation), and so microphysics can produce density perturbations at early times.

$\lambda_{ph} \leq H^{-1}$, microphysical processes can affect its evolution—such processes include: quantum mechanical effects, pressure support, free-streaming of particles, 'Newtonian gravity', etc. In this regime the evolution of the perturbation is very dynamical. When a perturbation is outside the physics horizon, $\lambda_{ph} \geq H^{-1}$, microphysical processes do not affect its evolution; in a very real sense its evolution is kinematic—it evolves as a wrinkle in the fabric of space–time.

In the standard cosmology, a given Fourier component crosses the horizon only once, starting outside the horizon and crossing inside at a time (see Fig. 3.9)

$$t \simeq (M/M_\odot)^{2/3} \text{ sec}$$

(valid during the radiation-dominated epoch). For this reason it is not possible to create adiabatic (more precisely, curvature) perturbations by causal microphysical processes which operate at early times[111,112]. In the standard cosmology, if adiabatic perturbations are present, they must be present ab initio. The smallness of the particle horizon at early times relative to the comoving volume occupied by the observable Universe today strikes again!

[It is possible for microphysical processes to create isothermal, more precisely isocurvature, perturbations. Once such perturbations cross inside the horizon they are characterized by a spectrum

$$(\delta\rho/\rho) \propto (M/M_H)^{-1/2}$$

or steeper. Here M_H is the horizon mass when the perturbations were created. Thus the earlier the processes operate, the smaller the perturbations are on interesting scales. By an appropriate choice of gauge it is possible to view these isothermal perturbations as adiabatic perturbations with a very steep spectrum, $\delta\rho/\rho \propto M^{-7/6}$; however, as must be the case, they cross the horizon with the amplitude mentioned above. For more details, see refs. 111, 112.]

Because the distance to the physics horizon remains approximately constant during inflation, the situation is very different in the inflationary Universe. All interesting scales start inside the horizon, cross outside the horizon and once again come inside the horizon (at the usual epoch); see Fig. 3.9. This means that causal microphysical processes can set up curvature perturbations on astrophysically-interesting scales. [This point seems to have been first appreciated by Press[113].]

Consider the evolution of a given Fourier component k. Early during the inflationary epoch $\lambda_{ph} \leq H^{-1}$, and quantum fluctuations in ϕ give rise to density perturbations on this scale. As the scale passes outside the horizon, say at $t = t_1$, microphysical processes become impotent, and $\delta\rho/\rho$ freezes out at a value,

$$(\delta\rho/\rho)_k \simeq O(\dot\phi H \Delta\phi/M^4),$$
$$\simeq O(\dot\phi H^2/M^4), \qquad (3.29)$$

as the scale leaves the horizon. From that point forward, the QM fluctuation is assumed to 'freeze in' and thereafter evolve classically. Note in the approximation that H and $\dot\phi$ are constant during the inflationary epoch the value of $\delta\rho/\rho$ as the perturbation leaves the horizon is independent of k. This scale independence of $\delta\rho/\rho$ when perturbations cross outside the horizon is of course traceable to the time translation invariance of deSitter space.

While outside the horizon the evolution of a perturbation is kinematical, independent of scale, and gauge dependent. There is a gauge independent quantity ($\equiv \varsigma$) which remains constant while the perturbation is outside the horizon, and which at horizon crossing ($t = t_1$ and t_H) is given by

$$\varsigma|_{\text{horizon crossing}} \simeq \delta\rho/(\rho + p),$$

$$\varsigma(t_1) = \varsigma(t_H)$$

$$\Rightarrow [\delta\rho/(\rho + p)]_{t=t_H} \simeq [\delta\rho/(\rho + p)]_{t=t_1}. \qquad (3.30)$$

(see refs. 108 and 114 for more details). When the perturbation crosses back inside the horizon: $(\rho + p) = n\rho (n = 4/3-$ radiation-dominated; $n = 1$, matter-dominated) so that up to a numerical factor $(\delta\rho/\rho)|_{t_H} \simeq [\delta\rho/(\rho + p)]|_{t_H}$. During inflation, however, $\rho + p = \dot{\phi}^2 \ll \rho \simeq M^4$ so that $(\delta\rho/\rho)|_{t_1} \simeq (\dot{\phi}^2/M^4)[\delta\rho/(\rho + p)]|_{t_1}$. Note, $M^4/\dot{\phi}^2$ is typically a very large number. Eqns(3.29, 3.30) then imply

$$(\delta\rho/\rho)_H \equiv (\delta\rho/\rho)_{t=t_H} \simeq (M^4/\dot{\phi}^2)(\delta\rho/\rho)_{t_1} \simeq H^2/\dot{\phi}, \qquad (3.31)$$

Note that in the approximation that $\dot{\phi}$ and H are are constant during inflation the amplitude of $\delta\rho/\rho$ at horizon crossing ($= (\delta\rho/\rho)_H$) is independent of scale. This fact is traceable to the time-translation invariance of the nearly-deSitter inflationary epoch and the scale-independent evolution of $(\delta\rho/\rho)$ while the perturbation is outside the horizon. The so-called scale-invariant or Zel'dovich spectrum of density perturbations was first discussed, albeit in another context, by Harrison[115] and Zel'dovich[116]. Scale-invariant adiabatic density perturbations are a generic prediction of inflation. [Because H and $\dot{\phi}$ are not precisely constant during inflation, the spectrum is not quite scale-invariant. However the scales of astrophysical interest, say $\lambda \simeq 0.1\text{Mpc} - 100\text{Mpc}$, cross outside the horizon during a very short interval, $\Delta N \simeq 6.9$, during which H, $\dot{\phi}$, and ϕ are very nearly constant. For most models of inflation the deviations are not expected to be significant; for further discussion see refs. 117, 118.] Although the details of structure formation are not presently sufficiently well understood to say what the initial spectrum of perturbations must have been, the Zel'dovich with an amplitude of about $10^{-4} - 10^{-5}$ is certainly a viable possibility.

Before moving on, let me be very precise about the amplitude of the inflation-produced adiabatic density perturbations. Perturbations which re-enter the horizon while the Universe is still radiation-dominated ($\lambda \leq \lambda_{eq} \simeq 13h^{-2}\text{Mpc}$), do so as a sound wave in the photons and baryons with amplitude

$$(\delta\rho/\rho)_H \equiv k^{3/2}|\delta_k|/(2\pi)^{3/2} \simeq H^2/(\pi^{3/2}\dot{\phi}) \qquad (3.32a)$$

Perturbations in non-interacting, relic particles (such as massive neutrinos, axions, etc.), which by the equivalence principle must have the same amplitude at horizon crossing, do not oscillate, but instead grow slowly ($\propto \ln R$). By the epoch of matter-radiation equivalence they have an amplitude of 2-3 times that of the initial baryon-photon sound wave, or

$$(\delta\rho/\rho)_{MD} \simeq (2-3)(\delta\rho/\rho)_H \simeq (2-3)H^2/(\pi^{3/2}\dot{\phi}) \qquad (3.32b)$$

It is this amplitude which must be of order $10^{-5} - 10^{-4}$ for successful galaxy formation.

Perturbations which re-enter the horizon when the Universe is already matter-dominated (scales $\lambda \geq \lambda_{eq} \simeq 13h^{-2}\text{Mpc}$) do so with amplitude

$$(\delta\rho/\rho)_H \simeq k^{3/2}|\delta_k|/(2\pi)^{3/2} \simeq (H^2/10)/(\pi^{3/2}\dot{\phi}) \qquad (3.33)$$

Once inside the horizon they continue to grow (as $t^{2/3}$ since the Universe is matter-dominated).

When the structure formation problem is viewed as an initial data problem, it is the spectrum of density perturbations at the epoch of matter domination which is the relevant input spectrum. The shape of this spectrum has been carefully computed by the authors of ref. 119. Roughly speaking, on scales less than λ_{eq} the spectrum is almost flat, varying as $\lambda^{-3/4} \propto M^{-1/4}$ for scales around the galaxy scale ($\simeq$ 1Mpc). On scales much greater than λ_{eq}, $(\delta\rho/\rho) \propto \lambda^{-2} \propto M^{-2/3}$ (in the synchronous gauge where adiabatic perturbations grow as t^n; $n = 2/3$ matter dominated, $n = 1$ radiation dominated. Since these scales have yet to re-enter the horizon they have not yet achieved their horizon-crossing amplitude).

In order to compute the amplitude of the inflation-produced adiabatic density perturbations we need to evaluate $H^2/\dot\phi$ when the astrophysically-relevant scales crossed outside the horizon. Recall, in the previous section we computed when the comoving scale corresponding to the present Hubble radius crossed outside the horizon during inflation—up to 'ln terms' $N \simeq 53$ or so e-folds before the end of inflation, cf., Eqn(3.23). The present Hubble radius corresponds to a scale of about 3000Mpc; therefore the scale λ must have crossed the horizon $\ln(3000\text{Mpc}/\lambda)$ e-folds later:

$$N_\lambda \simeq N_{HOR} - 8 + \ln(\lambda/\text{Mpc}) \simeq 45 + \ln(\lambda/\text{Mpc})$$
$$+ \frac{2}{3}\ln(M/10^{14}\text{GeV}) + \frac{1}{3}\ln(T_{RH}/10^{10}\text{GeV}).$$

Typically $H^2/\dot\phi$ depends upon N_λ to some power[117]; since N_λ only varies logarithmically ($\Delta N/N \simeq 0.14$ in going from 0.1Mpc to 3000Mpc), the scale dependence of the spectrum is almost always very minimal.

As mentioned earlier, a generic prediction of the inflationary Universe is that today Ω should be equal to one to a high degree of precision. Equivalently, that means

$$|(k/R^2)/(8\pi G\rho/3)| \ll 1$$

since

$$\Omega = 1/[1 - (k/R^2)/(8\pi G\rho/3)].$$

Therefore one might conclude that an accurate measurement of Ω would have to yield 1 to extremely high precision. However, because of the adiabatic density perturbations produced during inflation that is not the case. Adiabatic density fluctuations correspond to fluctuations in the local curvature

$$\delta\rho/\rho \simeq \delta(k/R^2)/(8\pi G\rho/3)$$

This means that should we be able to very accurately probe the value of Ω (equivalently the curvature of space) on the scale of our Hubble volume, say by using the Hubble diagram, we would necessarily obtain a value for Ω which is dominated by the curvature fluctuations on the scale of the present horizon,

$$\Omega_{obs} \simeq 1 + \delta(k/R^2)/(8\pi G\rho/3) \simeq 1 \pm O(10^{-4} - 10^{-5}),$$

and so we would obtain a value different from 1 by about a part in 10^4 or 10^5.

Finally, let me briefly mention that isothermal density perturbations can also arise during inflation. [Isothermal density perturbations are characterized by $\delta\rho = 0$, but $\delta(n_i/n_\gamma) \neq 0$ in some components. They correspond to spatial fluctuations in the local pressure due to spatial fluctuations in the local equation of state.] Such perturbations can arise from the deSitter produced fluctuations in other quantum fields in the theory.

The simplest example occurs in the axion-dominated Universe[120,121,122]. Suppose that Peccei-Quinn symmetry breaking occurs before or during inflation. Until instanton effects become important $(T \simeq \text{few } 100\,\text{MeV})$ the axion field $a = f_a \theta$ is massless and θ is in general not aligned with the minimum of its potential: $\theta = \theta_1 \neq 0$ (I have taken the minimum of the axion potential to be $\theta = 0$). Once the axion develops a mass (equivalently, its potential develops a minimum) θ begins to oscillate; these coherent oscillations correspond to a condensate of very cold axions, with number density $\propto \theta^2$. [For further discussion of the coherent axion oscillations see refs. 123-125.] During inflation deSitter space produced quantum fluctuations in the axion field gave rise to spatial fluctuations in θ_1:

$$\delta\theta \simeq \delta a/f_a \simeq H/f_a$$

Once the axion field begins to oscillate, these spatial fluctuations in the axion field correspond to fluctuations in the local axion to photon ratio

$$\delta(n_a/n_\gamma)/(n_a/n_\gamma) \simeq 2\delta\theta/\theta_1 \simeq 2H/(f_a\theta_1)$$

More precisely

$$(\delta n_a/n_a) = k^{3/2}|\delta a(k)|/(2\pi)^{3/2} = H/(2\pi^{3/2} f_\lambda \theta_1), \qquad (3.34)$$

where f_λ is the expectation value of f_a when the scale λ leaves the horizon (in some models the expectation value of the field which breaks PQ symmetry evolves as the Universe is inflating so that f_λ can be $< f_a$). It is possible that these isothermal axion fluctuations can be important for galaxy formation in an axion-dominated, inflationary Universe[121].

3.6. SPECIFIC MODELS–PART I. INTERESTING FAILURES

(1) 'Old Inflation'

By old inflation I mean Guth's original model of inflation. In his original model the Universe inflated while trapped in the $\phi = 0$ false vacuum state. In order to inflate enough the vacuum had to be very metastable; however, that being the case, the bubble nucleation probability was necessarily small—so small that the bubbles that did nucleate never percolated, resulting in a Universe which resembled swiss cheese more than anything else[87]. The interior of an individual bubble was not suitable for our present Universe either. Because he was not considering flat potentials, essentially all of the original false vacuum energy resided in bubble walls rather than in vacuum energy inside the bubbles themselves. Although individual bubbles would grow to a very large size given enough time, their interiors would contain very little entropy (compared to the 10^{88} in our observed Universe). In sum, the Universe inflated all right, but did not 'gracefully exit' from inflation back to a radiation-dominated Universe—close, Alan, but no cigar!

(2) Coleman–Weinberg SU(5)

The first model of new inflation studied was the Coleman–Weinberg SU(5) GUT[88,89]. In this model the field which inflates is the 24-dimensional Higgs which also breaks SU(5) down to SU(3) × SU(2) × U(1). Let ϕ denote its magnitude in the SU(3) × SU(2) × U(1) direction. The one-loop, zero-temperature Coleman–Weinberg[126] potential is

$$V(\phi) = 1/2 B\sigma^4 + B\phi^4\{\ln(\phi^2/\sigma^2) - 1/2\},$$
$$B = 25\alpha_{GUT}^2/16 \simeq 10^{-3} \quad (3.35)$$
$$\sigma \simeq 2 \times 10^{15} \text{GeV}$$

Due to the absence of a mass term $(m^2\phi^2)$, the potential is very flat near the origin (SSB arises due to one-loop radiative corrections[126]); for $\phi \ll \sigma$:

$$V(\phi) \simeq B\sigma^4/2 - \lambda\phi^4/4$$
$$\lambda \simeq |4B\ln(\phi^2/\sigma^2)| \simeq 0.1 \quad (3.36)$$

The finite temperature potential has a small temperature dependent barrier [height $O(T^4)$] near the origin [$\phi \simeq O(T)$]. The critical temperature for this transition is $O(10^{14} - 10^{15}\text{GeV})$, however the $\phi = 0$ vacuum remains metastable. When the temperature of the Universe drops to $O(10^9 \text{GeV})$ or so, the barrier becomes low enough that the finite temperature action for bubble nucleation drops to order unity and the $\phi = 0$ false vacuum becomes unstable[89]. In analogy with solid state phenomenon it is expected that at this the temperature of the Universe will undergo 'spinodal decomposition', i.e., will break up into irregularly shaped regions within which ϕ is approximately constant (so-called fluctuation regions). Approximately the potential by Eqn(3.36). It is easy to solve for the evolution of ϕ in the slow-rolling regime $[|V''| \leq 9H^2$ for $\phi^2 \leq \phi_e^2 \simeq \sigma^2(\pi\sigma^2/m_{pl}^2|\ln(\phi^2/\sigma^2)|)]$

$$(H/\phi)^2 \simeq \frac{2\lambda}{3} N(\phi), \quad (3.37)$$

$$H^2 \simeq \frac{4\pi}{3} \frac{B\sigma^4}{m_{pl}^2}, \quad (3.38)$$

where $N(\phi) \equiv \int_\phi^{\phi_e} H dt$ is the number of e-folds of inflation the Universe undergoes while ϕ evolves from ϕ to ϕ_e. Clearly, the number of e-folds of inflation depends upon the initial value of $\phi(\equiv \phi_0)$; in order to get sufficient inflation ϕ_0 must be $\simeq O(H)$. Although one might expect ϕ_0 to be of this order in the fluctuation regions since $H \simeq 5 \times 10^9$GeV $\simeq$ (temperature at which the $\phi = 0$ false vacuum loses its metastability), there is a fundamental difficulty. In using the semi-classical equations of motion to describe the evolution of ϕ one is implicitly assuming

$$\phi \simeq \phi_{cl} + \Delta\phi_{QM},$$
$$\Delta\phi_{QM} \ll \phi_{cl}$$

The deSitter space produced quantum fluctuations in ϕ are of order H. More specifically, it has been shown that[127,128]

$$\Delta\phi_{QM} \simeq (H/2\pi)(Ht)^{1/2}$$

Therein lies the difficulty—in order to achieve enough inflation the initial value of ϕ must be of the order of the quantum fluctuations in ϕ. At the very least this calls into question the semiclassical approximation.

The situation gets worse when we look at the amplitude of the adiabatic density perturbations:

$$(\delta\rho/\rho)_H \simeq (H^2/\pi^{3/2}\dot{\phi}) \qquad (3.39)$$
$$\simeq (3/\pi^{3/2})(H^3/\lambda\phi^3) \qquad (3.40)$$
$$(\delta\rho/\rho)_H \simeq (2/\pi)^{3/2}(\lambda/3)^{1/2}N^{3/2} \qquad (3.41)$$

For galactic-scale perturbations $N \simeq 50$, implying that $(\delta\rho/\rho)_H \simeq 30$! Again, its clear that the basic problem is traceable to the fact that during inflation $\phi \lesssim H$.

The decay width of the ϕ particle is of order $\alpha_{GUT}\sigma \simeq 10^{13}$GeV which is much greater than H (implying good reheating), and so the Universe reheats to a temperature of order 10^{14}GeV or so.

From Eqns(3.37, 3.41) it is clear that by reducing λ both problems could be remedied—however $\lambda \lesssim 10^{-13}$ is necessary[108]. Of course, as long as the inflating field is a gauge non-singlet λ is set by the gauge coupling strength. From this interesting failure it is clear that one should focus on weakly-coupled, gauge singlet fields for inflation.

(3) Geometric Hierarchy Model

The first model proposed to address the difficulty mentioned above, was a supersymmetric GUT[129,130]. In this model ϕ is a scalar field whose potential at tree level is absolutely flat, but due to radiative corrections develops curvature. In the model ϕ is also responsible for the SSB of the GUT. The potential for ϕ is of the form

$$V(\phi) \simeq \mu^4[c_1 - c_2\ln(\phi/m_{pl})] \qquad (3.42)$$

where $\mu \simeq 10^{12}$GeV is the scale of supersymmetry breaking, and c_1 and c_2 are constants which depend upon details of the theory. This form for the potential is only valid away from its SSB minimum ($\sigma \simeq m_{pl}$) and for $\phi \gg \mu$. The authors presume that higher order effects will force the potential to develop a minimum for $\phi \simeq m_{pl}$. Since $V' \propto \phi^{-1}$ the potential gets flatter for large ϕ—which already sounds good.

The inflationary scenario for this potential proceeds as follows. The shape of the potential is not determined near $\phi = 0$; depending on the shape ϕ gets to some initial value, say $\phi = \phi_0$, either by bubble nucleation or spinodal decomposition. Then it begins to roll. During slow roll which begins when $|V''| \simeq 9H^2$ and $\phi_s \simeq (c_2/24\pi c_1)^{1/2} m_{pl}$,

$$H^2 \simeq \frac{8\pi}{3m_{pl}^2} c_1 \mu^4 \qquad (3.43a)$$

$$(1 - \phi^2/m_{pl}^2) \simeq (c_2/4\pi c_1) N(\phi) \qquad (3.43b)$$

$$(\delta\rho/\rho)_H \simeq (H^2/\pi^{3/2}\dot\phi), \qquad (3.44a)$$

$$\simeq 8(8/3)^{1/2} (c_1^{3/2}/c_2) \mu^2 \phi/m_{pl}^3. \qquad (3.44b)$$

Note that during the slow roll $(\phi \geq \phi_s)$

$$\frac{\phi}{H} \geq \frac{\phi_s}{H} \simeq \frac{c_2^{1/2}}{c_1} \frac{1}{8\pi} \frac{m_{pl}^2}{\mu^2},$$

$$\simeq 10^{13} c_2^{1/2}/c_1 \gg 1,$$

thereby avoiding the difficulty encountered in the Coleman-Weinberg where $\phi \leq H$ was required to inflate. For $c_1 \simeq O(1)$, $c_2 \simeq 10^{-8}$—acceptable values in the model, $(\delta\rho/\rho)_H \simeq 10^{-5}$ and $N(\phi_s) \simeq 4\pi c_1/c_2 \simeq 10^9$. The number of e-folds of inflation is very large—10^9. This is quite typical of the very flat potentials required to achieve $(\delta\rho/\rho) \simeq 10^{-4} - 10^{-5}$.

Now for the bad news. In this model ϕ is very weakly coupled—it only couples to ordinary particles through gravitational strength interactions. Its decay width is

$$\Gamma \simeq O(\mu^6/m_{pl}^5), \qquad (3.45)$$

which is much less than H (implying poor reheating) and leads to a reheat temperature of

$$T_{RH} \simeq O[(\Gamma m_{pl})^{1/2}], \qquad (3.46a)$$

$$\simeq O(\mu^3/m_{pl}^2), \qquad (3.46b)$$

$$\simeq 10\,\text{MeV}. \qquad (3.46c)$$

Such a reheat temperature safely returns the Universe to being radiation-dominated before primordial nucleosynthesis, and produces a smooth patch containing an enormous entropy—for $c_2 \simeq 10^{-8}$, $c_1 \simeq 1$, $S_{patch} \simeq (m_{pl}^3/\mu^2 T_{RH}) e^{3N} \simeq 10^{35} \exp(3 \times 10^9)$, but does not reheat it to a high enough temperature for baryogenesis. Poor reheating is a problem which plagues almost all potentially viable models of inflation. Achieving $(\delta\rho/\rho)_H \lesssim 10^{-4}$ requires the scalar potential to be very flat, which necessarily means that ϕ is very weakly-coupled, and therefore $T_{RH}(\propto \Gamma^{1/2})$ tends to be very low.

(4) CERN SUSY/SUGR Models[131]

Early on members of the CERN theory group recognized that supersymmetry might be of use in protecting the very small couplings necessary in inflationary potentials from being overwhelmed by radiative corrections. They explored a variety of SUSY/SUGR models[131] (and dubbed their brand of inflation 'primordial inflation'). In the process, they encountered a difficulty which plagues almost all supersymmetric models of inflation based upon minimal supergravity theories.

It is usually assumed that at high temperatures the expectation value of an inflating field is at the minimum of its finite temperature effective potential (near $\phi = 0$); then as the Universe cools it becomes trapped there, and then eventually slowly evolves to the low temperature minimum (during which time inflation takes place). In SUSY models $\langle\phi\rangle_T$ is not necessarily zero at high temperatures. In fact in essentially all of their models $\langle\phi\rangle_T > 0$ and the high temperature minimum smoothly evolves into the low temperature minimum (as shown in Fig. 3.10)[132]. As a result the Universe in fact would never have inflated!

There are two obvious remedies to this problem: (i) arrange the model so that $\langle\varphi\rangle_T \leq 0$ (as shown in Fig. 3.10), then ϕ necessarily gets trapped near $\phi = 0$; or (ii) assume that ϕ is never in thermal equilibrium before the phase transition so that ϕ is not constrained to be in the high temperature minimum of its finite temperature potential at high temperatures. Variants of the CERN models[131] based on these two remedies have been constructed by Ovrut and Steinhardt[133] and Holman, Ramond, and Ross[134]

3.7. LESSONS LEARNED-A PRESCRIPTION FOR SUCCESSFUL NEW INFLATION

The unsuccessful models discussed above have proven to be very useful in that they have allowed us to 'write a prescription' for the kind of potential that will successfully implement inflation[117]. The following prescription incorporates these lessons, together with other lessons which have been learned (sometimes painfully). As we will see all but the last of the prescribed features, that the potential be part of a sensible particle physics model, are relatively easy to arrange.

(1) The potential should have an interval which is sufficiently flat so that ϕ evolves slowly (relative to the expansion timescale H^{-1})—that is, flat enough so that a slow-rollover transition ensues. As we have seen, that means an interval

$$[\phi_s, \phi_e]$$

where

$$|V''| \leq 9H^2,$$
$$|V'm_{pl}/V| \leq (48\pi)^{1/2}.$$

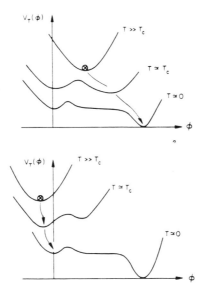

Figure 3.10: In SUSY/SUGR models $\langle\varphi\rangle_T$ is not necessarily equal to zero. If $\langle\varphi\rangle_T > 0$, then there is the danger that $\langle\varphi\rangle_T$ smoothly evolves into the zero temperature minimum of the potential, thereby eliminating the possibility of inflation (upper figure). A sure way of preventing this is to design the potential so that $\langle\varphi\rangle_T \le 0$ (lower figure).

(2) The length of the interval where ϕ evolves slowly should be much greater than $H/2\pi$, the scale of the quantum fluctuations, so that the semi-classical approximation makes sense. [Put another way the interval should be long enough so that quantum fluctuations do not quickly drive ϕ across the interval.] Quantitatively, this calls for

$$|\phi_e - \phi_s| \gg (H\Delta t)^{1/2}(H/2\pi)$$

where Δt is the time required for ϕ to evolve from $\phi = \phi_s$ to $\phi = \phi_e$. [More precisely, the semi-classical change in ϕ in a Hubble time, $\Delta\phi_{Hubble} \simeq -V'm_{pl}^2/8\pi V$, should be much greater than the increase in $\langle\phi^2\rangle_{QM}^{1/2} \simeq H/2\pi$, due to the addition of another quantum mode; see ref. 108.]

(3) In order to solve the flatness and homogeneity problems the time required for ϕ to roll from $\phi = \phi_s$ to $\phi = \phi_e$ should be greater than about 60 Hubble times

$$N \equiv \int_{\phi_s}^{\phi_e} H dt \simeq \int_{\phi_s}^{\phi_e} 3H^2 d\phi/(-V') \simeq 3H^2/V'' \geq 60$$

The precise formula for the minimum value of N is given in Eqn(3.23).

(4) The scalar field ϕ should be smooth in a sufficiently large patch (say size L) so that the energy density and pressure associated with the $(\nabla \phi)^2$ term is negligible:

$$1/2(\nabla \phi)^2 \simeq (\phi_0/L)^2 \ll V(\phi_0) \simeq M^4.$$

(Otherwise the $(\nabla \phi)^2$ term will dominate ρ and p, so that $R(t) \propto t$—that is, inflation does not occur.) Usually this condition is easy to satisfy, as all it requires is that

$$L \gg \phi_0/M^2 \simeq (\phi_0/m_{pl})H^{-1};$$

since ϕ_0 is usually $\ll m_{pl}$, $(\phi_0/m_{pl})H^{-1} \ll H^{-1}$—that is, ϕ only need be smooth on a patch comparable to the physics horizon H^{-1}. [I will discuss a case where it is not easy to satisfy—Linde's chaotic inflation.] Once inflation does begin, any initial inhomogeneities in ϕ are rapidly smoothed by the exponential expansion.

(5a) In order to insure a viable scenario of galaxy formation (and microwave anisotropies of an acceptable magnitude) the amplitude of the adiabatic density perturbations must be of order $10^{-5} - 10^{-4}$. [In a Universe dominated by weakly-interacting relic particles such as neutrinos or axions, $(\delta\rho/\rho)_{MD}$ must be a few $\times 10^{-5}$.] This in turn results in the constraint

$$\text{few} \times 10^{-5} \simeq (\delta\rho/\rho)_{MD} \simeq (2-3)(\delta\rho/\rho)_H \simeq (2-3)(H^2/\pi^{3/2}\dot\phi)_{Galaxy},$$

$$(H^2/\dot\phi)_{Galaxy} \simeq 10^{-4}$$

In general, this is by far the most difficult of the constraints (other than sensible particle physics) to satisfy and leads to the necessity of extremely flat potentials. I should add, if one has another means of producing the density perturbations necessary for galaxy formation (e.g., cosmic strings or isothermal perturbations), then it is sufficient to have

$$(H^2/\dot\phi)_{Galaxy} \lesssim 10^{-4}$$

(5b) Isothermal perturbations produced during inflation, e.g., as discussed for the case of an axion-dominated Universe, also lead to microwave anisotropies and possibly structure formation. The smoothness of the microwave background dictates that

$$(\delta\rho/\rho)_{ISO} \lesssim 10^{-4}$$

while if they are to be relevant for structure formation

$$(\delta\rho/\rho)_{ISO} \simeq 10^{-5} - 10^{-4}$$

In the case of isothermal axion perturbations this is easy to arrange to have $(\delta\rho/\rho)_{ISO} \ll 10^{-4}$ unless the scale of PQ symmetry is larger than about 10^{18} GeV.

(6a) Sufficiently high reheat temperature so that the Universe is radiation-dominated at the time of primordial nucleosynthesis ($t \simeq 10^{-2} - 10^2$ sec, $T \simeq 10$ MeV $- 0.1$ MeV). Only in the case of poor reheating is T_{RH} likely to be anywhere as low as 10 MeV, in which case $T_{RH} \simeq (\Gamma m_{pl})^{1/2}$ and the condition that T_{RH} be ≥ 10 MeV then implies

$$\Gamma \geq 10^{-23} \text{GeV} \simeq (6.6 \times 10^{-2} \text{ sec})^{-1}$$

(6b) The more stringent condition on the reheat temperature is that it be sufficiently high for baryogenesis. If baryogenesis proceeds in the usual way[68], the out-of-equilibrium decay of a supermassive particle whose interactions violate B, C, P conservation, then T_{RH} must be greater than about 1/10 the mass of the particle whose out-of-equilibrium decays are responsible for producing the baryon asymmetry. Assuming that this particle couples to ordinary quarks and leptons, its mass must be greater than 10^9 GeV or so to insure a sufficiently longlived proton, implying that the reheat temperature must be greater than about 10^8 GeV (at the very least). On the other hand if the baryon asymmetry can be produced by the decays of the ϕ particles themselves, then

$$n_B/s \simeq (0.75)(T_{RH}/m_\phi)\epsilon$$

and a very low reheat temperature may be tolerable

$$T_{RH} \simeq 10^{-10}\epsilon^{-1}m_\phi$$

where as usual ϵ is the net baryon number produced per ϕ-decay.

(7) If ϕ is not a gauge singlet field, as in the case of the original Coleman-Weinberg SU(5) model, one must be careful that 'ϕ rolls in the correct direction'. It was shown that for the original Coleman-Weinberg SU(5) models ϕ might actually begin to roll toward the SU(4) $\times$ U(1) minimum of the potential even though the global minimum of the potential was the SU(3) $\times$ SU(2) $\times$ U(1) minimum[135]. This is because near $\phi = 0$ the SU(4) $\times$ U(1) direction is usually the direction of steepest descent. This is the so-called problem of 'competing phases'. As mentioned earlier, the extreme flatness required to obtain sufficiently small density perturbations probably precludes the possibility that ϕ is a gauge non-singlet, so the problem of competing phases does not usually arise. [Although in SUSY/SUGR models ϕ is often complex and one has to make sure that it rolls in the correct direction.]

(8) In addition to the scalar density perturbations discussed earlier, tensor or gravitational wave perturbations also arise (these correspond to tensor perturbations in

the metric $g_{\mu\nu}$)[136]. The amplitude of these perturbations is easy to estimate. The energy density in a given gravitational wave mode (characterized by wavelength λ) is

$$\rho_{GW} \simeq m_{pl}^2 h^2/\lambda^2 \tag{3.47}$$

where h is the dimensionless amplitude of the wave. As each gravitational wave mode crosses outside the horizon during inflation deSitter space produced fluctuations lead to

$$(\rho_{GW})_{\lambda \simeq H^{-1}} \simeq H^4, \text{ or } h \simeq H/m_{pl}. \tag{3.48}$$

While outside the horizon the dimensionless amplitude h remains constant (h behaves like a minimally coupled scalar field), and so each mode enters the horizon with a dimensionless amplitude

$$h \simeq H/m_{pl}. \tag{3.49}$$

Gravitational wave perturbations with wavelength of order the present horizon lead to a quadrupole anisotropy in the microwave temperature of amplitude h. The upper limit to the quadrupole anisotropy of the microwave background ($\delta T/T \lesssim 10^{-4}$) leads to the constraint

$$H/m_{pl} \leq 10^{-4},$$
$$M \leq O(10^{17} \text{GeV}).$$

In turn this leads to a constraint on the reheat temperature (using $g_* \simeq 10^3$)

$$T_{RH} \leq g_*^{-1/4} M \leq \text{ few} \times 10^{16} \text{GeV}$$

(9) One has to be mindful of various particles which may be produced during the reheating process. Of particular concern are stable or very long-lived, NR particles (including other scalar fields which may be set into oscillation and thereafter behave like NR matter). Since $\rho_{NR}/\rho_R \propto R(t)$ and today $\rho_{NR}/\rho_R \simeq 3 \times 10^4$ or so one has to be careful that ρ_{NR}/ρ_R is sufficiently small at early times

$$\rho_{NR}/\rho_R \leq \begin{cases} 3 \times 10^4 & \text{today} \\ 10^{-8} & T = 1 \text{GeV} \\ 10^{-18} & T = 10^{10} \text{GeV} \end{cases}$$

which is not always easy—just ask any experimentalist about suppressing some effect by 18 orders-of-magnitude!

Of particular concern in supersymmetric models are gravitinos which, if produced, can decay shortly after nucleosynthesis and photodissociate the light elements produced (particularly D and ^7Li)[137]. [In fact, the constraint that gravitinos not be overproduced during the reheating process leads to the very restrictive bound: $T_{RH} \leq 10^9$GeV or so.] In supersymmetric models where SUSY breaking is done ala Polonyi[138], the Polonyi field can be set into oscillation[139] and these oscillations which behave like NR matter can come to dominate the energy density of

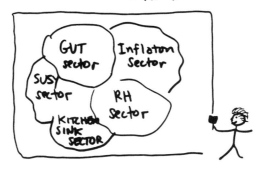

Figure 3.11: Constraint (11) in 'The Prescription for Successful Inflation'.

the Universe too early (leading to a Universe which if far too youthful when it cools to 3K) or even worse decay into dread gravitinos! In sum, one has to be mindful of the weakly-interacting, longlived particles which may be produced during reheating as they may eventually lead to an energy crisis.

(10) In SUSY/SUGR models where the scalar field responsible for inflation is in thermal equilibrium before the inflationary transition, one has to make sure that $\langle\varphi\rangle_T$ does not smoothly evolve into the zero temperature minimum of the potential. A sure way of doing this is to arrange to have

$$\langle\varphi\rangle_T \leq 0$$

this is the so-called thermal constraint[132].

(11) Last (in my probably incomplete list) but certainly not least, the scalar potential necessary for successful inflation should be but a part of a 'sensible, perhaps even elegant, particle physics model' (see Fig. 3.11). We do not want cosmology to be the tail that wags the dog!

These conditions are spelled out in more detail in ref. 117. In general they lead to a potential which is 'short and squat' and has a dimensionless coupling of order 10^{-15} somewhere. In order that radiative corrections not spoil the flatness, it is all but mandatory that ϕ be a gauge singlet field which couples very weakly to other fields in the theory.

[Suppose that ϕ has a nice, flat potential which will successfully implement inflation and has a ϕ^4 term whose coefficient $\lambda \simeq O(10^{-15})$ (as is usually the case). Now suppose that ϕ couples to another scalar field ψ or to a fermion field f through terms like: $\lambda'\psi^2\phi^2$ and $h\bar{f}f\phi$. One-loop corrections to the $\lambda\phi^4$ term in the scalar potential arise due to the coupling of ϕ to ψ of f: $(\lambda'^2 \ln + h^4 \ln)\phi^4$. In order that they not spoil the flatness of $V(\phi)$ by these 1-loop corrections, the couplings λ' and h must be small: $\lambda' \lesssim \ln^{-1/2} \lambda^{1/2}$; $h \lesssim \ln^{-1/4} \lambda^{1/4}$.]

To give an idea of the kind of potential which we are seeking consider

$$V = V_0 - a\phi^2 - b\phi^3 + \lambda\phi^4$$

The constraints discussed above are satisfied for the following sets of parameters

$$SET\ 1 \begin{cases} \lambda \leq 4 \times 10^{-16} \\ b \simeq 4 \times 10^7 \lambda^{3/2} m_{pl} \\ a \leq H^2/40 \simeq 10^{28} \lambda^3 m_{pl}^2 \\ \sigma \simeq 3 \times 10^7 \lambda^{1/2} m_{pl} \\ M \simeq V_0^{1/4} \simeq 3 \times 10^7 \lambda^{3/4} m_{pl} \\ \simeq \lambda^{1/4}\sigma \end{cases}$$

$$SET\ 2 \begin{cases} V = \lambda(\phi^2 - \sigma^2)^2 \quad (b=0,\ a = 2\lambda\sigma^2,\ V_0 = \lambda\sigma^4) \\ \sigma/m_{pl} = 1/2,\ 2,\ 3,\ 10 \\ \lambda = 2 \times 10^{-44},\ 5 \times 10^{-20},\ 10^{-15},\ 2 \times 10^{-15},\ 3 \times 10^{-16} \\ M \simeq \lambda^{1/4}\sigma \end{cases}$$

3.9. TWO SIMPLE MODELS THAT WORK

To date a handful of models that satisfy the prescription for successful inflation have been constructed[133,134,140-142,144,147-150]. Here, I will discuss two particularly simple and illustrative models. The first is an SU(5) GUT model proposed by Shafi and Vilenkin[141] and refined by Pi[142]. [Note, there is nothing special about SU(5); it could just as well be an E6 model.] I will discuss Pi's version of the model. In her model the inflating field $\vec{\phi}$ is a complex gauge singlet field whose potential is of the Coleman-Weinberg form[126]

$$V(\phi) = B[\phi^4 \ln(\phi^2/\sigma^2) + (\sigma^4 - \phi^4)/2]/4 \qquad (3.50)$$

where $\phi = |\vec{\phi}|$ and B arises due to 1-loop radiative corrections from other scalar fields in the theory and is set to be $O(10^{-14})$ in order to successfully implement inflation. (Note, in Eqn(3.50) I have only explicitly shown the part of the potential relevant for inflation.) Since the 1-loop corrections due to other fields in the model are of order $(\lambda^2 \ln)\phi^4$ (λ is a typical quartic coupling, e.g., $\lambda \phi^2 \psi^2$) the dimensionless couplings of ϕ to other fields in the theory must be of order 10^{-7} or so. In her model, $\vec{\phi}$ is the field responsible for Peccei-Quinn symmetry breaking; the vacuum expectation value of $|\vec{\phi}|$ breaks the PQ symmetry and the argument of $\vec{\phi}$ is the axion degree of freedom. In addition, the vacuum expectation value of $|\vec{\phi}|$ induces SU(5) SSB as it leads to a negative mass-squared term for the 24-dimensional Higgs in the theory which breaks SU(5) down to SU(3) × SU(2) × U(1). In order to have the correct SU(5) breaking scale, the vacuum expectation value of $|\vec{\phi}|$ must be of order 10^{18} GeV. In addition to the usual adiabatic density perturbations there are isothermal axion fluctuations of a similar magnitude[121]. The model reheats (barely) to a high enough temperature for baryogenesis. So far the model successfully implements inflation, albeit at the cost of a very small number ($B \simeq 10^{-14}$) whose origin is not explained and whose value is not stabilized (e.g., by supersymmetry).

The second model is a SUSY/SUGR model proposed by Holman, Ramond, and Ross[134] which is based on a very simple superpotential. They write the superpotential for the full theory as

$$W = I + S + G \quad (3.51)$$

where I, S, G pieces are the inflation, SUSY, and GUT sectors respectively. For the I piece of the superpotential they choose the very simple form

$$I = (\Delta^2/M)(\varphi - M)^2, \quad (3.52)$$

where $M = m_{pl}/(8\pi)^{1/2}$, Δ is an intermediate scale, and ϕ is the field responsible for inflation. Their potential has one free parameter: the mass scale Δ. This superpotential leads to the following scalar potential

$$V_I(\phi) = \exp(|\phi|^2/M^2)[|\partial I/\partial\phi + \phi^* I/M^2|^2 - 3|I|^2/M^2]$$
$$= \Delta^4 \exp(\phi^2/M^2)[\phi^6/M^6 - 4\phi^5/M^5 + 7\phi^4/M^4 - 4\phi^3/M^3 - \phi^2/M^2 + 1].$$

Expanding the exponential one obtains

$$V_I(\phi) = \Delta^4(1 - 4\phi^3/M^3 + 6.5\phi^4/M^4 - 8\phi^5/M^5 + ...), \quad (3.53a)$$
$$V_I' = \Delta^4(-12\phi^2/M^3 + 26\phi^3/M^4 - 40\phi^4/M^5 + ...) \quad (3.53b)$$

It is sufficient to keep just the first two terms in $V_I(\phi)$ to solve the equations of motion

$$\phi/M \simeq [12(N(\phi) + 1/3)]^{-1} \quad (3.54a)$$
$$H^2/\dot{\phi} \simeq (12\sqrt{3})^{-1}(\Delta/M)^2(\phi/M)^{-2} \simeq (12/\sqrt{3})(\Delta/M)^2 N^2, \quad (3.54b)$$

By choosing $\Delta/M \simeq 9 \times 10^{-5}$ density perturbations of an acceptable magnitude result (and about 2×10^6 e-folds of inflation!). Taking $\Delta/M \simeq 9 \times 10^{-5}$ corresponds to an intermediate scale in the theory of about $\Delta \simeq 2 \times 10^{14}$ GeV—a very suggestive value.

The ϕ field couples to other fields in the theory only by gravitational strength interactions and

$$\Gamma \simeq m_\varphi^3/M^2 \simeq \Delta^6/M^5, \qquad (3.55)$$

where $m_\varphi^2 \simeq 8e\Delta^4/M^2$.
The resulting reheat temperature is

$$T_{RH} \simeq (\Gamma m_{pl})^{1/2} \simeq (\Delta/M)^3 M \simeq 10^6 \text{GeV}. \qquad (3.56)$$

The baryon asymmetry in this model is produced directly by ϕ-decays ($\phi \to H_3 \bar{H}_3$; $H_3 \bar{H}_3 \to q's\, l's$; H_3 = color triplet Higgs

$$n_B/s \simeq (0.75\epsilon)T_{RH}/m_\phi$$
$$\simeq 10^{-1}\epsilon(\Delta/M)$$

Since $10^{-1}\Delta/M \simeq 10^{-5}$, a C, CP violation of about $\epsilon \simeq 10^{-5}$ is required to explain the observed baryon asymmetry of the Universe ($n_B/s \simeq 10^{-10}$).

Their model satisfies all the constraints for successful inflation except the thermal constraint. They argue that the thermal constraint is not relevant as the interactions of the ϕ field are too weak to put it into thermal equilibrium at early times. They therefore must take the initial value of ϕ ($\equiv \phi_0$) to be a free parameter and assume that in some regions of the Universe ϕ_0 is sufficiently far from the minimum so that inflation occurs ($\phi_0 \lesssim 10^{-3}M$). This model is somewhat ad hoc in that it contains a special sector of the theory whose sole purpose is inflation. Once again the model contains a small dimensionless coupling (the coefficient of the ϕ^4-term $\simeq 3 \times 10^{-16}$) or equivalently, a small mass ratio

$$(\Delta/M)^4 \simeq 10^{-16}$$

Since the model is supersymmetric that small number is stabilized against radiative corrections. Although the small ratio is not explained in their model, its value when expressed as a ratio of mass scales suggests that it might be related to one of the other small dimensionless numbers in particle physics (which also beg explanation)

$$(m_{GUT}/m_{pl}) \simeq 10^{-4}$$
$$(m_W/m_{pl}) \simeq 10^{-17}$$
$$g_e \simeq m_e/300\text{GeV} \simeq 10^{-6}$$

While neither of these models is particularly compelling and both have been somewhat contrived to successfully implement inflation, they are at the very least 'proof of existence' models which demonstrate that it is possible to construct a simple model which satisfies all the know constraints. Fair enough!

3.10. TOWARD THE INFLATIONARY PARADIGM

Guth's original model of inflation was based upon a strongly, first order phase transition associated with SSB of the GUT. The first models of new inflation were based upon Coleman-Weinberg potentials, which exhibit weakly-first order transitions. It now appears that the key feature needed for inflation is a very flat potential and that even potentials which lead to second order transitions (i.e., the $\phi = 0$ state is never metastable) will work just as well[143]. Most of the models for inflation now do not involve SSB, at least directly, they just involve the evolution of a scalar field which is initially displaced from the minimum of its potential. [There is a downside to this; in many models inflation is a sector of the theory all by itself.] Since the fields involved are very weakly coupled, thermal corrections can no longer be relied upon to set the initial value of ϕ. Inflation has become much more than just a scenario—it has become an early Universe paradigm!

On the horizon now are models which inflate, but are even more far removed from the original idea of a strongly-first order, GUT SSB phase transition; I'll discuss three of them here. Inflation—that is the rapid growth of our three spatial dimensions, appears to be a very generic phenomenon associated with early Universe microphysics.

(1) Chaotic Inflation

Linde[144] has proposed the idea that inflation might result from a scalar field with a very simple potential, say

$$V(\phi) = \lambda \phi^4$$

which due to 'chaotic initial conditions' (which thus far have not been well-defined) is displaced from the minimum of its potential—in this case $\phi = 0$ (see Fig. 3.12). With the initial condition $\phi = \phi_0$ this potential is very easy to analyze:

$$N(\phi_0) \equiv \int_{\phi_0}^{0} H dt \simeq \pi(\phi_0/m_{pl})^2, \qquad (3.57)$$
$$(\delta\rho/\rho)_H \simeq (H^2/\dot\phi) \simeq (32/3)^{1/2} \lambda^{1/2} N(\phi)^{3/2}.$$

In order to obtain density perturbations of the proper amplitude $(\delta\rho/\rho \simeq 10^{-4})$ λ must be very small

$$\lambda \simeq 10^{-14}$$

—as usual! In order to obtain sufficient inflation, the initial value of ϕ must be

$$N(\phi_0) \simeq \pi(\phi_0/m_{pl})^2 \geq 60$$
$$\Rightarrow \phi_0 \geq 4.4 m_{pl}$$

Both of these two conditions are rather typical of potentials which successfully implement inflation. However, when one talks about truly chaotic initial conditions one wonders if a large enough patch exists where ϕ is approximately constant. Remember the key constraint is that the gradient energy density be small compared to the potential energy

$$(\nabla\phi)^2/2 \ll \lambda \phi_0^4$$

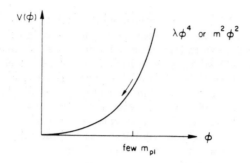

Figure 3.12: A potential for 'chaotic inflation'. In Linde's chaotic inflation, due to initial conditions, ϕ is displaced from the minimum of its potential ($\phi = 0$) and inflation occurs as it evolves to $\phi = 0$.

Labeling the typical dimension of the patch L, the above requirement translates to

$$L \gg \lambda^{-1/2}(m_{pl}/\phi_0)m_{pl}^{-1} \simeq 2(\phi_0/m_{pl})H^{-1}$$

which requires that L be rather large compared to the Hubble radius, therefore seeming to require rather special initial conditions. Still the simplicity of Linde's idea is very appealing.

[Note that the potential $V = \frac{1}{2}m^2\phi^2$ (corresponding to a massive scalar field) is also suitable for inflation. In this case

$$N(\phi_0) \simeq 2\pi(\phi_0/m_{pl})^2, \tag{3.58a}$$

$$(\delta\rho/\rho)_H \simeq H^2/\dot\phi \simeq 4(\pi/3)^{1/2}(m/m_{pl})N. \tag{3.58b}$$

Sufficient inflation requires: $\phi_0 \gtrsim 3m_{pl}$, and density perturbations of an acceptable magnitude requires: $(m/m_{pl}) \simeq 10^{-4}/(4N) \simeq 4 \times 10^{-7}$. This potential has been analyzed by I. Moss (private communication) and L. Jensen (private communication), and more recently by the authors of refs. 145.]

(2) Induced Gravity Inflation

Consider the Ginzburg-Landau theory of induced gravity based upon the effective Lagrangian[146]

$$\mathcal{L} = -\epsilon\phi^2 R/2 - \partial_\mu\phi\partial^\mu\phi/2 - \lambda(\phi^2 - v^2)^2/8, \tag{3.59}$$

where ϵ, λ are dimensionless couplings, R is the Ricci scalar, and $v \equiv \epsilon^{-1/2}(8\pi G)^{-1/2}$. The equation of motion for ϕ is

$$\ddot\phi + 3H\dot\phi + \dot\phi^2/\phi + [V' - 4V/\phi]/(1 + 6\epsilon) = 0 \tag{3.60a}$$

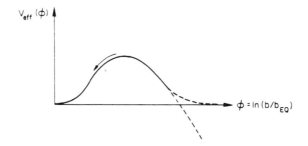

Figure 3.13: In theories with additional spatial dimensions there must be an effective potential associated with the size of the extra dimensions (shown here schematically). One might expect that very early on ($t \leq 10^{-43}$ sec) the size of the extra dimensions is displaced from its equilibrium value ($\equiv b_{eq}$), due to finite temperature corrections, initial conditions, or whatever. It is speculated that inflation might occur as the size of the extra dimensions evolves to its equilibrium value, thereby solving both the usual cosmological puzzles and the puzzle of why the extra dimensions are so small compared to our three familiar spatial dimensions.

supplemented by

$$H^2[1 + (2\dot\phi/\phi)/H] = (3\epsilon\phi^2)^{-1}[\dot\phi^2/2 + V(\phi)] \qquad (3.60b)$$

Successful inflationary scenarios can be constructed for $\phi_0 \ll v$ and for $\phi_0 \gg v$ ($\phi_0 =$ the initial value of ϕ), so long as $\epsilon \leq 10^{-2}$ and $\lambda \simeq O(10^{-12} - 10^{-16})$[147,148]. The small dimensionless coupling constant required in the scalar potential is by now a very familiar condition.

(3) The Compactification Transition

Ever increasing numbers of physicists are pursuing the idea that unification of the forces may require additional spatial dimensions (or as the optimist would say, unification of the forces is evidence for extra dimensions!), e.g., Kaluza-Klein theories, supergravity theories, and most recently, superstring theories. We know experimentally that these extra dimensions must be very small ($\ll 10^{-17}$cm) and indeed in most theories the extra dimensions form a compact manifold of characteristic size 10^{-34}cm or so. If our space-time is truly more than four dimensional, then we have yet another problem to add to our list of puzzling cosmological facts—the extreme smallness of the extra spatial dimensions, some $62 \simeq log(10^{28}\text{cm}/10^{-34}\text{cm})$ or so orders of magnitude smaller than the more familiar three spatial dimensions. The possible use of inflation to explain this largeness problem has not escaped the attention of researchers in this field.

In these theories there is a natural candidate for the 'inflating field' (which is also automatically a gauge singlet)—the radius of the extra dimensions. If there are extra dimensions there must be some dynamics which determine their present, equilibrium size ($\equiv b_{eq}$), and in principle one should be able to construct an effective potential associated with the size of the extra dimensions

$$V_{eff} = V(\phi),$$
$$\phi = \ln(b/b_{eq}).$$

(see Fig. 3.13). [The substitution $\phi = \ln(b/b_{eq})$ results in the usual kinetic term for ϕ when the higher dimensional Einstein equations are written down.] If the extra dimensions are displaced from their equilibrium value—an idea which seems not at all unreasonable since very early on ($t < 10^{-43}$ sec) one might expect all the dimensions to be on equal footing, then while they are evolving to their equilibrium value ($\phi = 0$) the Universe will be endowed with a large potential energy (and may very well inflate), thereby explaining the largeness of our three spatial dimensions as well as the usual cosmological puzzles. Inflationary models involving the compactification transition have already been investigated and the results are encouraging[149].

3.11. LOOSE ENDS

Inflation offers the possibility of making the present state of the Universe (on scales as large as our Hubble radius) insensitive to the initial data for the Universe. Since we have little hope of ever knowing what the initial data were this is a very attractive proposition. It has by no means yet achieved that lofty goal. There are a number of loose ends (and perhaps even a loose thread which may unravel the entire tapestry). I will briefly mention a few of them here.

(1) 'Who is ϕ?'

Inflationary models exist in which the scalar field ϕ: effects SSB of the GUT[140,141,142], effects SSB of SUSY[133], induces Newton's constant (in a Landau-Ginzburg model of induced gravity)[147,148], is $\sim \ln(r_X/r_{XEQ})$ (where r_X is the radius of compactified extra dimensions) in theories with extra dimensions which become compactified[149], is $\propto$ (scalar curvature)$^{1/2}$ in R^2 theories of gravity,[150] is just some 'random' scalar field[144], or is merely in the theory to effect inflation.[131,134] Given the number of different kinds of inflationary scenarios which exist, it seems as though inflation is generic to early Universe microphysics, occurring whenever a weakly-coupled scalar field finds itself displaced from the minimum of its potential. Clearly, a key question at this point is just how 'the inflation sector' of the theory fits into the Big Picture!

(2) What Determines the Initial Value of ϕ?

One thing is certain, and that is that ϕ must be very weakly-coupled, as quantified by its small dimensionless coupling constant. Because of this fact, it is almost certain that ϕ was not initially in thermal contact with the rest of the Universe and so the initial value of $\phi(\equiv \phi_0)$ is unlikely to be determined by thermal considerations (in the earliest models of new inflation, ϕ_0 was determined by thermal considerations, however these models resulted in density perturbations of an unacceptably large amplitude). At present ϕ_0 must be taken as initial data. Some have argued that ϕ_0 might be determined in an anthropic-like way, as regions of the Universe where ϕ_0 is sufficiently far displaced from equilibrium will undergo inflation and eventually occupy most of the physical volume of the Universe. Perhaps the wavefunction of the Universe approach will shed some light on the initial distribution of the scalar field ϕ. Or it could be that due to 'as-of-yet unknown dynamics' ϕ was indeed in thermal equilibrium at a very early epoch. It goes without saying that it is crucial that ϕ be initially displaced from its minimum.

(3) Validity of the Semi-Classical Equations of Motion for ϕ

While it may seem perfectly plausible that ϕ evolves according to its semiclassical equations of motion, the validity of this assumption has troubled inflationists from the 'dawn of new inflation'. While a full quantum field theory treatment of inflation is very difficult and has not been effected, a number of specific issues have been addressed. Several authors have studied the role of inhomogeneities in ϕ, and have found that for the very weakly-coupled fields one is dealing with, mode coupling is not important and the individual modes are quickly smoothed by the exponential expansion of their physical wavelengths[151]. I already mentioned the necessity of having ϕ smooth over a sufficiently large region so that the gradient terms in the stress energy do not dominate.

The effect of quantum fluctuations on the evolution of ϕ has been studied in some detail by Guth and Pi[143], Fischler et al.[152], Linde[127], Vilenkin and Ford[128], Semenoff and Weiss[153], and Evans and McCarthy[154]. The basic conclusion that one draws from the work of these authors is that the use of the semi-classical equations of motion is valid so long as $\phi_{cl} \gg \Delta\phi_{QM} \simeq N^{1/2}H/2\pi$, which is almost always satisfied for the very flat potentials of interest to inflationists (at least for the last 50 or so e-folds which affect our present Hubble volume). [More precisely, the semiclassical change in ϕ in a Hubble time, $\Delta\phi_{Hubble} \simeq -V'/3H^2 \simeq -V'm_{pl}^2/(8\pi V)$, should be much greater than the increase in $< \phi^2 >_{QM}^{1/2} \simeq H/2\pi$, due to the addition of another quantum mode; see Bardeen et al.[108].] At present the validity of the semi-classical equations of motion seems to be reasonably well established.

(4) No Hair Conjectures

While inflation has been touted from the very beginning as making the present state of the Universe insensitive to the initial spacetime geometry, not much has been done to justify this claim until very recently. As I mentioned earlier, inflation is nearly always analyzed in the context of a flat, FRW cosmological model, making such a claim somewhat dubious. However, it has now been shown that all of the homogeneous models (with the exception of the highly-closed models) undergo inflation, isotropize and remain isotropic to the present epoch providing that the model would have inflated the requisite 60 or so e-folds in the absence of anisotropy[155].

The proof of this result involves three parts. First, Wald[156] demonstrated that all homogeneous models with a positive cosmological term asymptotically approach deSitter (less the aforementioned highly-closed models which recollapse before the cosmological term becomes relevant). Wald's result follows because all forms of 'anisotropy energy density' decrease with increasing proper volume element, whereas the cosmological term remains constant, and so eventually triumphs. Of course, inflationary models do not, in the strictest sense, have a cosmological term, rather they have a positive vacuum energy as long as the scalar field is displaced from the minimum of its potential. Thus the dynamics of the scalar field comes into play: does ϕ stay displaced from the minimum of its potential long enough so that the vacuum energy comes to dominate? Due to the presence of anisotropy the expansion rate is *greater* than if there were only vacuum energy density, and so the friction felt by ϕ as it trys to roll (the $3H\dot\phi$ term) is greater and it takes ϕ longer to evolve to its minimum than without anisotropy. For this reason the Universe does become vacuum dominated before the vacuum energy disappears, and in fact the Universe inflates slightly longer in the presence of anisotropy (one or two e-folds)[157]. Finally, is the anisotropy reduced sufficiently so that the Universe today is still nearly isotropic? As it turns out, the requisite 60 or so e-folds needed to solve the other conundrums reduces the growing modes of anisotropy sufficiently to render them small today.

Allowing for inhomogeneous initial spacetimes makes matters much more difficult. Jensen and Stein-Schabes[158] and Starobinskii[159] have proven the analogue of Wald's theorem for spacetimes which are negatively-curved. Jensen and Stein-Schabes[158] have gone on to conjecture that spacetimes which have sufficiently large regions of negative curvature will undergo inflation, resulting in a Universe today which although not globally homogeneous, at least contains smooth volumes as large as our current Hubble volume.

Does this improve the situation that Collins and Hawking[63] discussed in 1973? While the work of Jensen and Stein-Schabes[158] seems to indicate that many inhomogeneous spacetimes undergo inflation and even leads one to speculate that the measure of the set of initial spacetimes which eventually inflate is non-zero, it is not possible to draw a definite conclusion without first defining a measure on the space of initial data. In fact, as Penrose[76] pointed out there is at least one way of defining the measure such that this is not the case. Consider the set of all Cauchy data at the present epoch; intuitively it is clear that those spacetime slices which are highly irregular are the rule, and those which are smooth in regions much larger than our current Hubble volume are the exception. Defining the measure today, it seems very reasonable that the smooth spacetime slices are a set of measure zero. Now evolve the spacetimes back to some initial epoch (for example $t = 10^{-43}$ sec). Using the seemingly very reasonable measure defined today and the mapping back to 'initial' spacetimes, one could argue that the set of initial data which inflate is still of measure zero. While I believe that this argument is technically correct, I also believe that it is silly. First, upon close examination of all of those initial spacetimes which led to spacetimes today without smooth regions as large as our present Hubble volume, one would presumably find that the scalar field responsible for inflation would be very close to the minimum of its potential (in order that they not inflate)–not a very generic initial condition. Secondly, if one adopts the point-of-view of an evolving Universe which has an 'initial epoch' (and not everyone does), then there is a preferred epoch at which one would define a measure—the 'initial epoch,' and at that epoch I believe any reasonably defined measure would lead to the set of initial spacetimes which inflate being of non-zero measure.

Although it is not possible *yet* to claim rigorously that inflation has resolved the problem of the seemingly special initial data required to reproduce the Universe we see today (at least within our Hubble volume), I think that any fairminded person would admit that it has improved the situation dramatically. Extrapolating from the solid results that exist, it seems to me that starting with a general inhomogeneous spacetime, there will exist regions which undergo inflation and which today are much larger than our present Hubble volume, thereby accounting for the smooth region we find ourselves in. From a more global perspective, one might expect that on scales $\gg H^{-1}$ the Universe would be highly irregular. [The evolution of a model universe which is isotropic and homogeneous except for one spherically-symmetric region of false vacuum (where $\phi \neq \sigma$) has been studied by the authors of ref. 160. The results are interesting in that they begin to address the problem of general initial conditions. The vacuum-dominated bubble becomes causally detached from the rest of the spacetime, becoming a 'child Universe' spawned by inflation.]

(5) The Present Vanishingly Small Value of the Cosmological Constant

Inflation has shed no light on this difficult and very fundamental puzzle (nor has anything else for that matter!). In fact, since inflation runs on vacuum energy so to speak, the fate of inflation hinges upon the resolution of this puzzle. For example, suppose there were a grand principle that dictated that the vacuum energy of the Universe is always zero, or that there were an axion-like mechanism which operated and ensured that any cosmological constant rapidly relaxed to zero; either would be a disaster to inflation, shorting out its source of power—vacuum energy. [Another possibility which has received a great deal of attention recently is the possibility that deSitter space might be quantum mechanically unstable[161]—of course, if its lifetime were at least 60 some e-folds that would not necessarily adversely affect inflation.]

3.12. INFLATION CONFRONTS OBSERVATION

No matter how appealing a theory may be, it must meet and pass the test of experimental verification. Experiment and/or observation is the final arbiter. One of the few blemishes on early Universe physics is the lack, thus far, of experimental/observational tests of the many beautiful and exciting predictions. That situation is beginning to change as the field starts to mature. Inflation is one of the early Universe theories which is becoming amenable to verification or falsification. Inflation makes the following very definite predictions (postdictions?):

(1) $\Omega = 1.0$ (more precisely, $R_{curv} = R(t)|k|^{-1/2} = H^{-1}/|\Omega - 1|^{1/2} \gg H^{-1}$)

(2) Harrison-Zel'dovich spectrum of constant curvature perturbations (and possibly isocurvature perturbations as well) and tensor mode gravitational wave mode perturbations

The prediction of $\Omega = 1.0$ together with the primordial nucleosynthesis constraint on the baryonic contribution, $0.014 \lesssim \Omega_B h^2 \lesssim 0.035 \lesssim 0.15$ (ref. 9), suggests that most of the matter in the Universe must be nonbaryonic. The simplest and most plausible possibility is that it exists in the form of relic WIMPs (Weakly-Interacting Massive Particles, e.g., axions, photinos, neutrinos; for a review, see ref. 162). Going a step further, these two original predictions then lead to testable consequences:

(3) $H_0 t_0 = 2/3$ (providing that the bulk of the matter in the Universe today is in the form of NR particles)

The observational data on both H_0 and t_0 are far from being definitive: $H_0 \simeq 40 - 100$km sec^{-1}Mpc^{-1} and $t_0 \simeq 12 - 20$Gyr, implying only that $H_0 t_0 \simeq 0.5 - 2.0$.

(4) $\Omega = 1.0$

All of the dynamical observations suggest that the fraction of critical density contributed by matter which is clumped on scales $\lesssim 10 - 30$Mpc is only about: $\Omega_{\lesssim 30} \simeq 0.2 \pm 0.1$ (± 0.1 is not meant to be a formal error estimate, but indicates the spread in the observations) (see refs. 11). If inflation is not to be falsified, that leaves but two options: (1) the observations are somehow misleading or wrong; or (2) there exists a component of energy density which is smoothly distributed on scales $\lesssim 10 - 30$Mpc (and therefore would not be reflected in the dynamical determinations). Candidates for the smooth component include: relic, light neutrinos, which by virtue of the large length scale ($\lambda_\nu \simeq 13 h^{-2}$Mpc) on which neutrino perturbations are damped by freestreaming, would likely still be smooth on these scales; relic relativistic particles produced by the recent decay of an unstable WIMP species;[163] a relic cosmological term;[164] 'failed galaxies,' referring to a population of galaxies which have the same mix of dark matter to baryons, but are more smoothly distributed and are too faint to observe (at least thus far);[165] relic population of light strings–either fast moving non-intercommuting strings or a tangled network of non-Abelian strings[166]. All of these smooth component scenarios have testable consequences[167]—their predictions for $H_0 t_0$ differ from $2/3$; the growth of perturbations is different; the evolution of the cosmic scale factor $R(t)$ is different from the matter-dominated model and various kinematic tests (magnitude-redshift, angular size-redshift, lookback time-redshift, proper volume element-redshift, etc.) can in principle differentiate between them.

(5) Microwave Fluctuations

Both the scalar and tensor metric perturbations lead to fluctuations in the CMBR on large angular scales ($\gg 1°$). On such large scales causal microphysical processes (such as reionization) cannot have erased the primordial fluctuations, and so if ever present, they must still be there. The scalar perturbations (if they have anything to do with structure formation) must be of amplitude $\gtrsim$ few $\times 10^{-6}$, which is within a factor of 10 or less of the current upper limits on these scales.

(6) Two Detailed Stories of Structure Formation

The simplest possibility, namely that most of the mass density is in relic WIMPs ($\Omega_{WIMP} = 1.0 - \Omega_B \simeq 0.9$) leads to two very detailed scenarios of structure formation: hot dark matter (the case where the dark matter is neutrinos) and cold dark matter (essentially any other WIMP as dark matter). At present, the numerical simulations of these scenarios are sufficiently definite that it is possible to falsify them–and in fact, both of these simplest scenarios have difficulties (see the recent review by White[168]). In the hot dark matter case it is forming galaxies early enough. The large-scale structure which evolves in this case (voids, superclusters, froth) qualitatively agrees with what is observed; however, in order to get agreement with the galaxy-galaxy correlation function, galaxies must form very recently (redshifts $\lesssim 1$) in contradiction to all the galaxies (redshifts as large as 3.2) and QSO's (redshifts as large as 4.0) which are seen at redshifts $\gtrsim 1$.

With cold dark matter the simulations can nicely reproduce galaxy clustering, most of the observed properties of galaxies (masses and densities, rotation curves, etc)[169]. However the simulations do not seem to be able to produce sufficient large-scale structure. In particular, they fail to account for the amplitude of the cluster-cluster correlation function (by a factor of about 3), large amplitude, large-scale peculiar velocities, and voids. [In fairness I should mention that our knowledge of large-scale structure of the Universe is still very fragmentary, with the first moderate sized ($\sim 10^4$), 3-dimensional surveys having just recently been completed.] In order to account for $\Omega = 1.0$, galaxy formation must be biased (i.e., only density-averaged peaks greater than some threshold, typically $2 - 3\sigma$, are assumed to evolve into galaxies which we see today, the more typical 1σ peaks resulting in 'failed galaxies' for some reason or another; see ref. 165).

[The situation with respect to large scale structure is becoming more interesting every moment. Several groups have now reported large-amplitude (600 – 1000km sec^{-1}) peculiar velocities on large scales ($\sim 50h^{-1}$Mpc) (Burstein et al.[170]; Collins et al.[171]). Such large peculiar velocities are very difficult, if not impossible, to reconcile with either hot or cold dark matter (or even smooth component models) and the Zel'dovich spectrum (see ref. 172). If these data hold up they may pose an almost insurmountable obstacle to any scenario with the Zel'dovich spectrum of density perturbations. The frothy structure observed in the galaxy distribution by de Lapparent et al.[173], galaxies distributed on the surfaces on large ($\sim 30h^{-1}$Mpc), empty bubbles, although somewhat more qualitative, also seems difficult to reconcile with cold dark matter.]

There are a number of observations/experiments which can and will be done in the next few years and which should really put the inflationary scenario to the test. They include improved sensitivity measurements of the CMBR anisotropy. The microwave background anisotropies predicted in the hot dark matter scenario are very close to the observational upper limits on angular scales of both 5 or so arcminutes and $\gtrsim$ few degrees[62]. With cold dark matter, the predictions are a factor of $3 - 10$ away from the observational limits (for the isocurvature spectrum, the quadrupole upper limit may actually rule out this possibility; see, Efstathiou and Bond[174]). An improvement in sensitivity to microwave anisotropies of the order of $3 - 10$ could either begin to confirm one of the scenarios or rule them both out, and is definitely within the realm of experimental reality (Wilkinson in ref. 8).

The relic WIMP hypothesis for the dark matter can also be tested. While it was once almost universally believed that all WIMP dark matter candidates were, in spite of their large abundance, essentially impossible to detect because of the feebleness of their interactions, a number of clever ideas have recently been suggested (and are being experimentally implemented) for detecting axions[175], photinos, sneutrinos, heavy neutrinos, etc.[176] Results and/or limits will be forth coming soon. With the coming online of the Tevatron at Fermilab, the SLC at SLAC, and hopefully the SSC it is possible that one of the candidates may be directly produced in the lab. Experiments to detect neutrino masses in the eV mass range also continue.

A geometric measurement of the curvature of the Universe (which uses the dependence of the comoving volume element as a function of redshift) has recently been made by Loh and Spillar[177]. Their preliminary results indicate $\Omega = 0.9^{+0.7}_{-0.5}$ (95% confidence) (for a matter-dominated model). This technique appears to have great cosmological leverage and looks very promising (especially the value!)—far more promising than the traditional approach of determining the density of the Universe through the deceleration parameter q_0.

Another area with great potential for improvement is 3d surveys of the distribution of galaxies. The largest redshift surveys at present contain only a few 1000 galaxies, yet have been very tantalizing, indicating evidence of voids and froth-like structure to the galaxy distribution[173]. The large, automated surveys which are likely to be done in the next decade could very well lead to a quantum leap in our understanding of the large scale features of the Universe and help to provide hints as to how they evolved.

The peculiar velocity field of the Universe is potentially a very valuable and direct probe of the density field of the Universe:

$$|\delta v_k| = |\dot{\delta}_k/k| \quad (= (\lambda H/2\pi)\delta_k \; for \; \Omega = 1) \tag{3.61}$$

$$(\delta v/c)_\lambda \simeq (\lambda/10^4 h^{-1} \text{Mpc})(\delta\rho/\rho)_\lambda \tag{3.62}$$

where δ_k and δv_k are the $k-th$ Fourier components of $\delta\rho/\rho$ and $\delta v/c$ respectively. The very recent measurements which indicate large amplitude peculiar velocities on scales of $\sim 50h^{-1}$Mpc are surprising in that they indicate substantial power on these scales, and are problematic to almost every scenario of structure formation. Should they be confirmed they will provide a very acute test of structure formation in inflationary models.

Of course, theorists are very accommodating and have already started suggesting alternatives to the simplest scenarios for structure formation. As I mentioned earlier, scenarios with a smooth component to the energy density have been put forward to solve the Ω problem. Cosmic strings present a radically different approach to structure formation with their non-gaussian spectrum of density fluctuations (for further discussion see refs. 183). [It is interesting to note that cosmic strings of the right 'weight' ($G\mu \simeq 10^{-6}$ or so, where μ is the string tension) seem to be somewhat incompatible with inflation, as they must necessarily be produced after inflation and require reheating to a temperature $\gtrsim \mu^{1/2} \simeq 10^{16}$GeV which seems difficult.] Somewhat immodestly I mention a proposal Silk and I recently made: 'double inflation'[178]. While the Harrison-Zel'dovich spectrum is a beautiful prediction both because of its geometric simplicity and its definiteness, it may well be in conflict with observation because it does not seem to allow enough power on large scales to account for the recent observations of froth and large amplitude peculiar velocities. In the variant we have proposed there are two (or more) episodes of inflation, with the final episode lasting only about 40 e-folds or so, so that the amplitudes of perturbations on large scales are set by the first episode and those on small scales by the second episode. That there might be multiple episodes of inflation seems quite plausible given the number of different microphysical scenarios which result in inflation. Arranging the most recent episode to last for only 40 or so e-folds so that some of the scales within our present Hubble volume crossed outside the horizon during an earlier episode of inflation is a more formidable task—but not an impossible or implausible one! If this can be arranged then it is possible to have very large amplitude perturbations on small scales (of order 10^{-1}) and larger than usual amplitude perturbations on large scales (nearly saturating the large scale microwave limits), thereby providing enough power for the large scale structure which the recent redshift surveys and peculiar velocity measurements indicate. The large amplitude perturbations on small scales allow for very early galaxy formation (and reionization of the Universe, thereby erasing the CMBR fluctuations on small angular scales). If the second episode of inflation proceeds via the nucleation of bubbles, they might directly explain the froth-like structure recently reported by de Lapparent et al.[173]

3.13. EPILOGUE

Despite the absence of a compelling model which successfully implements the inflationary paradigm, inflation remains a very attractive means of accounting for a number of very fundamental cosmological facts by microphysics that we have some understanding of: namely, scalar field dynamics at sub-planck energies. The lack of a compelling model at present must be viewed in the light of the fact that at present we have no compelling, detailed model for the 'Theory of Everything' and the fact that despite vigorous scrutiny there has yet to be a No-Go Theorem for inflation unearthed. It is my belief that the undoing of inflation (if it should come) will involve observations and not theory. At the very least The Inflationary Paradigm is still worthy of further consideration–and I hope that I have convinced you of that fact!

Due to space/time limitations my review of inflation has necessarily been incomplete, for which I apologize. I refer the interested reader to the more complete reviews by Linde[179]; by Abbott and Pi[180]; by Steinhardt[181]; by Brandenberger[182]; by Bonometto and Masiero[185]; and by Blau and Guth[184]. My prescription for successfully implementing inflation borrows heavily from the paper written by Steinhardt and myself[117]. This work was supported in part by the DoE (at Chicago) and by my Alfred P. Sloan Fellowship.

REFERENCES

1. M.S. Turner, in *Architecture of the Fundamental Interactions at Short Distances*, eds. P. Ramond and R. Stora (North-Holland, Amsterdam, 1987); S. Bonometto and A. Masiero, *La Vista del Nuovo Cimento* **9**, 3 (1986); *Inner Space/Outer Space*, eds. E.W. Kolb et al. (University of Chicago Press, Chicago, 1986); *Proceedings of the First Jerusalem Winter School*, eds. T. Piran and S. Weinberg (World Scientific, Singapore, 1986); *Proceedings of the Erice School on Gauge Theory and the Early Universe*, eds. P. Galeotti and D. Schramm (Reidel, Dordrecht, 1987); K. Olive, in *Grand Unification With and Without Supersymmetry* (World Scientific, Singapore, 1984); A. Dolgov and Ya.B. Zel'dovich, *Rev.Mod.Phys.* **53**, 1 (1981); G. Steigman, *Ann.Rev.Nucl.Part.Sci.* **29**, 313 (1979); E. Kolb in *Proceedings of the 1986 Santa Cruz TASI*, ed. H. Haber (World Scientific, Singapore, 1987).
2. M. B. Green and J. H. Schwarz, *Nucl.Phys.* **B181**, 502 (1981); **B198**, 252 (1982); **B198**, 441 (1982); *Phys.Lett.* **109B**, 444 (1982); M. B. Green, J. H. Schwarz, and L. Brink, *Nucl.Phys.* **B198**, 474 (1982); J. H. Schwarz, *Phys. Rep.* **89**, 223 (1982); M. B. Green and J. H. Schwarz, *Phys.Lett.* **149B**, 117 (1984); **151B**, 21 (1985); *Nucl.Phys.* **B243**, 475 (1984); L. Alvarez-Gaume and E. Witten, *Nucl.Phys.* **B243**, 475 (1984); G. Chapline and N. Manton, *Phys.Lett.* **120B**, 105 (1983); D. Gross, J. Harvey, E. Martinec, and R. Rohm, *Phys.Rev.Lett.* **54**, 502 (1984); P. Candelas, G. Horowitz, A. Strominger, and E. Witten, *Nucl.Phys.* **B256**, 46 (1985); M. B. Green, *Nature* **314**, 409 (1985); E. Kolb, D. Seckel, and M. S. Turner, *Nature* **314**, 415 (1985); K. Huang and S. Weinberg, *Phys.Rev.Lett.* **25**, 895 (1970).
3. M. Srednicki, *Nucl.Phys.* **B202**, 327 (1982); D. V. Nanopoulos and K. Tamvakis, *Phys.Lett.* **110B**, 449 (1982); J. Ellis, J. S. Hagelin, D. V. Nanopoulos, K. Olive, and M. Srednicki, *Nucl.Phys.* **B238**, 453 (1984); A. Salam and J. Strathdee, *Ann. Phys. (NY)* **141**, 316 (1982); P. G. O. Freund, *Nucl.Phys.* **B209**, 146 (1982); P. G. O. Freund and M. Rubin, *Phys.Lett.* **37B**, 233 (1980); E. Kolb and R. Slansky, *Phys.Lett.* **135B**, 378 (1984); C. Kounnas et al., *Grand Unification With and Without Supersymmetry* (World Scientific, Singapore, 1984); Q. Shafi and C. Wetterich, *Phys.Lett.* **129B**, 387 (1983); P. Candelas and S. Weinberg, *Nucl.Phys.* **B237**, 397 (1984); R. Abbott, S. Barr, and S. Ellis, *Phys.Rev.* **D30**, 720 (1984); E. Kolb, D. Lindley, and D. Seckel, *Phys.Rev.* **D30**, 1205 (1984).
4. S. Djorgovski, H. Spinrad, P. McCarthy, and M. Strauss, *Astrophys.J.* **299**, L1 (1985).
5. B.A. Peterson et al., *Astrophys.J.* **260**, L27 (1982); D. Koo, *Astron.J.* **90**, 418 (1985); C. Hazard, R-G. McMahon, and W.L.W. Sargent, *Nature* **322**, 38 (1986); *Nature* **322**, 40 (1986); S.J. Warren, et al., *Nature* **325**, 131 (1987).
6. P.J.E. Peebles, *Physical Cosmology* (Princeton Univ. Press, 1971); S. Weinberg, *Gravitation and Cosmology* (Wiley, NY, 1972), chapter 15; *Physical Cosmology*, eds. J. Audouze, R. Balian, and D.N. Schramm (North-Holland, Amsterdam,

1980).
7. J. Peterson, P. Richards, and T. Timusk, *Phys.Rev.Lett.* **55**, 332 (1985); G.F. Smoot et al., *Astrophys.J.* **291**, L23 (1985); D. Meyer and M. Jura, *Astrophys.J.* **276**, L1 (1984); D. Woody and P. Richards, *Astrophys.J.* **248**, 18 (1981).
8. D. Wilkinson, in *Inner Space/Outer Space*, eds. E. Kolb et al. (Univ. of Chicago Press, Chicago, 1986).
9. J. Yang, M.S. Turner, G. Steigman, D.N. Schramm, and K. Olive, *Astrophys.J.* **281**, 493 (1984); A. Boesgaard and G. Steigman, *Ann.Rev.Astron.Astrophys.* **23**, 319 (1985).
10. J. Huchra, and A. Sandage and G. Tammann, in *Inner Space/Outer Space*, eds. E. Kolb et al. (Univ. of Chicago Press, Chicago, 1986); J. Gunn and B. Oke, *Astrophys.J.* **195**, 255 (1975); J. Kristian, A. Sandage, and J. Westphal, *Astrophys.* **221**, 383 (1978); R. Buta and G. deVaucouleurs, *Astrophys.J.* **266**, 1 (1983); W.D. Arnett, D. Branch, and J.C. Wheeler, *Nature* **314**, 337 (1985).
11. S. Faber and J. Gallagher, *Ann.Rev.Astron.Astrophys.* **17**, 135 (1979); V. Trimble, *Ann.Rev.Astron.Astrophys.* **25**, in press (1987).
12. I. Iben, Jr., *Ann.Rev.Astron.Astrophys.* **12**, 215 (1974); A. Sandage, *Astrophys.J.* **252**, 553 (1982); D.N. Schramm, in *Highlights of Astronomy*, ed. R. West (Reidel, Dordrecht, 1983), vol.6; I. Iben and A. Renzini, *Phys.Rep.* **105**, 329 (1984).
13. G. Gamow, *Phys.Rev.* **70**, 572 (1946).
14. R. A. Alpher, J. W. Follin, and R. C. Herman, *Phys.Rev.* **92**, 1347 (1953).
15. P. J. E. Peebles, *Astrophys. J.* **146**, 542 (1966); R. V. Wagoner, W. A. Fowler, and F. Hoyle, *Astrophys. J.* **148**, 3(1967).
16. A. Yahil and G. Beaudet, *Astrophys. J.* **206**, 26 (1976).
17. R. V. Wagoner, *Astrophys. J.* **179**, 343 (1973).
18. K. Olive, D. N. Schramm, G. Steigman, M. S. Turner, and J. Yang, *Astrophys. J.* **246**, 557 (1981).
19. D. A. Dicus et al., *Phys.Rev.* **D26**, 2694 (1982).
20. J. Yang, M. S. Turner, G. Steigman, D. N. Schramm, and K. Olive, *Astrophys. J.* **281**, 493 (1984).
21. R. Epstein, J. Lattimer, and D. N. Schramm, *Nature* **263**, 198 (1976).
22. M. Spite and F. Spite, *Astron. Astrophys.* **115**, 357 (1982).
23. D. Kunth and W. Sargent, *Astrophys. J.* **273**, 81 (1983); J. Lequeux et al., *Astron. Astrophys.* **80**, 155 (1979).
24. S. Hawking and R. Tayler, *Nature* **209**, 1278 (1966); J. Barrow, *Mon. Not. r. Astron. Soc.* **175**, 359 (1976); K. Thorne, *Astrophys.J.* **148**, 51 (1967).
25. R. Matzner and T. Rothman, *Phys.Rev.Lett.* **48**, 1565 (1982).
26. Y. David and H. Reeves, in *Physical Cosmology*, see ref. 6; N. Terasawa and K. Sato, *Astrophys.J.* **294**, 9 (1985).
27. A. Boesgaard and G. Steigman, *Ann.Rev.Astron.Astrophys.* **23**, 319 (1985).
28. S. Faber and J. Gallagher, *Ann. Rev. Astron. Astrophys.* **17**, 135 (1979); V. Trimble, *Ann.Rev.Astron.Astrophys.* **25**, in press (1987); *Dark Matter in*

the *Universe* (IAU 117), eds. J. Kormendy and G. Knapp (Reidel, Dordrecht, 1987).
29. M. Davis and P. J. E. Peebles, *Ann. Rev. Astron. Astrophys.* **21**, 109 (1983).
30. M.S. Turner, in *Dark Matter in the Universe*, p.445, see ref. 28.
31. V. F. Shvartsman, *JETP Lett.* **9**, 184 (1969).
32. G. Steigman, D. N. Schramm, and J. Gunn, *Phys.Rev.Lett.* **43**, 202 (1977); J. Yang, D. N. Schramm, G. Steigman, and R. T. Rood, *Astrophys. J.* **227**, 697 (1979); J. Barrow and J. Morgan, *Mon.Not.r.Astron.Soc.* **202**, 393 (1983); K.A. Olive, D.N. Schramm, and G. Steigman, *Nucl.Phys.* **B180**, 497 (1981); A.S. Szalay, *Phys.Lett.* **101B**, 453 (1981); E.W. Kolb and R.J. Scherrer, *Phys.Rev.* **D25**, 1481 (1982); E.W. Kolb, M.S. Turner, and T.P. Walker, *Phys.Rev.* **D34**, 2197 (1986); R.J. Scherrer and M.S. Turner, *Astrophys.J.*, in press (1987); M.H. Reno and D. Seckel, *Phys.Rev.D*, in press (1987).
33. G. Arnison *et al.*, *Phys.Lett.* **126B**, 398 (1983).
34. D. N. Schramm and G. Steigman, *Phys.Lett.* **141B**, 337 (1984); D. Cline, D.N. Schramm, and G. Steigman, *Comments on Nucl.Part.Phys.*, in press (1987).
35. R. J. Scherrer and M. S. Turner, *Phys.Rev.* **D33**, 1585 (1986); **D34**, 3263(E) (1986).
36. S. Wolfram, *Phys.Lett.* **82B**, 65 (1979); G. Steigman, *Ann. Rev. Nucl. Part. Sci.* **29**, 313 (1979).
37. B. Lee and S. Weinberg, *Phys.Rev.Lett.* **39**, 169 (1977); E. W. Kolb, Ph.D. thesis (Univ. of Texas, 1978); also see, E. Kolb and K. Olive, *Phys.Rev.* **D33**, 1202 (1986).
38. M. S. Turner, in *Proceedings of the 1981 Int'l. Conf. on ν Phys. and Astrophys.*, eds. R. Cence, E. Ma, and A. Roberts, **1**, 95 (1981).
39. G. Steigman, *Ann. Rev. Astron. Astrophys.* **14**, 339 (1976).
40. F. Stecker, *Ann. NY Acad. Sci.* **375**, 69 (1981).
41. T. Gaisser, in *Birth of the Universe*, eds. J. Audouze and J. Tran Thanh Van (Editions Frontiers: Gif-sur-Yvette, 1982).
42. A. Sakharov, *JETP Lett.* **5**, 24 (1967).
43. M. Yoshimura, *Phys.Rev.Lett.* **41**, 281; **42**, 746(E) (1978).
44. D. Toussaint, S. Treiman, F. Wilczek, and A. Zee, *Phys.Rev.* **D19**, 1036 (1979).
45. S. Dimopoulos and L. Susskind, *Phys.Rev.* **D18**, 4500 (1978).
46. A. Ignatiev, N. Krasnikov, V. Kuzmin, and A. Tavkhelidze, *Phys.Lett.* **76B**, 486 (1978).
47. J. Ellis, M. Gaillard, and D. V. Nanopoulos, *Phys.Lett.* **80B**, 360; **82B**, 464(E) (1979).
48. S. Weinberg, *Phys.Rev.Lett.* **42**, 850 (1979).
49. E. Kolb and S. Wolfram, *Nucl.Phys.* **172B**, 224; *Phys.Lett.* **91B**, 217 (1980); J. Harvey, E. Kolb, D. Reiss, and S. Wolfram, *Nucl.Phys.* **201B**, 16 (1982).
50. J. N. Fry, K. A. Olive, and M. S. Turner, *Phys.Rev.* **D22**, 2953; 2977; *Phys.Rev.Lett.* **45**, 2074 (1980).
51. D. V. Nanopoulos and S. Weinberg, *Phys.Rev.* **D20**, 2484 (1979).

52. S. Barr, G. Segrè and H. Weldon, *Phys.Rev.* **D20**, 2494 (1979).
53. G. Segrè and M. S. Turner, *Phys.Lett.* **99B**, 339 (1981).
54. J. Ellis, M. Gaillard, D. Nanopoulos, and S. Rudaz, *Phys.Lett.* **99B**, 101 (1981).
55. E. W. Kolb and M. S. Turner, *Ann.Rev.Nucl.Part.Sci.* **33**, 645 (1983).
56. I. Affleck and M. Dine, *Nucl.Phys.* **B249**, 361 (1985).
57. V.A. Kuzmin, V.A. Rubakov, and M.E. Shaposhnikov, *Phys.Lett.* **155B**, 36 (1985).
58. F. Accetta, P. Arnold, E.W. Kolb, L. McLerran, and M.S. Turner, *Phys.Rev.D*, submitted (1987); E.W. Kolb and M.S. Turner, *Mod.Phys.Lett.A*, in press (1987).
59. C.B. Collins and M.J. Perry, *Phys.Rev.Lett.* **34**, 1353 (1975).
60. G.F.R. Ellis, *Ann.NY.Acad.Sci.* **336**, 130 (1980).
61. R. Sachs and A. Wolfe, *Astrophys.J.* **147**, 73 (1967).
62. N. Vittorio and J. Silk, *Astrophys.J.* **285**, L39 (1984); J.R. Bond and G. Efstathiou, *Astrophys.J.* **285**, L44 (1984); J. Silk, in *Inner Space/Outer Space*, eds. E.W. Kolb et al. (University of Chicago Press, Chicago, 1986).
63. C.B. Collins and S. Hawking, *Astrophys.J.* **180**, 317 (1973).
64. See, e.g., P.J.E. Peebles, *Large-Scale Structure of the Universe* (Princeton Univ. Press, Princeton, 1980); G. Efstathiou and J. Silk, *Fund.Cosmic Phys.* **9**, 1 (1983); S.D.M. White, in *Inner Space/Outer Space*, eds. E. Kolb et al. (Univ. of Chicago Press, Chicago, 1986).
65. R.H. Dicke and P.J.E. Peebles, in *General Relativity: An Einstein Centenary Survey*, eds. S.W. Hawking and W. Israel (Cambridge University Press, Cambridge, 1979).
66. G. Steigman, *Ann.Rev.Astron.Astrophys.* **14**, 339 (1976).
67. G.G. Ross, *Grand Unified Theories* (Benjamin/Cummings, Menlo Park, 1984).
68. E.W. Kolb and M.S. Turner, *Ann.Rev.Nucl.Part.Sci.* **33**, 645 (1983).
69. G. 't Hooft, *Nucl.Phys.* **B79**, 276 (1974); A.M. Polyakov, *JETP Lett.* **20**, 194 (1974).
70. T.W.B. Kibble, *J.Phys.* **A9**, 1387 (1976); J. Preskill, *Phys.Rev.Lett.* **43**, 1365 (1979); Ya. B. Zel'dovich and M. Yu. Khlopov, *Phys.Lett.* **79B**, 239 (1978).
71. J. Preskill, *Ann.Rev.Nucl.Part.Sci.* **34**, 461 (1984); M.S. Turner, in *Monopole '83*, ed. J. Stone (Plenum Press, NY, 1984).
72. A. Guth and S-H. H. Tye, *Phys.Rev.Lett.* **44**, 631 (1980).
73. A. Guth, *Phys.Rev.* **D23**, 347 (1981).
74. S. Bludman and M. Ruderman, *Phys.Rev.Lett.* **38**, 255 (1977); Ya. B. Zel'dovich, *Sov.Phys.Uspekhi* **11**, 381 (1968).
75. C.W. Misner, *Astrophys. J.* **151**, 431 (1968); in *Magic Without Magic*, ed. J. Klauder (Freeman, San Francisco, 1972); R. Matzner and C.W. Misner, *Astrophys. J.* **171**, 415 (1972).
76. R. Penrose, in *General Relativity: An Einstein Centenary Survey*, eds. S.W. Hawking and W. Israel (Cambridge University Press, Cambridge, 1979).
77. B. deWitt, *Phys.Rev.* **90**, 357 (1953).
78. L. Parker, *Nature* **261**, 20 (1976).

79. Ya.B. Zel'dovich, *JETP Lett.* **12**, 307 (1970).
80. A.A. Starobinskii, *Phys.Lett.* **91B**, 99 (1980).
81. P. Anderson, *Phys.Rev.* **D28**, 271 (1983).
82. J.M Hartle and B.-L Hu, *Phys.Rev.* **D20**, 1772 (1979).
83. M.V. Fischetti, J. Hartle, and B.-L. Hu, *Phys.Rev.* **D20**, 1757 (1979).
84. J.M. Hartle and S.W. Hawking, *Phys.Rev.* **D28**, 2960 (1983).
85. B.J. Carr and M.J. Rees, *Nature* **278**, 605 (1979); J.D. Barrow and F. Tipler, *The Anthropic Cosmological Principle* (Oxford University Press, Oxford, 1986).
86. D. Lindley, Fermilab preprint (1985).
87. A. Guth and E. Weinberg, *Nucl.Phys.* **B212**, 321 (1983); S. Hawking, I. Moss, and J. Stewart, *Phys.Rev.* **D26**, 2681 (1982).
88. A. Linde, *Phys.Lett.* **108B**, 389 (1982).
89. A. Albrecht and P.J. Steinhardt, *Phys.Rev.Lett.* **48**, 1220 (1982).
90. S. Coleman and F. de Luccia, *Phys.Rev.* **D21**, 3305 (1980); S. Coleman, *Phys.Rev.* **D15**, 2929 (1977); C.G. Callan and S. Coleman, *Phys.Rev.* **D16**, 1762 (1977).
91. S.W. Hawking and I.G. Moss, *Phys.Lett.* **110B**, 35 (1982); L. Jensen and P.J. Steinhardt, *Nucl.Phys.* **239B**, 176 (1984); K. Lee and E.J. Weinberg, *Nucl.Phys.* **B267**, 181 (1986).
92. A. Albrecht, P.J. Steinhardt, M.S. Turner, and F. Wilczek, *Phys.Rev.Lett.* **48**, 1437 (1982).
93. L. Abbott, E. Farhi, and M. Wise, *Phys.Lett.* **117B**, 29 (1982).
94. A. Dolgov and A. Linde, *Phys.Lett.* **116B**, 329 (1982).
95. P.J. Steinhardt and M.S. Turner, *Phys.Rev.* **D29**, 2162 (1984).
96. J.R. Gott, *Nature* **295**, 304 (1982); J.R. Gott and T.S. Statler, *Phys.Lett.* **136B**, 157 (1984).
97. M.S. Turner, *Phys.Lett.* **115B**, 95 (1982); J. Preskill, in *The Very Early Universe*, eds. G. Gibbons, S. Hawking, and S. Siklos (Cambridge Univ. Press, Cambridge, 1983); G. Lazarides, Q. Shafi, and W.P. Trower, *Phys.Rev.Lett.* **49**, 1756 (1982).
98. G. Lazarides and Q. Shafi, in *The Very Early Universe, ibid*; K. Olive and D. Seckel, in *Monopole '83*, ed. J. Stone (Plenum Press, NY, 1984).
99. M.S. Turner, *Phys.Rev.* **D28**, 1243 (1983).
100. J.D. Barrow and M.S. Turner, *Nature* **292**, 35 (1981); J.D. Barrow and D.H. Sonoda, *Phys.Rep.* **139**, 1 (1986).
101. D. Nanopoulos, K. Olive, and M. Srednicki, *Phys.Lett.* **127B**, 30 (1983).
102. R.J. Scherrer and M.S. Turner, *Phys.Rev.* **D31**, 681 (1985).
103. T. Bunch and P.C.W. Davies, *Proc.Roy.Soc.London* **A360**, 117 (1978); also see, G. Gibbons and S. Hawking, *Phys.Rev.* **D15**, 2738 (1977).
104. L. Abbott and M. Wise, *Nucl.Phys.* **B244**, 541 (1984).
105. S. Hawking, *Phys.Lett.* **115B**, 295 (1982).
106. A.A. Starobinskii, *Phys.Lett.* **117B**, 175 (1982).
107. A. Guth and S.-Y. Pi, *Phys.Rev.Lett.* **49**, 1110 (1982).
108. J. Bardeen, P. Steinhardt, and M.S. Turner, *Phys.Rev.* **D28**, 679 (1983).

109. R. Brandenberger, R. Kahn, and W. Press, *Phys.Rev.* **D28**, 1809 (1983); W. Fischler, B. Ratra, and L. Susskind, *Nucl.Phys.* **B259**, 730 (1985); R. Brandenberger, *Rev.Mod.Phys.* **57**, 1 (1985).
110. E.M. Lifshitz and I.M. Khalatnikov, *Adv.Phys.* **12**, 185 (1963); E.M. Lifshitz, *Zh.Ek.Teor.Fiz.* **16**, 587 (1946).
111. W. Press and E.T. Vishniac, *Astrophys.J.* **239**, 1 (1980).
112. J.M. Bardeen, *Phys.Rev.* **D22**, 1882 (1980).
113. W. Press, in *Cosmology and Particles*, eds. J. Audouze et al. (Editions Frontieres, Gif-sur-Yvette, 1981), p. 137.
114. J.A. Frieman and M.S. Turner, *Phys.Rev.* **D30**, 265 (1984); R. Brandenberger and R. Kahn, *Phys.Rev.* **D29**, 2172 (1984).
115. E. Harrison, *Phys.Rev.* **D1**, 2726 (1970).
116. Ya.B. Zel'dovich, *Mon.Not.Roy.Astron.Soc.* **160**, 1p (1972).
117. P.J. Steinhardt and M.S. Turner, *Phys.Rev.* **D29**, 2162 (1984).
118. M.S. Turner and E.T. Vishniac, in preparation (1985).
119. J. Bond and A. Szalay, *Astrophys.J.* **276**, 443 (1983); P.J.E. Peebles, *Astrophys.J.* **263**, L1 (1982); J. Bond, A. Szalay, and M.S. Turner, *Phys.Rev.Lett.* **48**, 1036 (1982).
120. M. Axenides, R. Brandenberger, and M.S. Turner, *Phys.Lett.* **120B**, 178 (1983); P.J. Steinhardt and M.S. Turner, *Phys.Lett.* **129B**, 51 (1983).
121. D. Seckel and M.S. Turner, *Phys.Rev.* **D32**, 3178 (1985).
122. A.D. Linde, *JETP Lett.* **40**, 1333 (1984); *Phys.Lett.* **158B**, 375 (1985).
123. J. Preskill, M. Wise, and F. Wilczek, *Phys.Lett.* **120B**, 127 (1983); L. Abbott and P. Sikivie, *Phys.Lett.* **120B**, 133 (1983); M. Dine and W. Fischler, *Phys.Lett.* **120B**, 137 (1983).
124. M.S. Turner, F. Wilczek, and A. Zee, *Phys.Lett.* **125B**, 35, 519(E) (1983); J. Ipser and P. Sikivie, *Phys.Rev.Lett.* **50**, 925 (1983).
125. M.S. Turner, *Phys.Rev.* **D33**, 889 (1986).
126. S. Coleman and E. Weinberg, *Phys.Rev.* **D7**, 1888 (1973).
127. A.D. Linde, *Phys.Lett.* **116B**, 335 (1982).
128. A. Vilenkin and L. Ford, *Phys.Rev.* **D26**, 1231 (1982).
129. S. Dimopoulos and S. Raby, *Nucl.Phys.* **B219**, 479 (1983).
130. A. Albrecht, S. Dimopoulos, W. Fischler, E. Kolb, S. Raby, and P. Steinhardt, *Nucl.Phys.* **B229**, 528 (1983).
131. J. Ellis, D. Nanopoulos, K. Olive, and K. Tamvakis, *Phys.Lett.* **118B**, 335 (1982); D.V. Nanopoulos, K. Olive, M. Srednicki, and K. Tamvakis, *Phys.Lett.* **123B**, 41 (1983); J. Ellis, K. Enqvist, D. Nanopoulos, K. Olive, and M. Srednicki, *Phys.Lett.* **152B**, 175 (1985); K. Enqvist and D. Nanopoulos, *Phys.Lett* **142B**, 349 (1984); C. Kounnas and M. Quiros, *Phys.Lett.* **151B**, 189 (1985); for a review of the CERN SUSY/SUGR models see D.V. Nanopoulos, *Comments on Astrophysics* **X**, 219 (1986).
132. B. Ovrut and P.J. Steinhardt, *Phys.Lett.* **133B**, 161 (1983); L. Jensen and K. Olive, *Nucl.Phys.* **B263**, 731 (1986).

133. B. Ovrut and P.J. Steinhardt, *Phys.Rev.Lett.* **53**, 732 (1984); *Phys.Lett.* **147B**, 263 (1984).
134. R. Holman, P. Ramond, and G.G. Ross, *Phys.Lett.* **137B**, 343 (1984); C.D. Coughlan, R. Holman, P. Ramond, and G.G. Ross, *Phys.Lett.* **140B**, 44 (1984); **158B**, 47 (1985).
135. J. Breit, S. Gupta, and A. Zaks, *Phys.Rev.Lett.* **51**, 1007 (1983).
136. A.A. Starobinskii, *JETP Lett.* **30**, 682 (1979); V. Rubakov, M. Sazhin, and A. Veryaskin, *Phys.Lett.* **115B**, 189 (1982); R. Fabbri and M. Pollock, *ibid* **125B**, 445 (1983); L. Abbott and M. Wise, *Nucl.Phys.* **B244**, 541 (1984).
137. J. Ellis, J. Kim, and D. Nanopoulos, *Phys.Lett.* **145B**, 181 (1984); L.L. Krauss, *Nucl.Phys.* **B227**, 556 (1983); M.Yu. Khlopov and A.D. Linde, *Phys.Lett.* **138B**, 265 (1984).
138. J. Polonyi, Budapest preprint KFKI 1977-93, unpublished (1977).
139. C. Coughlan, W. Fischler, E. Kolb, S. Raby, and G.G. Ross, *Phys.Lett.* **131B**, 54 (1983).
140. S. Gupta and H.R. Quinn, *Phys.Rev.* **D29**, 2791 (1984).
141. Q. Shafi and A. Vilenkin, *Phys.Rev.Lett.* **52**, 691 (1984).
142. S.-Y. Pi, *Phys.Rev.Lett.* **52**, 1725 (1984).
143. A. Guth and S.-Y. Pi, *Phys.Rev.* **D32**, 1899 (1985).
144. A.D. Linde, *Phys.Lett.* **129B**, 177 (1983).
145. V.A. Belinsky, L.P. Grischuk, I.M. Khalatnikov, and Ya.B. Zel'dovich, *Phys.Lett.* **155B**, 232 (1985); T. Piran and R.M. Williams, *ibid* **168B**, 331 (1985).
146. S. Adler, *Rev.Mod.Phys.* **54**, 729 (1982); L. Smolin, *Nucl.Phys.* **B160**, 253 (1979); A. Zee, *Phys.Rev.Lett.* **42**, 417 (1979).
147. F. Accetta, D. Zoller, and M.S. Turner, *Phys.Rev.* **D31**, 3046 (1985).
148. B.L. Spokoiny, *Phys.Lett.* **147B**, 39 (1984).
149. Q. Shafi and C. Wetterich, *Phys.Lett.* **129B**, 387 (1983); **152B**, 51 (1985).
150. A.A. Starobinskii, *Phys.Lett.* **91B**, 99 (1986); M.B. Mijic, M.S. Morris, and W.-M. Suen, *Phys.Rev.* **D34**, 2934 (1986).
151. A. Albrecht and R. Brandenberger, *Phys.Rev.* **D31**, 1225 (1985); *ibid* **D32**, 1280 (1985).
152. W. Fischler, B. Ratra, and L. Susskind, *Nucl.Phys.* **B259**, 730 (1985).
153. G. Semenoff and N. Weiss, *Phys.Rev.* **D31**, 689 (1985).
154. M. Evans and J. McCarthy, *Phys.Rev.* **D31**, 1799 (1985).
155. M.S. Turner and L. Widrow, *Phys.Rev.Lett.* **57**, 2237 (1986); L. Jensen and J. Stein-Schabes, *Phys.Rev.* **D34**, 931 (1986).
156. R.M. Wald, *Phys.Rev.* **D28**, 2118 (1983).
157. G. Steigman and M.S. Turner, *Phys.Lett.* **128B**, 295 (1983).
158. L. Jensen and J. Stein-Schabes, *Phys.Rev.*, in press (1987).
159. A.A. Starobinskii, *JETP Lett.* **37**, 66 (1983).
160. S.K. Blau, E.I. Guendelman, and A.H. Guth, *Phys.Rev.D*, in press (1987); K. Sato, M. Sasaki, H. Kodama, and K. Maeda, *Phys.Lett.* **108B**, 103 (1982); *Prog.Theor.Phys.* **65**, 1443 (1981); *ibid* **68**, 1979 (1982); *ibid* **66**, 2287 (1981).

161. N. Myhrvad, *Phys.Lett.* **132B**, 308 (1983); E. Mottola, *Phys.Rev.* **D31**, 754(1985); **D33**, 2136; L. Parker, *Phys.Rev.Lett.* **50**, 1009 (1983); L. Ford, *Phys. Rev.* **D31**, 710 (1985); P. Anderson, *Phys.Rev.* **D32**, 1302 (1985); J. Traschen and C.T. Hill, *Phys. Rev.* **D33**, 3519 (1986).
162. M.S. Turner, in *Dark Matter in the Universe*, eds. J Kormendy and G. Knapp (Reidel, Dordrecht, 1987), p. 445.
163. M.S. Turner, G. Steigman, and L.L. Krauss, *Phys.Rev.Lett.* **52**, 2090 (1984); D.A. Dicus, E.W. Kolb, and V. Teplitz, *Phys.Rev.Lett.* **39**, 168 (1977); K.A. Olive, D. Seckel, and E.T. Vishniac, *Astrophys.J.* **292**, 1 (1985).
164. P.J.E. Peebles, *Astrophys.J.* **284**, 439 (1984); M.S. Turner, et al. in ref. 111.
165. N. Kaiser, *Astrophys.J.* **273**, L17 (1983); J.M. Bardeen, J. Bond, N. Kaiser, and A. Szalay, *Astrophys.J.* **304**, 15 (1986).
166. A. Vilenkin, *Phys.Rev.Lett.* **53**, 1016 (1984).
167. J.C. Charlton and M.S. Turner, *Astrophys.J.* **313**, 495 (1987).
168. S.D.M. White, in *Inner Space/Outer Space*, eds. E.W. Kolb et al. (University of Chicago Press, Chicago, 1986).
169. G. Blumenthal, S. Faber, J. Primack, and M. Rees, *Nature* **311**, 517 (1984); M. Davis, G. Efstathiou, C. Frank, and S.D.M. White, *Astrophys.J.* **292**, 371 (1985).
170. D. Burstein, et al., in *Galaxy Distances and Deviations from Universal Expansion*, eds. B. Madore and R. Tully (Reidel, Dordrecht, 1987), p. 123.
171. C.A. Collins, et al., *Nature* **320**, 506 (1986).
172. N. Vittorio and M.S. Turner, *Astrophys.J.* **316**, in press (1987).
173. V. deLapparent, M. Geller, and J. Huchra, *Astrophys.J.* **302**, L1 (1986).
174. G. Efstathiou and J.R. Bond, *Mon.Not.r.Astron.Soc.* **218**, 103 (1986).
175. P. Sikivie, *Phys.Rev.Lett.* **51**, 1415 (1983).
176. M. Goodman and E. Witten, *Phys.Rev.* **D31**, 3059 (1985).
177. E. Loh and E. Spillar, *Astrophys.J.* **307**, L1 (1986); *Phys.Rev.Lett.* **57**, 2865.
178. J. Silk and M.S. Turner, *Phys.Rev.* **D35**, 419 (1986); M.S. Turner, et al., *Astrophys.J.*, in press (1987).
179. A. Linde, *Rep.Prog.Phys.* **47**, 925 (1984); *Comments on Astrophysics* **10**, 229 (1985); *Prog.Theo.Phys. (Suppl.)* **85**, 279 (1985).
180. *The Inflationary Universe*, eds. L. Abbott and S.-Y. Pi (World Publishing, Singapore, 1986).
181. P.J. Steinhardt, *Comments on Nucl.Part.Phys.* **12**, 273 (1984).
182. R. Brandenberger, *Rev.Mod.Phys.* **57**, 1 (1985).
183. A. Vilenkin, *Phys.Rep.* **121**, 263 (1985); A. Albrecht and N. Turok, *Phys.Rev.Lett.* **54**, 1868 (1985); N. Turok, *Phys.Rev.Lett.* **55**, 1801 (1985); R. Scherrer, *Astrophys.J.*, in press (1987); A. Melott and R. Scherrer, *Nature*, in press (1987).
184. S.K.Blau and A.H. Guth, in *300 Years of Gravitation*, eds. S.W. Hawking and W. Israel (Cambridge University Press, Cambridge, 1987).
185. S. Bonometto and A. Masiero, *La Rivista del Nuovo Cimento* **9**, 3 (1986).

BABY UNIVERSES*

Andrew Strominger

Department of Physics

University of California

Santa Barbara, CA 93106

---

* Lectures presented at the TASI summer school, Brown University, June, 1988.

I. INTRODUCTION

The subject of baby universes and their effects on spacetime coupling constants [1,2,3,4,5,6,7] is in its infancy and rapidly developing. The subject is based on the non-existent (even by physicists' standards) Euclidean formulation of quantum gravity, and it is therefore necessary to make a number of assumptions in order to proceed. Nevertheless, the picture which has emerged is quite appealing: all spacetime coupling constants become dynamical variables when the effects of baby universes are taken into account. This fact might even solve the puzzle of the cosmological constant [8,9,10]. The subject therefore seems worth further investigation. Several important, as yet incompletely answered, questions are

1. *How does one describe, even formally, a system of interacting universes?* An ordinary quantum mechanical system is described by a set of initial data along with laws governing time evolution. Since each universe has its own time, such a description is not directly applicable here. Consistent laws of physics and interpretational rules for the many-universe system must be found.

2. *How can the dynamics of the many-universe system be approximated?* The description of the many-universe system will involve a formal sum-over-four-geometries and topologies which is intractable for a variety of reasons. In order to better understand its properties, and certainly to make any physical predictions, approximation methods must be developed. One aspect of this is that, since we appear to live in a single universe, it is important to understand when the single universe approximation is valid.

3. *To what extent can low-energy couplings be predicted independently of the precise details of Planck-scale physics?* It has been argued that one effect of baby universes is to shift the low-energy cosmological constant to zero, and that this result does not depend on details of Planck scale physics. Why is this possible? Can other low-energy couplings be determined as well?

In these lectures we will address all three of these questions. We will have something fairly definite to say concerning the first two. The last question appears unresolved at present [11,12,13,14,15,16], so our comments will be incomplete.

The organization is as follows. In Section II we consider a toy model of topology change in a no-space one-time dimensional universe. This model will be used to illustrate, in a simple context, the proposal of "third quantization" as a means of defining a system of interacting universes. We show how the third quantized equation of motion becomes a dynamical equation for the second quantized coupling constants in the one-dimensional universe. In Section III it is explained how third quantization is applied by analogy to 3+1 dimensional universes. A new feature, third quantized gauge invariance, arises and is discussed. In Sections IV, V and VI small expansion parameters and approximation methods for describing this third quantized field theory are discussed. IV describes the parent-baby universe approximation, which involves an expansion in the ratio of low-energy scales to the Planck or baby universe scale. V contains a mini-review of instantons, beginning with quantum mechanics and discussing in some detail the subject of gravitational instantons. In VI we discuss the axion model, for which the third quantized system can be quite explicitly described in the instanton approximation. In VII we discuss the possibility that baby universes are responsible for the vanishing of the cosmological constant.

The material covered in these notes is roughly the same as what I covered in TASI in June, 1988, but the presentation differs. The original presentation was more along the lines of references [6,5,7,10] and followed somewhat the (recent!) historical development of the subject. I hope that the presentation given here is a little more systematic. In particular, the discussion of the 0+1 dimensional system has not been given previously and provides a simple context for discussing many of the basic concepts.

I would like to thank my collaborator, Steve Giddings, for many discussions on baby universes. Much of the material herein is part of the collective folklore, and should not be attributed to Giddings and me. I have otherwise attempted to give proper references. Sidney Coleman and Stephen Hawking have in particular had a deep influence on many aspects of the subject.

II. TOPOLOGY CHANGE AND THIRD QUANTIZATION IN 0+1 DIMENSIONS

In this section we are going to consider, as a toy model, a quantum field theory in a universe (or universes) with zero space and one time dimension. In the absence of topology change, this is mathematically equivalent to a quantum mechanical particle(s) moving in a potential. If topology change is allowed, space is not described as a point, but as a set of points. The interacting many-universe system is no longer equivalent to a quantum mechanics model. We shall see that it becomes equivalent to a 'third quantized' many-particle quantum field theory on a larger space, and that the third quantized equation of motion is an equation for effective coupling constants in the one dimensional universe.

Mathematically, this section contains nothing new. We are merely repeating the well-known steps leading from first quantized, single-particle quantum mechanics to a second quantized, many-particle quantum field theory. However, we shall say different words as we take these steps, in order to explain their relevance to the problem of topology change and interacting universes.

The reader should be forewarned that, while there are many features common to the problems of topology change in one and four (the real case of interest) dimensions, there are also important qualitative differences. For example, in four dimensions the spatial topology is described by some (possibly disconnected) three-manifold. It has been shown [17] that all three-manifolds are cobordant. This means that given any two three-manifolds, there always exists a smooth interpolating four-manifold whose boundary is the two given three-manifolds. As we shall see in the next section, this leads to a very natural expression for the quantum transition between the two topologies in terms of a functional integral on the interpolating four-manifolds. In contrast, in one spacetime dimension topology change never occurs smoothly in this sense. The spatial topology is just a set of disconnected points, and n points is cobordant only to n points. There is no smooth one-manifold which interpolates between one point and two points. It is therefore much more natural to consider topology change in four (in fact two or higher) dimensions than in one dimension. In one dimension the topology-changing interaction must be introduced in a rather artifical way, and its form is rather unconstrained. Nevertheless, it does provide an instructive example.*

2.1 Third Quantization of Free One-dimensional Universes

Consider a one dimensional universe described by the second quantized action (see e.g. [19])

$$S = \int_{\tau_i}^{\tau_f} d\tau \left(e^{-1} \dot{X}^\nu \dot{X}^\mu g_{\mu\nu} - em^2 \right) \tag{2.1}$$

where the summation convention is implied for the index $\mu = 1, 2, ..D$. The second quantized fields are the einbein e and the D "matter" fields X^μ, although e turns out to be non-dynamical. We refer to the fields X^μ and e and the action S as 'second quantized' in keeping with our aim to stress the analogy with four dimensions. e is related to the one-dimensional (timelike) line element by $ds^2 = -e^2 d\tau^2$. m^2 is the one-dimensional cosmological constant. $g_{\mu\nu}$ is a constant Lorentzian metric. (We are interested in the case $g_{\mu\nu} = \eta_{\mu\nu}$ but we consider this more general action because we will later need to analytically continue $\eta_{\mu\nu}$ to $\delta_{\mu\nu}$ to define a path integral.) They are one-dimensional versions of the usual second quantized objects in a four dimensional spacetime. S is invariant under local diffeomorphisms of the world line

$$\delta\tau = \epsilon(\tau) \tag{2.2}$$

$$\delta X^\mu(\tau) = \epsilon(\tau) \dot{X}^\mu(\tau) \tag{2.3}$$

---

* In some ways a better example is provided by string theory, which was in fact a guiding example for third quantization of four dimensional universes [10,9,18]. However the analogy is somewhat obscured by the important role played by Weyl invariance in string theory, which does not have an obvious four dimensional analog.

$$\delta e = \dot{\epsilon}(\tau)e(\tau) + \epsilon(\tau)\dot{e}(\tau) \tag{2.4}$$

$$\delta S = 0 \tag{2.5}$$

provided $\epsilon(\tau_i) = \epsilon(\tau_f) = 0$.

The diffeomorphism invariance must be fixed in order to define the quantum theory. One choice is synchronous gauge

$$\dot{e}(\tau) = 0 \tag{2.6}$$

which implies

$$e(\tau) = N \tag{2.7}$$

for some constant N. N is not a gauge degree of freedom since the proper length of the world line is $|N|(\tau_f - \tau_i)$. Since the metric involves the square of e, there are two choices of N (plus or minus) which describe the same geometry. We restrict N to be non-negative in order to avoid double-counting (i.e. to give a single cover of moduli space). This gauge choice leaves unfixed the global translation $\tau \to \tau + constant$. This symmetry is generated by the Hamiltonian

$$H = g^{\mu\nu} P_\mu P_\nu + m^2 \tag{2.8}$$

where $P_\nu = \frac{1}{N} \dot{X}^\mu g_{\mu\nu}$ or, in the quantum theory, $P^\mu = -i\frac{\partial}{\partial X^\mu}$. Invariance of the quantum theory under the (unfixed) global translation symmetry is then obtained by the constraint

$$H\phi(X^\mu) = 0 \tag{2.9}$$

which must be satisfied by physical second quantized states ϕ. (2.9) is the famous Wheeler-DeWitt equation [20]. In the absence of topology change this equation encodes all the dynamics of the theory. (Mathematically, (2.9) is equivalent to the time-independent Schroedinger equation for a free particle wave function in D dimensions with energy $E = -m^2$.)

In this simple model, $\phi(X^\mu)$ is the second quantized "wave function of the universe." Since the metric is trivial in zero space dimensions, ϕ is a function only of the D matter fields X^μ. It gives the probability amplitude for observing the field configuration of the universe to be X^μ.

To third quantize this free second quantized field theory, we write an action which is a functional of the second quantized wave function $\phi(X^\mu)$:

$$\begin{aligned} S &= \tfrac{1}{2}\int d^D X \sqrt{-g}\, \phi H \phi \\ &= \tfrac{1}{2}\int d^D X \sqrt{-g}\, (g^{\mu\nu}\nabla_\mu \phi \nabla_\nu \phi + m^2 \phi^2) \end{aligned} \quad (2.10)$$

Variation of this S with respect to ϕ leads directly to the Wheeler-DeWitt equation. This equation contains all the information of the second quantized theory, so the two formulations are equivalent. At this level, third quantization is rather trivial. We shall see later on that allowing topology change in the second quantized theory leads to an interacting third quantized theory.

To facilitate the discussion of topology change, we would like to describe this system in terms of a path integral "sum-over-one-geometries." Such expressions are well defined only in Euclidean space. In order to obtain a convergent path integral, we must Wick rotate both the one dimensional and ten dimensional metrics in (2.1), i.e.

$$e \to -ie \quad (2.11)$$

$$\eta_{\mu\nu} \to \delta_{\mu\nu} \quad (2.12)$$

or, equivalently

$$\tau \to -i\tau \quad (2.13)$$

$$X^0 \to iX^0 \quad (2.14)$$

The Lorentzian path integral is then obtained by analytic continuation (in both variables) from the Euclidean expression.

The path integral "sum-over-one-geometries" is defined as the weighted sum over all gauge inequivalent field configurations on the line interval with initial (final) value $X_i^\mu (X_f^\mu)$. We choose the coordinate system to run from $\tau_i = 0$ to $\tau_f = 1$. One then has in synchronous gauge,

$$G_E(X_i^\mu, X_f^\mu) = \int_{X_i^\mu}^{X_f^\mu} DX^\mu(\tau) \int_0^\infty dN\, e^{-S_E} \qquad (2.15)$$

The integration over the "modular" parameter N is the one remnant of the integration over the einbein which survives the gauge fixing. The motivation for the notation "G_E" will be evident shortly. In synchronous gauge, the Euclidean action is given by

$$S_E = \int_0^1 d\tau (\frac{1}{N}\dot{X}^\mu \dot{X}^\nu g_{\mu\nu} + N m^2) \qquad (2.16)$$

where $g_{\mu\nu}$ here is a Euclidean metric.

It is then a standard problem in quantum mechanics to show that, [21]

$$G_E(X_i^\mu, X_f^\mu) = \int_0^\infty d\tau K_E(X_i^\mu, X_f^\mu; \tau) \qquad (2.17a)$$

where

$$K_E(X_i^\mu, X_f^\mu; \tau) = <X_f^\mu | e^{-H\tau} | X_i^\mu > \qquad (2.17b)$$

K is the kernel of a quantum mechanical system with Hamiltonian H. It obeys the (Euclidean) Schroedinger equation

$$HK_E = -\frac{\partial}{\partial \tau} K_E \qquad (2.18)$$

This in turn implies the desired result that G_E is a Green function for H:

$$HG_E = \delta^D(X_i^\mu - X_f^\mu) \qquad (2.19)$$

The fact that the Euclidean sum-over-geometries with fixed boundaries gives a Green function for the Wheeler-DeWitt operator is also true for higher dimensional universes. It was verified for the two dimensional universes described by string theory in [22], for four dimensional minisuperspace models by Halliwell in [23] and was argued more generally to be the correct interpretation of the sum-over-four-geometries in [24,10].

It is also possible to obtain a third quantized formula for the quantity G_E. The answer is

$$G_E(X_i^\mu, X_f^\mu) = \int D\phi\, \phi(X_i^\mu)\phi(X_f^\mu)\, e^{-S_E[\phi]} \qquad (2.20)$$

where S_E is the Euclidean version of (2.10). Normalization of the right hand side to divide out disconnected vacuum diagrams is implicit. This formula is obvious since (2.19) states that G_E is the inverse of the kinetic operator H appearing in S_E. Hence the notation G_E. G_E is a Euclidean Green function of the third quantized field theory.

The real time Lorentzian Green function is then obtained by Wick rotation either of (2.20) or (2.15). This one dimensional problem is actually simple enough that the entire analysis can be done without Wick rotation of either X_0 or τ. This is also possible for the two dimensional universes desribed by string theory, where the real time (light-cone) and imaginary time (Polyakov) methods give the same results. However in four dimensions only the Euclidean methods are understood. For that reason it is useful to discuss the Euclidean formalism in one dimension. In particular, our treatment of the indefiniteness of the four dimensional action (due to the conformal factor) will be analogous to the treatment given here (and in string theory) of the indefinite action of the field X_0.

2.2 Third Quantization of Interacting One-Dimensional Universes

Now we are going to introduce topology changing interactions. For the moment, we will just discuss the construction and properties of the sum-over-one geometries describing topology change without ascribing any physical interpretation to it. After the construction has been understood, we will be in a better position to state our interpretation.

Topology change is introduced in the path integral by allowing the basic process illustrated in Figure 1 and all its iterations to occur with an arbitrary weighting λ. This process clearly represents one universe splitting into two, or two joining into one. We impose the natural boundary condition that the values of $X^\mu(\tau)$ on each of the three world lines are equal to X_0^μ where they meet, and then integrate over all values of X_0^μ.*

The three point function of Figure 1 can be easily expressed in terms of products of two point functions. The result is

$$G_E(X_1^\mu, X_2^\mu, X_3^\mu) = -\lambda \int d^D X_0^\mu \sqrt{g} G_E(X_1^\mu, X_0^\mu)$$
$$G_E(X_2^\mu, X_0^\mu) G_E(X_3^\mu, X_0^\mu) \qquad (2.21)$$

Iterating this process leads to arbitrarily complicated diagrams, as illustrated in Figure 2. The rule for computing the n-point function is to include every diagram with n external lines, and to count every inequivalent geometry on that diagram once and only once.

A third quantized formula can be obtained for these interacting Green functions just as was done for the free Green function in (2.20). We require an action $S_E[\phi]$ with the property that the correlator of n field operators weighted by S_E exactly reproduces the second quantized Euclidean sum-over-one-geometries with n boundary points. Such an action is given by

$$S_E = \tfrac{1}{2} \int d^D X^\mu \sqrt{g} (g^{\mu\nu} \nabla_\mu \phi \nabla_\nu \phi + m^2 \phi^2 + \frac{\lambda}{3} \phi^3) \qquad (2.22)$$

It is then a matter of combinatorics to verify that

$$G_E(X_1^\mu, X_2^\mu \cdots X_n^\mu) = \int \mathcal{D}\phi \, e^{-S_E} \phi(X_1^\mu) \phi(X_2^\mu) \cdots \phi(X_n^\mu). \qquad (2.23)$$

---

* While this is a natural boundary condition, many others are possible. We could multiply by functions of X^μ or its τ derivatives before integrating, have four or more world lines meeting, or attach group theory factors. This would lead to inequivalent third quantized field theories, in the latter case, gauge theories. This ambiguity is eliminated in higher dimensions by requiring smooth geometries.

Green functions for scattering Lorentzian universes governed by the free action (2.1) are then obtained by Wick rotation.

We stress that the third quantized action is *defined* by the requirement that it reproduce the sum-over-one-geometries. This definition carries over to higher dimensions.

Let us now turn to the physical interpretation of these diagrams. This is a highly non-trivial issue, and one on which there is not general agreement. As far as I can tell, the physical interpretation of the sum-over-geometries including topology change cannot be derived from the theory without topology change. Rather, one must simply postulate the interpretation, and then check to see if it is consistent or sensible. It is not even ruled out (though it seems unlikely) that there is more than one sensible interpretation. Interpretations which are apparently inequivalent to that presented here can be found in [25,8,26].

Clearly a larger Hilbert space is needed to describe a theory with topology change. The states must have support on configurations with all possible numbers of universes. How can such a state be extracted from the sum-over-one-geometries? We postulate:

The many universe system is described by a Schroedinger state $\Psi[\phi(X^i), X^0]$ *of the third quantized Hilbert space.*

Here $i = 1, \cdots D-1$, and X^0 is a second quantized field operator which has been chosen to serve as a third quantized "time" coordinate. The state $|\Psi>$ obeys the third quantized Schroedinger equation:

$$\mathcal{H}|\Psi> = i\frac{\partial}{\partial X^0}|\Psi> \qquad (2.24)$$

where $\mathcal{H}$ is the Hamiltonian of the third quantized action (not to be confused with the Wheeler-DeWitt operator H).

What is the interpretation of the state $|\Psi>$? $|\Psi>$ can be decomposed into components with definite universe number in the following sense. Let $\mathcal{N}$ be the universe number operator defined in the standard manner from the action S. ($\mathcal{N}$ of course does not commute with the full $\mathcal{H}$.) We may then define orthonormal eigenspaces

$$\mathcal{N}|n> = n|n> \qquad (2.25)$$

and decompose $|\Psi>$ at some moment X_0:

$$|\Psi> = \sum_n \Psi_n(X_0)|n> \qquad (2.26)$$

$\Psi_n(X^0)$ is then the "probability amplitude for n universes at time X^0" or the "probability amplitude for n universes with field values X^0".

More generally, eigenstates of some complete set of observables, such as $\phi(X^i)$, can be constructed. These represent coherent states (of indefinite universe number) of universes with wave functions $\phi(X^i)$. The state $|\Psi>$ may then be decomposed at time X^0 in terms of these eigenstates, and describes a general many-universe state.

Note that the specification of an auxilliary variable playing the role of time (in this case X^0) is necessary to make sense of the question *"How many universes are there?"*. Since each universe has its own intrinsic time τ, and these times are unrelated to one another, the intrinsic time τ is not a suitable auxilliary variable. Instead, we use the second quantized field X^0 as a "time" variable. We can then ask the sensible question *"How many universes are there on which the matter field variable X^0 takes the specified value?"*

In more general models, there may not be a variable such as X^0 which can play the role of time, or a corresponding Hamiltonian formulation of the third quantized theory. In this case other descriptions of the theory might be developed, as for example briefly described in [10,9]. However, in many cases of interest a time variable is available, and for simplicity we restrict our attention to these.

2.3 The Single-Universe Approximation and Dynamical Determination of Coupling Constants

The third quantized state $|\Psi>$ in general describes a system in which interactions between universes are not small and the single-universe approximation is not valid. It is certainly true that the single-universe approximation is at some level valid for the universe we inhabit (these investigations were not motivated by any experimental observation). One therefore is especially interested in when the single-universe approximation is valid. This is a dynamical issue.

In four dimensions, we shall see that the dynamics associated to the Einstein action (plus axions) leads to universes at two widely separated scales. These are small (roughly Planck-scale) baby universes and large (roughly Hubble-scale) parent universes. One can then compute the effect of baby universes on parent universe dynamics, and ask when the latter is well approximated by single-universe dynamics.

In one dimension, all universes have the same size–they are spatially just points. Therefore there can be no separation of scales as in four dimensions. It is nevertheless possible to construct a model which mimics some (but not all) of the features encountered in four dimensions, as follows.

Consider two species of one dimensional universes described by the actions

$$S_P = \int d\tau (e^{-1}\dot{X}^2 - e\, m_P^2) \qquad (2.27a)$$

and

$$S_B = \int d\tau (e^{-1}\dot{X}^2 - e\, m_B^2) \qquad (2.27b)$$

where $m_P \neq m_B$. S_P describes a 'parent' and S_B a baby universe. Note that we have set $D = 1$. Now include topology changing interactions of the form parent-parent-baby and baby-baby-baby as illustrated in Figure 3. As before the values of $X(\tau)$ must all equal X_0 at the junction of the world lines, and all values of X_0 are integrated over.

The resultant sum-over-one-geometries is generated by a third quantized action, but now there are two fields (ϕ_P and ϕ_B) which create and annihilate the two species of universes. Since there is only one X, the third quantized field theory is one dimensional. It is described by the action

$$S[\phi] = \frac{1}{2g^2} \int dX (-(\nabla \phi_P)^2 + m_P^2 \phi_P^2$$
$$- (\nabla \phi_B)^2 + m_B^2 \phi_B^2 + \kappa \phi_P^2 \phi_B + \frac{\lambda}{3} \phi_B^3) \tag{2.28}$$

A factor of g^2 has been scaled out of the action to facilitate discussion of the (third quantized) semi-classical limit.

Let us now consider the limit of very large m_P. There is then a clear energy gap between the sectors with zero and one parent universe. The couplings conserve parent universe number mod 2, so the one-parent universe state cannot decay into a state with no parent universes. The large value of m_P suppresses pair production of parent universes. It is therefore consistent to restrict attention to the one parent universe sector.

The single parent universe propagates in a plasma of baby universes. A typical process in its evolution is depicted in Figure 4. We wish to determine to what extent the parent universe dynamics, including baby universe effects, can be described by an ordinary second quantized effective action on the parent universe. To this end, we introduce a hybrid description which treats the baby universes in a third quantized manner, the single parent universe in a second quantized manner and includes parent-baby interactions by means of a mixed interaction Lagrangian S_I. The utility of this alternate description will be evident shortly. S_I will take the general form

$$S_I = \int d\tau e \sum_i \mathcal{L}_i(\tau) \phi_B^i \tag{2.29}$$

where $\mathcal{L}_i(\tau)$ are as yet unspecified local second quantized operators on the parent universe and ϕ_B^i are modes of the third quantized baby universe field operator. The physical meaning of this expression is that the creation or annihlation of a baby universe in the mode ϕ_B^i by the parent universe at τ is accompanied by a 'jolt' to the parent universe equivalent to an operator insertion of $\mathcal{L}_i(\tau)$.

The precise form of S_I is determined by the requirement that the hybrid formulae reproduce the sum-over-one-geometries. Consider for example the (connected) parent universe two point function in the Schroedinger state $|\Psi_B, 0>$ with no parent universe and baby universes in the state $|\Psi_B>$, given by the third quantized expression

$$G(X_f, X_i) = \int_{\Psi_B, 0} \mathcal{D}\phi_B \mathcal{D}\phi_P e^{iS} \phi_P(X_f) \phi_P(X_i)$$
$$= <\Psi_B, 0|\phi_P(X_f)\phi_P(X_i)|\Psi_B, 0> \qquad (2.30a)$$

The boundary conditions on the functional integral here are determined by the third quantized state, and the bra (ket) are states at time $X = \infty$ ($X = -\infty$). The restriction to connected diagrams is implicit throughout. The third quantized functional integral for the parent universe propagator in a baby universe background may be reexpressed as a second quantized path integral. This leads to the hybrid expression

$$G(X_f, X_i) = -ig^2 \int_{X_i}^{X_f} \mathcal{D}X(\tau) \int_0^\infty dN \int_{\Psi_B} \mathcal{D}\phi_B e^{iS_P + iS_I + iS_B}$$
$$= -ig^2 \int_{X_i}^{X_f} \mathcal{D}X(\tau) \int_0^\infty dN <\Psi_B|e^{iS_P + iS_I}|\Psi_B> \qquad (2.30b)$$

with S_P given by (2.27a), and in synchronous gauge

$$S_I = \kappa \int_0^1 d\tau \phi_B(X(\tau)) \qquad (2.31)$$

In terms of the Fourier transformed field

$$\tilde{\phi}_B(k) = \frac{1}{2\pi} \int_{-\infty}^\infty dX\, e^{ikX} \phi_B(X) \qquad (2.32)$$

S_I is of the general form (2.29) with the continuous variable k replacing the index i. It can further be demonstrated that analogous hybrid formulae for all n point functions in arbitrary states using S_I gives the same answer as the original sum-over-one-geometries.

Let us now consider the semiclassical limit of the third quantized theory, namely $g^2 \to 0$ in (2.28). In this limit the field operators all commute, and, as pointed out by Coleman [6], it is possible to diagonalize ϕ_B in terms of (real time) third quantized baby universe eigenstates $|\alpha(X)>$:

$$\phi_B(X)|\alpha(X)> = \alpha(X)|\alpha(X)> \qquad (2.33)$$

The eigenvalues $\alpha(X)$ are constrained to obey the baby universe field equation

$$(\nabla^2 + m_B^2)\alpha(X) + \frac{1}{2}\lambda\alpha^2(X) = 0 \qquad (2.34)$$

in the absence of parent universe sources. In such a state, the baby universe field operator may be replaced by its eigenvalue. The parent universe two point function becomes

$$G(X_i, X_f) = -ig^2 \int_{X_i}^{X_f} \mathcal{D}X(\tau) \int_0^\infty dN \, e^{iS_P + iS_I[\alpha]} \qquad (2.35)$$

where

$$S_P + S_I[\alpha] = \int d\tau (\frac{1}{N}\dot{X}^2 - Nm^2 - N\kappa\alpha(X)) \qquad (2.36)$$

We neglect here the back reaction of the parent universe on the baby universe state which is justified for small parent-baby coupling κ.

The effective action $S_P + S_I$ looks like an ordinary second quantized action for a one dimensional universe. Comparing (2.35) with formula (2.15), which involves no topology change, we see that the effects of the baby universes have been entirely summarized by the addition of an ordinary potential $\alpha(X)$ into the field theory. This key observation is due, in a slightly different context, to Coleman [6] and reference [7].

Note that this potential is subject to the dynamical constraint (2.34) [9,10]. It must obey the equation of motion of a particle in a cubic potential. Surely the inhabitants of the parent universe, unaware of the existence of the baby universes, would be mystified by the shape of the potential! It is hoped that just such a dynamical constraint might explain the mysterious vanishing of the cosmological constant in our universe, as will be discussed in Sections VI and VII.

2.4 The Third Quantized Uncertainty Principle

In the previous subsection it was argued that if the baby universes were in an eigenstate $|\alpha(X)>$, their sole effect was to generate parent universe couplings. However in general there is no reason to suppose that the baby universes are in such an eigenstate. Suppose, for example, that they are in a linear superposition of eigenstates

$$|\alpha, \alpha'> = \beta|\alpha(X)> + \beta'|\alpha'(X)> \quad (2.37)$$

where $|\beta|^2 + |\beta'|^2 = 1$. To understand the behavior of a parent universe in this state, it is useful to insert an ideal clock into the parent universe. One may then discuss correlation functions of n field operators at times $\tau_1, \cdots \tau_n$. These are given by

$$<X(\tau_1)\cdots X(\tau_n)>_{\alpha,\alpha'} = <\alpha, \alpha'|\int DX(\tau)e^{iS_P+iS_I}X(\tau_1)\cdots X(\tau_n)|\alpha, \alpha'> \quad (2.38)$$

Since the α-states are orthogonal, this separates into two pieces:

$$<X(\tau_1)\cdots X(\tau_n)>_{\alpha,\alpha'} = |\beta|^2 \int DX(\tau)e^{iS_P+iS_I[\alpha]}X(\tau_1)\cdots X(\tau_n)$$
$$+ |\beta'|^2 \int DX(\tau)e^{iS_P+iS_I[\alpha']}X(\tau_1)\cdots X(\tau_n) \quad (2.39)$$

Each of these pieces looks like an ordinary correlation function, but in universes with different coupling constants α and α'. This separation occurs because second quantized operators in a single universe (correponding to physical observables) do not affect the baby universe state and so cannot connect the states $|\alpha>$ and $|\alpha'>$. Thus an observer who measures the value α can never talk to one who measures α'. Given that the result of some measurements which indicate that the coupling constants are α (α') all future measurements will agree that the coupling constants are α (α'), as argued by Coleman [6].

This result may be rephrased using the Copenhagen interpretation of quantum mechanics [7]. Initially the coupling constants of the universe are not well defined, rather they are governed by a probability distribution. However performing a measurement collapses the wave function into the state $|\alpha>$ ($|\alpha'>$) with probability $|\beta|^2$($|\beta'|^2$). (This was shown explicitly in [7]). All future measurements are then consistent with some definite value of the coupling constants.

An important feature which naturally emerges here is the idea of probability distributions for coupling constants. While all coupling constants are subject to shifts by baby universe effects, given some initial conditions or other criteria for choosing a baby universe state, it may be possible to determine their most probable values.

States of the form (2.37) are still far from the most general baby universe state. In general, away from the semiclassical limit $g^2 = 0$, one has

$$[\phi_B(X), \phi_B(X')] = \mathcal{O}(g^2) \tag{2.40}$$

Since ϕ_B does not commute with itself for different values of X, we cannot possibly diagonalize ϕ_B for all X, and the α-states do not exist.

How then, can one interpret such a system? The answer to this question is not obvious, and is not generally agreed upon. We advance here the following interpretation of the formulae for nonzero g^2, which will reduce to the preceding interpretation in the limit $g^2 \to 0$.

In the limit $g^2 \to 0$, the value of the parent universe potential $\alpha(X)$ and its first derivative at one value of X uniquely determine, along with the third quantized equation of motion, the values of $\alpha(X)$ at all other values of X. In principle this could lead to definite predictions for values of coupling constants.

If $g^2\neq 0$, the baby universe state is subject to quantum mechanical fluctuations, and definite predictions for values of unmeasured coupling constants are no longer possible. Instead, one must speak of conditional probability amplitudes for the results of various measurements. For example, given that the potential at X_1 and X_2 have been measured to be

$$\alpha_1 = \alpha(X_1) \tag{2.41}$$

and

$$\alpha_2 = \alpha(X_2) \tag{2.42}$$

we may then ask for the conditional probability amplitude that the potential at an intermediate point X_3 is measured to have the value α_3 This is given by

$$A(\alpha_3) = C \int D\phi(X) e^{iS_B[\phi]} \tag{2.43}$$

where S_B is the third quantized baby universe action and the path integral is over all paths obeying

$$\alpha_1 = \phi(X_1)$$
$$\alpha_2 = \phi(X_2)$$
$$\alpha_3 = \phi(X_3) \tag{2.44}$$

and is suitably normalized so that the probabilities sum to one.

It should be stressed that this does *not* mean that coupling constants do not take definite values. Once a coupling constant has been measured to lie in a certain range, the wave function is collapsed and it retains that value regardless of what other measurements are made. What is becoming "uncertain" here is the predictions implied by the third quantized equation of motion for relations among coupling constants.

Our interpretation does, however, imply that in practice it will be difficult or impossible to actually obtain precise measurements of all couplings. For example, if the field X runs over an infinite range, the probability amplitude $A(\alpha_3)$ is zero for all values of α_3 when it is correctly normalized. This corresponds to the well known quantum mechanical fact that in order to measure the position of a particle *exactly* (as we have done in this case at X_1 and X_2) you must give it so much momentum that you will never find it again.

Even if X is somehow restricted to lie in a finite range, there will still be difficulties in measuring the first derivative of the potential. Suppose that, as would be the case in practice, the potential has not been measured exactly at X_1 and X_2, but has been determined to within a Gaussian of width λ around the values α_1 and α_2. Then, given this measurement, the conditional probability amplitude for measuring the first derivative of the potential at X_3 to take the value

$$\frac{\partial \alpha}{\partial X}(X_3) = \alpha' \qquad (2.45)$$

is given by the Fourier transform of $A(\alpha_3)$. For X_3 very near X_2 this is

$$A(\alpha') = \sqrt{2\pi\lambda}e^{-\alpha'^2/2\lambda} \qquad (2.46)$$

In particular as the difference between the two field values X_1 and X_3 and the uncertainty λ of the measurement of the potential at X_1 go to zero, the spread in α' goes to infinity. This is again equivalent to the statement that the momentum spread of a quantum mechanical particle is very large shortly after a precise measurement of its position.

In practice, a real detector can only measure the derivative of the potential within some finite range. Thus if we obtain a very precise measurement of the potential at X_1, it will be impossible to obtain a precise measurement of its derivative at X_3 as X_3 gets arbitrarily close to X_1. This obstruction to obtaining precise measurements of coupling constants was referred to in [10] as the uncertainty principle for spacetime couplings.

III. THIRD QUANTIZATION IN 3+1 DIMENSIONS

The idea that third quantization might be useful for describing interacting four dimensional universes has been periodically proposed in published [27,28,29,30,31,32,33,9,10,18,34] and unpublished works for a variety of motivations. It's utility in the context of baby universes was advocated in [9,10,35,36,18].

.The basic idea of third quantization is to treat the many-universe system as a quantum field theory on superspace. The following table illustrates the analogy:

Second Quantization	Third Quantization
Particle	Universe
Interaction Vertex	Topology Change
Field	Third Quantized Field
Spacetime	Superspace of Three Geometries
Free Laplacian	Wheeler-DeWitt Operator
Vacuum	Void

Third quantization provides an interpretation and definition of a system of interacting four dimensional universes just as it does for the one dimensional universes (particles) described by ordinary quantum field theory discussed in the previous section. In ordinary quantum field theory, each particle has its own proper time along its world line. Dynamics therefore cannot be defined with respect to this proper time, rather the arena is spacetime. Analogously in third quantization each universe has its own time. Dynamics cannot be defined with respect to this time, rather the arena is superspace. The third quantized field operators act on the 'void', i.e. the third quantized state with no universes, and create second quantized states in the field theory of a single universe. In the absence of interactions these operators, and hence the single universe states, obey the Wheeler-DeWitt equation. Interactions then generalize this equation to a non-linear form. The relation between third quantized fields and second quantized couplings (discussed in Section 2.3 for the one dimensional case and in the next section for the four dimensional case) then implies that the non-linear Wheeler-DeWitt equation is an equation for spacetime couplings. This will be seen in detail in Sections IV and VI.

While there are many similarities between the four and one dimensional cases, several important differences also arise:

1. The resultant third quantized field theory is a gauge theory [37], and account must be taken of the third quantized gauge symmetries in construction of the action.

2. The third quantized field theory lives on an infinite dimensional space (superspace) and is therefore ill-defined because of non-renormalizability and other issues.

3. The topology-changing interactions are naturally described by a sum-over-smooth-four-geometries, with fixed boundaries. This is unlike the one dimensional case, where the interactions are introduced by hand.

3.1 The Gauge Invariant Action

The complete construction of the third quantized action is an unsolved problem, even at a formal level. The main difficulty is related to the third quantized gauge symmetry. While interesting in its own right, the complete construction does not at present appear essential for studies of baby universes. The reason for this is that the problem of gauge symmetries (or the problem of non-renormalizability) does not arise in the approximations used in treating baby universes.

Nevertheless, let us sketch on how such a construction might proceed. One route

[37,10] is based on BRST second quantization of gravity (see Schleich [38] for a detailed discussion) and follows the approach taken for construction of gauge invariant string field theory actions [39].

The second quantized BRST charge is constructed from the metric and ghost and antighost fields c^μ and $\bar{c}_\mu$. It has the important properties that

$$Q^2 = 0 \tag{3.1}$$

and

$$\{Q, \bar{c}_\mu(x)\} = H_\mu(x) \tag{3.2}$$

where the constraints $H_\mu(x)$ generate diffeomorphisms. A physical state $|\phi>$ is identified as a cohomology class of Q; i.e., a solution of

$$Q|\phi> = 0 \tag{3.3}$$

modulo an exact state of the form $Q|\epsilon>$. Diffeomorphism invariance of matrix elements between two physical states $|\phi>$ and $|\phi'>$ is then a consequence of

$$<\phi'|H_\mu(x)|\phi> = 0 \tag{3.4}$$

which follows directly from (3.2) and (3.3).

Note that (3.4) is weaker than the usual Wheeler-DeWitt [20] physical state condition $H_\mu(x)|\phi> = 0$. However, only the weaker matrix element condition can be demanded on physical grounds, and in string theory imposing the stronger Wheeler-DeWitt condition would eliminate all states.

As for third quantization of one-dimensional universes, the equation of motion of the free third quantized action should reproduce the second quantized physical state condition. The action which accomplishes this is*

$$S_2 = <\phi|Q|\phi> \tag{3.5}$$

_____

* This may suffer from the "doubling problem" of closed string field theory in that there may be multiple copies of the physical spectrum at differing ghost numbers.

This action has a gauge invariance generated by

$$\delta|\phi> = Q|\epsilon> \qquad (3.6)$$

so we see quite generally that the third quantized field theory is a gauge theory. This gauge symmetry must be fixed in order to co..pute correlation functions. Presumably by doing so one can reproduce the second quantized sum-over-four-geometries, although this has not been done explicitly.

The third quantized gauge invariance is preserved by interactions between universes. This can be deduced (as in string theory) from the second quantized sum-over-four-geometries from which the third quantized interactions are derived. Consider the sum-over-four geometries with n fixed three-boundaries. The sum of the BRST charges on the n boundaries vanishes by BRST conservation. This implies that the coupling of a BRST exact or "pure-gauge" state to $n-1$ physical states vanishes. The diagrams thus obey Ward identities which are equivalent to gauge invariance of the interacting third quantized field theory. The nature and consequences of this third quantized gauge invariance remain to be understood.

3.2 Relation to Other Formalisms

Third quantization is a specific form of the non-linear generalizations of quantum mechanics which have been discussed in the literature. Before the incorporation of topology change, the quantum mechanical wave function of the universe obeys a linear equation such as the Wheeler-DeWitt equation or (3.3). Including the effects of topology change amounts to adding non-linear terms to this equation [32]. In [40], Weinberg has discussed experimental signals of and constraints on non-linear quantum mechanics models. Part of that discussion may be relevant here.

An apparently different way of defining a many-universe system has been proposed by Hartle and Hawking [25,41]. They compute the n-universe wave function as the sum-over-four-geometries with n fixed boundary components, and demand that each universe separately obeys the linear Wheeler-DeWitt equation. This contrasts with third quantization, in which this same sum-over-four-geometries is an off-shell Green function and accordingly obeys a non-linear Schwinger-Dyson equation. In the Hartle-Hawking program, it is hoped that an appropriate complex contour for integration over the conformal part of the metric will insure that the Wheeler-DeWitt equation is obeyed [42].

The discussions of this and the previous sections are illustrations of the apparently general phenomona that generalizing a quantum field theory by including topology change leads to a quantum field theory on a diferent (usually bigger) space. This notion was pursued in the direction opposite to that taken here by Green [43] who suggested that the two dimensional quantum field theory of the string world sheet itself arises in this manner. The idea was taken to its logical extreme by Srednicki [37] who suggested that topology change occurs at all levels, and the universe is actuall described by an infinite sequence of quantum field theories.

IV. PARENT AND BABY UNIVERSES

A third quantized field theory in general allows the joining and splitting of universes of all sizes. However, the joining or splitting of a macroscopic universe from our own would lead to rather dramatic effects which have not been observed. We are, therefore, particularly interested in theories where such processes are dynamically suppressed. This suppression indeed occurs in the semiclassical approximation to theories governed by the Einstein action at long distances. By dimensional analysis, the action associated to nucleation of a universe of radius R is $R^2 M_p^2$. The nucleation of 'baby' universes large relative to the Planck length is, therefore, exponentially suppressed.† The third quantized description of baby universe phenomona is discussed by Banks [9] and in reference [10].

We then have two widely separated scales: the baby universe scale, denoted μ, and the parent universe scale, $\frac{\Lambda^{\frac{1}{2}}}{M_p}$. The construction of the third quantized theory simplifies in an expansion in the small parameter $\frac{\Lambda}{\mu^2 M_p^2}$. (Later it will be seen that validity of the instanton approximation also requires $\frac{\mu}{M_p} << 1$). To leading order in this approximation we have only the diagrams of Figure 5 representing wormholes connecting parent universes to themselves and to one another. There are no diagrams representing bifurcation of parent universes, this is assumed to be exponentially small in $\frac{M_p^4}{\Lambda}$. Bifurcation of baby universes is also negligible relative to the interaction depicted of three baby universes coupling via a parent universe, since the latter process is enhanced by a phase space factor of the cube of the parent universe volume in Planck units.

The processes in Figure 5 resemble Feyman diagrams with the wormholes representing propagators and the parent universe vertices. Thus the baby universes couple to one another via interaction with the parent universes, whose second quantized couplings are

---

† This assumes the sign of the action is positive. In fact for the nucleation of one universe from nothing, the sign is negative and production of large universes may be enhanced. Further discussion of this important sign issue can be found in Sections V and VII.

in turn determined by the state of the baby universes. This provides an unusual feedback mechanism between long and short distance physics–the baby universe system knows about the long distance couplings. This circumstance forces us to reexamine the usual lore that values of low energy couplings - such as the cosmological constant - can be understood independently of short distance physics. This will be further discussed in Sections VI and VII.

4.1 The Hybrid Action

To analyze the parent-baby universe interactions, it is convenient to construct a hybrid representation of these diagrams using second quantized parent universe variables and third quantized baby universe variables as was done for the one dimensional parent-baby universe model in Section II. This hybrid formula will reproduce the Euclidean sum-over-four-geometries.

We simply state the answer, various pieces of which are derived in [6,7,10,9,13,44] and leave it to the reader to verify the combinatorics. Let ϕ_i be a mode of the third quantized baby universe field operator ϕ which creates or annihilates a baby universe of type i from a parent universe

$$\phi(^3g) = \sum_i \phi_i f_i[^3g] \tag{4.1}$$

where the f_i are a set of orthonormal functions on the space of (small) three metrices on S^3. The effect of nucleation of a small baby universe (of type i) on an observer in the parent universe are equivalent to the insertion of a local operator, denoted $\mathcal{L}_i(x)$, at the nucleation event x [3,4,5,6,7]. Let G_{ij} be the propagator defined by the sum-over-four-geometries from a baby universe of type i to type j, i.e.,

$$G_{ij} = \int \mathcal{D}^3 g \int \mathcal{D}^3 g' f_i[^3g] f_j[^3g'] G_E(^3g,^3g') \tag{4.2}$$

The second quantized sum-over-four-geometries is then reproduced by the third quantized functional integral [10,13]

$$Z = \int D\phi \, e^{-S[\phi_i]} \qquad (4.3)$$

where

$$S[\phi_i] = e^{2\gamma} \int_{S^3} D^3 g_s \left[\tfrac{1}{2}\phi_i G_{ij}^{-1}\phi_j - \int_{S^4} D^4 g_l e^{(\phi_i - \lambda_i)S_i}\right] \qquad (4.4)$$

and

$$S_i = \int d^4x \sqrt{g}\, \mathcal{L}_i(x) \qquad (4.5)$$

Summation over the (potentially continuous) indices i,j is implied. γ here is the coefficient of the Euler character which appears in the second quantized action. For the topologies we consider, this counts the number of closed loops of universes. $e^{-2\gamma}$ therefore plays the role of a third quantized Planck's constant. The functional integral $\int D^3 g_s$ denotes an integration over small three geometries (corresponding to baby universes) as well as possible additional matter fields. ϕ is a function on this space. $\int D^4 g_l$ is an integration over large four geometries (plus possibly other matter fields) on the parent universe. λ_i are the fundamental coupling constants. We see from this formula that the effective parent-universe coupling constants are $\lambda_i - \phi_i$ and are shifted by the baby universes.

Without going through all the details, it can be seen qualitatively how (4.3)-(4.5) reproduces the sum-over-four-geometries. The parent universes act as vertices, and the integral over large four geometries in the interaction term of S corresponds to the fact that there is one vertex for each configuration of the parent universe. The use of G^{-1} as the kinetic operator then produces the desired factor of G with each wormhole propagator.

An important issue, which awaits further clarification, is the relation between the action (4.4) for Euclidean universes and the action for Lorentzian universes. Formally the two are related by factors of i (arising from the rotation of the lapse function) exactly as in the case of one dimensional universes discussed in Section II. However this issue cannot be entangled from the problem of the conformal factor, which renders the action indefinite in Euclidean space. A proper understanding of this issue is essential to further progress in the subject.

4.2 Baby Universe Field Operators and Spacetime Couplings

To understand the relation between the third quantized baby universe operators ϕ_i and second quantized coupling constants, it is convenient (but probably not essential) to assume the existence of a Hilbert space representation of the third quantized field theory. The ϕ_i are then operators on this Hilbert space, and the state $|\psi>$ of the baby universe system is described as a state in this Hilbert space. Measurements on a single parent universe are then represented as correlation functions of second quantized operators $\mathcal{O}_i(X_i)$ in this state

$$< \mathcal{O}_1(X_1)\cdots\mathcal{O}_n(X_n) >_\psi = <\psi| \int_{S^4} D^4 g_l e^{-(\lambda_i - \phi_i) S_i} \mathcal{O}_1(X_1)\cdots\mathcal{O}_n(X_n) |\psi> \qquad (4.6)$$

Let us now suppose that the third quantized theory is classical in the sense that the loop expansion parameter $e^{-2\gamma}$ is taken to zero. In that case, the ϕ's all commute and may be diagonalized in terms of coherent α–states $|\{\alpha_i\}>$ obeying

$$\phi_j |\{\alpha_i\}> = \alpha_j |\{\alpha_i\}>$$
$$<\{\alpha_i\}|\{\alpha_i'\}> = \Pi_i \delta(\alpha_i - \alpha_i') \qquad (4.7)$$

If the third quantized system is in an α–state, the correlation function (4.7) becomes, after normalization,

$$<\mathcal{O}_1(X_1)\cdots\mathcal{O}_n(X_n) >_{\{\alpha_i\}} = \int_{S^4} D^4 g_l e^{-(\lambda_i - \alpha_i) S_i} \mathcal{O}_1(X_1)\cdots\mathcal{O}_n(X_n) \qquad (4.8)$$

We see from this expression that the second quantized parent universe couplings (below the wormhole scale) are shifted by the eigenvalues α_i of the third quantized field operators ϕ_i [6,3,7,4,5,45].

In general, there is at least one species of baby universe for every local operator $\mathcal{L}_i(x)$ so, in the absence of a symmetry forbidding the coupling of the baby universe, one expects that all low energy couplings will be shifted by the α parameters.

The field ϕ_i is subject to the equation of motion of the (possibly loop corrected) third quantized action [9,10]. This in turn becomes an equation of motion or dynamical constraint for second quantized couplings. In the classical limit, this equation is

$$G_{ij}^{-1}\alpha_j - \int_{S^4} D^4 g_l e^{-(\lambda_i - \alpha_i)S_i} S_j = 0 \qquad (4.9)$$

Obviously this equation is in general intractable. Later we shall discuss approximations within which it becomes tractable.

A special case of formula (4.7) for correlation functions was introduced by Hartle and Hawking [25,26,41] and used by Coleman [8] in his analysis of the cosmological constant. Consider the case of one operator in the baby universe ground state

$$<\mathcal{O}(X)> = <0|\int_{S^4} D^4 g_l e^{-(\lambda_i - \alpha_i)S_i} \mathcal{O}(X)|0> \qquad (4.10)$$

Taking the baby universe ground state is equivalent to summing over all vacuum bubbles. This may be expressed in a second quantized functional integral (neglecting terms of order $\frac{M_p^2 \mu^2}{\Lambda}$):

$$<\mathcal{O}(X)> = \sum_{\text{topologies}} \int D^4 g e^{-\lambda_i S_i} \mathcal{O}(X) \qquad (4.11)$$

Coleman attempted to derive this formula from the Hartle-Hawking wave function of the universe. We see here that it arises in computing expectation values in the third quantized baby universe ground state.

In general the state $|\psi>$ of the baby universes may not be one of the $|\{\alpha_i\}>$ eigenstates, as discussed in Section 2.4. For example suppose it is in a linear superposition of two eigenstates

$$|\psi> = \beta|\{\alpha_i\}> + \beta'|\{\alpha'_i\}> \qquad (4.12)$$

where $|\beta|^2 + |\beta'|^2 = 1$. The correlator of n parent universe operators then becomes

$$< \psi|O_1(X_1)\cdots O_n(X_n)|\psi >$$
$$= |\beta|^2 < O_1(X_1)\cdots O_n(X_n) >_{\{\alpha_i\}}$$
$$+ |\beta'|^2 < O_1(X_1)\cdots O_n(X_n) >_{\{\alpha'_i\}} \qquad (4.13)$$

The correlator is split into two separate pieces because neither second quantized operators nor the full third quantized Hamiltonian connect different α–states. There is a superselection rule which prevents us from interfering different α–states. This means that an observer who measures α will never know about the one who measures α'. Given that the results of a set of measurements indicate a set of couplings $\{\alpha_i\}$, all future measurements will be governed by dynamics with the couplings $\{\alpha_i\}$ [6].

This generalizes to states involving superpositions of all possible eigenvalues of the form

$$|f> = \Pi_i \int_{-\infty}^{\infty} d\alpha_i f(\{\alpha_i\})|\{\alpha_i\}> \qquad (4.14)$$

where

$$\Pi_i \int_{-\infty}^{\infty} d\alpha_i f^* f = 1 \qquad (4.15)$$

$f(\{\alpha_i\})f^*(\{\alpha_i\})$ is then the probability that the universe is governed by the set of couplings $\{\alpha_i\}$. Allowed values of $\{\alpha_i\}$ are constrained by the third quantized equation of motion. It is natural to consider general states of the form (4.15) because there is no particular reason that the baby universes are in an α–state. For example the third quantized ground state is in general not an α–state. Predictions for physical couplings are possible if the baby universe wave function is highly peaked on a subspace of the $\{\alpha_i\}$.

The preceeding discussion has assumed the existence of a set of coherent states obeying $\phi_i|\{\alpha_j\}> = \alpha_i|\{\alpha_j\}>$. Such states exist only in the semiclassical limit for which

$[\phi_i, \phi_j] = 0$. Away from that limit $[\phi_i, \phi_j]$ is in general order $e^{-2\gamma}$, and the α-states do not exist as states in the third quantized Hilbert space. The eigenvalues of ϕ_i are then subject to the third quantized uncertainty principle [10]. This appears to imply that spacetime couplings cannot be measured to arbitrary accuracy, as was discussed in the one dimensional context in Section 2.4., but the full implications are not yet understood.

V. INSTANTONS–FROM QUANTUM MECHANICS TO QUANTUM GRAVITY

Quantum mechanics often allows the occurrence of processes which are classically forbidden. In favorable circumstances instantons can be used to calculate the rate of such processes as a systematic expansion in a small parameter. Spatial topology change in three space dimensions is just such a classically forbidden process, and we shall see that there are models in which, given some reasonable assumptions, its effects can be systematically calculated as an expansion in a small parameter. The applications of instantons in quantum field theory follows from analogy to their applications in quantum mechanics. This section reviews the relevant features of instanton methods in quantum mechanics and quantum gravity. A classic, and more detailed review can be found in [46].

5.1 Quantum Mechanics

The propagation of a quantum mechanical particle is described by the kernel [21] in the path integral presentation

$$K(X_f, X_i; \tau_f - \tau_i) = < X_f | e^{-iH(\tau_f - \tau_i)} | X_i >$$
$$= \int_{X_i}^{X_f} DX(\tau) e^{iS[X]/g^2} \qquad (5.1)$$

where the Minkowskian action is

$$S = T - V = \int_{\tau_i}^{\tau_f} d\tau (\tfrac{1}{2}\dot{X}^2 - V(X)) \qquad (5.2)$$

and we have scaled out a factor of $\frac{1}{g^2}$. K is the probability amplitude that a particle in the potential V observed at X_i at time τ_i will be observed at X_f at time τ_f. K can be evaluated by analytic continuation in the variable τ.

$$K_E(X_f, X_i; \tau_f - \tau_i) = <X_f|e^{-H(\tau_f - \tau_i)}|X_i>$$
$$= \int_{X_i}^{X_f} DX(\tau) e^{-S_E/g^2} \tag{5.3}$$

where now

$$S_E = T + V = \int_{\tau_i}^{\tau_f} d\tau (\tfrac{1}{2}\dot{X}^2 + V(X)) \tag{5.4}$$

In this form, K_E can be conveniently calculated for small g^2 using the saddle point approximation. From the manner in which g^2 enters the exponent of the functional integral (5.3), it is evident that this is the same thing as a semiclassical expansion in $\hbar$. Such an expansion is sensible only if g^2 is small or, equivalently, if quantum fluctuations are weak for the system in question.

The action is expanded around a trajectory $\overline{X}(\tau)$ which extremizes the action

$$\left.\frac{\delta S_E}{\delta X}\right|_{X=\overline{X}} = 0 \tag{5.5}$$

and obeys the boundary conditions imposed on the functional integral:

$$\overline{X}(\tau_i) = X_i \tag{5.6}$$

$$\overline{X}(\tau_f) = X_f \tag{5.7}$$

As shall be elaborated momentarily, an instanton describes tunneling between two semiclassical WKB states. The pre (post) tunneling position of the particle is X_i (X_f). Specification of a semiclassical state requires the velocity as well as the position. This is obtained from the time derivative of $\overline{X}$ at the boundary. If this time derivative is non-zero in Euclidean space, it will Wick rotate to an imaginary value in Minkowski space. In order that the instanton describe a real tunnneling process in Minkowski space, we require the additional boundary condition:

$$\dot{\overline{X}}(\tau_i) = \dot{\overline{X}}(\tau_f) = 0. \tag{5.8}$$

S_E may be Taylor expanded around $\overline{X} = X - \hat{X}$

$$S_E(X) = S_E(\overline{X}) + \tfrac{1}{2}\int d\tau\, \hat{X} D(\overline{X}) \hat{X} + \cdots \tag{5.9}$$

where

$$D(\overline{X}) = -\partial_\tau^2 + V''(\overline{X}) \tag{5.10}$$

For small g^2, the exponential is highly peaked around $\overline{X}$, and higher order terms in $S(X)$ may be ignored. K_E can then be evaluated by a Gaussian integration:

$$K_E(X_f, X_i; \tau_f - \tau_i) = e^{-\overline{S}/g^2} \det^{-\frac{1}{2}} D \left(1 + O(g^2)\right) \tag{5.11}$$

where $\overline{S}$ is the instanton action. We shall not have much to say about the determinant of the operator $D(\overline{X})$ appearing in (5.11), see [46]. It is defined as the (renormalized) product of the eigenvalues of D, and is often quite difficult to compute. It must be computed to obtain quantitative results, but qualitative results can often be obtained without computing it. The only time it is qualitatively important is if it vanishes or is imaginary. It can vanish as a result of fermion zero modes, certain processes will then be suppressed. It can be imaginary as a result of a negative mode (note the square root in (5.11)). We shall see that this is interpreted as evidence of an instability.

There are two main types of tunneling processes which can be described using instantons. The first is (using the language of quantum field theory) the decay of a particle from a "false", metastable vacuum into the true vacuum as for example occurs in the potential $V(X)$ illustrated in Figure 6. Comparing the Minkowskian action (5.2) with the Euclidean action (5.4) we see that solving the Euclidean equations of motion is equivalent to solving the Minkowskian equations for a particle in this potential $-V$, as illustrated in Figure 7. It is then evident that there is a "bounce" solution for which the particle begins at $\tau = -\infty$

in the bottom of the false vacuum at X_F, rolls towards the true vacuum until it reaches the turning point X_T, whereupon it bounces back and asymptotically approaches X_F as $\tau \to +\infty$.

This solution has one negative mode, *i.e.*, one negative eigenvalue for the operator $\mathcal{D}(\overline{X}) = -\partial_\tau^2 + V''(\overline{X})$. Intuitively this can be understood as follows. The bounce solution $\overline{X}$ can be continuously deformed to the constant solution $X(\tau) = X_F$. It is straightforward to see that the constant solution has less action than the bounce solution. It follows that either there exists a deformation of $\overline{X}$ to X_F for which the action decreases continuously, or there is another saddle point in between which we don't know about. The latter alternative is implausible. The former (correct) alternative implies that the solution $\overline{X}$ has a negative mode.

As mentioned earlier, since the path integral involves the square root of the product of the eigenvalues, the existence of a negative mode (or any odd number of negative modes) implies that the path integral is imaginary. The meaning of this is revealed by using the relation for the vacuum energy F in terms of the Euclidean path integral

$$F = \lim_{T \to \infty} -\frac{1}{T} \ln < X_F | e^{-HT} | X_F > \qquad (5.12)$$

(5.12) can be evaluated by summing over saddle points. To leading non-trivial order in the small parameter $e^{-\overline{S}/g^2}$ we must include the zero and one-bounce solutions. These contribute to the matrix element in (5.12) an amount

$$1 + KTe^{-\overline{S}/g^2} \qquad (5.13)$$

where K is the one loop determinant and the factor of T arises from integrating over the location of the center of the bounce from 0 to T. The free energy is then given by, to leading order

$$F = -Ke^{-\overline{S}/g^2} \qquad (5.14)$$

Since K is imaginary, the energy F of the state $|X_F>$ is also imaginary. This imaginary part is directly related to the decay rate of the particle from X_F into the true vacuum [46].

The validity of this derivation, and the instanton approximation in general, depends on the smallness of $e^{-\overline{S}/g^2}$, which in turn follows from small g^2. The relevance of small $e^{-\overline{S}/g^2}$ can be seen by noting that the instanton density, $<\frac{N}{T}>$, is given by

$$<\frac{N}{T}> = g^2 \frac{dF}{d\overline{S}} = Ke^{-\overline{S}/g^2} \ll 1. \tag{5.15}$$

When this parameter is not small, the instantons are close together and interactions between instantons are important. Ignoring these interactions, as we have done here, is known as the dilute gas or dilute instanton approximation.

The form of the bounce solution also tells us what $|X_F>$ decays into. There is a moment τ_S of time symmetry where $\dot{\overline{X}}(\tau_S) = 0$ and $\overline{X}(\tau_S) = X_T$. Cutting the instanton in half at this moment, one obtains a saddle point which contributes to the matrix element

$$<X_T|e^{-HT}|X_F> \tag{5.16}$$

After the particle tunnels to X_T, it then oscillates around the (true) bottom of the potential. The instanton thus describes tunnelling between two classical solutions of the same energy.

Note that the entire history of the particle can be described semiclassically. Before tunneling it is approximated by a classical particle at the stable minimum. It then undergoes semiclassical tunneling. After tunneling its evolution is again classically described as oscillating motion. The initial data determining the post tunnelling behavior is obtained from the instanton solution. Such a semiclassical description of instanton processes in quantum field theory and quantum gravity is also possible.

The second type of instanton which arises in quantum mechanics exists for the potential of Figure 8. This double well potential has two degenerate minima at X_+ and X_-. To find the instanton, the potential should be turned upside down. It is then evident that there is a kink solution $\overline{X}(\tau)$ which begins at X_- for $\tau = -\infty$ and asymptotically approaches X_+ at $\tau = +\infty$. Unlike the previous example, it does not bounce back. Because of the boundary conditions $\overline{X}(\pm\infty) = X_\pm$, this solution can not be continuously deformed to either of the trivial solutions $X(\tau) = X_\pm$. Correspondingly, there is no negative mode. The instanton gives a real contribution to the matrix elements

$$ln < X_+|e^{-HT}|X_- > = ln < X_-|e^{-HT}|X_+ >$$
$$= KTe^{-\overline{S}/g^2} \qquad (5.17)$$

as computed in the dilute instanton approximation. This implies that the states $|X_\pm >$ do not diagonalize the Hamiltonian. Rather H is diagonalized by the coherent states $|X_+ > \pm |X_- >$, and the energy splitting between these states can be computed using (5.17). The lesson to learn here is that the existence of an instanton with no negative modes in general signifies that the quantum vacuum is constructed as a coherent superpostion of classical vacua.

5.2 Quantum Field Theory

We now turn to some examples in four dimensional quantum field theory. Consider the action

$$S = -\int d^4x \left(\tfrac{1}{2}\nabla^\mu\phi\nabla_\mu\phi + V(\phi)\right) \qquad (5.18)$$

where the potential V is the same function appearing in the first example above with vacuum decay. Vacuum decay occurs here as well. There is a Euclidean solution of the form illustrated in Figure 9, with a round bubble of the true vacuum inside a sea of the false vacuum. This solution can be continuously deformed to the false vacuum, and correspondingly has one negative eigenvalue. It therefore represents a decay process.

The state to which the false vacuum decays can be found as in the quantum mechanics example, by cutting the instanton in half at the moment of time symmetry. This reveals a bubble of true vacuum in the sea of false vacuum, and all time derivatives of fields vanish. Despite the lower energy of the true vacuum, this tunneling process conserves energy exactly because there is energy in the bubble wall. The nucleated bubble then begins exponential acceleration, as governed by the classical equation of motion, and grows indefinitely.

Now suppose V is as in Figure 8, corresponding to degenerate vacua. Tunneling from one minima (ϕ_+) to another (ϕ_-) can not proceed via a bubble as in the preceeding example because energy could not be conserved. Instead all of space must tunnel at the same time. But the action of such an instanton diverges like the volume of space. Tunnelling is therefore exponentially suppressed and vanishes in the infinite volume limit.

In Yang-Mills theory, instantons mediate vacuum topology change. Classical vacua of Yang-Mills theory with $F_{\mu\nu} = 0$ are labelled by integers

$$n = tr \int_{R^3} A \wedge A \wedge A \qquad (5.19)$$

and may be denoted

$$|n> \qquad (5.20)$$

A Yang-Mills instanton with non-zero integral $tr \int F \wedge F = 1$ tunnels between two such classical vacua. There is then a non-zero matrix element

$$<n+1|e^{-HT}|n> \neq 0 \qquad (5.21)$$

so that the $|n>$ vacua do not diagonalize the Hamiltonian. The Hamiltonian is diagonalized instead by the θ vacua

$$|\theta> = \sum_n e^{in\theta}|n> \qquad (5.22)$$

Thus, as in the degenerate double well, the quantum vacua are coherent superpositions of classical vacua.

5.3 Quantum Gravity

Classically, topology change is forbidden in general relativity. If we pick initial data on a space with arbitrary topology and evolve using Einstein's equations, the topology does not change (unless there is a singularity, in which case general relatively itself breaks down).

Quantum mechanically topology change might occur in principle, but there is no real argument that it must or must not occur. Our attitude is simply to assume that it does occur, try to find models where the effects are calculable, and then to see whether these effects are interesting or even measurable.

We shall also assume that the Euclidean path integral and instanton techniques are applicable for computation of topology change. That is the transition amplitude between two manifolds M_I and M_F is given by

$$< M_F | e^{-HT} | M_I > = \sum_{topologies} \int_{M_I}^{M_F} D^4 g e^{-S} \tag{5.23}$$

for some action S. There are several immediate problems with this formula:

1) In general, there is no well defined time 'T' to use on the right hand side. This formula only makes sense when M_I and M_F are three manifolds which bound a four-manifold that has an asymptotically flat region. T is then the Euclidean time as measured in this asymptotic region. This will not work if M_I and M_F are compact. In that case, the Euclidean sum over-four-geometries instead gives a Green function of the third quantized field theory, as discussed in Sections II and III, and instantons provide an approximation to this Green function.

2) The sum-over-four-topologies in (5.23) is problematic because four-manifolds are unclassifiable. Demonstrating the equivalence of two four-manifolds is a Goedel unsolvable problem [47]. In practice we simply restrict ourselves to some subset. If this subset is closed under composition of the functional integral, the theory thereby obtained is at least naively self-consistent.

3) The Einstein action is unrenormalizable. This problem can be remedied by simply imposing a cutoff μ on loop integrations. The semiclassical loop expansion is then an expansion in $\frac{\mu}{M_p}$, where M_p is the Planck mass. Validity of this expansion therefore requires $\frac{\mu}{M_p} \ll 1$. The physical idea behind this cutoff is that some new physics (such as string theory) which does not have the divergence problems of the Einstein action is relevant above the scale μ. We assume here (as suggested by string theory) that a generally covariant cutoff procedure exists.

4) The most serious problem, in my view, is that the Einstein action is unbounded from above and below, so expression (5.23) is not well defined even with a cutoff. To see this let us fix the sign of the action so that a transverse traceless graviton around flat Euclidean space (h_{TT}) has positive action

$$S_E = -\frac{M_p^2}{16\pi} \int d^4x \sqrt{g} R(\delta + h_{TT}) \geq 0 \tag{5.24}$$

where δ is the flat metric. With this sign, a conformal transformation of the metric decreases the action

$$S_E(\Omega^2 g) = -\frac{M_p^2}{16\pi} \int d^4x \sqrt{g} \Omega^2 R(g) - \frac{3M_p^2}{8\pi} \int d^4x \sqrt{g} g^{ab} \nabla_a \Omega \nabla_b \Omega \tag{5.25}$$

So for example a sphere of volume V, which is related to flat space by a (singular) conformal transformation, has action

$$S(V) = -M_p^2 \sqrt{\frac{3V}{2}} \tag{5.26}$$

which grows in magnitude with the size of the sphere. This bizarre fact is ultimately the origin of the Hawking factor [48] in his analysis of the cosmological constant, as discussed in Section VII. However, there do not seem to be many examples in quantum field theory with indefinite actions, and we don't really understand yet how to deal with such systems. My approach in these lectures will be to treat the indefinite modes in much the same way that the indefinite mode X_0 is treated in the one and two dimensional cases, as mentioned at the end of Section (2.1). In the following this amounts to simply ignoring the fact that the functional integral is unbounded, and obtaining results by formal manipulations.

There is a good physical reason why the functional integral for gravity is unbounded, and understanding this may ultimately lead to a resolution of the indefiniteness problem. The Euclidean path integral with periodic boundary conditions is equal to the canonical partition function at a temperature related to the periodicity. However the canonical partition function is, and should be, divergent for gravity because of negative specific heat associated with the attractive force.

Even setting all these problems aside, it is exceedingly difficult to find gravitational instantons of the Einstein action which contribute to any physical process. Let us first consider tunneling from flat R^3 to N where N is a connected but topologically nontrivial three manifold. The tunneling process always conserves energy, so N must have zero energy just as does flat R^3. We now encounter the following theorem of Schoen and Yau [49]:

Theorem: *There are no asymptotically flat solutions of Einsteins equations with zero energy except flat space.*

This rules out any such tunneling processes. We might then consider tunneling

$$R^3 \to R^3 \tag{5.27}$$

which might conceivably be accompanied by topology change of a spin structure. However, we are foiled by another theorem of Schoen and Yau [49]:

Theorem: *There are no asymptotically flat four geometries which are Ricci flat except flat R^4.*

Now we might try to make M^3 disconnected e.g., $M^3 = R^3 \oplus S^3$. The extrinsic curvature induced on M^3 from the interpolating four-manifold must vanish by the analog of the boundary condition (5.8). The interpolating manifold could then be as depicted in Figure 10. Then we encounter the theorem of Cheeger and Grommol (restated) [50]:

Theorem: *Given an asymptotically flat four geometry with $n > 1$ compact interior boundaries with vanishing extrinsic curvature, the Ricci tensor always has a negative eigenvalue somewhere.*

This rules out such instantons for pure gravity, or for gravity coupled to a scalar field for which the Ricci tensor obeys $R_{\mu\nu} = \nabla_\mu \phi \nabla_\nu \phi$ and has everywhere positive eigenvalues.

Because of these theorems, instances in which physical applications of gravitational instantons are possible are very rare. I will now mention a few that I know of.

a.) **Instability of Hot Flat Space**

One way of avoiding the above theorems is to change the boundary conditions. In finite temperature field theory the asymptotic boundary conditions are on $S^2 \times S^1$, rather than S^3, because time is compactified. Euclidean Schwarzchild is a Ricci-flat geometry with just such asymptotic behavior, and was argued by Gross, Perry and Yaffe [51] to represent quantum nucleation of black holes at finite temperature. A similar process also occurs in de Sitter space [52].

b.) **Instability of the Kaluza-Klein Vacuum**

The five dimensional Euclidean Schwarzchild has an $S^3 \times S^1$ boundary with an asymptotically flat metric. The geometry used for five dimensional Kaluza-Klein compactification on a circle is flat Minkowski space cross a circle $(M^4 \times S^1)$ and its Euclideanization $(R^4 \times S^1)$ therefore also has an asymptotically (in fact everywhere) flat $S^3 \times S^1$ boundary. Witten [53] used the fact that these two boundary conditions are the same to interpret the five dimensional Euclidean Schwarzchild as a Kaluza-Klein instanton. Euclidean Schwarzchild has one transverse traceless negative mode. * The one loop determinant is therefore imaginary and this represents a decay of flat space.

What does flat space $(R^3 \times S^1)$ decay into in Witten's example? The instanton has a discrete Z_2 time symmetry. It therefore also has a surface ($"T = 0"$) running through the center of the Schwarzchild "bolt" on which the time derivative of the metric vanishes or, stated more geometrically, the extrinsic curvature vanishes. The topology of this surface is $R^2 \times S^2$. The instanton therefore represents tunnelling from flat $R^3 \times S^1$ to $R^2 \times S^2$. A more detailed analysis shows that the $R^2 \times S^2$ geometry is a hole in space which proceeds to explode.

c.) Kaluza-Klein Charge Confinement

The four dimensional Kaluza-Klein (Euclideanized) vacuum has an asymptotically flat $S^2 \times S^1$ boundary. Gross [54] showed that the boundary conditions are such that Taub-Nut is an instanton for this theory. It leads to confinement of the $U(1)$ electric charge arising from isometries of the internal S^1.

d.) Nucleating a Closed Universe from the Void

Another way to avoid the no-go theorems is to change the action. The simplest

---

* Of course there are an infinite number of pure trace negative modes, corresponding to the fact that the conformal modes all have the wrong sign. Flat space also has an infinite number of such negative modes, but no transverse traceless ones. The working hypothesis is therefore to ignore the trace zero modes, but a careful justification has not been given. This is a manifestation of problem (4) above.

way to change the action is to add a cosmological constant:

$$S = \int d^4x \sqrt{g} \left(-\frac{M_p^2}{16\pi} R + \Lambda\right) \qquad (5.28)$$

There is then an obvious instanton: the round metric on S^4. As argued by Vilenkin, [55] if the S^4 is cut in half along the equator, one obtains an instanton with one S^3 boundary. The extrinsic curvature on this boundary vanishes, so the half-sphere represents tunnelling from nothing to S^3. The geometry on this S^3, which provides the initial data for post-tunneling classical evolution, corresponds to de Sitter space at its minimum radius.

e.) Nucleating an Open Universe from the Void

Myers [56] has recently proposed a novel interpretation of the Eguchi-Hanson instanton [57]. An instanton with an asymptotically flat S^3 boundary contributes to the process $R^3 \to R^3$. This can be seen by dividing the boundary into three regions; the past R^3, the future R^3 and a timelike cylinder at spatial infinity. Now consider an S^3/Z_2 boundary, where the Z_2 identifies antipodal points. The past and future R^3's are identified, and the timelike boundary closes on itself. Such boundary conditions seem appropriate for the process of nothing going into R^3. These are the boundary conditions obeyed by the Eguchi-Hanson instanton, which therefore contributes to this tunnelling.

f.) Wormholes

Recently there has been much interest in spacetime wormholes. These were first discussed in [2,58] as instantons of the positive-definite gravitational action

$$S = \int d^4x \sqrt{g} \, C_{\mu\nu\alpha\beta} C^{\mu\nu\alpha\beta} \qquad (5.29)$$

where $C_{\mu\gamma\alpha\beta}$ is the Weyl curvature tensor. These instantons are asymptotically flat, and therefore contribute to the $R^3 \to R^3$ vacuum-vacuum amplitude. They have a "wormhole" handle, whose cross sections are T^3 (three tori). There is a Z_2 time reflection symmetry, and the moment of time symmetry slices the handle in half. The boundary of the half-instanton sliced along the moment of time symmetry thus contains two disconnected pieces: a non-compact, asymptotically flat portion which turns out to be topologically T^3 minus a point (space with a knot, which turns out to have fermionic character! [59]) and a compact T^3 portion where the handle is sliced. This instanton therefore represents nucleation of a small toroidal baby universe.

There is no negative mode, so this is not a decay process. Rather it implies that, as in the double well, the quantum vacuum must be constructed as a coherent superposition of classical vacua. In this case the quantum vacuum was shown to contain configurations with all numbers of baby universes [2].

More recently, wormhole configurations have been discussed by Hawking [3] and Lavreshvili, Rubakov, and Tinyakov [4] but instantons were not found in these works. However, another alteration of the Einstein action which avoids the no-go theorems has recently been found [5]: the addition of axions. This also leads to wormhole instantons, and will be the subject of the next subsection.

In addition to the above examples, there are many solutions of the Euclidean Einstein equations that have been discussed in the literature which have no known physical interpretations. This occurs when the boundary conditions do not correspond to those imposed on the functional integral representing any physical process. An example is K^3, which has no boundary at all.

5.4 Axionic Instantons

The coupling of gravity to axions may be described by the Euclidean action

$$S = \frac{M_p^2}{16\pi} \int d^4x \sqrt{g}(-R + H_{\mu\nu\lambda}H^{\mu\nu\lambda}) \qquad (5.30)$$

where the three form

$$H = dB \qquad (5.31)$$

is the axion field strength. In general surface terms are also required, but these are unimportant in the present context, see [60] for a discussion. The equations of motion following from (5.32) are

$$R_{\mu\nu} = -6 \,^*H_\mu \,^* H_\nu \qquad (5.32)$$

$$d^*H = 0 \qquad (5.33)$$

where

$$^*H_\mu = \frac{1}{3!}\epsilon_\mu{}^{\nu\alpha\beta} H_{\nu\alpha\beta} \qquad (5.34)$$

defines the dual of H. Notice that the Ricci tensor is *negative definite*. The precepts of the no-go theorems are violated and instantons are possible. In fact, there does exist [5] an instanton of the form depicted in Figure 10. The line element is given by

$$ds^2 = \frac{|q|}{2\pi^2 M_p^2} \cosh 2x (dx^2 + d\Omega^2) \qquad (5.35)$$

while the axion field strength is

$$H = \frac{q\epsilon}{M_p^2} \qquad (5.36)$$

where the three form ϵ is the volume element for surfaces of constant x normalized to integrate to one. It is easy to see that the two regions $x \to \pm\infty$ are asymptotically flat. The solution is invariant under $x \to -x$, so the extrinsic curvature vanishes at $x = 0$. Considering the coordinate region $x > 0$, we then have the half-wormhole instanton of Figure 10. There is one asymptotic S^3 region at $x = \infty$ with an asymptotically flat metric. The induced metric on the small S^3 boundary at $x = 0$ is the round metric, and has vanishing extrinsic curvature as required by (5.7). The H field on the S^3 boundary is proportional to the volume element on the boundary. The data on this surface corresponds to initial data for a Robertson-Walker cosmology with axionic matter. Dividing the asymptotically flat boundary into past and future regions, this instanton is seen to represent tunneling from $R^3 \to R^3 + S^3$ (or $R^3 + S^3 \to R^3$), *i.e.*, the nucleation (or annihlation) of a baby Robertson-Walker universe. The precise application of this instanton will be discussed in the next section.

The current $^*H_\mu$ is conserved according to (5.31). We can therefore associate a charge,

$$q = M_p^2 \int H \tag{5.37}$$

known as the Peccei-Quinn charge, with any homology class of three surfaces. There is a one parameter family of instantons labeled by the charge q running through the wormhole. Their action is

$$S_q = \frac{3|q|}{8} \tag{5.38}$$

The charge is related to the radius, R of the baby universe (at the moment of nucleation) by the equations of motion

$$|q| = 2\pi R^2 M_p^2 \tag{5.39}$$

Thus the nucleation of baby universes large relative to the Planck mass is exponentially suppressed. This follows on dimensional grounds.

At large distances, the effects of the mediation of a baby universe can be approximated by the insertion of a local operator at the nucleation event. Since the total Peccei-Quinn charge (on all universes) is conserved, and the baby universe carries off charge q, the local operator insertion must itself carry charge $-q$. Such operators are awkward to express in terms of the potential B. It is convenient to introduce a pseudoscalar field a defined by

$$^*H_\mu = \frac{4\pi}{3}\nabla_\mu a \tag{5.40}$$

which is always locally possible by virtue of (5.33). Under a Peccei-Quinn symmetry transformation, a is shifted by a constant. A local operator $\mathcal{L}_q(x)$ of charge q should acquire a phase under this shift. The unique operator is therefore

$$\mathcal{L}_q(x) = e^{iqa(x)} \tag{5.41}$$

Higher order corrections to this formula are discussed in [16].

A process closely related to baby universe creation is tunneling from a Robertson-Walker to a DeSitter universe. Instantons describing this process in Einstein gravity with a cosmological constant and axions were discussed in [61], and in Einstein gravity with axions and a scalar field in [35].

If one begins with an action defined in terms of the pseudoscalar a, rather than the field strength H, the equations of motion are identical to those of an ordinary scalar field and instantons naively appear not to exist. However, the analyses of references [62,63,64] show that, if careful account is taken of the boundary conditions appropriate for tunneling between charge eigenstates, tunnelling nevertheless can be seen to occur.

5.5 The Small Expansion Parameter

The advantage of the instanton approximation to the functional integral, as opposed to e.g. a minisuperspace approximation, is that in favorable circumstances it is the first term in an expansion in a small parameter of the exact functional integral. We then have good reason to believe that our results are both quantitatively and qualitatively accurate. Let us discuss when the instanton approximation may be accurate in quantum gravity.

On dimensional grounds, an instanton of Einstein gravity will generically have action

$$S \simeq R^2 M_p^2 \tag{5.42}$$

where R is the scale of the instanton. The instanton density per Planck four-volume is then of order

$$\frac{N}{M_p^4 V} \simeq e^{-R^2 M_p^2} \tag{5.43}$$

For $R \gg \frac{1}{M_p}$, the instantons are dilute. However, for $R \sim \frac{1}{M_p}$, the instanton separation is of order their size. Interactions between instantons become important, and the dilute gas approximation breaks down. Therefore, one prerequisite for the validity of the instanton approximation is that the instantons are large relative to the Planck length.

In the axion model, Euclidean wormhole solutions exist for all values of the radius R. However, we must impose a cutoff μ since the theory is non-renormalizable. New physics arises at the scale μ, and the Einstein action, or its extrema, are not relevant above that scale. This means that the integration over instanton sizes should be cutoff at μ. The most probable instantons are then of size $R \sim \frac{1}{\mu}$ and have a density

$$\frac{N}{M_p^4 V} \simeq e^{-M_p^2/\mu^2} \tag{5.44}$$

which is very small for $\frac{\mu}{M_p} \ll 1$.

$\frac{\mu}{M_p}$ is also the loop expansion parameter in quantum gravity. If it is small, both the interactions between instantons and the corrections to the saddle point approximation are small.

An unsatisfactory feature of using a small cutoff to justify the instanton approximation is that we cannot discuss what happens as the cutoff is taken away. Also there may be other effects related to the new physics at the cutoff scale of the same size as the instanton effects. Thus the problem of computing wormhole effects cannot in this context be clearly separated from the problem of finding a consistent quantum theory of gravity at short distances.

A better way to justify the instanton approximation, peculiar to the axion model, is the following. Strings can be coupled to axions via a coupling of the form

$$\frac{T}{2}\int d^2\sigma\, dX^\mu \wedge dX^\nu B_{\mu\nu} \tag{5.45}$$

to the string world sheet. This results in a quantization condition on the charge q, analogous to electric charge quantization due to monopoles [65,66]

$$q = n\frac{M_p^2}{T} \tag{5.46}$$

for an integer n. If we now adjust T so that

$$T \ll M_p^2 \tag{5.47}$$

we find that the minimum value of q is very large. This corresponds to a minimum radius of

$$R^2 = \frac{1}{2\pi T} \tag{5.48}$$

and instanton density

$$\frac{N}{M_p^4 V} = e^{-3M_p^2/8T} \tag{5.49}$$

The instanton approximation is then justified, and for small T the results are insensitive to the manner in which quantum gravity solves its short distance problem.

This result was pointed out by Kim and Lee [67] in the language of pseudoscalars. If the pseudoscalar is a periodic angular variable, the charge is quantized and there is a minimum radius wormhole.

VI. THE AXION MODEL AND THE INSTANTON APPROXIMATION

In Section III we discussed the general problem of third quantization of four dimensional universes. In Section IV the general formulae were simplified in the parent-baby universe approximation. All of the formulae presented were quite formal because, among other reasons, they involved divergent functional integrals over four geometries. In the instanton approximation, which is the first term in a semiclassical expansion in $\frac{\mu}{M_p}$, the intractable sum-over-four-geometries is reduced to a tractable integration over saddle points. It is thereby possible to obtain concrete expressions and give a more precise meaning to our formalism [10].

To leading order in the semiclassical approximation, the third quantized action is defined by the requirement that it reproduce the sum-over-instantons. Consider the axion model with a small cosmological constant Λ_0 whose action is

$$S_0 = \int d^4x \sqrt{g} \left(\frac{M_p^2}{16\pi}(-R + H_{\mu\nu\lambda}H^{\mu\nu\lambda}) + \Lambda_0 \right) - \gamma\chi \qquad (6.1)$$

χ is the Euler character of the manifold, and the corresponding topological coupling constant γ has been added here. This theory has two known types of instantons: the small wormhole instantons of Section 5.4 and large Einstein metrics on S^4 with scale governed by the long-distance effective cosmological constant. The first (small) instanton provides an approximation to the kinetic term of the third quantized action, while the second (large) instanton provides an approximation to the third quantized potential. Thus the assumption of widely separated scales used in the parent-baby universe approximation is justified in the axion model by the semiclassical expansion in $\frac{\mu}{M_p}$.

The general action which reproduces the sum-over-parent and baby universes was given in (4.4) as

$$S = e^{2\gamma} \int_{S_3} D^3 g_s [\tfrac{1}{2}\phi_i G_{ij}^{-1} \phi_j - \int_{S^4} D^4 g_l \, e^{\phi_i S_i - S_0}] \tag{6.2}$$

where (6.1) has been substituted for $\lambda_i S_i$. Each term in this action is simplified in the instanton approximation. In the next few pages we will show that this field theory in fact reduces to a quantum mechanics model with action given by (6.8). It is instructive to derive this model beginning with (6.2) but the final answer can be checked by verifying that it reproduces the sum-over-instantons. The reader who is willing to take this for granted may skip directly to equation (6.8). *

In the axion model, baby universes are labeled by the charge q. The label i on the baby universe field operator should then run over the possible values of the charge, so ϕ_i becomes $\tilde{\phi}(q)$. Similarly the functional integral $\int_{S^3} D^3 g_s$ becomes $\int_{-\infty}^{\infty} dq$. The inverse propagator $\tilde{G}^{-1}(q, q')$ is the instanton approximation to the sum-over-four-geometries carrying charge q. Since in this approximation only one geometry contributes, this is simply given by the exponential of the instanton action

$$\tilde{G}^{-1}(q, q') = \delta(q + q') \, e^{S_q} \tag{6.3}$$

with S_q given in (5.40). We ignore here, and in the following, factors arising from the instanton determinant and normalization of the zero modes and measure. These lead to powers of q (or powers of V in the following) and do not qualitatively affect the answer.

The local operator $\mathcal{L}(q)$ associated to the insertion of a wormhole end carrying charge q is, from (5.43)

$$\mathcal{L}(q) = e^{iqa(x)} \tag{6.4}$$

---

* The following derivation, which has not appeared before, was developed in conversations with Steve Giddings.

where a is the pseudoscalar axion field. The combination $\phi_i S_i$ then becomes

$$\int_{-\infty}^{\infty} dq \, e^{iqa(x)} \tilde{\phi}(q) = \phi(a(x)) \tag{6.5}$$

where ϕ is the Fourier transform of $\tilde{\phi}$. Putting this all together, redefining ϕ by a minus sign and Fourier transforming with respect to q, we obtain

$$S = e^{2\gamma} \int_{-\infty}^{\infty} d\tau [\tfrac{1}{2} \phi G^{-1} \phi - \int_{S^4} \mathcal{D}^4 g_l \mathcal{D}\hat{a} \, e^{-S_0 - \int d^4 x \sqrt{g} \phi(a(X))}] \tag{6.6}$$

where G^{-1} is the Fourier transform of $\tilde{G}^{-1}$ and is non-local in τ-space. $\hat{a} = a - \tau$ is the part of the axion field a orthogonal to the zero mode τ on S^4. In rewriting the integration over the axion field in terms of a (rather than B) this zero mode is omitted. However the $\mathcal{D}\hat{a}$ integration combines with the overall τ integration to give an integration over the full axion field.

The integral over large four metrics and $\hat{a}$ can be evaluated in the saddle point approximation, for which $R_{\mu\nu} = g_{\mu\nu}(\Lambda + \phi(\tau))$ and $\hat{a} = 0$. One then obtains our final expression for the third quantized action

$$S = e^{2\gamma} \int_{-\infty}^{\infty} d\tau [\tfrac{1}{2} \phi G^{-1} \phi - e^{\frac{3M_p^4}{8(\Lambda + \phi)}}] \tag{6.7}$$

If the charge q is quantized, τ becomes a periodic variable.

In the semiclassical limit $e^{-2\gamma} \to 0$, coherent states can be constructed which are eigenstates of $\phi(\tau)$ for all τ

$$\phi(\tau)|\alpha(\tau)> = \alpha(\tau)|\alpha(\tau)> \tag{6.8}$$

$\alpha(\tau)$ is constrained to be a solution of the classical equation of motion of S. It represents the trajectory of a particle in the potential $e^{\frac{3M_p^4}{8(\Lambda+\phi)}}$. Comparing with (4.8) and using equation (6.4), we then see that the effective second quantized action for a parent universe interacting with baby universes in the state $|\alpha(\tau)>$ is

$$S_{\text{eff}} = \int d^4 x \sqrt{g} (-\frac{M_p^2}{16\pi} R + \frac{8\pi M_p^2}{3} \nabla_\mu a \nabla^\mu a + \Lambda + \alpha(a)) \tag{6.9}$$

In conclusion: *The spacetime axion potential is given by the classical trajectory of the particle governed by the third quantized action (6.7)*.

One interesting consequence is that, because of the singular nature of the third quantized potential at $\Lambda + \phi = 0$, the axion potential appears to always have a minimum where the bare cosmological constant Λ is just cancelled. To the low energy observer, this would appear to be a fine tuning of coupling constants. Here it is a conseqence of the third quantized equation of motion. This vanishing of the cosmological constant is more general than our specific model, and is further discussed in a more general context in the next section.

VII. THE COSMOLOGICAL CONSTANT.

Recently there has been much discussion of the possibility that non-perturbative effects in quantum gravity might account for the observed vanishing of the cosmological constant. The basic argument appears in a 1984 paper of Hawking's [48] which received little attention at the time. This paper was slightly preceded by Baum [68] which contains some of the important ideas. We begin by restating their argument in a slightly more modern form.

7.1 The Hawking-Baum Argument.

Consider the following formula for the correlation function of n operators

$$< \mathcal{O}_1(X_1)\cdots\mathcal{O}_n(X_n) > = \int_{S^4} \mathcal{D}g\mathcal{D}A \ e^{-S}(\mathcal{O}_1(X_1)\cdots\mathcal{O}_n(X_n)) \tag{7.1}$$

where S is the Einstein action plus matter

$$S = \int d^4x \sqrt{g}(-\frac{M_P^2}{16\pi}R + \Lambda + \mathcal{L}(A)) \tag{7.2}$$

and the functional integral is over matter fields A and metrics g on S^4. Formulae of the type (7.1) have been discussed by Hartle and Hawking [25,26,41] in the context of the wave function of the universe. (7.1) may also arise in an approximation to spacetime correlation functions in a third quantized theory; alternately it may simply be postulated. However the relation between the Euclidean formulae (7.1) and real time Lorentzian expectation values is poorly understood at present.

Hawking's mechanism exists whenever the cosmological constant becomes a dynamical variable. To take a familiar case, let us suppose that A is a Yang-Mills field and there is a corresponding θ angle. The vacuum state is then in general given by

$$|f> = \int d\theta f(\theta)|\theta> \tag{7.3}$$

In a θ eigenstate, the effective cosmological constant is well known to be

$$\Lambda(\theta) = \Lambda + M^4 \cos\theta \qquad (7.4)$$

for an appropriate mass M of order the Yang-Mills confinement scale. Correlation functions in the state $|f>$ involve an integral over θ. This is equivalent to integrating over the cosmological constant. The formula for correlation functions in the state $|f>$ is

$$< O_1(X_1) \cdots O_n(X_n) >_f$$
$$= \int d\theta \int_{S^4} DgDA |f(\theta)|^2 \, e^{-S(\theta)} (O_1(X_1) \cdots O_n(X_n)) \qquad (7.5)$$

where

$$S(\theta) = \int d^4x \sqrt{g} \left(-\frac{M_p^2}{16\pi} R + \Lambda(\theta) + \mathcal{L}(A) \right) \qquad (7.6)$$

Let us now approximate the integral over the metric by its saddle point:

$$< O(X_1) \cdots O(X_n) >_f$$
$$\simeq \int d\theta |f(\theta)|^2 \int DA \, e^{-S[A] + \frac{3M_p^4}{8\Lambda(\theta)}} (O(X_1) \cdots O(X_n)) \qquad (7.7)$$

We assume that $\Lambda < M^4$ (otherwise this mechanism doesn't work), which might for example be explained by supersymmetry. The θ integral has an essential singularity at $\Lambda(\theta) = 0$ or $\theta = \arccos(\frac{\Lambda}{M^4})$ where the total effective cosmological constant vanishes.

Regulating the functional integral with an infrared cutoff and normalizing, the essential singularity can be replaced by a delta function $\delta(\Lambda(\theta))$. As long as $f(\theta)$ has non-zero support at $\theta = \arccos(\frac{\Lambda}{M^4})$, the θ angle adjusts itself to cancel the cosmological constant.

Several points are worth noting about the mechanism. First, it is not specific to baby universes. While baby universes provide a natural and appealing mechanism for making the cosmological constant (as well as all other coupling constants) dynamical, any such mechanism will do. There are certainly far less exotic mechanisms than baby universes for cancelling the cosmological constant, such as the θ parameter mentioned here.

The real reason that the cosmological constant is forced to vanish is the sign of the Euclidean action. On dimensional grounds, one expects the magnitude of the action to grow with the scale. Ordinarily this would mean that large configurations have large action and are suppressed. The preferred size for the universe would then be the Planck size. However, since large Einstein metrics on the four sphere have negative action such configurations are enhanced (in fact infinitely) rather than suppressed.

As discussed previously, large Einstein four spheres are not the only configurations with negative action. In general the action of any configuration can be arbitrarily decreased by conformal transformations of the metric. The action of a conformally rescaled metric is given by the expression

$$S_E(\Omega^2 g) = \frac{M_p^2}{16\pi} \int d^4x \sqrt{g} \left(-\Omega^2 R(g) - 6\nabla_\mu \Omega \nabla^\mu \Omega\right) \tag{7.8}$$

Note in particular that the action can be made arbitrarily negative with rapidly varying conformal factors. Thus while Hawking's argument predicts zero cosmological constant, it might also predict that the conformal factor should be rapidly varying on all length scales.

Of course a key difference is that Einstein metrics on S^4 are not just configurations with large negative action, they are extrema. One may thus hope that a careful treatment, based on a semiclassical analysis, can justify Hawking's analysis and choice of sign. Clearly this point needs further clarification. We need a systematic derivation of the vanishing of the cosmological constant which does not also predict rapidly varying conformal factors.

7.2 Baby Universes and Coleman's Argument

In the last section we saw in the axion model that the axion potential appears to always have a minimum for which the effective cosmological constant vanishes. This is obviously closely related to Hawking's mechanism.

To see the connection, recall from Sections IV that in the parent-baby universe approximation spacetime correlation functions in the baby universe ground state are given by

$$< O_1(x_1)\cdots O_n(x_n) > = \int D\phi \, e^{-S[\phi]} \int_{S^4} Dg \, e^{-S_0 - \phi_i S_i} O_1(x_1)\cdots O_n(x_n) \qquad (7.9)$$

ϕ here is the baby universe field and S its third quantized action. This is eqivalent to the second quantized formula [25,26,41]

$$< O_1(x_1)\cdots O_n(x_n) > = \sum_{topologies} \int D^4 g \, e^{-S_0} O_1(x_1)\cdots O_n(x_n) \qquad (7.10)$$

which in turn is simply (7.1) summed over different topologies.

(7.9) or (7.10) can be easily evaluated in the axion model if if we suppress the non-constant part of the baby universe field ϕ. This is equivalent to ignoring the charge collective coordinate for the instanton. The third quantized action then has only a potential term

$$S = -e^{\frac{3M_p^4}{8(\Lambda_0 + \phi)}} \qquad (7.11)$$

With this ansatz, correlation functions are given by

$$< O_1(x_1)\cdots O_n(x_n) > = \int d\phi \, e^{e^{\frac{3M_p^4}{8(\Lambda_0 + \phi)}}} \int_{S^4} Dg \, e^{-S_0 - V\phi} O_1(x_1)\cdots O_1(x_n) \qquad (7.12)$$

and the effective cosmological constant is $\Lambda_0 + \phi$. This formula is identical in form to (7.5). The effect of baby universes (with this ansatze) is simply to introduce a weighted integration over the cosmological constant. The prefactor $e^{e^{\frac{3M_p^4}{8(\Lambda_0 + \phi)}}}$ is the probability distribution for ϕ and is obviously highly peaked where the effective cosmological constant vanishes.

This result is of course due to Coleman [8]. We have rephrased his analysis here as a computation of the potential of the third quantized action.

The factor $e^{e^{\frac{3M_p^4}{8(\Lambda_0 + \phi)}}}$ is different from Hawking's factor $e^{\frac{3M_p^4}{8(\Lambda_0 + \phi)}}$. This difference arises simply because Hawking calculates probabilities for Λ_{eff} for one universe, while Coleman calculates probabilities for a dilute gas of universes. At the level of the discussion given here, both analyses give a delta function for vanishing cosmological constant, and appear equivalent. However, more refined considerations appear to distinguish the two analyses. Coleman [8] argues that by computing corrections to (7.11), peaked probability distributions for *all* coupling constants can be derived. The viability of this proposal is currently under active investigation [11,12,13,15,16], but as of this writing the verdict is not in. It has also been argued that baby universes are essential for understanding how the universe can have zero cosmological constant *and* hot matter [13].

The physical picture behind Coleman's analysis appears quite different from that of Hawking and Baum. Spacetime coupling constants are baby universe vacuum parameters, and are dynamically determined by the baby universe interactions. This provides a totally new framework for understanding low energy coupling constants. It is a tantalizing possibility that this framework may explain values of coupling constants in our universe.

ACKNOWLEDGEMENTS

I am grateful to T. Banks, S. Coleman, S. Giddings, G. Horowitz, B. Keay, R. Myers, S.-J. Rey and M. Srednicki for useful discussions and comments, to the students at TASI for their interest, and to A. Jevicki and C-I Tan for organizing a stimulating school. This work was supported in part by DOE Outstanding Junior Investigator Grant DE-AT03-76ER70023 and an A. P. Sloan Foundation Fellowship.

REFERENCES

1) K. Sato, H. Kodama, M. Sasaki and K. Maeda, Phys. Lett. 108B **103**, (1982).

2) A. Strominger, *Vacuum Topology and Incoherence in Quantum Gravity*, Phys. Rev. Lett. **52**, 1733 (1984).

3) S. W. Hawking, *Coherence Down the Wormhole*, Phys. Lett. **195B**, 337 (1987); *Wormholes in Spacetime*, Phys. Rev. D **37**, 904 (1988).

4) G. V. Lavrelashvili, V. A. Rubakov and P. G. Tinyakov, JETP Lett. **46** 167 (1987); Nucl. Phys. B **299**, 757 (1988).

5) S. B. Giddings and A. Strominger, *Axion-Induced Topology Change in Quantum Gravity and String Theory*, Nucl. Physics B, **306**, 890 (1988).

6) S. Coleman, *Black Holes as Red Herrings: Topological Fluctuations and the Loss of Quantum Coherence*, Nucl. Phys. B, **307**, 864 (1988).

7) S. B. Giddings and A. Strominger, *Loss of Incoherence and Determination of Coupling Constants in Quantum Gravity*, Nucl. Physics B, **307**, 854 (1988).

8) S. Coleman, *Why There Is Nothing Rather Than Something: A Theory of the Cosmological Constant*, Harvard preprint HUTP–88/A022.

9) T. Banks, *Prolegomena to a Theory of Bifurcating Universes: A Nonlocal Solution to the Cosmological Constant Problem, or Little Lambda Goes Back to the Future*, Santa Cruz preprint SCIPP 88/09.

10) S. B. Giddings and A. Strominger, *Baby Universes, Third Quantization and the Cosmological Constant*, Harvard preprint, HU TP-88/A036 (1988). (to appear)

11) B. Grinstein and M. Wise, Cal Tech preprint CALT-68-1505.

12) V. Kaplunovsky, unpublished.

13) I. Klebanov, L. Susskind and T. Banks, *Wormholes and the Cosmological Constant* SLAC preprint. (to appear)

14) A. Gupta and M. B. Wise, Caltech preprint CALT-68-1520 (1988).

15) R. C. Myers and V. Periwal, *Constants and Correlations in the Wormhole Calculus*, ITP preprint (1988).

16) S. J. Rey, *The Axion Dynamics in Wormhole Background*, UCSB preprint (1988).

17) J. Milnor and J. O. Stasheff, Characteristic Classes, (Princeton University Press: New Jersey (1974)).

18) M. McGuigan, *On the Second Quantization of the Wheeler–DeWitt Equation* Report No. DOE/ER/40325-38-Task-B (1988). (to appear)

19) J. B. Hartle and K. V. Kuchar, Phys. Rev. **D34**, 2323 (1986).

20) B. S. DeWitt, Phys. Rev. **160**, 1113 (1967); J. A. Wheeler, Batelle Recontres, C. DeWitt and J. A. Wheeler, eds. (Benjamin: (1968)).

21) R. P. Feynman and A. Hibbs, Quantum Mechanics and Path Integrals, (McGraw-Hill: (1965)).

22) A. Cohen, G. Moore, P. Nelson, J. Polchinski, Nucl. Phys. B **267**, 143 (1986).

23) J. J. Halliwell, *Derivation of the Wheeler-DeWitt Equation from a Path Integral for Minisuperspace Models*, ITP preprint NSF–ITP–88–25.

24) C. Teitelboim, Phys. Rev. D **25**, 3159 (1982).

25) S. W. Hawking and J. B. Hartle, Phys. Rev. D **28**, 2960 (1983).

26) J. B. Hartle, *Gravitation and Astrophysics*, proceedings of the Carg/'ese 1986 Summer Institu J. B. Hartle and B. Carter, eds. (Plenum: (1987)), and in *High Energy Physics 1985*, M. Bowick and F. Gürsey, eds. (World Scientific: (1986)), and references therein.

27) K. Kuchar, J. Math. Phys. **22** 2640 (1981); Quantum Gravity 2, C. J. Isham, R. Penrose, and D. W. Sciama, eds. (Clarendon Press: (1981)).

28) A. Jevicki, Frontiers in Particle Phys. '83; Dj. Sijacki et. al., eds. (World Scientific: (1984)).

29) N. Caderni and M. Martellini, Inst. J. Theor. Phys. **23**, 23 (1984).

30) I. Moss in Field Theory, Quantum Gravity and Strings II, eds. H. J. deVega and N. Sanchez, (Springer: Berlin (1987)).

31) A. Anderson, *Changing Topology and Non-Trivial Homotopy*, University of Maryland preprint 88–230.

32) C. Hill, *Non-Linear Quantum Mechanics as a Relaxation Method for the Cosmological Constant*, CERN preprint TH.4908/87 (1988).

33) V. Gates, E. Kangaroo, M. Roachcock, and W. C. Gall, *The Super G-String*, in Unified String Theories, M. Green and D. Gross, eds. (World Scientific: (1986)).

34) A. Hosoya and M. Morikawa, *Quantum Field Theory of Universe*, RITP Hiroshima University preprint RKK88-20, 1988.

35) G. V. Lavrelashvili, V. A. Rubakov and P. G. Tinyakov, *Destruction of Quantum Coherence Due to Topological Changes: A Toy Model*, in Proc. Inst. Seminar "Quarks–1988," Tbilisi, May 1988 (World Scientific, Singapore). (to appear)

36) V. A. Rubakov and Institute for Nuclear Research of the Academy of Sciences of the USSR, *On the Third Quantization and the Cosmological Constant*, preprint (1988).

(to appear)

37) M. Srednicki, *Infinite Quantization*, UCSB preprint 88-07.

38) K. Schleich, Phys. Rev. D **36**, 2342 (1987).

39) E. Witten, Nucl. Phys. B **268**, 253 (1986).

40) S. Weinberg, *Particle States the as Realizations of Spacetime Symmetries*, Austin preprint, UTTG-15-88 (1988).

41) J. B. Hartle, *Simplicial minisuperspace I. General discussion*, J. Math Phys. **26**, 804 (1985).

42) J. J. Halliwell and J. B. Hartle, in progress.

43) M. B. Green, *World Sheets for World Sheets*, Nucl. Phys. B **293**, 593 (1987).

44) A. Hosoya, *A Diagrammatic Derivation of Coleman's Vanishing Cosmological Constant*, RITP Hiroshima University preprint RRK 88-28, 1988.

45) S. W. Hawking and R. LaFlamme, *Baby Universes and the Non-Renormalizability of Gravity* Cambridge DAMPT preprint 88-0290.

46) S. Coleman, Aspects of Symmetry: Selected Erice Lectures, (Cambridge University Press: NY (1985)), pp. 265-351.

47) A. A. Markov, Proceedings of the International Congress of Mathematicians, (Cambridge University Press: Cambridge, England (1958)).

48) S. W. Hawking, *The Cosmological Constant is Probably Zero*, Phys. Lett. **134B**, 403 (1984).

49) R. Schoen and S. T. Yau, Commun. Math. Phys. **79**, 231 (1982).

50) J. Cheeger and D. Grommol, *On the Structure of Complete Manifolds with Non-Negative Ricci Curvature*, Ann. Math. **96(3)**, 413 (1972).

51) D. J. Gross, M. Perry, and L. Yaffe, Phys. Rev. **D25**, 330 (1982).

52) P. Ginsparg and M. Perry, *Semi-classical Perdurance of De-Sitter Space*, Nucl. Phys. B **22**, 245 (1983).

53) E. Witten, *Instability of the Kaluza-Klein Vacuum*, Nucl. Phys. B **195**, 481 (1982).

54) D. J. Gross, Nucl. Phys. B **236**, 349 (1984).

55) A. Vilenkin, Phys. Lett. B, **117** 25 (1983); Phys. Rev. D, **27** 2848 (1983).

56) R. Myers, private communication.

57) T. Eguchi and A. S. Hanson, Phys. Lett. B **74**, 249 (1978).

58) G. Horowitz, M. Perry and A. Strominger, *Instantons in Conformal Gravity*, Nucl. Phys. **B238**, 653 (1984).

59) J. L. Friedman and R. D. Sorkin, Phys. Rev. Lett. **44**, 1100 (1980).

60) S. W. Hawking, General Relativity an Einstein Centenary Survey, S. W. Hawking and W. Israel, eds. (Cambridge University Press: (1979)).

61) R. Myers, *New Axionic Instantons in Quantum Gravity*, ITP preprint NSF–ITP–88–54–(1988).

62) K. Lee, *Wormholes and Goldstone Bosons*, Fermilab–PUB–88/27–T (1988).

63) J. E. Kim, University of Michigan preprint, UM-TH-88-09.

64) C. P. Burgess and A. Kshirsagar, *Wormholes and Duality*, McGill University (unpublished).

65) R. Rohm and E. Witten, Ann. Phys. **170**, 454 (1986).

66) C. Teitelboim, Phys. Lett. B **176**, 69 (1986).

67) J. E. Kim and K. Lee, *The Scale Problem in Wormhold Physics*, Fermilab–Pub-88/95-T, (1988). (to appear)

68) E. Baum, Phys. Lett. **B133**, 185 (1983).

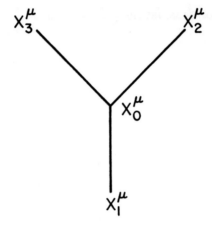

Figure 1: The bifurcation of a one dimensional universe. The values of the field X^μ agree at the meeting point of the three universes.

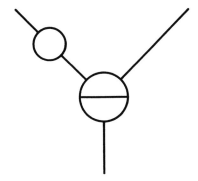

Figure 2: Iterating the basic joining-splitting interaction leads to arbitrarily complicated many-universe processes.

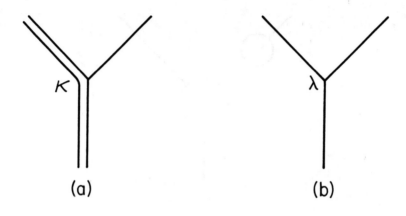

Figure 3: A double line represents a parent universe and a single line a baby universe. The two basic interactions are a) nucleation (or annihlation) of a baby by a parent universe and b) bifurcation of a baby universe.

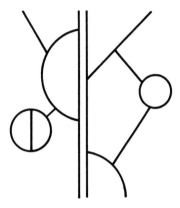

Figure 4: A typical process describing a parent universe propagating in a bath of baby universes.

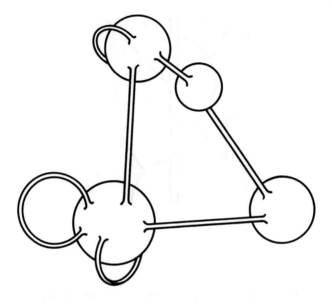

Figure 5: The large spheres represent parent universes, and the thin tubes baby universes. This is a typical contribution to the void-void amplitude. In the parent-baby universe approximation, baby universes interact only via coupling to the parent universe.

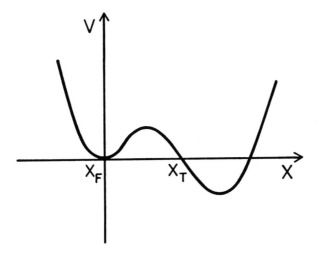

Figure 6: A quantum mechanical potential with a metastable and a true minimum.

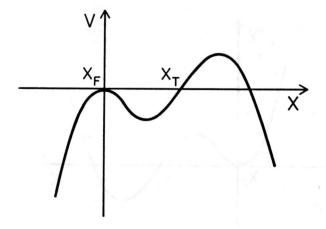

Figure 7: The Euclidean bounce solution is found by solving the equation of motion for a particle in the 'upside down' potential illustrated.

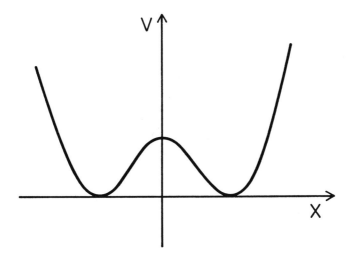

Figure 8: The degenerate double well potential.

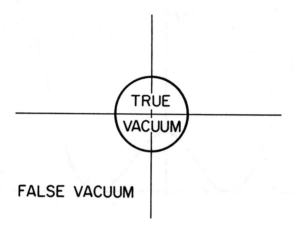

Figure 9: A bubble of true vacuum nucleating in the false vacuum.

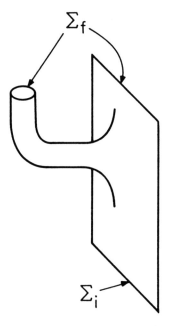

Figure 10: An instanton desribing tunneling from a topologically R^3 initial geometry Σ_i to a topologically $R^3 + S^3$ final geometry Σ_f.

Phase Transitions as the Origin of Large Scale Structure in the Universe

Neil Turok[1]

NASA/Fermilab Astrophysics Center
Fermi National Accelerator Laboratory, Batavia, Illinois 60510

Abstract

A review of the formation of large scale structure through gravitational growth of primordial perturbations is given. This is followed by a discussion of how symmetry breaking phase transitions in the early universe might have produced the required perturbations, in particular through the formation and evolution of a network of cosmic strings. Finally the statistical mechanics of string networks, for both cosmic and fundamental strings is discussed, leading to some more speculative ideas on the possible role of fundamental strings (superstrings or heterotic strings) in the very early universe.

August, 1988

Lectures presented at the Theoretical Advanced Studies Institute, Brown University, June 1988.

[1] Address after September 1st: Joseph Henry Laboratories, Princeton University, Princeton NJ08544

1 Very Large Scale Structure and Gravity

The standard big bang model of the early universe is both simple and successful. It assumes that at early times the universe was a homogeneous and isotropic thermal bath of all the particle species see today. From this and Einstein's equations it correctly predicts the Hubble expansion, microwave background radiation and the abundances of the light elements (for a review, see [1]).

Nevertheless it is obviously only a first approximation - the universe is very clearly not homogeneous and isotropic today and so there must have been some inhomogeneity at early times. In fact, as we shall see in this section, a number of recent observations point to the existence of structure in the universe on scales much larger than the distances matter has moved (under gravity) since the big bang. This is exciting - the new very large scale observations are giving us a more or less direct 'window' on the primordial fluctuations.

Where could the large scale structure in the universe have come from?

A beautiful idea has emerged from fundamental particle physics in the last few years. It is based on the notion of spontaneous symmetry breaking, which has emerged as fundamental in unified theories of particle interactions. What could be simpler than that the mechanism which broke the symmetry of particle interactions also broke the spatial symmetry of the universe?

This happens through the dynamics of the symmetry breaking process, as I shall discuss in section 2. At very high temperatures the symmetry is unbroken but as the universe cools and expands a phase transition occurs, forming defects as the universe goes out of thermal equlibrium which survive to much later epochs and seed the formation of structure.

One particular class of defects shows particularly interesting behavior in an expanding universe. Cosmic strings[2] are linelike defects formed in symmetry breaking phase transitions, and have the property that they evolve into a 'scaling solution', where the strings remain a fixed fraction of the total density throughout the radiation dominated era. This results in a highly predictive theory of galaxy formation, an alternative to the more popular theory based on quantum fluctuations produced during a period of inflation. I will concentrate on cosmic strings here but will also compare and contrast the two scenarios in section 2.

Of course we still do not know what the right fundamental theory is. Some grand unified theories predict strings and others do not. Over the last few years there has been increased hope that superstring theories may offer an 'ultimate' unified theory, including gravity. Some 'superstring-inspired' models also predict cosmic strings, but so far these are so contrived that it is hard to take them seriously.

In the final section I will describe an interesting connection between cosmic strings and fundamental strings. The latter may be used as an approximate description of the former, and for fundamental strings the statistical mechanics is relatively straightforward. I hope to give a clear and simple account of this in section 3, and discuss what insight this gives us into both cosmic and fundamental strings in the early universe.

1.1 Observations and Motivations

Cosmology is one of the most rapidly growing areas in physics for two very good reasons.

First the observational situation is qualitatively better than just a few years ago and improving all the time. Advances in telescope technology have made large scale complete surveys of the three dimensional galaxy distribution feasible. There are also now very stringent and improving limits on the microwave background anisotropy and gravity wave background, which put strong constraints on the magnitude of the primordial inhomogeneities - typically the magnitude of the density fluctuation must have been less than a part in ten thousand.

What makes the large scale observations interesting is that the distribution of galaxies is very clearly not random on very large scales. Early three dimensional surveys revealed giant 'filaments' (roughly linear overdense regions in the galaxy distribution) about $200h_{50}^{-1}$ Megaparsecs long and $10h_{50}^{-1}$ Mpc across [3], large 'voids' $120h_{50}^{-1}$ Mpc in diameter [4]. A complete 'slice of the universe' out to $300h_{50}^{-1}$ Mpc shows the galaxy distribution to be 'frothy' i.e. many galaxies lying on the surfaces of almost empty 'bubbles' $40-60h_{50}^{-1}$ Mpc across [5], while deep 'needle' surveys out to further than $2000h_{50}^{-1}$ Mpc have shown a shell-like distribution of galaxies on a scale of as much as $200h_{50}^{-1}$ Mpc [11]. The rich clusters of galaxies, called Abell clusters [12], are (roughly) defined observationally to be regions $3h_{50}^{-1}$ Mpc in radius containing more than 50 bright galaxies. These seem to be significantly clustered on scales of at least $200h_{50}^{-1}$ Mpc [13]. For comparison the present Hubble radius H_0^{-1} is $6000h_{50}^{-1}$ Mpc, the mean galaxy separation (i.e. the inverse cube root of the number density) is $10h_{50}^{-1}$ Mpc [14] and the mean Abell cluster separation is $110h_{50}^{-1}$ Mpc [13]. h_{50} is of course Hubbles constant H_0 in units of $50 kms^{-1}$, observationally $1 < h_{50} < 2$.

Statistics to describe these structures are still being developed - the simplest is the 2-point correlation function $\xi(r)$, defined to be the excess probability over random of finding a galaxy at a distance r from another galaxy [7]. There are different ideas relating the primordial fluctuations to the final $\xi(r)$ we see today.

We now believe that most of the mass in the universe is dark. There is no strong reason why the shining matter distribution should exactly trace the dark matter distribution. What processes could end up making the matter in one part of the universe shine and leave another part dark? One popular idea is that of *threshold biasing* - that shining matter might only be formed at the *peaks* of the underlying matter distribution [21,22]. With Gaussian noise, the peaks are more correlated than the underlying distribution, and this is used crucially to get the standard inflation plus CDM model (which I shall call the ICDM model) to agree with the galaxy-galaxy correlation function and pair velocity distribution. However this idea, at least as applied in the ICDM model, does not appear to work particularly well as far as fitting large scale structure[79].

Another idea, and one that is at least partly realised in the cosmic string scenario, is that the distribution of galaxies and clusters may reflect a perturbation distribution with a *scale invariant* correlation function i.e. objects of mean separation d have a correlation function $\xi(r/d)$ with ξ the same function for all objects. This is consistent with the observational fact that clusters poorer than Abell clusters [16], the richer Abell clusters [13] and even superclusters[15] have clustering apparently consistent with a single scale invariant correlation function. In fact ξ is approximately equal to $0.2(r/d)^{-2}$. Written the same way galaxies appear slightly *more* correlated: $\xi_{gg} \approx (r/d)^{-2}$[18], consistent with the view that gravity has enhanced the correlations on small scales.

Statistics on these observations are still not good and the ideas for explaining them clearly preliminary, but the observations are improving all the time and becoming increasingly constraining. Clearly data like this are going to play a crucial role in deciding which theories of the origin of large scale structure are viable.

The fact that the distribution of galaxies does not necessarily reflect the distribution of the mass (and in particular the dark matter) in a simple way represents a profound problem for matching cosmological theories to observation. Which is why recent determinations of the actual peculiar velocity field (the velocities of galaxies relative to the overall Hubble flow) are of such great importance. These seem to indicate a very large scale flow field centred on a region ('the great attractor') $80h_{50}^{-1}$ Mpc from us with magnitude about $500 kms^{-1}$ at our galaxy [17]. I will discuss below how this velocity field is related, in the linear regime, to the mass perturbation. If these observations are correct, they practically rule out the ICDM model.

So what if there is large scale structure in the universe? The most striking thing about it is that matter could not have moved these distances under gravity since the big bang. In the linear regime (and on these scales the structure is not yet nonlinear) there is a precise relation between the distance a particle has moved relative to the Hubble flow since the big bang δr and its present peculiar velocity δv: $\delta r = H_0^{-1} \delta v$ (see below). Now galaxies are very rarely observed to have velocities of greater than $600 km s^{-1} = 2.10^{-3} c$ relative to the observed structures. This means they can have moved no more than $12 h_{50}^{-1} \mathrm{Mpc}$ since the big bang. So when we see much larger structures we are seeing the primordial density perturbation field itself! This is an encouraging thought.

The second good reason for a growing interest in cosmology is that only in the last few years have any well-defined ideas as to the *origin* of large scale structure come forward. In particular cosmic strings and quantum fluctuations in inflationary models are well defined theories with clear predictions. For example, the quantum fluctuation scenario with hot dark matter is close to being ruled out by its excessively large predicted anisotropy in the microwave background [19].

1.2 Dark Matter

The greatest single obstacle to the interpretation of the observations of large scale structure is the by now firmly established fact that most of the matter in the universe is dark (For a nice review, see [9]). The shining matter - gas and stars - is only the 'tip of the iceberg'. The best evidence for dark matter comes from observations of the rotation curves of spiral galaxies[20]. To keep a particle in orbit around a galaxy requires an acceleration of v^2/r where v and r are the orbital velocity and radius. Assuming that this is provided by the gravitational pull of the mass $M_{<r}$ inside the orbit one finds

$M_{<r} = \frac{v^2 r}{G}$. Observations indicate that v^2 remains constant with increasing r out to distances several times the optical radius of a galaxy. Thus most of the mass is dark. Estimates of the mass of apparently virialised clusters of galaxies confirm this - and extrapolating the observed mass to light ratios in clusters to the whole universe leads to an estimate of Ω typically between 0.1 and 0.3. This may well be an underestimate, particularly if galaxies form preferentially in denser regions like clusters, making the mass to light ratio larger outside clusters.

Nucleosynthesis calculations[8] in fact indicate that even the *baryons* are mostly dark! They yield the bounds $.04 < \Omega_{baryons} h_{50}^2 < 0.14$ whereas the observed $\Omega_{luminous}$ i.e. the density in stars and gas is less than 0.02. So we can't even see the baryons!

However things are worse than this, according to most theorists. There is a strong prejudice that $\Omega = 1$. This is because in a matter or radiation dominated universe $\Omega = 1$ is an unstable fixed point of Einstein's equation. If Ω deviates from 1 then the deviation grows rapidly in time. Given that Ω is today fairly close to unity it would seem crazy that we should apparently be here precisely around the time when it started to deviate from unity. Inflation 'solves' this problem (called the flatness problem) - during a period of inflation $\Omega = 1$ becomes an *attractor* in Einstein's equation so that in a typical region of the universe which underwent inflation Ω would still be very close to 1.

1.3 Newtonian Approximation

For many purposes the Newtonian approximation to Einstein's equation is entirely adequate. In particular it is a good approximation when dealing with scales (a) far inside the horizon so velocities are small and (b) where gravity is fairly weak so the dimensionless gravitational potential is much less than 1. It is also a lot simpler than full relativistic perturbation theory and brings out the main points about the growth of density perturbations under gravity.

So let us begin with Newton's equations. A particle at 'physical' coordinate $\vec{r}(t)$ obeys

$$\ddot{\vec{r}} = -\vec{\nabla}_r \Phi \tag{1}$$

where the Newtonian potential Φ is given by

$$\vec{\nabla}_r^2 \Phi = 4\pi G \rho \tag{2}$$

with ρ the density. Before discussing perturbations, let us describe the homogeneous matter dominated background solution to (1) and (2). We set $\rho = \rho_b$, independent of $\vec{r}$, and find in spherical coordinates about some point chosen as the origin (where $r = \Phi = 0$),

$$\begin{aligned} \Phi &= \frac{2\pi G \rho_b r^2}{3} \\ \ddot{\vec{r}} &= -\frac{4\pi G \rho_b \vec{r}}{3} \\ \rho_b r^3 &= const \end{aligned} \tag{3}$$

the last equation being matter conservation. Integrating we find

$$\frac{\dot{r}^2}{2} - \frac{4\pi G \rho_b r^2}{3} = const \tag{4}$$

and choosing the constant to be zero (i.e. Ω to be 1) we find

$$r \propto t^{\frac{2}{3}}$$
$$\rho_b = \frac{1}{6\pi G t^2} \qquad (5)$$

the standard Friedmann matter dominated universe. This solution is true for every particle, and we write it as $\vec{r} = a(t)\vec{x}$ where $a(t)$ is the scale factor. We can choose $a(t)$ equal to 1 initially in which case $\vec{x}$ is the initial coordinate of the particle (the 'comoving coordinate'). The Newtonian approximation gives the correct result as a consequence of homogeneity and Birkhoff's theorem[6] - it is only the contribution of the matter *inside* a spherical shell which is important, and this shell may be chosen small enough so that fields are weak and velocities small so Newtons equations apply. By homogeneity the same rate of expansion must apply on all scales.

1.4 Linear Perturbations in Cold Dark Matter

For cold dark matter we also have mass conservation in a comoving volume

$$\int \rho dV_{comoving} = const \qquad (6)$$

Now imagine we give each particle a small perturbation at some early time and follow it thereafter

$$\vec{r} = a(t)(\vec{x} + \vec{\psi}(\vec{x},t)) \qquad (7)$$

$\vec{\psi}$ is called the comoving displacement. Mass conservation tells us the resulting density perturbation: if the initial unperturbed density is $\rho_i = \rho_b a^3$ then

$$\rho_i d^3\vec{x} = \rho_b a^3 d^3\vec{x}$$

$$= \rho(\vec{r})d^3\vec{r} \quad (8)$$

so the total density

$$\rho(\vec{r}) = \frac{a^3 \rho_b}{Det(\frac{\partial \vec{r}}{\partial \vec{x}})}$$
$$\approx \rho_b(1 - \vec{\nabla}_x.\vec{\psi}) \quad (9)$$

where we expand the determinant to first order in the perturbation using $Det(1+A) = 1 + Tr(A) + o(A^2)$. Thus the fractional density perturbation is given by

$$\frac{\delta\rho}{\rho}(\vec{r}) \equiv \frac{\rho(\vec{r}) - \rho_b}{\rho_b}$$
$$= -\vec{\nabla}_x.\vec{\psi} \quad (10)$$

At this point it is convenient to Fourier transform $\vec{\psi}(\vec{x})$ and decompose it into two pieces;

$$\vec{\psi}(\vec{k}) \equiv \int d^3\vec{x} e^{i\vec{k}.\vec{x}} \vec{\psi}(\vec{x})$$
$$\equiv \psi_{\parallel}(\vec{k})\frac{\vec{k}}{k} + \vec{\psi}_{\perp}(\vec{k}) \quad (11)$$

$\psi_{\parallel}$ (parallel to $\vec{k}$) is the *compressional* piece and $\psi_{\perp}$ (perpendicular to $\vec{k}$) is the *rotational* piece. Now from (2) and (9) only the *compressional* piece causes a perturbation in the potential $\delta\Phi$. Fourier transforming and solving (2) we find

$$\vec{\nabla}_r \delta\Phi = -4\pi G \rho_b a \vec{\psi}_{\parallel} \quad (12)$$

where to lowest order we use $\vec{\nabla}_r = a^{-1}\vec{\nabla}_x$. Now (1) decomposes into

$$\ddot{\vec{\psi}}_{\parallel} + 2\frac{\dot{a}}{a}\dot{\vec{\psi}}_{\parallel} = 4\pi G \rho_b \vec{\psi}_{\parallel}$$
$$\ddot{\vec{\psi}}_{\perp} + 2\frac{\dot{a}}{a}\dot{\vec{\psi}}_{\perp} = 0 \quad (13)$$

These equations have a simple interpretation. The left hand side describes a test particle moving in comoving coordinates - setting it to zero results in $\dot{\vec{\psi}} \propto a^{-2}$ so the peculiar velocity (the physical velocity $\dot{\vec{r}}$ minus the Hubble flow $\frac{\dot{a}}{a}\vec{r}$) $\vec{v}_p = a\dot{\vec{\psi}} \propto a^{-1}$. In comoving coordinates a test particle slows down. This is all that happens to the rotational piece. The right hand side of (13) describes the feedback of the compressional perturbation on itself - the gravitational instability first discovered by Jeans. If the background were not expanding, $\vec{\psi}$ would grow exponentially but in a matter dominated universe with $a \propto t^{\frac{2}{3}}$ we find

$$\ddot{\vec{\psi}}_\| + \frac{4}{3t}\dot{\vec{\psi}}_\| = 4\pi G \rho_b \vec{\psi}_\| = \frac{2}{3t^2}\vec{\psi}_\|$$
$$\ddot{\vec{\psi}}_\perp + \frac{4}{3t}\dot{\vec{\psi}}_\perp = 0$$
$$\to \vec{\psi}_\| = \vec{A}(\vec{x})t^{\frac{2}{3}} + \vec{B}(\vec{x})t^{-1}$$
$$\vec{\psi}_\perp = \vec{C}(\vec{x})t^{\frac{-1}{3}} + \vec{D}(\vec{x}) \qquad (14)$$

with $\vec{A}, \vec{B}, \vec{C},$ and $\vec{D}$ arbitrary functions of x. From (10) $\delta\rho/\rho$ behaves just as $\vec{\psi}_\|$, so in the growing mode it grows like $a(t)$.

More generally the right hand side of (13) has an extra term $-a^{-1}\vec{\nabla}_r \Phi_s(\vec{r})$ if there is a source driving the perturbation producing a potential Φ_s. For example for a point mass it reads $-Gm\vec{x}/(a^3 x^3)$. Likewise in the equation for $\delta\rho/\rho$ the source term is $4\pi G \rho_s(\vec{x})$, where $\rho_s(\vec{x})$ is the density of the source. Notice that with cold dark matter the evolution is *purely local* in comoving coordinates.

In general if one has a theory for the origin of the perturbations one wants to evolve the perturbations through the radiation era into the matter era. Now the density ρ_b on the right hand side of (13) is only the *matter* density, at early times much less than the total density. Setting it to zero and using $a \propto t^{\frac{1}{2}}$, the radiation dominated result, one finds the solution $\psi = E + F \ln(t)$ with E and F constants i.e. slow logarithmic growth. In fact solving (13) through the transition one finds at late times ($a \gg a_{eq}$, the scale factor at equal matter and radiation density)

$$\vec{\psi} = \vec{A}(\vec{x})(1 + \frac{3}{2}\frac{a}{a_{eq}}) \quad (15)$$

where $\vec{A}(\vec{x})$ is the perturbation at early times [7]. Thus there is very little growth in the radiation era and one can think of the perturbation growth as beginning at a_{eq}.

1.5 Local Nonlinearity 'doesn't matter'

Whilst these equations are correct only in the linear regime they can be used perfectly well on a surface surrounding a nonlinear mass concentration as long as the density perturbation is always small *on the surface*. If one has a comoving surface S surrounding a volume V then the change in the volume enclosed by S is

$$\delta V = a^3 \int_S \vec{\psi}.\vec{dS} \quad (16)$$

and the change in the mean density ρ inside S defined by $\rho_{av}(V + \delta V) \equiv \rho_i V_i = \rho_b V$ where $V = V_i a^3$. Now we find

$$\begin{aligned}\frac{\delta \rho}{\rho} &\equiv \frac{(\rho_{av} - \rho_b)}{\rho_b} \\ &= -\frac{\delta V}{V} \\ &= -\frac{\int_S \vec{\psi}.\vec{dS}}{V_i}\end{aligned} \quad (17)$$

which of course grows just as (14). Thus linear theory has a wide range of applicability.

1.6 Peculiar velocities

Linear theory also gives information about peculiar velocities in a very direct way - we see from $\vec{v}_p = a\dot{\vec{\psi}}$ and (14) that peculiar velocities grow as $t^{\frac{1}{3}}$ in the compressional mode. We also see using $\vec{\psi} \propto a$ in the growing mode that $\vec{v}_p = \dot{a}\vec{\psi} = H\delta\vec{r}$, the relation used in the introduction.

In particular geometries v_p is related to $\delta\rho/\rho$ - using (17) we find that defining the Hubble flow $v_H = Hr$ we get

$$\begin{aligned}\frac{v_p}{v_H} &= \frac{\delta\rho}{\rho} \text{ for planar collapse} \\ &= \frac{1}{2}\frac{\delta\rho}{\rho} \text{ for cylindrical collapse} \\ &= \frac{1}{3}\frac{\delta\rho}{\rho} \text{ for spherical collapse} \end{aligned} \quad (18)$$

Observations of peculiar velocities such as those recently reported[17] thus give a unique window on $\delta\rho/\rho$. If they are correct, we are moving in part of a very roughly spherical collapse pattern centred on the 'great attractor', 80 h_{50}^{-1} Mpc away from us. Our velocity towards it is approximately $500 km s^{-1}$. This corresponds to a value of $\delta\rho/\rho$ of approximately 0.4 on this scale.

1.7 Sources

As discussed above gravitational sources (like loops of cosmic string) add an additional driving term to the right hand side of (13). It may then be solved fairly simply with a Green's function[65],[70]. Since the equation is completely local $\delta\rho/\rho\,(x)$ simply evolves in proportion to $\delta\rho_S\,(x)$, the source density. In linear theory therefore a point mass always remains a point mass its mass simply grows as $a(t)$. In the linear regime (13) with a source which 'turns on' in the matter era at t_i is easily solved with a Green's function

$$\delta\rho/\rho(x,t) = \int_{t_i}^{t} dt' G(t,t') 4\pi G \rho_s(\vec{x},t')$$
$$G(t,t') = \frac{3}{5}(t^{\frac{2}{3}} t'^{\frac{1}{3}} - t^{-1} t'^2) \tag{19}$$

Extended sources like large cosmic string loops trace out an extended 'trumpet - shaped' structure in comoving coordinates leading to a density contrast of the same pattern[70], quite similar to the giant 'filament' mentioned above[3]!

1.8 Hot Dark Matter

If the dark matter particles have appreciable thermal velocities at t_{eq}, when perturbations start to grow, then the above approach must be modified - it is hard to follow the particles! Instead one must follow the particle trajectories in phase space and I will discuss this in the next section.

However the main effect is simple to understand. Roughly speaking, particles move a distance $d \approx vt$ in an expansion time. where v is their velocity. Structure on smaller scales than this is washed out by particle diffusion ('free streaming') while structure on larger scales evolves unaffected, just as with cold dark matter. This length scale corresponds to a mass scale, often called the 'neutrino Jeans mass'. Since peculiar velocities are damped as we discussed above, the thermal velocity v of the particles decays as a^{-1}. The comoving scale $d_c = d/a \propto t^{-\frac{1}{3}} \propto a^{-\frac{1}{2}}$ corresponds to a mass scale

$M_J \propto a^{-\frac{3}{2}}$. With neutrinos for example, of mass 25 h_{50}^2 eV such as would be required to make $\Omega = 1$ while respecting the nucleosynthesis bound, M_J is of order 10^{15} $M_\odot$ at matter-radiation equality and about 10^9 $M_\odot$ today.

1.9 Linearising Liouville

Clustering of hot dark matter can be treated quite precisely by linearising the equation for the phase space density, the Liouville equation. As before I will remain strictly Newtonian.

The phase space density f is defined so that the number of particles in the phase space volume $d^3\vec{r}d^3\vec{p}$ at time t is $f(\vec{r},\vec{p},t)d^3\vec{r}d^3\vec{p}$. Thus the ordinary density is given by $\rho = m \int d^3\vec{p}\, f$, where m is the mass of the particles. Conservation of particle number yields the continuity equation

$$\frac{\partial f}{\partial t} + \vec{\nabla}.(f\vec{v}) = 0 \qquad (20)$$

where the divergence and velocity are in the full phase space: $\vec{\nabla} = (\vec{\nabla}_r, \vec{\nabla}_p)$, $\vec{v} = (\dot{\vec{r}}, \dot{\vec{p}})$. Recalling Liouvilles theorem that $\vec{\nabla}.\vec{v} = \vec{\nabla}_r.\dot{\vec{r}} + \vec{\nabla}_p.\dot{\vec{p}} = \vec{\nabla}_r.\vec{\nabla}_p H - \vec{\nabla}_p.\vec{\nabla}_r H = 0$ where the total Hamiltonian is H, we get

$$\frac{\partial f}{\partial t} + \vec{v}.\vec{\nabla} f = 0 \qquad (21)$$

which is the statement that the function f is constant along particle trajectories. We will return to this in the next section. Next we use the fact that $(\dot{\vec{r}}, \dot{\vec{p}}) = (\vec{p}/m, -m\vec{\nabla}_r\Phi)$ with Φ the Newtonian potential. It is convenient to change coordinates to $\vec{x} = \vec{r}/a, \vec{q} = a\vec{p} - ma\dot{\vec{r}}$, comoving position and peculiar momentum, and $d\tau = dt/a$, conformal time. This yields

$$\frac{1}{a}\frac{\partial f}{\partial \tau} + \frac{\vec{q}}{ma^2}.\vec{\nabla}_x f - m\ddot{a}\vec{r}.\vec{\nabla}_q f - m\vec{\nabla}_x\Phi.\vec{\nabla}_q f = 0 \qquad (22)$$

(Φ is independent of $\vec{q}$). Here $\ddot{a}\vec{r} = \frac{d^2a}{dt^2}\vec{r} = -\vec{\nabla}_x\Phi_0$ from Newton's equations for the background as discussed above, where Φ_0 is the zeroth order part of the potential. The homogeneous, zeroth order solution to this is

$$f_0 = \frac{2}{e^{\frac{q}{T_\nu a}}+1}$$

$$\to \rho_0 = \frac{m}{a^3}\int d^3\vec{q}\,\frac{2}{e^{\frac{q}{T_\nu a}}+1} = \frac{2mT_\nu^3}{\pi^2}\eta_3 \qquad (23)$$

where $\eta_3 \approx 1.803$ and $T_\nu \propto a^{-1}$ is the neutrino temperature, equal to $(4/11)^{\frac{1}{3}}$ the temperature of the microwave background radiation. We use the relativistic phase space density of one species of neutrino (and antineutrino) when they decoupled as initial conditions, and this is preserved as the universe expands even as the neutrinos go out of thermal equilibrium and go nonrelativistic, as we learn from (22). As an excercise the reader can calculate the required mass of the neutrino needed to make $\Omega = 1$, stated earlier.

It is fairly straightforward to linearise and solve (23) through the matter-radiation transition including sources just as for cold dark matter[24,25]. The interested reader may refer to [26] for a pedagogical account.

1.10 Phase Space Constraints for Hot Dark Matter

The fact that for neutrino (and more generally fermionic) dark matter the initial phase space density (23) has a maximum, namely $\frac{1}{2}$ per neutrino or antineutrino, has a nice consequence first pointed out by Tremaine and Gunn [27]. Because it is constant along particle trajectories, it cannot increase beyond this maximum. This puts a strong constraint, valid even in the nonlinear regime, on the clustering of neutrinos, relevant to the issue of whether neutrinos can form the dark matter in galactic halos.

Crudely, the phase space density corresponding to a virialised lump of N neutrinos with volume r^3 and momentum p_ν is just $Nr^{-3}p_\nu^{-3}$ and by the constraint, this cannot be larger one. However for a virialised lump we have $v^2 \approx GM/r \approx GNm/r$. Solving for N and substituting back into the previous formula we obtain $m_\nu^4 > 1/(vr^2 G)$ (I use units where $\hbar$ is 1). A more detailed treatment[27] yields the bound

$$m_\nu > 30eV (\frac{100\ km\ s^{-1}}{\sigma})^{\frac{1}{4}} (\frac{15\ kpc}{r_c})^{\frac{1}{2}} \qquad (24)$$

assuming the collapsed object is an isothermal sphere with core radius r_c and one dimensional velocity dispersion σ. Observations of dwarf galaxies with dark matter halos can therefore put strong constraints on the neutrino scenario.

1.11 Summary

In this section I have given a flavour of the new observations and what is encouraging about them. I have also given a brief introduction to the kind of calculation necessary to follow the gravitational clustering of matter around primordial perturbations. In the next section I will give an account of one theory as to where these perturbations came from, the theory of cosmic strings.

2 Phase Transitions, Defects and Fluctuations

In this section I will discuss the physics of symmetry breaking phase transitions predicted to occur in grand unified theories in the very early universe[28,29,2]. The discussion will necessarily be qualitative and crude - it is difficult to do precise calculations in this area! Nevertheless according to our present view the precise details are not crucial in most cases. This is because the phase transitions concerned happened very early in the life of the universe, typically at times of 10^{-34} seconds. Transitions producing domain walls or monopoles would have been disastrous fairly quickly - the defects quickly coming to dominate the universe. Those producing string however, which I shall concentrate on as the most interesting case, lead to a 'scaling solution' for the string network, in which the strings remain a fixed fraction of the total energy density in the universe. As I shall discuss, we have good reason to believe that the late time evolution of the scaling solution is rather insensitive to the precise details of the initial conditions.

For the moment I will ignore the possibility that the phase transitions were inflationary - indeed it is very difficult to arrange for inflation in a gauge symmetry-breaking transition - but will return to the question of how cosmic strings relate to inflation later on.

2.1 Physical picture of a phase transitions

According to our current and successful understanding of high energy physics, the underlying field theory describing nature has a high degree of symmetry but most of this symmetry is broken at low energy.

As the simplest example of how this works, consider the theory of a scalar field ϕ invariant under $\phi \rightarrow -\phi$. If we give the field a potential $V(\phi)$ as in Figure 2.1 then the potential shares this symmetry, so the classical ground state of the theory is degenerate - in vacuo ϕ can sit at $+$ or at $-$. The chosen configuration breaks the symmetry[1]

At high temperature however the symmetry is 'restored' simply because thermal excitations are energetic enough to 'kick' the field ϕ over the potential barrier. Any masses become irrelevant at high temperatures and so by dimensions one must have $<\phi>^2 \approx T^2$. Thus ϕ explores a wide range of the potential. As the system cools the field explores a smaller and smaller range until at low temperatures it flops down into one of the two minima. In some parts of space it chooses the $+$ minimum and in other parts the $-$ minimum. Because of this defects are inevitably formed - in going from a $+$ region to a $-$ region ϕ must cross over the potential barrier in $V(\phi)$. Where it does so there is a localised lump of potential and gradient energy - a *domain wall* separating the $+$ and $-$ regions. It is important to realise that these defects are, at low temperatures, a *nonequilibrium* phenomenon - they are Boltzmann suppressed. To understand them one needs to understand how the fields go out of equilibrium as the universe expands.

According to the most naive picture, at high T the system is characterised by thermal oscillations with $k \approx T$. Shorter wavelength modes are

---

[1]This idea is *classical* -quantum mechanically of course tunnelling can occur between the $+$ and $-$ states. The true ground state is a symmetric combination of $+$ and $-$. In field theory however, in macroscopic volumes the tunnelling rate is very tiny so the classical picture is reasonable - there is simply a *classical* probability of being in $+$ or $-$.

suppressed. Thus the crudest picture of the ϕ field is that it is smooth on a scale of the thermal wavelength. On might therefore expect that the size of the $+$ and $-$ domains would, as long as one stayed in equilibrium, be just this. However what is really relevant is not the shortest wavelength modes - these oscillate vigorously and simply average out to zero as the system cools. Rather it is the *long* wavelength modes, with their more sluggish oscillations, which determine how the system goes out of equilibrium.

2.2 Gaussian Approximation

Let us try to understand this in a little more detail. To begin with assume that Φ has a long wavelength component $\bar{\Phi}$ which we shall initially treat as constant. We write $\Phi = \bar{\Phi} + \delta\Phi$ where $\delta\Phi$ is the short wavelength component. I will be more precise about the division into short and long wavelengths later. We then calculate the partition function for Φ by first summing over the fluctuations $\delta\Phi$ about $\bar{\Phi}$ in the Gaussian approximation (i.e. keeping only quadratic terms in $\delta\Phi$ in the energy).

The small fluctuations are just massive fields whose mass is determined by the value of $\bar{\Phi}$. For $V(\Phi) = \frac{\lambda}{8}(\Phi^2 - \eta^2)^2$ (shown in Figure 2.1) the mass squared of $\delta\Phi$ is $m^2 = V''(\bar{\Phi}) = \frac{\lambda}{2}(3\bar{\Phi}^2 - \eta^2)$. In general if Φ is a Higgs field it gives a mass to other fields as well - in particular fermion and gauge fields. Their contribution may be included in exactly the same way.

Now recall that a massive boson field has a partition function

$$Z_b = Tr(e^{-\beta H})$$
$$= \sum_{\{n_{\vec{k}}\}} e^{-\beta \sum_{\vec{k}} E_{\vec{k}}}$$

$$= \prod_{\vec{k}} \sum_{n_{\vec{k}}=0}^{\infty} e^{-\beta n_{\vec{k}} \omega_k}$$

$$= \prod_{\vec{k}} \frac{1}{(1 - e^{-\beta \omega_k})} \qquad (25)$$

where $\vec{k}$ labels each momentum mode and $n_{\vec{k}}$ are the occupation numbers for each mode. ω_k is just $\sqrt{\vec{k}^2 + m^2}$, and $\beta = 1/T$. For fermions one has instead

$$Z_f = \prod_{\vec{k}} (1 + e^{-\beta \omega_k}) \qquad (26)$$

The free energy contributed by a massive field is (recall $Z \equiv e^{-\beta F}$)

$$F_b = \frac{V}{\beta} \int d^3k \ln(1 - e^{-\beta \omega_k})$$

$$\approx V(-\frac{\pi^2 T^4}{90} + \frac{m^2 T^2}{24} + ...) \quad \frac{m}{T} \ll 1 \qquad (27)$$

at high temperature. A Dirac fermion contributes

$$F_f = 4\frac{V}{\beta} \int d^3k \ln(1 + e^{-\beta \omega_k})$$

$$\approx V(-\frac{7\pi^2 T^4}{180} + \frac{m^2 T^2}{12} + ...) \quad \frac{m}{T} \ll 1 \qquad (28)$$

similarly. Here V is the volume : we replace $\sum_{\vec{k}}$ by $V \int d^3k \equiv V \int d^3k/(2\pi)^3$. The high T expansion is obtained by scaling out the T dependence and Taylor expanding the integrand in m/T about $m/T = 0$.

It is easily seen that (27) and (28) are dominated by modes with $k \approx T$, just as we expected above. Thus for $m/T \ll 1$ we can regard (27) and (28) as the contribution to the full partition function of only summing over the high k modes, leaving the low k modes (i.e. $k \ll T$) not summed over. We then have an effective theory for the low k modes with a potential energy given by adding the contributions of (27) and (28) to the zero temperature potential for $\bar{\Phi}$

$$\begin{aligned} V_{eff}(\bar{\Phi}) &\approx V(\bar{\Phi}) + m^2(\bar{\Phi})\frac{T^2}{24} \\ &\approx \frac{\lambda}{8}\bar{\Phi}^4 - \frac{\lambda}{4}\eta^2\bar{\Phi}^2 + \frac{\lambda}{16}\bar{\Phi}^2 T^2 \\ &\approx \frac{\lambda}{8}(\bar{\Phi}^2 - \eta_{eff}^2)^2 \end{aligned} \qquad (29)$$

ignoring $\bar{\Phi}$ independent terms. Here

$$\begin{aligned} \eta_{eff}^2 &= \eta^2(1 - \frac{T^2}{T_c^2}) \\ T_c &= 2\eta \end{aligned} \qquad (30)$$

To repeat, the extra temperature dependent terms come from summing over the short wavelength fluctuations about the long wavelength 'background'. This has a very important effect - at high temperature it changes the sign of the $\bar{\Phi}^2$ term and pushes the minimum to $\bar{\Phi} = 0$, the 'unbroken' phase. The transition occurs when the $\bar{\Phi}$ field becomes effectively massless, at $T = T_c = 2\eta$, called the critical temperature. This is shown in Figure 2.2. If the Φ field couples to other fields then it also gives them a mass and their contribution may be included in V_{eff} similarly. For example Φ generally gives gauge fields a mass squared $\approx g^2\Phi^2$ giving a correction term in (29) of $\approx g^2\Phi^2 T^2$ and we find $T_c \approx \sqrt{\lambda/g^2}\eta$.

What is the cause of this 'restoration' of symmetry? In the partition function we are really finding the most likely configuration for the system plus heat bath (see Lecture 3). In this context this amounts to minimising the free energy with respect to $\bar{\Phi}$. The most important effect of $\bar{\Phi}$ being nonzero and giving the other modes a mass is to cut down the number of states available (at given T) to that mode. This is an *entropy* effect, and at high T as we have seen it tends to favour $\bar{\Phi} = 0$.

One other point must be made. In the Gaussian approximation the exact expressions in (27) and (28) do not make sense for all $\bar{\Phi}$: for $\bar{\Phi}^2 < \eta^2/3$ we have $m^2 = \frac{\lambda}{2}(3\bar{\Phi}^2 - \eta^2) < 0$. This means that ω_k and hence F are not even real! This is clearly a problem with the approximation. Nevertheless the qualitative behaviour exhibited by the small m/T expansion is quite believable.

The function $V_{eff}(\bar{\Phi})$ is often called the finite temperature effective potential. In quantum mechanics, by considerations similar to those I explained in the context of symmetry breaking, this should never have two minima - $\bar{\Phi}$ can tunnel through from one minimum to another. However we are interested in an evolving macroscopic system, and we are approximating it as a *classical* $\bar{\Phi}$ plus a quantum fluctuation. Ignoring quantum tunnelling should be reasonable and so $V_{eff}(\bar{\Phi})$ is the relevant quantity. However as far as I am aware none of this has been made rigourous and I would certainly encourage those of you feeling queasy with it to try and do better yourselves[30]!

2.3 Spatial Distribution of the Fields near T_c

The interpretation of $V_{eff}(\bar{\Phi})$ as an effective potential for the long wavelength modes means that we can use it to discuss their distribution at high T. Clearly $\bar{\Phi}$ will not be spatially constant at finite T. To understand its spatial distribution consider a massive field in equilibrium. As we approach T_c from above $\bar{\Phi}$ looks like a massive field with m decreasing to zero. For T below

T_c, in the neighbourhood of the minima $\bar{\Phi} = \pm\eta_{eff}$ of $V_{eff}(\bar{\Phi})$, $\bar{\Phi}$ looks like a massive field with mass squared $m_{eff}^2 = V_{eff}''(\bar{\Phi}) = \lambda\eta_{eff}^2 = \lambda\eta^2(1-(T/T_c)^2)$. Now in thermal equilibrium a massive field has spatial correlation function

$$<\Phi(r)\Phi(0)> = \frac{Tr(\Phi(r)\Phi(0)e^{-\beta H})}{Tr(e^{-\beta H})} \qquad (31)$$

The trace over all states is conveniently calculated in an occupation number basis just as the partition function in (25) was. With the mode expansion for Φ one finds

$$<n_{\vec{k}} \mid \Phi(r)\Phi(0) \mid n_{\vec{k}}> \;=\; \int \frac{d^3k}{2\omega_k} e^{i\vec{k}\cdot\vec{r}}(2n_{\vec{k}}+1) \qquad (32)$$

and so summing over all states

$$<\Phi(r)\Phi(0)> \;=\; \int \frac{d^3k}{2\omega_k} e^{i\vec{k}\cdot\vec{r}}(\frac{2}{(e^{\beta\omega_k}-1)}+1) \qquad (33)$$

the second piece being the usual vacuum piece. For $r >> m^{-1} >> T^{-1}$ the integral in (33) is dominated by long wavelength modes and gives

$$<\Phi(r)\Phi(0)> \;\approx\; \frac{T}{4\pi r}e^{-mr} \qquad (34)$$

This tells us that $\bar{\Phi}$ is uncorrelated on scales larger than $\xi \equiv m_{eff}^{-1}$, which is called the correlation length. If we defined $\bar{\Phi}_R$ as the average of $\bar{\Phi}$ over a region of size $R >> \xi$ then it is straightforward to check that (34) implies the same result for $<\bar{\Phi}_R^2>$ as that obtained if $\bar{\Phi}_R$ were the result of averaging over $(R/\xi)^3$ volumes within which $\bar{\Phi}$ took on a randomly chosen values $\pm\sqrt{m_{eff}T}$. This serves as a rough picture of the distribution of $\bar{\Phi}$. Near the transition point the magnitude of the fluctuation, $\sqrt{m_{eff}T}$, is much less than T but the range, m_{eff}^{-1}, is much greater than T^{-1}. The energy in a typical fluctuation $\approx$ *potential* $\times$ *volume* $\approx m_{eff}^2\Phi^2\xi^3 \approx T$, as expected.

In fact if the system remains in equilibrium the correlation length goes to infinity at T_c, when $m_{eff} = 0$. In any real system with a finite rate of cooling, this never occurs. Instead the maximum correlation length is determined by the rate of cooling. I leave it as an exercise to find the maximum value of ξ by using the equilibrium value, with $T \propto a^{-1}$ and demanding that its rate of increase $\dot{\xi}/\xi$ not exceed the Hubble expansion rate $\dot{a}/a$.

2.4 Out of Equilibrium: Defect Formation

As the universe cools below a symmetry breaking phase transition, even though V_{eff} has the symmetry breaking form thermal fluctuations are initially still large enough to erase whole domains - in the example above a whole + domain may be 'flipped' to a − domain. Domains are really 'frozen in' very soon after the temperature falls below the energy required to kick them over the potential barrier. This is because below this temperature 'flipping' becomes exponentially (Boltzmann) suppressed, and the finite expansion rate of the universe means that it rapidly becomes negligible. The temperature at which domains are still frequently 'flipped' is given by

$$T \approx \xi^3 \Delta V_{eff} \approx \xi^3 \frac{\lambda}{8} \eta_{eff}^4 \tag{35}$$

being the energy required to take a domain ξ^3 over the potential barrier separating the two minima of V_{eff}. Using the equilibrium expressions right down to this temperature we find

$$T = \frac{\eta}{8\sqrt{\lambda}}(1 - (\frac{T}{T_c})^2)^{\frac{1}{2}}$$
$$\to T^{-1} = \sqrt{\frac{64\lambda}{\eta^2} + \frac{1}{T_c^2}}$$

$$\approx T_c \qquad (36)$$

for small λ. This temperature, the temperature of domains freezing in, is called the *Ginzburg* temperature T_G. Below this temperature the domains and defects rapidly go out of thermal equilibrium[2].

At T_G we have $\xi = m_{eff}^{-1} = \eta_{eff}^{-1}/\sqrt{\lambda} \approx 1/(8T_c\lambda)$ which for λ not too small and $T_c \approx m_{Gut} \ll m_{Planck}$ is smaller than the horizon scale at this time. For very small λ the assumption of thermal equilibrium would be dubious. However λ must be quite small for the small m/T expansion to be reasonable.

I have given a very simplistic account of what is a very complex subject. For one recent attempt to do better see [31].

2.5 Defects and Topology

In the previous section we discussed the simplest example of a symmetry, the discrete Z_2 symmetry $\Phi \to -\Phi$. In unified theories we are interested in more complicated symmetries including continuous symmetries. Nevertheless defect formation can be understood in a simple and general way.

Say we have a theory invariant under a symmetry group G which is spontaneously broken by a Higgs field Φ to a subgroup H, defined as $H = \{g \in G : g\Phi = \Phi\}$. Now the whole theory and in particular the Higgs potential $V(\Phi)$ is invariant under $\Phi \to g\Phi$. Thus if we have one point η on the minimum V_{min} of V we can obtain a whole set of values of Φ degenerate in energy by applying elements of G to η. In fact if there is no 'accidental' degeneracy in the theory the resulting set will be the entire space V_{min}. V_{min} is not however equal to G because if we apply two different grroup elements g and g' we may obtain the same point on V_{min}. This happens if $g\Phi = g'\Phi$ which is true iff $g^{-1}g' \in H$ so that g and g' are related by right multiplication in H. V_{min} is therefore obtained by *identifying* elements in G related by right multiplication in H - it is a *coset space*

$$V_{min} = \frac{G}{H} \qquad (37)$$

Thus the topology of V_{min} is simply a group theoretic entity, not depending on details of the Higgs potential or its coupling constants. In our simple example, $G = Z_2$ was broken to $H = 1$ by Φ and so from (37) $V_{min} = Z_2$, the two points $+$ and $-$.

As another simple example, consider a theory with a continuous $G = U(1)$ broken by a complex scalar field Φ transforming under $g = e^{i\theta}$ as $\Phi \to e^{iQ\theta}\Phi$. Here $H = 1$ so from (37) $V_{min} = U(1)$ which is just a circle. The corresponding quartic potential, $V = \lambda(|\Phi|^2 - \eta^2)^2$ is the 'Mexican hat' potential. As a more complicated example if $G = SO(3)$ is broken by a 3 - component vector $\hat{\Phi}$ to a subgroup $H = SO(2)$ (rotations about the axis defined by $\hat{\Phi}$) then we find $V_{min} = SO(3)/SO(2) = S_2$, a two sphere.

The topology of V_{min} has very important physical consequences. In the 'domain' picture explained above, $\bar{\Phi}$ falls onto different points of V_{min} in different domains. Moving through space, Φ will wind around V_{min} and this will in general result in defects just as in Section 2.1. In the simplest case, V_{min} has two points $+$ and $-$. In going along any path in space from a domain where $\Phi = +$ to one where $\Phi = -$, Φ has to climb over the potential barrier in between. This means that somewhere there must be a localised wall of energy density separating the two domains, a 'domain wall'. In the second example above, as we traverse a loop in space Φ will in general wind around the circle V_{min}. This means that somewhere inside the loop Φ must vanish. In fact Φ must vanish somewhere on every surface bounded by the loop. Where it vanishes, there is a nonzero energy density. The lowest energy configuration compatible with Φ winding around V_{min} on the loop is a *line* of energy density - a string - passing through the loop. Neither walls nor strings can have edges or ends - they must either be closed or infinite. In the third example above, traversing a two- sphere in space one would find in general that Φ would wind around the S^2 of V_{min}. This results in a localised *point* defect - a magnetic monopole (see Figure 2.3).

2.6 A Useful Theorem

A useful theorem in the case of strings is that the group of loops in $V_{min} = G/H$ not continuously deformable into one another (the homotopy group $\pi_1(G/H)$) is given by

$$\pi_1(\frac{G}{H}) = \pi_0(H) \tag{38}$$

in the case where G is simply connected. $\pi_0(H)$ is just the number of disconnected components of H. A pictorial 'proof' of this statement is given in Figure 2.4. This formula is still useful when G is simple but not simply connected because in this case one has $G/H = \bar{G}/\bar{H}$ where $\bar{G}$ is the simply connected covering group of G and $\bar{H}$ is the subgroup of $\bar{G}$ leaving Φ invariant. For example $SO(3)$ can be broken by *two* 3-vectors to 1. This produces strings because the simply connected covering group of $SO(3)$ is $\bar{G} = SU(2)$ and it would be broken to a subgroup $\bar{H} = Z_2$, because the 3-vectors are in the adjoint representation of SU(2), and are left invariant by the two elements of its centre. Thus we would have $\pi_1(V_{min}) = Z_2$. These strings are more complicated than the $U(1)$ strings discussed above - in this case a string running in one direction is topologically equivalent to one running in the opposite direction. Strings of this type form in many simple grand unified theories, for example in the symmetry breaking transitions

$$SO(10) \to {}^\backprime SU(5)XZ_2'$$
$$E(6) \to {}^\backprime SO(10)XZ_2' \qquad (39)$$

the first being mediated by a Higgs **126** and the second by a **351**. The quotation marks indicate that the full global structure of H is more complicated, as explained in [32].

In low energy models 'inspired' by superstrings cosmic strings of the simplest U(1) type as well as those of the more complicated Z_n type occur[33].

As an excercise I leave the reader to construct the 'proof' analogous to that in Figure 2.4 that $\pi_2(\frac{G}{H}) = \pi_1(H)$ assuming $\pi_2(G)$ and $\pi_1(G)$ are trivial. The latter are always true for simple Lie groups. Since H must ultimately contain the $U(1)$ of electromagnetism, if the grand unified theory is based on a simple group then magnetic monopoles are inevitable. This is the basis of the famous 'magnetic monopole problem' [34].

2.7 Cosmic Strings

Strings are the most interesting topological defects because their existence is compatible with the observed universe and may even explain its large scale structure. Their possible existence in relativistic field theories was first pointed out by Nielsen and Olesen [45]. They are simple generalisations of the flux lines found in superconductors. For a review, see [35].

The microscopic structure of a string is illustrated in Figure 2.5 for the simplest case, a $U(1)$ string. Outside the string, Φ winds around V_{min} as we go around the string. At the centre of the string Φ goes to zero. The 'magnetic' gauge field of the $U(1)$ group that is broken in forming the string is nonzero and points along the string in one direction in the minimum energy configuration. The energy per unit length μ of a static straight string is obtained from the energy functional

$$\begin{aligned} \mu \equiv E/L &= \int d^2x [(\vec{D}\Phi)^2 + \frac{1}{2}\vec{B}^2 + V(\Phi)] \\ &= 2\pi\eta^2 f(\frac{\lambda}{g^2}) \approx m_{Gut}^2 \\ V(\Phi) &= \frac{\lambda}{2}(|\Phi|^2 - \eta^2)^2 \\ \vec{D}\Phi &= (\vec{\nabla} - ig\vec{A})\Phi \end{aligned} \quad (40)$$

where f is a slowly varying function equal to unity when $\lambda = g^2$ [38,39]. These strings generally carry no electric charges and their strongest coupling to other matter is through their gravitational field. The gravitational coupling is thus the only parameter relevant to the (ordinary) cosmic string theory of galaxy formation. It is given by the dimensionless number $G\mu \approx (m_{Gut}/m_{Planck})^2 \approx 10^{-6}$ typically in those grand unified theories predicting strings.

In some theories the string can couple to long range fields other than gravity. *Global* strings occur where the $U(1)$ symmetry broken by Φ is not gauged. Here there is a massless Goldstone boson which couples to the string and is radiated from it. This kind of string occurs in axion models[40]. *Su-*

perconducting strings form in theories where Φ couples to another electrically charged field χ in such a way as to force χ to be nonzero on the string. These strings may carry a net electric charge or current[41]. With the assumption of a primordial magnetic field (which is hard to justify) these strings are the basis for the OTW scenario of explosive galaxy formation[42]. I will not discuss this scenario in these lectures.

2.8 Simulating String Formation

Cosmic string formation was originally understood numerically by simply throwing down values for the Higgs field on V_{min} at random on a lattice of domains of size ξ, with a prescription for smoothly varying the Higgs field phase in going from one domain to the next. This is of course a crude model and ignores any interactions. The scale ξ is the correlation length at the Ginzburg temperature as discussed in section 2.4.

In more detail, for U(1) strings V_{min} is approximated by three points assigned the values 0,1 or 2. For a cubic lattice of domains, any edge is surrounded by four domains. The rule for going from one domain to the next is that Φ takes the shortest path along V_{min}. This is illustrated in Figure 2.6. The reader can easily check that at any vertex on the lattice the same number of strings leave as enter. With periodic or reflecting boundary conditions this means that the strings have no ends - they are all in the form of closed loops.

The numerical simulations first done by Vachaspati and Vilenkin [56] have been checked by many people on a variety of lattices and are in good agreement [49,46,52]. Most (about 80%) of the string is in strings as large as the box in which the simulation is performed ('infinite' strings). The remainder is in the form of closed loops. These have a *scale invariant* distribution - the number of loops of size (a *course* measure like r.m.s. radius) greater than R per unit volume, $n_>(R) \propto R^{-3}$. The term scale invariant simply means that the result depends only on R and not on the domain size ξ. Dimensional analysis then determines the power law. Both the 'infinite' strings and the loops are in the form of Brownian random walks. Recently David Mitchell and myself have shown how these results may be obtained by counting states in the quantised closed bosonic string, an intriguing connection I shall explain in Lecture 3.

A picture of the string network at formation is shown in Figure 2.7.

2.9 Evolution of Strings

After the strings form we have to evolve them. At first sight this appears horribly complicated - they are inherently nonlinear entities. However two effects straighten out the strings very rapidly. In the early stages (until the temperature falls to $\approx (G\mu)^{\frac{1}{2}} m_{Gut}$) the strings are heavily damped by collisions with particles[35]. Secondly the expansion of the universe stretches the string out. The typical radius of curvature of the string increases rapidly while the width of the string remains fixed.

Quite quickly it becomes a very good approximation to treat the strings as infinitely thin relativistic lines or 'Nambu' strings, the action for which is simply the area of the two dimensional worldsheet they sweep out in spacetime [43]

$$S = -\mu \int dA = -\mu \int d^2\sigma \sqrt{-\gamma}$$
$$\gamma = det(\gamma_{\mu\nu}) \quad \gamma_{\mu\nu} = \partial_\mu x^a(\sigma)\partial_\nu x^b(\sigma)g_{ab}(x(\sigma)) \quad (41)$$

where $g_{ab}(x)$ is the metric of the background spacetime. $x^a(\sigma)$ are the four spacetime coordinates of the string worldsheet which is parametrised by the two worldsheet coordinates $\sigma^\mu = (\tau, \sigma)$. The string worldsheet has one spacelike coordinate σ and one timelike coordinate τ.

Let me explain how (41) arises. For a $U(1)$ theory producing strings the action is

$$S = \int d^4y \sqrt{-g}((D\Phi)^2 - V(\Phi) - \frac{1}{4}F^2)$$
$$V(\Phi) = \frac{\lambda}{2}(|\Phi|^2 - \eta^2)^2$$
$$D_\mu \Phi = (\partial_\mu + igA_\mu)\Phi$$
$$F_{\mu\nu} = \partial_\mu A_\nu - \partial_\nu A_\mu \quad (42)$$

We seek an approximate classical solution to the full field theory equations in the form of a curved and moving string. Locally it should look like a Lorentz boosted version of the static cylindrically symmetric solutions obtained by minimising the energy functional (40). It should also be built around a curved worldsheet $x^a(\sigma)$.

First we imagine such a worldsheet is given and construct two normal vectors to it, $n_A^a(\sigma), A = 1, 2$. These obey $n_A^a x^b_{,\mu} = 0$ and are orthonormal $n_A^a n_B^b g_{ab} = -\delta_{ab}$. Now for any point in space y^a closer to the string than its radius of curvature R we can associate two worldsheet coordinates σ^μ and two radial coordinates ρ^A uniquely

$$y^a = x^a(\sigma) + \rho^A n_A^a(\sigma) \tag{43}$$

This is illustrated in Figure 2.8. We will use this relation to change variables from the four y^a coordinates to the four $\xi^\alpha = (\sigma^\mu, \rho^A)$ coordinates. Our ansatz for the field configuration corresponding to this worldsheet is the $\Phi(y) = \Phi_s(\rho)$ and $A^a(y) = n_A^a(\sigma) A_s^A(\rho)$ where the subscript s refers to the static cylindrically symmetric solution. The reader can check that this ansatz gives the exact solution for a straight moving string (i.e. just the Lorentz boosted version of the static straight solution). We expect that it will provide us with an approximate solution if the radius of curvature of the string R is much greater than its width W. Now we simply change coordinates in the action (42) from y^a to ξ^α. The Jacobian is most easily calculated by noting that

$$\sqrt{-g}\,det(\frac{\partial y}{\partial \xi}) = \sqrt{-det(g_{ab}\frac{\partial y^a}{\partial \xi^\alpha}\frac{\partial y^b}{\partial \xi^\beta})} \tag{44}$$

and

$$\begin{aligned}M_{\alpha\beta} &= g_{ab}\frac{\partial y^a}{\partial \xi^\alpha}\frac{\partial y^b}{\partial \xi^\beta} \\ &= diag(\gamma_{\mu\nu}, -\delta_{AB}) + o(\rho/R)\end{aligned} \tag{45}$$

where worldsheet derivatives of x^a are down by order ρ/R. Thus $detM = det\gamma$ to order ρ/R. We also have to the same order $(D\Phi)^2 = M^{\alpha\beta}D_\alpha\Phi D_\beta\Phi = -(D_A\Phi_s)^2$ and $F^2 = F_s^2$ so when the $d^2\rho$ integral is performed the Lagrangian reduces to minus the energy per unit length, μ. The terms of order ρ are down by W/R after integrating ρ across the string profile, where W is the string width. Thus to lowest order we find the action (41). For loops seeding galaxies R is of the order of 10 parsecs and W is of order $m_{gut}^{-1} \approx 10^{-33} cm$ so W/R is indeed pretty small! However in the very small regions of the string worldsheet like 'cusps' (see below) where the worldsheet becomes highly curved we expect the corrections to be important.

In fact the $o(W/R)$ corrections to (41) vanish by symmetry - the first nontrivial corrections arise at order $(W/R)^2$. They are

$$\Delta S = \mu \int d^2\sigma (\alpha K^2 - \beta \omega^2) \qquad (46)$$

where K is the extrinsic curvature of the worldsheet and w the 'twist' along the string [44]. α and β are positive constants of order W^2. The sign of the extrinsic curvature term corresponds to a *negative* rigidity - cosmic string likes to be crinkled rather than straight. This might lead to interesting effects in the late time evolution of loops, where presumably it will be counteracted by the gravitational radiation acting to reduce high frequency modes on the string.

The nicest feature of the Nambu action is that it is completely geometrical. The parameter μ does not enter into the equations of motion at all. The motion of the string just depends on the background spacetime metric. Waves propagate along the string with the speed of light.

Of course our discussion breaks down where two strings collide. In this case one has to solve the full nonlinear field equations. However it really is a very local process (the string fields approach the vacuum exponentially rapidly outside the string) and can be done perfectly well in flat spacetime. Numerical calculations by Shellard [36](for global strings) and Matzner [37] (for local strings) show that when two strings collide they always reconnect the other way (see Figure 2.9) for velocities less than .95 in the centre of mass frame, i.e. essentially always. This is a very nice result because the string interactions are therefore also fixed and cannot be adjusted. For strings with more complicated internal structure like superconducting strings this may not be the case.

2.10 Free Strings in an Expanding Universe

We know from present observation that at early times Ω was very close to unity and that the universe was radiation dominated and quite homogeneous. Thus the background metric in which we have to evolve the strings is completely specified. The metric is

$$ds^2 = dt^2 - a^2(t)d\vec{x}^2 = a^2(\eta)(d\eta^2 - d\vec{x}^2) \tag{47}$$

with $a(t) \propto t^{\frac{1}{2}} \propto \eta$ and η is the *conformal* time, in which the metric is conformally related to the Minkowski metric. The action (41) is invariant under *reparametrisations* of the string worldsheet. This means that to obtain a well defined initial value problem one has to choose a gauge. A convenient gauge is

$$\begin{aligned} x^0(\tau,\sigma) &= \tau \\ \partial_\tau \vec{x}.\partial_\sigma \vec{x} &= 0 \end{aligned} \tag{48}$$

(i.e. choosing τ to be the time and choosing σ at each instant so that the motion of the point labelled by σ is perpendicular to the string). Upon varying (41) one finds in this gauge the equations[47]

$$\ddot{\vec{x}} + 2\frac{\dot{a}}{a}\dot{\vec{x}}(1-\dot{\vec{x}}^2) = \frac{1}{\epsilon}\partial_\sigma(\frac{\vec{x}'}{\epsilon})$$

$$\dot{\epsilon} = -2\frac{\dot{a}}{a}\epsilon\dot{\vec{x}}^2$$

$$\epsilon = \sqrt{\frac{\vec{x}'^2}{1-\dot{\vec{x}}^2}} \qquad (49)$$

where dots refer to τ derivatives and primes to σ derivatives. The quantity ϵ has a simple meaning - the string energy per parameter σ length (see Section 2.14 below) is given by $\mu a(\eta)\epsilon$. The numerator in ϵ is just the comoving length of the string per unit σ, corresponding to a potential energy μ times (physical length). The denominator contributes the kinetic energy of the string.

In (49) there are two competing forces. On the left hand side is a damping term due to the expansion of the universe. On the right hand side is the curvature term which accelerates the string. The result of the competition between these terms is that for curvature scales R on the string much *smaller* than the Hubble radius $R_H = H^{-1} = a/\frac{da}{dt}$ the curvature wins and the string evolves as in flat spacetime. One finds in the case [47]

$$E \approx const \to \rho \propto a^{-3} \quad R << R_H \qquad (50)$$

Highly curved strings behave like *matter*. For curvature scales much *greater* than the Hubble radius the damping term wins so $\ddot{\vec{x}} \approx 0$ and the string is *conformally* stretched, its shape remaining fixed but its length growing like a. Thus

$$E \propto a \to \rho \propto a^{-2} \quad R >> R_H \qquad (51)$$

Either of these two behaviours would be cosmologically disastrous - the string would quickly come to dominate the radiation background (for which $\rho \propto a^{-4}$). However this is not the end of the story. Loops smaller than the Hubble radius are not a problem because they slowly radiate themselves away into gravity waves for which $\rho \propto a^{-4}$. As we shall see the long strings which would cause problems if there were no string-string interactions gradually chop themselves up into loops which then radiate themselves away.

2.11 Inside the horizon: Free Strings in Flat Spacetime

In flat spacetime the string equations are particularly simple and can be solved analytically as follows [105,48]. From (49) we have $\dot{\epsilon} = 0$. If we initially parametrise the string so that $\epsilon(\sigma)$ is equal to 1 along the string then it will remain so for all time. We then have

$$\ddot{\vec{x}} = \vec{x}''$$
$$\dot{\vec{x}} \cdot \vec{x}' = 0 \quad \dot{\vec{x}}^2 + \vec{x}'^2 = 1 \tag{52}$$

The general solution is given by

$$\vec{x} = \frac{1}{2}(\vec{a}(\sigma + \tau) + \vec{b}(\sigma - \tau))$$
$$\vec{a}'^2 = \vec{b}'^2 = 1 \tag{53}$$

Here $\vec{a}'$ and $\vec{b}'$ are arbitrary curves on the suface of a unit sphere. For a closed loop in its centre of mass frame $\vec{a}'$ and $\vec{b}'$ are in addition constrained to have $\int d\sigma \vec{a}' = \int d\sigma \vec{b}' = 0$. Some simple nonintersecting Fourier mode solutions [55] are shown on Figure 2.10 (courtesy A.Stebbins).

It is easy to check that a closed loop has period $L/2$ where L is the parameter length in this gauge. Some of the trajectories in Figure 2.10 have 'cusps' - points where $|\vec{x}'|= 0$ and the string reverses back on itself. From (52) these points have $|\dot{\vec{x}}|= 1$ - they are instantaneously moving at the speed of light![55]. Gravitational radiation is 'beamed' out from these points[55,58].

2.12 Interacting String networks in an Expanding Universe: the Scaling Solution

Now we proceed to the full problem of the evolution of an interacting string network in an expanding background. We simply take the initial configuration as shown in Figure 2.7 and evolve it with the equations (49) and the rule for interactions that whenever two strings cross we reconnect them the other way as in Figure 2.9.

A. Albrecht and myself have been developing a numerical code to do this for several years [49]. We have recently made qualitative improvements over our early code and brought the results to what (for us!) is a rather convincing state. Bennett and Bouchet [52] have also developed a code and our latest code agrees at least in part with theirs.

A picture of part of one of our simulations is shown in Figure 2.11 . We begin with initial conditions as in Figure 2.7 in a 24^3 or 30^3 box, with 10 points on the string for every initial 'correlation length' ξ. We then evolve (49) numerically. The only parameter to vary in the simulation is the ratio of ξ to the Hubble radius $R_H \equiv a/\frac{da}{dt} = 2t$. The initial string density is approximately $\rho_s = \mu/\xi^2$, so starting at large or small R_H/ξ corresponds to starting at high or low string density relative to the background radiation density ($\rho_b = \rho_{radiation} = 3/(32\pi G t^2) = 3/(8\pi G R_H^2)$).

It is useful to separate the strings into 'long string', strings longer than the Hubble radius, and 'loops', strings shorter than the Hubble radius. The precise dividing line is not very important.

Figure 2.12 shows how the density in long string (defined as loops for whom E/μ is greater than $2R_H$) behaves for differing initial string densities. A good way to think of this is that if there is a fixed number of long strings of length $\approx R_H$ per Hubble volume R_H^3 then the long string density scales with time as R_H^{-2} so the ratio ρ_s/ρ_b is constant. As can be seen this ratio does indeed appear to approach a constant and $\rho_s \approx 200/R_H^2$.

Why is this? The main idea of the 'scaling solution' [49,50,51] is that one should define a length scale ξ by

$$\rho_L = \frac{\mu}{\xi^2} \qquad (54)$$

where ρ_L is the density in long strings. As long as reconnection keeps the network 'random' then ξ should also be the typical curvature scale on the string. Now as long as $\xi << R_H$, the strings should evolve as in flat space so that the energy in the long strings (ignoring chopping off loops) scales like matter. Also, the timescale for the long strings to chop off a fraction of their length into loops is simply ξ (recall the characteristic velocity of waves on the string is just the speed of light which we set equal to 1). Thus ignoring reconnections of loops we find

$$\dot{\rho}_L = -3\frac{\dot{a}}{a}\rho_L - c\frac{\rho_L}{\xi} \qquad (55)$$

with c a constant (here dot refers to $\frac{d}{dt}$ not $\frac{d}{d\eta}$). The first term is simply a result of the expansion, the second the loss of energy into loops. Ignoring reconnections is reasonable in the light of the results of the next lecture, where I will show that in flat space chopping off is heavily favoured by phase space over reconnection .

Equation (55) has a simple solution: for a radiation dominated background $a \propto t^{\frac{1}{2}}$ it is given by

$$\rho_{sc} = \frac{\mu}{c^2 R_H^2} = \frac{\mu}{c^2(2t)^2} \qquad (56)$$

This is called the *scaling solution*.

In fact the scaling solution is a stable fixed point of (55) as may be seen by defining $y = (\rho_L/\rho_{sc})$ and calculating

$$\dot{y} = \frac{y}{R_H}(1 - \sqrt{y}) \qquad (57)$$

If y is greater than unity then it decreases, but if it is less than unity it increases. Thus $y = 1$ is a stable fixed point. Taking the initial scale factor a_i to be unity (57) may be solved for an initial value y_i to yield

$$y = \frac{y_i \sqrt{a}}{(1 - y_i(1 - \sqrt{a}))} \qquad (58)$$

Thus y approaches unity rather slowly.

The dashed lines on Figure 2.12 show the predicted approach to scaling from (58) with y_i chosen soon after the initial time (there is a short initial transient due to the fact that we start our strings off stationary in comoving coordinates). As can be seen the results fit (58) rather well. Figure 2.11 shows the string configuration in a carved out Hubble volume (actually $(R_H/2)^3$) in the scaling solution.

2.13 Loops

The distribution of loops produced in the scaling solution plays a very important role in the cosmic string theory of large scale structure formation. The importance of loops was first emphasised in [53] by Vilenkin, and the simple identification of one loop with one object today proposed in [54,64,66].

The energy lost from long strings in (55) goes into loops. Defining $\rho_l(e)$ to be the density contributed by loops of energy e to $e + de$ we have an equation

$$\dot{\rho}_l(e) = -3\frac{\dot{a}}{a}\rho_l + \frac{1}{R_H^4}f(e,t) \tag{59}$$

which defines the dimensionless *loop production function* f. In the scaling solution there is only one length and time scale in the problem, the Hubble radius $R_H = 2t$ in the radiation era. Thus f being dimensionless can only be a function of $e/(\mu R_H)$. Some of the loops produced will further selfintersect. I will ignore this complication here by regarding f as the loop production function for non-self-intersecting loops. Integrating (59) we find

$$\begin{aligned}\rho_l(e)de &= (\frac{\mu R_H}{e})^{\frac{3}{2}}\frac{\lambda de}{R_H^3} \\ \lambda &= \frac{1}{2}\int_0^{x_C}dx x^{\frac{1}{2}}f(x)\end{aligned} \tag{60}$$

for loops with $e/(\mu R_H) << 1$ and we define $f(x)$ to be zero for $x > x_C \approx 1$ - such loops are simply part of the long string. Similarly the number of loops per unit volume with energy e to $e + de$ is

$$n_l(e)de = (\frac{\mu R_H}{e})^{\frac{3}{2}}\frac{\lambda de}{e R_H^3} \tag{61}$$

Our original simulations found $\lambda \equiv \nu(2\beta)^{\frac{3}{2}}$ (in the old notation; see [66]) approximately equal to 0.8. The results of our latest simulations will shortly be revealed [49]. Equation (60) shows that the loop density scales like matter - the small loops have been around longest so they dominate the total density in strings. This would be a disaster if the loops were totally stable. However they do slowly decay by emitting gravity waves, at a rate which is exactly calculable for simple loop trajectories [55,57,59]. The rate of energy loss is given by $\dot{e} = -\Gamma G\mu^2$ independent of the size of the loops, with Γ generically of order 50-100 for simple loops. The energy of a loop thus decreases slowly with time: $e(t) = e_i - \Gamma G\mu^2(t - t_i)$ where e_i and t_i are the initial values. For $t \gg t_i$ this yields the final loop distribution

$$n_l(e,t)de = (\frac{\mu}{R_H})^{\frac{3}{2}} \frac{\lambda de}{(e + \Gamma G\mu t)^{\frac{5}{2}}} \qquad (62)$$

These loops act as gravitating 'seeds' which start accreting matter at around t_{eq} in the manner described in Lecture 1. The mass a loop accretes is proportional to it's mass (with cold dark matter and as long as it was produced before t_{eq}, see below). Smaller loops are more numerous and give rise to smaller objects like galaxies. Larger loops are rarer and give rise to larger objects like clusters of galaxies. Thus (62) gives a simple prediction for number density versus mass for objects in todays universe (though this is in general altered by merging).

2.14 Strings as Density Perturbations

Cosmic strings of the simplest kind interact with matter mainly through gravity. As discussed in Lecture 1 density perturbations only start growing appreciably when the universe becomes matter dominated. To see how strings cause matter perturbations to grow in their vicinity we need to calculate their gravitational field.

For simplicity I will only discuss strings in a flat background in the weak field limit (the gravitational potential is of order $G\mu \ll 1$ so this is a very good approximation most of the time for cosmic strings. It may fail at 'cusps'). The string stress tensor is given from the Nambu action as

$$T^{ab}(t,\vec{x}) = -\frac{2}{\sqrt{-g}} \frac{\delta S}{\delta g_{ab}} |_{g_{ab}=\eta_{ab}}$$
$$= \mu \int d\sigma \begin{pmatrix} 1 & \dot{x}^i \\ \dot{x}^j & \dot{x}^i\dot{x}^j - x'^i x'^j \end{pmatrix} \delta(\vec{x} - \vec{x}(t,\sigma)) \qquad (63)$$

in the gauge (48) where η_{ab} is the Minkowski metric and $\vec{x}(t,\sigma)$ the string trajectory in space. For a straight static string along the z axis, $\vec{x}(t,\sigma) = (0,0,\sigma)$, we find $T_{00} = -T_{33} = \mu\delta^2(\vec{x})$ with other components zero. The sign of the T_{33} term corresponds to a *negative* pressure i.e. a tension along the z axis.

To calculate the metric perturbation produced by the string we write the metric as $g_{ab} = \eta_{ab} + h_{ab}$ with $|h| \approx G\mu \ll 1$. In the harmonic gauge $g^{ab}\Gamma^c_{ab} = 0$ the linearised Einstein equations become [6]

$$\partial^2 h_{ab} = -16\pi G(T_{ab} - \frac{1}{2}\eta_{ab}T^c_c) \qquad (64)$$

The usual Newtonian potential is just $h_{00}/2$: for our straight string it is easily seen that the source for h_{00} vanishes exactly! The only regular solution is $h_{00} = 0$ everywhere, so there is no Newtonian potential outside the string. h_{33} also vanishes but there is a nonzero h_{11} and h_{22};

$$\nabla^2 h_{11} = \nabla^2 h_{22} = 16\pi G\mu\delta(\vec{x})$$
$$\Rightarrow h_{11} = h_{22} = 8\pi G\mu ln(r/W) \qquad (65)$$

choosing $h_{11} = h_{22} = 0$ at the string width W. Inside the string the true stress energy is not singular like the 'Nambu' approximation (63) and this removes the apparent singularity at $r = 0$ [63]. Clearly the weak field approximation breaks down at very large radii as well. Ignoring this for the moment we have in cylindrical coordinates

$$ds^2 = dt^2 - dz^2 - (1 - 8G\mu ln(r/W))(dr^2 + r^2 d\phi^2) \qquad (66)$$

We can remove the prefactor to the dr^2 term by defining a new radial variable $dr' = dr(1 - 4G\mu ln(r/W))$ or $r' = r(1 + 4G\mu(1 - ln(r/W)))$. Neglecting terms of order $(G\mu)^2$, consistent with our approximation, we then find

$$\begin{aligned}ds^2 &= dt^2 - dz^2 - dr'^2 - r'^2(1 - 8G\mu)d\phi^2 \\ &= dt^2 - dz^2 - dr'^2 - r'^2 d\phi'^2\end{aligned} \qquad (67)$$

upon defining $\phi' = (1 - 4G\mu)\phi$. This is not only regular everywhere, it is flat space! However the new angular variable ϕ' only runs from 0 to $2\pi - 8\pi G\mu$ the geometry is that of 'flat space minus a wedge', or a cone [61,60]. This is illustrated in Figure 2.13. The only difference is at radii of order W, the string width, where the apex of the cone is smoothly rounded out[62,63].

What about loops? These *do* produce a Newtonian potential - one finds

$$\partial^2 h_{00} = -16\pi G\mu \int d\sigma \dot{\vec{x}}^2 \delta^3(\vec{x} - \vec{x}(\sigma, t)) \qquad (68)$$

For an oscillating loop the right hand side oscillates peiodically in the centre of mass frame of the loop. This leads to a peculiar pulsating gravitational field[54]. It's time average however is simple, because of the following identity. If the length of the loop is L and its period T we have

$$\begin{aligned}\int \frac{d\sigma}{L} \int \frac{dt}{T} \dot{\vec{x}}^2 &= \int\int -\vec{x}.\ddot{\vec{x}} = \int\int -\vec{x}.\vec{x}'' \\ &= \int\int \vec{x}'^2 = \int\int (1-\dot{\vec{x}}^2) \\ &\Rightarrow \int\int \dot{\vec{x}}^2 = \frac{1}{2}\end{aligned} \quad (69)$$

where we used integration by parts and the equations (52). The average velocity squared for a closed loop is thus $\frac{1}{2}$. Now if we time average (68) we find the time averaged long range field of a loop is that of a shell of Newtonian dust with suface density proportional to $\dot{\vec{x}}^2$ and with total mass equal to μL, the mass of the loop. At distances from the loop much greater than its size we can treat it to a good approximation as a point mass.

Vachaspati [58] has tried to use the time variation of h_{ab} and the *beaming* effect mentioned earlier [55] to argue that an accreting loop will eject matter into an interesting shape. However the actual amount of mass which is in the region where h is so large that it moves the matter substantially during the existence of a cusp is tiny, so most of the matter around the loop simply sees the time averaged field, which shows no beaming effect [55].

2.15 Accretion by Loops

Accretion by moving loops has been looked at in some detail in [65] and [70]. In linear theory, the equation for the density perturbation $\delta\rho/\rho(\vec{x})$ is purely local in comoving coordinates, as noted in Lecture 1. This means that the surface 'traced out' by the loop as it moves through space will appear in the density perturbation it produces. This is only true while the perturbation is linear - as soon as collapse and virialisation occur the structure will be smoothed out and the object become more and more spherical. The shape traced out by a loop is in fact like a 'trumpet' [70]. This is seen as follows. The peculiar velocity of a loop $v_{pec} = a(t)\dot{x}$ decays [1] as a^{-1}. This means that, in the matter dominated era, the comoving distance a loop moves is given by

$$\Delta x = \int_{t_i}^{t} v_i(\frac{a_i}{a})^2 dt = 3v_i t_i (1 - (\frac{a_i}{a})^{\frac{1}{2}}) \tag{70}$$

where v_i is the initial velocity, at scale factor a_i and time t_i. Thus a loop slows down and stops in comoving coordinates in the matter era (in the radiation era Δx increases logarithmically with time). However the radius of the loop in comoving coordinates r_c shrinks (the *physical* radius is approximately constant, ignoring oscillations) so

$$r_c = r_i \frac{a_i}{a} \tag{71}$$

So defining the distance from where the loop ends up as $z = 3v_i t_i - \Delta x$ we see that the shape traced out by the loop is

$$r_c \propto z^2 \tag{72}$$

the surface of revolution of a parabola - a 'trumpet' shape. As stated above, nonlinearity smears this out, making the accreted lump in the end rather spherical [65]. For galaxies this would be the case today. However larger clusters should still exhibit this shape - in fact the Perseus-Pisces filament [3] does look rather 'trumpet' like! Bingelli has made an attempt to measure the eccentricity for Abell clusters - this will serve as an interesting test in the future[10].

---

[1] More precisely the momentum $\int d\sigma \epsilon \dot{x}$ decays as a^{-1} as can be seen from (49), but if its velocity is substantially less than 1 and it is well inside the horizon then the energy is roughly constant.

2.16 Loops as seeds for galaxies and clusters

The string plus cold dark matter (SCDM) scenario is the easiest to understand. It is clear from Section 1 that the *total mass* accreted by a loop around at t_{eq} is today roughly $M_{loop}(1 + Z_{eq})$. In fact a precise calculation in the spherical collapse model yields [67]

$$m_{loop}(1 + Z_{eq}) = \frac{\xi}{5}M(\frac{\rho}{\rho_b})^{\frac{1}{3}} \qquad (73)$$

where M is the mass of the accreted object, ρ/ρ_b is the overdensity today and ξ is a 'loss of growth' factor, equal to 1 for loops laid down well before t_{eq}, equal to 4 for loops laid down at t_{eq} and greater for loops laid down still later.

It is now simple to calculate the required string tension. First we pick a class of objects, for example Abell clusters. Their number density is $1/(110h_{50}^{-1}Mpc)^3$. Let me assume that the loops seeding them are all produced in the radiation era for simplicity (they are actually produced at or just after the matter-radiation transition according to our latest simulations). Now we take the loop distribution, given in equation (62), and evolve it through the matter - radiation transition to the present day. We then ask for what cutoff in energy will the integrated number density be equal to $1/(110h_{50}^{-1}Mpc)^3$. Doing this[66,67], one finds that the length of loops producing clusters is $L = e/\mu \approx 20kpc$. We can then demand that loops of this size are massive enough to accrete clusters by today, from (73) and the observed mass and overdensities of clusters [67]. One finds

$$G\mu = 2.10^{-6}h_{50}^{-1}\sigma_{700}^{\frac{5}{3}}d_{110}^{-2}\lambda^{-\frac{2}{3}}\xi_4 \qquad (74)$$

Here σ_{700} is the velocity dispersion in a cluster in units of 700 kms^{-1}, d their mean separation in units of 110 h_{50}^{-1}Mpc, λ the constant in (62) and ξ the growth factor mentioned before. There are clearly large uncertainties in the normalisation of $G\mu$, probably of an order of magnitude either way. In particular there are several effects which might reduce $G\mu$. If clusters form by mergers of several loops, or directly from curved sections of the long strings, this would decrease the required value. This is one of the main motivations for the more detailed calculations now being performed in which we are directly evolving the matter distribution around a string network produced in the simulations[71].

Independently of $G\mu$ however, the *spatial distribution* of the loops and long strings should have testable consequences for the spatial distribution of galaxies and clusters, particularly clear cut on scales larger than those where gravitational clustering has been efficient (i.e. greater than 12 h_{50}^{-1} Mpc or so, see Lecture 1). Unlike theories like the ICDM model, there are no parameters to adjust here - the evolution of the strings is independent of the string tension. An early calculation [64] showed that remarkably enough the two-point correlation function of loops produced in a cosmic string simulation (in the radiation era) matched the observed cluster-cluster correlation function very closely. Subsequent simulations by Bennett and Bouchet, and those by Albrecht and myself using our improved code, have confirmed this correlation of loops when they are produced. The more detailed calculations using improved codes and including the radiation-matter transition and the effect of the long strings will show whether this apparently miraculous agreement with observation is confirmed. [2]

2.17 Cosmic String and Hot Dark Matter

It has recently been shown that cosmic string could solve the usual problems of the standard Hot Dark Matter (IHDM) scenario [25,83] (see also the earlier work of [112]). I will call this the SHDM scenario. This is nice - the massive neutrino is still the best motivated dark matter candidate. Of course exactly why $m_\nu = 30eV$ as is required is still a mystery (as are the other fermion masses!).

Without going into details, the reason for the improvement is that loops survive the free streaming era and begin accreting baryons slowly at around recombination. The baryon/loop lumps then start accreting neutrinos efficiently when the neutrino Jeans mass falls to galaxy scales. Flat rotation curves are obtained automatically (once heralded as a success of the standard ICDM scenario). The loss of growth on galaxy scales means that a higher value of $G\mu$ is needed - this brings the theory more into line with the observed large scale velocity fields [78]. Clusters and their correlations are similar to SCDM[25,26,83].

---

[2]It has been argued, in particular by Bennett and Bouchet[68], that if one extrapolates the loop positions for several expansion times then the correlations are washed out by loop velocities. I believe this is a red herring - loops forming clusters are produced at or just after t_{eq} - accretion begins as soon as the loops are produced. Thus the *initial* positions of the loops is most relevant. Second, some (e.g. Scherrer and Press[69]) have argued that loop fragmentation produces fragments with fairly high average velocities $\approx 0.6c$, so one can use randomly distributed loops to calculate structure formation. However it is the larger loops around (and even those that will subsequently self-intersect) at or after t_{eq} that will form clusters, and these have much lower velocities [49]. String plus matter simulations will hopefully resolve the issue.

2.18 Cosmic String and Inflation?

How do cosmic strings fit in with inflation? Many cosmologists today take the idea of inflation very seriously (some even refer to it as a 'paradigm') because it 'explains' why the universe is so large, so homogeneous and so flat today.

However inflation is not a gospel. Many people including myself feel rather unhappy with the 'explanations' it offers. At a general level, it is certainly true that inflation will turn a wider class of initial conditions into a large homogeneous universe like our own. However it certainly will not turn *arbitrary* initial conditions into universes like ours and until it is possible to say precisely 'how much better' inflation makes things, one really cannot judge the idea.

It also seems very likely that it will be quite some time before we can say anything very definite about the initial conditions of the universe that is in any way testable or provable. The whole discussion may well simply remain in the realm of philosophy rather than physics.

In cosmology today there is a strong case for being conservative in our ambitions. It has always been easy in cosmology to get into the realms of untestable speculation. What is new today is that there is a rapidly increasing data base upon which theories can be tested. And there are now at least two rather precise theories to test. This is the most hopeful aspect of the subject. Now the aspect of inflation that *is* certainly a major advance from this point of view is that the fluctuation spectrum it 'predicts' (at least in the simplest version [73]) has definite consequences for the galaxy distribution, the microwave background fluctuations and so on. Cosmic strings are from this point of view also a very good theory, with clear testable predictions.

In fact one can make the case that cosmic strings are a much better justified extrapolation of low energy physics. The underlying successful idea in high energy physics has been that of gauge symmetry and spontaneous symmetry breaking. Cosmic strings are fairly generically (though certainly not always) produced in symmetry breaking phase transitions, and the precise details of the symmetry breaking (i.e. whether it occurs dynamically via a condensate or through a scalar 'Higgs field') are not crucial. On the other hand it is very difficult to make inflation work in a gauge symmetry breaking phase transition. Instead one has to use a gauge *singlet* inflaton field which 'allows' one to tune the coupling constant to very small values. This raises real questions about the initial distribution of the inflaton field - it was almost certainly not in thermal equilibrium. It also makes the subject very ad hoc.

I am often faced with the question 'well how do cosmic strings explain the flatness and homogeneity of the universe?' to which I reply 'they don't'. It is perfectly legitimate in seeking to explain the large scale structure in the universe to use the fact that we *know* the universe was homogeneous and flat at early times, from the microwave background. This is exactly what we do in nucleosynthesis calculations for example.

Inflation does not go well with cosmic strings (as you might have guessed from my critical comments!). During inflation the string density would fall exponentially so that after inflation there would be far fewer than one string per horizon volme. Indeed that is how inflation solves the monopole problem (to which there are other solutions [74,75,76]). For the reasons I have given, this does not bother me too much.

If you want to use inflation to solve the monopole problem, but have cosmic strings provide the fluctuations later on, you have to 'stand on your head'. You must first form the monopoles, inflate them away and then reheat to a temperature high enough to form strings but too low to form monopoles. This is tricky - in most inflationary models the universe heats up to temperatures well below the 10^{16} GeV scale typically needed to form strings. However in cosmic strings the mass per unit length $\mu \approx \eta^2$ is only weakly dependent on couplings whereas the phase transition temperature $T_c \approx \sqrt{\lambda/g^2}\eta$ can be made much lower by decreasing λ or imposing discrete symmetries which only allow high powers of Φ in the Higgs potential which effectively does the same thing [77].

Just as none of the models for inflation alone is compelling, none of the models for inflation plus strings is particularly inspiring. In fact they are quite ugly [81,80].

2.19 Cosmic String versus Inflation

Let me now turn to comparing the predictions of cosmic strings with those of 'inflation'. By inflation I mean the simplest prediction of quantum fluctuations from inflation - the scale-free Harrison- Zel'dovich spectrum[84,85,86].

The theories are really very different. In the case of inflation it makes sense to think of the density perturbation field as a linear superposition of plane waves with random phases

$$\frac{\delta\rho(\vec{x})}{\rho} = \sum_{\vec{k}} \delta_{\vec{k}} e^{i\vec{k}.\vec{x}} \qquad (75)$$

with the probability distribution for each $\delta_{\vec{k}}$ being a Gaussian : if $\delta_{\vec{k}} = r_{\vec{k}} e^{i\theta_{\vec{k}}}$ then each mode is independent (except for the reality condition which forces $\delta_{-\vec{k}} = \delta_{\vec{k}}^*$) with probability measure

$$\prod_{\vec{k}} \frac{d\theta_{\vec{k}}}{2\pi} \frac{dr_{\vec{k}} r_{\vec{k}}}{P_k} e^{-\frac{r_{\vec{k}}^2}{2P_k}} \qquad (76)$$

This is a result of the inflaton field being essentially a free massless field during inflation. The power spectrum $P_k = <|\delta_{\vec{k}}|^2>$ completely defines the theory. The fact that the different $\vec{k}$ modes are decoupled means that any quantity like the excess mass in a ball of radius R

$$\delta M = \int_0^R d^3 r \delta\rho(\vec{r}) \qquad (77)$$

is also Gaussian distributed, by the central limit theorem. This makes large fluctuations away from the 'typical' fluctuations very rare.

With strings the situation is quite different, and

$$\delta\rho(\vec{x}) = \sum_{loops, longstrings} \delta\rho(\vec{x}) \qquad (78)$$

is often the simplest way to look at the density perturbations. Many different $\vec{k}$ modes are correlated in a loop for example. The probability distribution is nonGaussian as can be seen from (62) - in a volume V (ignoring correlations) the number of loops expected with mass M to $M + dM$ is

$$N(M)dM \propto M^{-\frac{5}{2}} dM \qquad (79)$$

When this number is much less than one, it is just the probability of finding a loop of mass M in the volume. Thus $P(\delta M) \propto (\delta M)^{-\frac{5}{2}}$ at large δM. Any one realisation of cosmic string looks like a collection of spikes and lines in the density distribution - small loops and long strings cause very high density perturbations locally. Fluctuations from inflation however look like a superposition of plane waves that is everywhere linear. This should make it easier in principle to identify objects with initial density perturbations in the cosmic string theory - the distinction between a lump and the surrounding matter is more clear-cut. In inflation there are mass fluctuations on every scale right down to the smallest scales. This makes the identification of 'galaxies' quite hazardous. Of course merging of lumps, which is important in the cosmic string plus CDM scenario[87] tends to destroy the identification of one loop with one object.

The 'typical' mass fluctuation on a given scale δ_M is actually quite similar in both theories - $\delta M/M \propto R^{-1.25}$ approximately for scales R corresponding to the mean separation of galaxies [72] for inflation, which leads to the successful prediction of the galaxy-galaxy correlation function. For strings, the mass m of the largest 'typical' loop in a volume R^3 is given from the loop distribution (62) by $n_>(m)R^3 \approx 1$ so $\delta M \propto R^2$ and $\delta M/M \propto R^{-1}$ instead. So the gross mass fluctuation spectrum on galaxy scales is not very different.

One of the virtues of the prediction from inflation is that it is much easier to calculate with. The linear phase of the evolution is relatively trivial. With strings however it is necessary to do a computer simulation of the strings even before including the growth of perturbations. This should not of course mistakenly be seen as a reason why one theory is 'better'.

The successes of the ICDM theory are mostly on small scales where clustering has gone nonlinear. With the correct *biassing* factor (initially put in by hand but now argued by some to emerge naturally [23]) the ICDM model matches the galaxy correlation function, pair velocity distribution, and gives flat rotation curves in good agreement with observation[21]. This is impressive but it must be remembered that the nonlinear dissipative baryonic physics has not yet been put in, and this is actually in determining whether the postulated bias actually happens.

In contrast, the ICDM model has fared rather poorly in matching the large scale observations. These are, as I argued in Section 1 the most unambiguous testing ground for theories of the origin of structure. The basic problem of the ICDM model is that it has very little power on large scales - the predicted cluster cluster correlation function for example goes through zero at only 80 h_{50}^{-2} Mpc[22]. With suitable *biassing* it is claimed that it can produce 'bubbles' such as those seen in the 'slice of the universe'[21]. However with this biassing it fails by a long way to match the observed velocity field [17,79]. While the theory is not yet clearly ruled out it certainly has enough problems to make the investigation of alternatives such as cosmic strings worthwhile. In particular the SHDM model offers particular promise.

Cosmic strings are as I have said harder to calculate with. However there are already indications that cosmic strings have a lot of power on large scales. Early calculations of the correlation function for loops showed remarkable agreement with that for Abell clusters[64]. The later simulations have confirmed these correlations are present and have the correct r^{-2} form. However they have also found a much higher scaling density and this makes the long strings relatively more important. The detailed calculation of large scale structure produced by long strings is underway. This will also be important in accurately normalising the theory.

2.20 Other Observational Tests for Cosmic Strings

Cosmic strings have several fairly direct observational consequences which give us some confidence that if they exist, and were massive enough to seed the observed large scale structure, they should be observable fairly soon.

The most clear-cut effect, as with other theories, is the predicted anisotropy of the microwave background. Observations of intrinsic anisotropies would be a major advance in cosmology - the statistics of the fluctuations would tell us a lot about the kind of processes that could have caused them.

With cosmic strings the effect is quite dramatic. Moving strings lead to linear discontinuity across the sky in the temperature of the microwave background[88]. This is a result of the conical metric discussed in Section 2.14. In the string's restframe the there is a difference of $8\pi G\mu$ in the angle at which light from the same point on the last scattering surface of the microwave radiation is recieved by the observer. The observer thus appears to have a small component $8\pi G\mu v_s$ *towards* the last scattering surface behind the string, where v_s is the string's velocity perpendicular to the line of sight. This results in a Doppler shift of the radiation temperature observed of

$$\frac{\delta T}{T} = 8\pi G\mu v_s. \tag{80}$$

of the order 10^{-5} if $G\mu = 10^{-6}$. This is not far below present sensitivities. The difficulty is that strings are actually quite rare on the last scattering surface, so one has to cover quite a large angle in order to be likely to see one at all. Recently Bouchet et al. have calculated the expected pattern from a string simulation [89] - they deduce the limit

$$G\mu < 5 \times 10^{-6} \tag{81}$$

from the experiment of Melchiorri et. al. [90].

Another unique signature of cosmic strings is their gravitational lensing effect. The conical metric leads, in the case of a straight string, to two identical images for objects behind the string with angular separation

$$\delta\theta = 8\pi G\mu \frac{d_o}{d_o + d_s} \tag{82}$$

where d_s and $d_o > d_s$ are the distances from us to the string and the object. This is of the order of a few arc seconds if $G\mu = 10^{-6}$. What is unique is that a string would produce a line of double images. In fact a candidate event involving four double galaxies was reported recently by Cowie and Hu [91]. Deeper observations in the same area of the sky to see whether galaxies behind the putative string connecting the objects are also lensed are currently underway.

A less direct but still powerful test of the cosmic string theory comes from the gravitational radiation produced by the strings during the radiation dominated era. The fact that the millisecond pulsar is so 'quiet' imposes strong limits on the level of gravitational background that can be around today. The expected contribution to Ω_{GW} today from cosmic strings with periods of the order of a year or so (the millisecond pulsar observations put the stringest constraint on waves with this period) is[92,93,106,95]

$$\Omega_{GW} \approx 2 \times 10^{-7} \lambda (\frac{\Gamma}{50})^{-\frac{1}{2}} (\frac{G\mu}{10^{-6}})^{\frac{1}{2}} h_{50}^{-2} \Omega^{-1} \qquad (83)$$

whereas the observations already constrain $\Omega_{GW} < 4 \times 10^{-7}$ [96]. Things are already getting tight! This limit is improving, and will rapidly improve with time should another quiet millisecond pulsar be discovered, and should be able to convincingly rule out the scenario with $G\mu > 10^{-8}$ in 10 years time.

The other major effect of the gravity waves produced by strings in the radiation era is the disruption of standard nucleosynthesis. This gives the bound [97,51,95,98]

$$G\mu < 10^{-5} (\frac{\Gamma}{50}) \lambda^{-2} \qquad (84)$$

The strength of these bounds can only be assessed when we have an accurate calculation of the normalisation of $G\mu$, as discussed in Section 2.16. One cautionary remark about the gravitational radiation bounds must be made. The calculations of the rate of loss of radiation were made using the Nambu action and equations. If there is more complicated dynamics, particularly in the late history of a loop, as I discussed in connection with corrections to the Nambu action, or if loops continue self-intersect late in their lifetime Γ could increase, and there could be other types of radiation emitted. Both of these effects would weaken the bounds (83) and (84). Finally there are many other large scale structure predictions waiting to be made and tested - higher correlation functions, void statistics, velocity field predictions and so on.

3 Fundamental Strings and Cosmology

In this lecture I will run the risk of confusing you by discussing both cosmic strings (i.e. topological defects in spontaneously broken gauge theories) and fundamental strings which I have not so far mentioned at all. The latter are *quite different* in many important ways - they are the elementary excitations in string theories which attempt to unify gravity with the other forces.

However fundamental strings are similar in some important respects to cosmic strings. In particular the 'Nambu-Goto' action which provides an *approximate* description of the motion of cosmic string as long as the radius of curvature is much greater than the string width is the *exact* action for free fundamental strings. Provided we keep in mind the circumstances where the 'Nambu-Goto' action breaks down for cosmic strings, we can use fundamental strings as an approximation to cosmic strings. We shall also see that our numerical results for cosmic strings in turn provide considerable insight into the behaviour of fundamental strings at high density (at least in the weak coupling limit).

The more immediate goal of this lecture will be to understand and explain some of the numerical results on string networks mentioned above. In particular we would like to understand the distribution of strings at formation (Section 2.8) where

a) most (roughly 80%) of the string is in 'infinite' string

b) the remainder is in a scale invariant distribution of loops.

c) both types of string are in the form of 'Brownian' walks.

We would also like to understand why the chopping off of loops from long string is favoured over reconnection, which was very important in leading to the scaling solution (55). Our understanding of this below will lead to a simple argument that the string network cannot come to dominate the universe and that the scaling solution is, barring very long time transients, more or less inevitable.

Of course we will only obtain precise results for strings in equilibrium in Minkowski space. These will nevertheless be relevant to strings in an expanding background in the limiting regime where ξ is much smaller than the Hubble radius.

3.1 Counting String States : the Quantised String

In studying string statistical mechanics we are simply interested in finding the configurations which maximise the number of states available to a certain length[3] of string in a box. Assuming ergodicity, the system will end up in these, the most probable configurations. This is the idea behind the *microcanonical ensemble* and is completely independent of notions of temperature, heat baths and so on.

But how do we count states for the strings? Classically, a short string has just as many configurations as a long string. Thus one might imagine that the greatest number of states would be obtained by putting the string into the smallest possible loops. However this argument is incorrect - in counting states one has to put a *measure* on configuration space, and this brings in a

---

[3] I shall use length L and energy E interchangeably for strings, since $E = \mu \int d\sigma$ in the gauge $\epsilon = 1$. $\int d\sigma$ can he taken as the definition of the length, related to the naive one $L_N = \int d\sigma \, | \vec{x}' |$ by the factor $| \vec{x}' |$. Since $<| \vec{x}' |^2> = \frac{1}{2}$ for a closed loop by an argument analogous to (69) the two definitions are quite similar.

length scale, roughly the size of the smallest wiggles allowed on the string. A good way to do this is actually to *quantise* the strings - then the states form a well defined discrete set and the scale of the wiggles is given in terms of the string tension. In fact the number of states for a large string is exponentially greater than that for a small string as we shall see.

The free quantised string is a beautiful example of a nonlinear but exactly solvable quantum theory. For a thorough review see [102]. Here I will only summarise the main formulae we need.

We only wish to count the physical degrees of freedom of the strings so we use the 'light-cone' gauge of [105]. In this gauge the transverse coordinates of the quantised string (i= 2,3..d-1 in d spacetime dimensions) obey

$$x^i(\sigma,\tau) = q^i + \frac{p^i}{2\pi\mu} + \frac{i}{2\sqrt{\pi\mu}} \sum_{n \neq 0} \frac{\alpha_n^i}{n} e^{-2in(\tau-\sigma)} + \frac{\tilde{\alpha}_n^i}{n} e^{-2in(\tau-\sigma)} \quad (85)$$

$$\left[\alpha_n^i, \alpha_m^j\right] = n\delta_{n+m,0}\delta^{ij} \quad (86)$$

and similarly for the $\tilde{\alpha}$ oscillators. The string has been parametrised so that σ runs from 0 to π. In addition we have the rest mass formula and constraints

$$m^2 = 4\pi\mu(N + \tilde{N} - 2)$$
$$N = \tilde{N}$$
$$N = \sum_{n>0} \alpha_{-n}^i \alpha_n^i \quad (87)$$

and similarly for $\tilde{N}$. Here N and $\tilde{N}$ are the level number operators for the left and right moving oscillators respectively.

I shall ignore the tachyon altogether in what follows. This is the state with $N = \tilde{N} = 0$ and has negative mass squared. For cosmic strings, we shall not trust the string spectrum right down to the lowest mass states since the Nambu approximation is not valid for small loops. For heterotic strings or superstrings there are no tachyons anyway. I will also for simplicity ignore fermionic or winding number modes - including these is straightforward and does not affect the main conclusions.

A complete set of states for a single string is obtained by applying the α^i_{-n} creation operators to the vacuum, which is labelled by the centre of mass momentum [102]. For example at level number 0 we have the tachyon, with $m^2 = -8\pi\mu$. At level 1 we have the tensor states $\alpha^i_{-1} \tilde{\alpha}^j_{-1} \mid 0 >$, with $m^2 = 0$ and so on. In general the level number is obtained by adding the subscripts of the α or $\tilde{\alpha}$ oscillators (which from (87) must give the same result). Thus the number of states at level n involves (apart from the left-right degeneracy and the transverse indices) just the number of different ways of writing n as a sum of smaller positive integers. In mathematics this is called the partition function, and it was first realised by the mathematicians Hardy and Ramanujan that there is a simple way to calculate it for large n[106].

The method is as follows. One considers the expression

$$P(z) = Tr(z^N)$$
$$= (1 + z + z^2 + ..)^{d-2}(1 + z^2 + z^4 + ..)^{d-2}(1 + z^3..)^{d-2} \quad (88)$$

where the trace is over the entire Fock space of one set of oscillators (say the left moving oscillators) and factorises into the product of traces for each oscillator separately. In fact

$$P(z) = (\prod_{n=1}^{\infty} \frac{1}{(1-z^n)})^{d-2}$$

$$= \sum_{n=0}^{\infty} d_n z^n \qquad (89)$$

where d_n is the degeneracy for the left moving oscillators at level n because the trace in (88) adds one term z^n for every state at level n.

Now we invert (89) for d_n

$$d_n = \oint \frac{dz}{2\pi i z} z^{-n} P(z) \qquad (90)$$

where the contour runs around the origin in the complex z plane. Now the integrand is easily seen to have a saddle point for z just less than 1 on the real axis - for real z the function $P(z)$ blows up as z approaches 1 and conversely z^{-n} blows up as z decreases away from 1. We therefore distort the contour in (90) to run over the saddle point. To find the saddle point we calculate $P(z)$ near $z = 1$;

$$\begin{aligned} P(z) &= e^{-(d-2)\sum_{n=1}^{\infty} \ln(1-z^n)} = e^{(d-2)\sum_{n,m=1}^{\infty} \frac{z^{mn}}{m}} \\ &= e^{(d-2)\sum_{m=1}^{\infty} \frac{z^m}{m(1-z^m)}} \sim e^{\frac{d-2}{1-z}\sum_{m=1}^{\infty} \frac{1}{m^2}} \\ &= e^{\frac{\pi^2(d-2)}{6(1-z)}} \end{aligned} \qquad (91)$$

We then perform the Gaussian saddle point integral to obtain

$$d_n \sim e^{\pi\sqrt{\frac{2n(d-2)}{3}}} \qquad (92)$$

as the leading behaviour. A more detailed calculation gives the prefactor [103,99]

$$d_n \approx C n^{-\frac{d+1}{4}} e^{\pi\sqrt{\frac{2n(d-2)}{3}}} \qquad (93)$$

with C a constant. The total degeneracy at level n is simply the square of (93) - d_n ways for the left moving oscillators times d_n ways for the right moving oscillators.

Using the mass formula $m^2 \approx 8\pi\mu n$ we find for the number of states for a string of mass m to $m + dm$

$$\rho(m)dm \approx cm^{-d}e^{bm}dm$$
$$b = \sqrt{\frac{\pi(d-2)}{3\mu}} \qquad (94)$$

Thus as stated above the number of states available to a string grows exponentially with its mass. c is a constant depending on μ with dimensions of inverse volume.

3.2 The Fractal Dimension of Big Strings

As we have seen there is an exponentially large number of states available to a string of mass m. What do these states look like? This is surprisingly easy to answer[99]. Consider the position operator (85) at $\tau = 0$. Using it we can construct the operator which measures the average squared radius of the string

$$\Delta r^2 = (d-1)\int \frac{d\sigma}{\pi}(x^i - q^i)^2 = \frac{d-1}{2\pi\mu}(R + \tilde{R} + 2\sum_{n=1}^{\infty}\frac{1}{n})$$
$$R = \sum_{n=1}^{\infty} \frac{\alpha^i_{-n}\alpha^i_n}{n^2}$$
$$\tilde{R} = \sum_{n=1}^{\infty} \frac{\tilde{\alpha}^i_{-n}\tilde{\alpha}^i_n}{n^2} \qquad (95)$$

where i is not summed over - a single transverse index is used and the answer obtained by isotropy. The last term is a logarithmic divergence obtained by 'normal ordering' the oscillators. For cosmic string there is a cutoff corresponding to modes whose wavelength is of order the string width - for fundamental string this is a delicate issue but I will take the view that there should also always be an ultraviolet cutoff in any 'measurement' of the size of the string (short wavelength modes diffract light more for example). With any reasonable cutoff the logarithm will turn out to be negligible compared to the first two terms.

Ignoring it, we find for example for the two extreme states at level n

$$\Delta r^2 (\alpha_{-1})^n \mid 0 > \; \propto \; n(\alpha_{-1})^n \mid 0 >$$
$$\Delta r^2 \alpha_{-n} \mid 0 > \; \propto \; n^{-1} \alpha_{-n} \mid 0 > \qquad (96)$$

The first set of states may be thought of as a long and 'straight' loops with a size Δr proportional to their length m/μ. The second set of states are increasingly tightly wound loops whose size actually decreases with increasing m.

Now we wish to calculate the average of Δr^2 for *all* states at level n - this will tell us what typical string configurations look like. It is easily seen that this is obtained by

$$\sum_{\mid \psi > \text{ level } n} < \psi \mid \Delta r^2 \mid \psi > \;=\; \frac{d-1}{\mu \pi} \frac{d}{dy} Tr(x^N y^R) \mid_{y=1} \mid_{coeff \; of \; x^n}$$
$$= \; \frac{d-1}{\mu \pi} \frac{1}{d-2} P(x) ln(P(x)) \mid_{coeff \; of \; x^n} \qquad (97)$$

Evaluating the right hand side by a contour integral just as we did for d_n in the previous section, and dividing the answer by d_n, the degeneracy of level n, we find

$$\Delta r^2 = \frac{d-1}{\mu \pi} \sqrt{\frac{n}{6(d-1)}} \propto m \qquad (98)$$

Since the mass m is proportional to the length, this means that the typical string trajectory has a fractal dimension of 2, the same as a Brownian random walk. This also applies to the point-to-point separation on a string [99].

3.3 Failure of Canonical Ensemble

We are now ready to discuss the statistical mechanics of strings in general. Since string interactions involve the splitting and joining of strings, the number of strings is not conserved. The problem we wish to address is simple - given a box of volume V containing string with energy E to $E + dE$, which configurations are most probable? We shall make the usual assumption that all states are *a priori* equally likely so the probability distribution is just given by counting states.

Historically this problem was first addressed by Hagedorn, Frautschi and Carlitz [107,108,109] who were motivated by the desire to understand hadronic matter at high densities. Nowadays we believe that the string model of hadrons is only good at low energies, and at high densities hadronic matter behaves more like free quarks and gluons. Nevertheless many of the important conclusions were reached by these early works. For some recent papers see [104]. Here I will present a simplified account of the work of D. Mitchell and myself on the subject [99](see also [100]).

Let me begin by reviewing the basics of canonical statistical mechanics. One starts from the partition function

$$Z(\beta) = \int_0^\infty dE \, \Omega(E) e^{-\beta E} \qquad (99)$$

where $\Omega(E)$ is the number of states for the system to have energy in the range E to $E + dE$. β is $1/T$ as usual. The Boltzmann factor $e^{-\beta E}$ may be thought of as the decrease in the number of states for a heat bath if we take energy E from it into the system: the integrand of $Z(\beta)$ is the E dependent part of the total number of states for system plus heat bath. For 'normal' macroscopic systems, $\Omega(E)$ grows rapidly with E, typically as E^n where n is the number of particles in the system. In fact this is where the $e^{-\beta E}$ for the heat bath comes from. Assume that for the heat bath the number of states $\Omega_{hb}(E) \propto E^n$. From the entropy, defined as $S = ln(\Omega)$ we calculate the temperature via $\frac{1}{T} \equiv \frac{dS}{dE} = \frac{n}{E}$. The definition of a heat bath is that we hold T fixed and let n and E go to infinity.

Now if our system is in contact with the heat bath but the total energy E_{tot} conserved, the energy of the heat bath is given by $E_{tot} - E$ where E is the energy in the system. Thus the number of states for the heat bath is given up to constants by $ln(\Omega_{hb}(E_{tot} - E)) = n ln(E_{tot} - E) \propto -E/T$ expanding to first order in E/E_{tot} which becomes small as n goes to infinity. Thus the dependence of Ω_{hb} on E is given by the Boltzmann factor, as stated.

For 'normal' systems this dependence on E is strong enough to kill the rapid growth of $\Omega(E)$ as long as the heat bath is much larger than the system. The result is that the integrand in (99) is very sharply peaked. Its maximum is obtained by differentiating the exponent i.e. solving $\frac{dS}{dE}(E) = \beta$. Physically this corresponds to the value of E for which there is an equilibrium between the system and the heat bath - the total number of states does not change to first order if one moves energy from the system to the heat bath or vica versa. The sharpness of the peak in the integrand is essential if we are to calculate the expected value of properties of the system at fixed E by simply inserting those quantities into the integrand - in other words there must be a very tight correspondence between E and T if we are to use the canonical ensemble to answer questions about the isolated system. To see how sharply peaked the integrand actually is we expand $S(E)$ to second order about the equilibrium value $\overline{E}$

$$\Omega(E) \propto e^{\frac{1}{2}\frac{d^2S}{dE^2}(\overline{E})(E-\overline{E})^2+...} \tag{100}$$

The integrand has in this approximation a Gaussian peak with width $\delta E^2 = -\frac{d^2S}{dE^2}^{-1} = -\frac{dE}{d\beta} = T^2\frac{dE}{dT} \equiv T^2 c_V$ where c_V is the specific heat. Thus the fractional energy fluctuation is $\delta E^2/E^2 = c_V T^2/\overline{E}^2 \propto V^{-1}$ for normal systems, where the specific heat is positive and proportional to the volume. Thus as we take the volume to infinity the energy fluctuations go to zero.

I have reviewed these basic assumptions because strings will turn out to violate them. The cause of this violation is very simple. We saw in Section 3.1 that the number of states available to a string grows *exponentially* with energy (we can put all the energy into the rest mass). Thus the integrand in (99) contains the term $e^{bE-\beta E}$. For $\beta < b$ the integral diverges: this happens for

$$T > T_H \equiv b^{-1} = \sqrt{\frac{3\mu}{\pi(d-2)}} \tag{101}$$

Thus canonical thermodynamics simply does not make sense at temperatures above the Hagedorn temperature T_H. As I will emphasise this is not a fundamental inconsistency in any sense but simply means that the usual assumptions involved in the canonical approach break down.

3.4 Superstrings: $Z=\infty$ for $T > T_H$

Surprisingly this simple fact has been ignored in some of the recent superstring literature on strings at high temperature. For example it has been claimed[110] that for heterotic strings there is a 'duality' relation $Z(\beta) =$

$Z(\pi/\beta\mu)$ which is clearly inconsistent with Z being finite below T_H and infinite above it. What has gone wrong is that some of the resummations involved in the 'proofs' of this formula are illegitimate. The far-reaching conclusions drawn from this work by Atick and Witten[111] for example seem to me to be on very shaky ground. Let me give a brief and rigorous proof that Z is indeed infinite for superstrings above T_H.

Just as in field theory we can calculate the partition function for a gas of free strings, as in Section 2.2. The only difference is that there is an extra sum over the internal mass levels for the strings;

$$Z = e^{-\beta F}$$
$$-\beta F = \sum_{m=0}^{\infty} d_m V \int d^{d-1}k(-ln(1 - e^{-\beta\omega_k}) + ln(1 + e^{-\beta\omega_k})) \quad (102)$$

where d_m is the degeneracy of the mth mass level and the two terms come from bosons and fermions. Supersymmetry of course gives them identical degeneracies. Expanding the logarithms, we obtain

$$-\beta F = \sum_{m=0}^{\infty} d_m \sum_{n=1}^{\infty}(1+(-1)^{n+1})V \int d^{d-1}k \frac{e^{-n\beta\omega_k}}{n}$$
$$> 2\sum_{m=0}^{\infty} d_m V \int d^{d-1}k e^{-\beta\omega_k} \quad (103)$$

But for the massive modes $\omega_k = \sqrt{\vec{k}^2 + m^2} < m + \vec{k}^2/(2m)$ so we obtain the further inequality

$$-\beta F > 2\sum_{m=1}^{\infty} d_m e^{-\beta m} V \int d^{d-1}k e^{-\frac{\beta \vec{k}^2}{2m}}$$
$$= 2V \sum_{m=1}^{\infty} d_m e^{-\beta m}(\frac{m}{2\pi\beta})^{\frac{d-1}{2}} \quad (104)$$

Since $d_m \approx e^{bm}$ the sum is clearly divergent above T_H. Supersymmetry therefore makes little difference - the canonical partition function does not exist above T_H.

Now it might reasonably be argued that string interactions have so far been ignored and that these might well ameliorate the divergence in Z above T_H. Indeed this is precisely what happens with cosmic strings. In that case, as T approaches T_c, the critical temperature of the string - forming phase transition, the mass of the Higgs field goes to zero and thus the strings become very fat. Interactions between strings become important and the Nambu description breaks down. Both effects act to reduce the number of states available below that of free Nambu strings. The same is true of hadronic strings as one approaches the QCD phase transition.

The same phenomena could also occur in string theory but then string theory would probably not be a good name to call it! In any case we would obviously need more than free string calculations to understand it and this is really all that has been done so far. For small string coupling constant (the value of the string coupling constant depends upon the dilaton expectation value and cannot be fixed at present from low energy physics) the free string picture I will present below should indeed be reasonable. As we shall see it is not the strings but the canonical approach to thermodynamics that breaks down.

3.5 Low Temperatures and Densities: The canonical approach

At low temperatures and densities canonical thermodynamics provides a good description of a gas of strings. It is also easier than the microcanonical approach. As in the previous section we can write down the partition function for a gas of closed oriented bosonic strings

$$Z = e^{-\beta F}$$
$$-\beta F \approx -\int_{m_0}^{\infty} dm \rho(m) V \int d^{d-1}k \ln(1 - e^{-\beta \omega_k}) \qquad (105)$$

where we use $\rho(m) = c e^{bm} m^{-d}$ from (94) and we include a cutoff m_0 which is a characteristic low mass in the spectrum - we shall not believe any answers precisely when they involve m_0. For fundamental closed strings the mass m of the first massive level is $\sqrt{8\pi\mu}$ and so $bm = \sqrt{\frac{8\pi^2(d-2)}{3}} >> 1$ and we should take m_0 to be at least a few times this in order for the form of $\rho(m)$ to be valid. Thus we have $bm_0 >> 1$. This will be useful.

For $\beta m_0 > bm_0 >> 1$ we can expand the logarithm and keep only the first term, and approximate w_k by its nonrelativistic value - $w_k \approx m + \vec{k}^2/(2m)$. The fact that this is a good approximation (which is easily checked) means that the configurations that dominate are those where the loops are nonrelativistic - roughly speaking the internal degrees of freedom are more important than the centre of mass degrees of freedom. This also means that we can use energy and mass more or less interchangeably. We now perform the Gaussian $\vec{k}$ integral to obtain the thermodynamic quantities

$$-\beta F \approx \frac{cV}{(2\pi\beta)^{\frac{d-1}{2}}} \int_{m_0}^{\infty} dm \frac{e^{(b-\beta)m}}{m^{\frac{d+1}{2}}} \qquad (106)$$

$$\overline{E} = \frac{\partial}{\partial \beta}(\beta F) \approx \frac{cV}{(2\pi\beta)^{\frac{d-1}{2}}} \int_{m_0}^{\infty} dm \, \frac{e^{(b-\beta)m}}{m^{\frac{d-1}{2}}} \tag{107}$$

$$c_V \propto \overline{(E-\overline{E})^2} = \frac{\partial^2}{\partial \beta^2}(-\beta F) \approx \frac{cV}{(2\pi\beta)^{\frac{d-1}{2}}} \int_{m_0}^{\infty} dm \, \frac{e^{(b-\beta)m}}{m^{\frac{d-3}{2}}} \tag{108}$$

$$\tag{109}$$

where I have dropped terms down by βm_0. The density (108) corresponds to a distribution of string loops with the number of loops of mass m to $m + dm$ per unit volume

$$n(m) = \frac{c}{(2\pi\beta)^{\frac{d-1}{2}}} \frac{dm}{m^{\frac{d+1}{2}}} e^{(b-\beta)m} \tag{110}$$

which is just a Boltzmann distribution at temperature β^{-1}.

The first consequence is therefore that large loops are suppressed at low temperatures. The corresponding density is approximately given from (107) as

$$\rho \approx \frac{2\pi c}{(2\pi\beta)^{\frac{d-1}{2}}} \frac{e^{(b-\beta)m_0}}{m_0^{\frac{d-1}{2}}} \tag{111}$$

we can solve this to find the temperature corresponding to a given density;

$$T \approx \frac{T_H}{1 + \frac{1}{bm_0}ln(\frac{\rho_0}{\rho})}$$

$$\rho_0 \equiv m_0 c \tag{112}$$

recalling that $b^{-1} = T_H$. This shows that the thermodynamic temperature of a system of strings depends only very weakly on its density. For example for a network of cosmic strings in todays universe, $T_H \approx \mu^{\frac{1}{2}} \approx 10^{16} GeV$. Taking $bm_0 \approx 10$ and $ln(\rho_0/\rho) \approx 10^2$ we find $T \approx 10^{15}$ GeV! Of course this does not mean that the strings are literally hot but shows they are very far from thermal equilibrium with the surrounding radiation today, and in fact become so quite soon after they form. Thus, unfortunately, we can learn very little about the distribution of cosmic string in the late universe from equilibrium calculations.

What we do learn however is that a system of strings allowed to equilibrate at low density is very different from the system of strings at formation I discussed in section 2.8. Large loops are exponentially suppressed. The phase space is dominated by configurations where essentially all the string is in the smallest possible loops.

This is consistent with what we have learned from the numerical simulations where we saw that chopping off of loops from long strings is favoured over reconnection. It also indicates that it would be inconsistent to have a string dominated universe in the following sense [99]. Imagine tha $G\mu$ was very much less than unity. If the density of long strings grew relative to the background then the scale ξ on the network would become much smaller than the Hubble radius R_H (see Section 2). If ξ is much less than R_H however the strings evolve as in flat space. The timescale for the string to approach thermal equilibrium is however given by ξ, as I discussed. In a time of order $\xi \ll R_H$, the expansion time, therefore, the long strings would lose a fraction of their length into loops (because this is favoured by phase space) and the density of the long strings would fall relative to the background density. Thus the scaling solution is really inevitable, barring really long time transients which would make the string system quite unique among statistical systems (a gas in a box for example equilibrates in a few mean collision times).

The result for the loop distribution in flat spacetime was calculated in [99] and shown to agree well with the flat spacetime simulations of Smith and Vilenkin [112]. In the simulations the strings are discretised and the scale b enters as the spacing between points on the string. m_0 enters as the cutoff in the smallest loops allowed. The predictions from [99] and the simulation results [112] are shown in Fig 3.1.

3.6 The Microcanonical Approach

It is clear from (108) that the canonical approach will fail at high densities for $d \geq 4$ because the canonical density tends to a finite limit

$$\rho_H \approx \frac{2}{d-3} \frac{cm_0}{(2\pi b m_0)^{\frac{d-1}{2}}} \tag{113}$$

as the temperature approaches T_H. It is also clear that it will fail for $d = 4$ because the fluctuations are large at T_H. Thus for every d of interest it is essential to use the more fundamental microcanonical ensemble.

In the microcanonical ensemble one uses the density of states $\Omega(E)$ directly. One simply finds the configurations that dominate the total number of states, for energy fixed in the range E to $E + dE$ and for fixed volume V.

Now from (99) the partition function $Z(\beta)$ is simply the Laplace transform of $\Omega(E)$. Therefore one way to obtain $\Omega(E)$ is to invert the Laplace transform. Starting from (105), then as long as βm_0 is small we can expand the logarithm and keep only the first term. This yields

$$Z(\beta) \approx \sum_{n=1}^{\infty} \frac{1}{n!} (V \int d^{d-1}k \int_{m_0}^{\infty} dm \, \rho(m) e^{-\beta \omega_k})^n \tag{114}$$

Now since the inverse Laplace transform of an exponential is just a delta function we find

$$\Omega(E) = \sum_{n=1}^{\infty} \frac{1}{n!} \prod_{i=1}^{n} (V \int d^{d-1}k_i \int_{m_0}^{\infty} dm_i \, \rho(m_i)) \delta(E - \sum_{i=1}^{n} E_i) \qquad (115)$$

with $E_i = \sqrt{m_i^2 + \vec{k}_i^2}$. This is just the *classical* formula for the total number of states for an arbitrary number n strings of total energy E in a box. The $1/n!$ is just the usual 'classical statistics' overcounting factor for identical particles. It is only strictly valid when the average occupancy of each state is much less than unity, which is true here since we always have $e^{-\beta E_i} < e^{-\beta m_0} < e^{-b m_0} \ll 1$.

Equation (115) is the starting point for the microcanonical approach first followed by Frautschi[108] and Carlitz[109].

3.7 High Density

In reference [99] Mitchell and myself started from (115) and found the configurations that dominate the integrals directly. This is a hard calculation but there is a simpler approach that leads to the same result, based on examining the nature of the divergences in canonical expectation values at T_H which I shall explain here.

We can actually determine a lot about the function $\Omega(E)e^{-bE}$ (the integrand of the partition function at T_H) from its moments (107 - 109) and higher moments considered as T approaches T_H. This then tells us about the integrand $\Omega(E)e^{-\beta E}$ at arbitrary β. Consider the probability distribution

$$p(E)dE = \frac{\Omega(E)e^{-\beta E}dE}{Z(T_H)} \qquad (116)$$

We shall be interested in its limit as $\beta \to b$. Physically this is the probability for finding the system with energy E when it is in contact with a heat bath at T_H. The expectation value of the energy in this distribution is given by (108) at T_H;

$$\overline{E} = \frac{cV}{(2\pi b)^{\frac{d-1}{2}}} \int_{m_0}^{\infty} \frac{dm}{m^{\frac{d-1}{2}}} = \frac{2}{d-3} \frac{cVm_0}{(2\pi bm_0)^{\frac{d-1}{2}}} \qquad (117)$$

which is finite in $d \geq 4$ and corresponds to a loop distribution

$$n(m) = \frac{c}{(2\pi b)^{\frac{d-1}{2}}} \frac{dm}{m^{\frac{d+1}{2}}} \qquad (118)$$

For $d \geq 6$ the fluctuations around this value are small;

$$\overline{\delta E^2} = \frac{2}{d-5} \frac{cVm_0^2}{(2\pi bm_0)^{\frac{d-1}{2}}} \qquad (119)$$

so the fractional fluctuation $\delta E^2/E^2 \propto V^{-1}$ which tends to zero in the infinite volume limit. However for $d < 6$ the fluctuations are large.

In $d \geq 6$ near $\overline{E}$ the probability distribution is a sharply peaked Gaussian. However (and this is how strings differ from 'normal' systems) at large E $p(E)$ is not Gaussian at all.

We can see this by calculating higher moments of E just as in (107 - 109) for β close to b. The first divergent moment is

$$\overline{\delta E^{\frac{4}{3}}} = \frac{cV}{(2\pi\beta)^{\frac{d-1}{2}}} \int_{m_0}^{\infty} \frac{dm}{m^{\frac{1}{2}}} e^{(b-\beta)m}$$
$$\approx \frac{cV}{(2\pi\beta)^{\frac{d-1}{2}}} \frac{\sqrt{\pi}}{(\beta-b)^{\frac{1}{2}}} \qquad (120)$$

as $\beta \to b$ which tells us that for large E and at $\beta = b$

$$p(E) \approx \frac{cV}{(2\pi b)^{\frac{d-1}{2}}} \frac{1}{E^{\frac{d+1}{2}}} \qquad (121)$$

so

$$\Omega(E) \approx Z(T_H)\frac{cV}{(2\pi b)^{\frac{d-1}{2}}}\frac{e^{bE}}{E^{\frac{d+1}{2}}}$$

$$Z(T_H) = e^{-bF_H}$$

$$-bF_H = \frac{cV}{(2\pi b)^{\frac{d-1}{2}}}\int_{m_0}^{\infty}\frac{dm}{m^{\frac{d+1}{2}}} \qquad (122)$$

The second term in Ω is simply the number of states available to a single string of energy E:

$$\Omega_1(E) = cV\int d^{d-1}p\int_{m_0}^{\infty}dm\,\frac{e^{bm}}{m^d}\delta(E-\sqrt{\vec{p}^2+m^2}) \qquad (123)$$

where one uses the nonrelativistic solution to perform the m integral: $m \approx E - \vec{p}^2/(2E)$ and thus the momentum integral to find

$$\Omega_1(E) \approx \frac{cV}{(2\pi b)^{\frac{d-1}{2}}}\frac{e^{bE}}{E^{\frac{d+1}{2}}} \qquad (124)$$

Thus as one increases the density beyond ρ_H the energy just goes into a single long string. Now the first term in (122) is just the partition function at T_H

$$Z(T_H) \approx e^{-b\overline{E}}\Omega_{loops}(\overline{E})\Delta E \qquad (125)$$

in $d \geq 6$ because the integral is strongly peaked around $\overline{E}$, and we take it's width to be ΔE. Ω_{loops} is the number of states corresponding to the loop distribution (118) at T_H. Thus assuming $E >> \overline{E}$ (123) is just

$$\Omega(E) \approx \Omega_{loops}(\overline{E})\Omega_1(E-\overline{E})\Delta E \qquad (126)$$

which is the number of states where $\bar{E}$ of the string is in loops and the remainder is in one long string. This is therefore what happens at very high density, $\rho > \rho_H$. The configurations that dominate consist of a density independent loop distribution and a very large string which takes up a greater and greater fraction of the total energy as the density is increased.

In fact using our previous result (98) that $m \propto r^2$ we find the size distribution of loops

$$n(r)dr \propto \frac{dr}{r^d} \qquad (127)$$

precisely the scale invariant distribution of loops found in simulations of cosmic string formation. The remainder of the energy is in one long string, whose energy equals

$$E_{long\ string} = E - \rho_{loops} V \qquad (128)$$

where ρ_{loops} is just the energy density corresponding to the loop distribution (118).

All these results (random walks, loop distribution and predominance of long string) agree with what is found in the simulations of cosmic string formation discussed above. This is reassuring, but it must be remembered that all these results were derived for *free* strings. The cosmic string formation simulations used instead the assumption of *free Higgs field phases* and found the typical configurations. Both these calculations neglect interactions - as I have emphasised in Section 2 it is very hard to do better analytically. Happily the final scaling solution for the string network is rather insensitive to the details of the initial conditions (though it does require long strings to be present).

The predictions derived above and first made in [99] were checked in flat spacetime simulations by Sakellariadou and Vilenkin [113], extending the work of Smith and Vilenkin cited earlier. Their results are shown in Figure 3.1. As can be seen from the Figure, the transition to long string domination is very sharp.

3.8 Microcanonical Temperatures above T_H

It is interesting to examine the transition to long string domination a little more closely. In particular let us imagine that our system is in thermal contact with a heat bath. For superstrings this would simply be the massless modes in the theory - radiation, gravity waves etc. Now the microcanonical temperature

$$\frac{1}{T_\mu} = \frac{dS}{dE} = \frac{d}{dE} ln\Omega(E) \qquad (129)$$

is well defined at arbitrarily high density. It also has a simple physical meaning. It is the temperature of a heat bath which would be in equilibrium (to first order) with the system of strings (see Section 3.3). For $d \geq 6$ as we have seen the canonical approach is valid right up to T_H in the large volume limit. Thus to high accuracy we have $T_\mu(E) = T_{can}(E)$. Let me just emphasise that these two quantities are *a priori* different: to calculate T_μ one fixes E and calculates T from (129). In the canonical approach one does just the reverse, fixing T and calculating the expected value of $E(T)$. $T_{can}(E)$ is the canonical temperature required to obtain energy E.

$T_{can}(E)$ and $T_\mu(E)$ are compared in Figure 3.2. Of course T_{can} does not make sense above T_H. However at high E we can calculate T_μ from (122)

$$\frac{1}{T_\mu} = \frac{d}{dE} ln\Omega(E) = \frac{1}{T_H} - \frac{d+2}{2E} \qquad (130)$$

which yields $T_\mu > T_H$! However the difference in temperature is tiny in the large volume limit - the $E - T$ curve turns vertical at T_H in the large volume limit. Differentiating (130) we would find a negative specific heat, but $c_V \propto V^2$, an infinite specific heat per unit volume.

We have come a long way in the analysis of the Hagedorn transition to long string domination. The main conclusion is simple. In the large volume limit there are more states available to a system of strings than there are to a heat bath at any temperature above T_H. Thus as one squeezes a box full of string plus radiation there comes a density above which all the energy will flood into the long string.

This may have important consequences for fundamental strings in the early universe, or in black holes, where very high densities are reached. As long as equilibrium is attained and the effects of gravity sufficiently weak to validate the above analysis, above a certain density all the energy in the radiation will flood into long strings which would fill the universe. It is important to note that the *radiation* density is always bounded above, basically to T_H^d. Thus the problem of understanding singularities in string theory involves the massive string modes crucially. This makes it so different from the usual problems with ordinary matter or radiation that it may even have a solution!. Some discussion of the quantum mechanics of the massive modes in the very early universe was given in [101] where it was argued that they may actually lead to an inflationary phase (see also [100]). Some recent work on calculating the interaction rates for massive string modes may be found in [114,115] - this is important in eventually deciding how reasonable the assumption of thermal equilibrium is in the case of fundamental strings.

Acknowledgements

I would like to thank the organisers and participants of TASI for making the school such a stimulating one. I thank A. Albrecht and D. Mitchell for very enjoyable continuing collaboration, and J. Atick and C. -I. Tan for bringing some heat to the discussion of string statistical mechanics! This work was supported in part by the DOE and by the NASA at Fermilab.

References

[1] E. Kolb and M.S. Turner, *The Early Universe*, Addison Wesley, 1988.

[2] T.W.B. Kibble, J. Phys. A9 (1976) 1387; Physics Reports 67 (1980) 183.

[3] R. Giovanelli and M. P. Haynes, Astrophysical Journal, 87 (1982) 1355.

[4] R.F. Kirschner, A. Oemler, P.L. Shecter and S.A. Schectman, in 'Early Evolution of the Universe and its Present Structure', I.A.U. Symposium 104 (1983) 197.

[5] V. De Lapparent, M. Geller and J. Huchra, Astrophysical Journal 302 (1986) L11.

[6] S. Weinberg, *Gravitation and cosmology: Principles and Applications of the General Theory of Relativity*, John Wiley, New York, 1972.

[7] P.J.E. Peebles, *The Large Scale Structure of the Universe*, Princeton University Press, 1980.

[8] J. Yang, M.S. Turner, G. Steigman, D.N. Schramm and K. Olive, Ap. J. 281 (1984) 493.

[9] J. Primack, *Dark Matter in The Universe*, lectures presented at the International School of Physics, Varenna 1984, SLAC-PUB-3387.

[10] B. Bingelli, Astron. Astrophys. 107 (1982) 338.

[11] D. Koo, R. Kron, and A. Szalay, 13th Texas Symposium on Relativistic Astrophysics, Ed. P. Ulmer, World Scientific, 1986.

[12] G.O. Abell, Astrophysical Journal Supp. 3 (1958) 211; for a nice review see J. Oort, Ann. Rev. Astron. Astrophys. 21 (1983) 373.

[13] N.A. Bahcall and R.M. Soneira, Astrophysical Journal 270 (1983) 20.

[14] M. Davis and J. Huchra, Astrophysical Journal 254 (1982) 437.

[15] N.A. Bahcall and W.S. Burgett, Astrophysical Journal 300 (1986) L35.

[16] S. Schectman, Astrophysical Journal Supp. 57 (1985) 77.

[17] D. Lynden-Bell, S.M. Faber, D. Burstein, R.L. Davies, A. Dressler, R.J. Turlevich and G. Wegner, Astrophysical Journal, March 1988 in press.

[18] A. Szalay and D.N. Schramm, Nature 314 (1985) 718.

[19] J.R. Bond and G. Efstathiou, Astrophysical Journal 285 (1984) L45.

[20] D. Burstein and V.C. Rubin, Astyrophysical Journal 297 (1985) 423.

[21] M. Davis, G. Efstathiou, C. Frenk and S. White, Ap. J. 292 (1984) 371; S.D.M. White, in *Nearly Normal Galaxies, From the Planck Time to the Present*, Ed. S.M. Faber, Springer-Verlag, 1987.

[22] J.M. Bardeen, J.R. Bond, N. Kaiser, A. Szalay, Ap. J. 304 (1986) 15.

[23] C. Frenk, S. White, M. Davis and G. Efstathiou, Ap. J. 327 (1988) 507.

[24] J.R. Bond and A. Szalay, Ap. J. 274 (1984) 443.

[25] R. Brandenberger, N. Kaiser, D. Schramm and N. Turok, Phys. Rev. Lett. 59 (1987) 2371

[26] R. Brandenberger, N. Kaiser and N. Turok, Phys. Rev. D36 (1987) 335.

[27] S. Tremaine and J. Gunn, Phys. Rev. Lett. 42 (1979) 407.

[28] D.A. Kirzhnits and A.D. Linde, Phys. Lett. 42B (1972) 745.

[29] L. Dolan and R. Jackiw, Phys. Rev. D9 (1974) 3320; S. Weinberg, Phys. Rev. D((1974) 3357.

[30] A more detailed discussion is contained in K. Wilson and J. Kogut, Phys. Rep. 12 (1974) 75.

[31] E. Copeland, D. Haws and R. Rivers, Fermilab preprint 1988.

[32] T.W.B. Kibble, G. Lazarides and Q. Shafi, Phys. Rev. D26 (1982) 435; D. Olive and N. Turok, Phys. Lett. 117B (1982) 193.

[33] E. Witten, Phy. Lett. 149B (1985) 351.

[34] J. Preskill, Phy. Rev. Lett. 43 (1979) 1365.

[35] A. Vilenkin, Phys. Rep. 121 (1985) 263.

[36] E.P.S. Shellard Nuc. Phys. B283 (1987) 624.

[37] R. Matzner, University of Texas report, 1987.

[38] E.B. Bogomolny, Sov. J. Nucl. Phys. 24 (1976) 449; E.B. Bogomolny and M.S. Marinov, Sov. J. Nucl. Phys. 23 (1976) 355.

[39] L. Jacobs and C. Rebbi, Phys. Rev. B19 (1979) 4486.

[40] R. Davis, Phys. Rev. D32 (1985) 3172, A. Vilenkin and T. Vachaspati, Phys. Rev. D35 (1987) 1138.

[41] E. Witten, Nuc. Phys. B249 (1985) 557.

[42] J. Ostriker, C. Thompson and E. Witten, Phys. Lett. 180B (1986) 231.

[43] Y. Nambu, Proceedings of International Conference on Symmetries and Quark Models (Wayne State University, 1969); Lectures at the Copenhagen Summer Symposium, 1970

[44] K. Maeda and N. Turok, Phys. Lett. 202B (1988) 376.

[45] K.B. Nielsen and P. Olesen, Nuclear Physics B61 (1973) 45.

[46] J. Frieman, R. Scherrer, Phys. Rev. D33 (1986) 3556.

[47] N. Turok and P. Bhattacharjee, Phys. Rev. D33 (1984) 1557.

[48] N. Turok and T.W.B. Kibble, Phys. Lett. 116B (1982) 141.

[49] A. Albrecht and N. Turok, Phy. Rev. Lett. 54 (1985) 1868; in preparation, 1988.

[50] T.W.B. Kibble, Nuc. Phys. B252 (1985) 227.

[51] D. Bennett, Phys. Rev. D33 (1986) 872.

[52] D. Bennett and F. Bouchet, Phys. Rev. Lett. 60 (1988) 257.

[53] A. Vilenkin, Phys. Rev. Lett. 46 (1981) 1169, 1496(E).

[54] N. Turok, Phys. Lett. 123B (1983) 387.

[55] N. Turok, Nucl. Phys. B242 (1984) 520.

[56] T. Vachaspati and A. Vilenkin, Phys. Rev. D30 (1984) 2036; D. Garfinkle and T. Vachaspati, Phys. Rev. D36 (1987) 2229.

[57] T. Vachaspati and A. Vilenkin, Phys. Rev. D31 (1985) 3052.

[58] T. Vachaspati, Phys. Rev. D35 (1987) 1767; Gravity Research Foundation Prize Essay, 1987.

[59] C. Burden Phy. Lett. 164B (1985) 277.

[60] A. Vilenkin, Phys. Rev. D23 (1981) 852.

[61] An older reference is L. Marder, Proc. Roy. Soc. Lond. A252 (1959) 45.

[62] J.R. Gott, Ap. J. 288 (1985) 422.

[63] D. Garfinkle, Phys. Rev. D32 (1985) 1323.

[64] N. Turok, Phys. Rev. Lett. 55 (1985) 1801.

[65] E. Bertschinger Ap. J. 316 (1987) 489.

[66] N. Turok and R. Brandenberger, Phys. Rev. D33 (1986) 2175.

[67] N. Turok, in *A Unified View of the Macro- and Micro-Cosmos*, Ed. A De Rujula, D.V. Nanopoulos and P.A. Shaver, World Scientific 1987.

[68] F. Bouchet, LBL preprint, 1988.

[69] R. Scherrer and W. Press, Smithsonian preprint, 1988.

[70] N. Turok, in *Nearly Normal Galaxies, From the Planck Time to the Present*, Ed. S.M. Faber, Springer-Verlag, 1987.

[71] A. Albrecht, A. Stebbins and N. Turok, in preparation, 1988.

[72] P.J.E.Peebles, Ap. J. 263 (1982) L1.

[73] L. Kofman and A.D. Linde, Nuc. Phys. B282 (1987) 555.

[74] P. Langacker and S.-Y. Pi, Phys. Rev. Lett. 45 (1980) 1.

[75] A. Everett, T. Vachaspati and A. Vilenkin, Phys. Rev. D31 (1985) 1925.

[76] E. Copeland, D. Haws, T. Kibble, D. Mitchell and N. Turok, Nuc. Phys. B298 (1988) 445.

[77] G. Lazarides C. Panagiotakopoulos and Q. Shafi, Phys. Lett. 183 (1987) 289.

[78] R. Brandenberger, N. Kaiser, E.P.S. Shellard and N. Turok, Phys. Rev. D36 (1987) 335.

[79] J.R. Bond, in *Nearly Normal Galaxies, From the Planck Time to the Present*, Ed. S.M. Faber, Springer-Verlag, 1987.

[80] Q. Shafi and A. Vilenkin, Phys. Rev. D29 (1984) 1870.

[81] E. Vishniac, K. Olive and D. Seckel, Nuc. Phys. B289 (1987) 717.

[82] A. Vilenkin and Q. Shafi, Phys. Rev. Lett. 51 (1983) 1716.

[83] E. Bertschinger and P. Watts Ap. J. 328 (1988) 23.

[84] E. Harrison, Phys. Rev. D1 (1980) 2726.

[85] Ya. B. Zel'dovich, M. N. R. A. S.,160 (1982) 1P.

[86] S. Hawking, Phys. Lett. 115B (1982) 295; A. Guth and S.-Y. Pi, Phys. Rev. Lett. 49 (1982) 1110; J. Bardeen, P. Steinhardt and M. S. Turner, Phys. Rev. D28 (1983) 679.

[87] R. Brandenberger and E.P.S. Shellard, Brown preprint, 1988.

[88] N. Kaiser and A. Stebbins, Nature 310 (1984) 391; A. Stebbins, Ap. J. 327 (1988) 584.

[89] F. Bouchet, D. Bennett and A. Stebbins, Fermilab preprint, 1988.

[90] F Melchiorri, B. Melchiorri, C. Ceccarelli and L. Pietranera, Ap. J. Lett. 250 (1981) L1.

[91] L.L. Cowie and E.M. Hu, Ap. J. 318 (1987) L33.

[92] A. Vilenkin, Phys. Lett. 107B (1981) 47.

[93] E. Witten, Phys. Rev. D30 (1984) 272.

[94] C. Hogan and M. Rees, Nature 311 (1984) 109.

[95] R. Brandenberger, A. Albrecht and N. Turok, Nuc. Phys. B277 (1986) 605.

[96] L.A. Rawley, J.H. Taylor, M.M. Davis and D.W. Allan, Science 238 (1987) 761.

[97] R. Davis, Phys. Lett. 161B (1285) 285; D. Bennett, Phys. Rev. D33 (1986) 872.

[98] H. Hodges and M. Turner, Fermilab preprint, 1988.

[99] D. Mitchell and N. Turok, Phys. Rev. Lett. 58 (1987) 1577; Nucl. Phys. B294 (1987) 1138.

[100] Y.Aharonov, F. Englert and J. Orloff, Phys. Lett. B 199 (1987) 366.

[101] N. Turok, Phys. Rev. Lett. 60 (1988) 552.

[102] M. Green, J. Schwarz and E. Witten, *Superstring Theory*, Cambridge University Press, 1987.

[103] K. Huang and S. Weinberg, Phys. Rev. Lett. 25 (1970) 895.

[104] M. Bowick and L.C.R. Wijewardhana, Phys. Rev. Lett. 54 (1985) 2485; B. Sundborg, Nucl. Phys. B254 (1985) 583; E. Alvarez, Phys. Rev. D31 (1985) 418, Nucl. Phys. B269 (1986) 596; P. Salomonson and B.-S. Skagerstam, Nucl. Phys. B268 (1986) 349.

[105] P. Goddard, G. Goldstone, C. Rebbi and C. Thorn, Nucl. Phys. B 56 (1973) 109.

[106] G. H. Hardy and S. Ramanujan, Proc. Lond. Math. Soc. 17 (1918) 75.

[107] R. Hagedorn, Nuovo Cimento Suppl. 3 (1965) 147.

[108] S. Frautschi, Phys. Rev. D3 (1971) 2821.

[109] S. Carlitz, Phys. Rev. D5 (1972) 3231.

[110] K. H. O'Brien and C.-I.Tan, Phys. Rev. D36 (1987) 1184.

[111] J. Atick and E. Witten, IAS preprint IAS-HEP-88/14.

[112] G. Smith and A. Vilenkin, Phys. Rev. D36 (1987) 987.

[113] M. Sakellariadou and A. Vilenkin, Phys. Rev. D37 (1988) 885.

[114] J. Polchinski, Texas preprint UTTG-07-8.

[115] D. Mitchell, N. Turok, R. Wilkinson and P. Jetzer, Fermilab preprint, 1988.

Figure Captions

Figure 2.1

The potential $V(\Phi) = \frac{\lambda}{8}(\Phi^2 - \eta^2)^2$ used in Section 2.

Figure 2.2

The effective potential for the long wavelength modes $V_{eff}(\Phi)$ calculated in Section 2.2, at $T = 0$, $T = T_c$, and $T = \sqrt{2}T_c$.

Figure 2.3

The three simplest types of defects. On the left is the spatial distribution of the fields. Arrows show the location of the fields on the minimum of the potential V_{min}, which is shown on the right.
a) a domain wall
b) a string
c) a magnetic monopole.

Figure 2.4

A pictorial demonstration that $\pi_1(G/H) = \pi_0(H)$ when G is simply connected. On the left, H has two components, the component connected to the identity, H_e, and the disconnected component H'. When G/H is constructed from G one identifies all the elements of H with the identity. This can be pictured as squeezing all the elements of H together, forming a 'tunnel' as shown on the right. This clearly has a noncontractible loop.

Figure 2.5

The profile of a cosmic string. The solid line is $|\Phi|/\eta$ where the angular dependence (in cylindrical coordinates) is simply $\Phi = |\Phi| e^{i\theta}$. The dashed line is $A_\theta gr$, with G the gauge coupling. The gauge potential A_θ asymptotically tends to $1/(gr)$ so the total 'magnetic' flux down the string is $\int \vec{B}.\vec{dS} = \int \vec{A}.\vec{dl} = 2\pi/g$. The radius is in units of $1/(g\eta)$, and the solution is shown for $\lambda = g^2$, the case where the Higgs particle mass equals the gauge particle mass in the broken phase (Section 2.7).

Figure 2.6

The approximation used in simulations of the formation of the simplest 'U(1)' cosmic strings (Section 2.8).
a) The space V_{min} (the circle) is approximated by 3 points. Each domain is assigned a value 0,1 or 2 corresponding to one of these points. The rule for

the phase of the Higgs field is to move along the shortest path on V_{min} when going from one domain to the next.
b) an edge with four surrounding domains - this edge is a string.
c) this edge is not a string.

Figure 2.7

A box of strings at formation. The strings are formed on a cubic lattice with the algorithm described in Figure 2.6. The box is tilted and the string width decreases away from the nearest side of the box to give an impression of perspective. Periodic boundary conditions are used so all of the string is in the form of closed loops- most of it is actually in a single very long loop.

Figure 2.8

Coordinates used in the derivation of the Nambu action given in Section 2.9 . It is assumed that the radius of curvature of the string, R, is much greater than it's width W.

Figure 2.9

When two strings collide they reconnect the other way (Section 2.9).

Figure 2.10

A selection of simple loop trajectories in flat spacetime which do not self-intersect at any time. The solutions plotted are given by
$\vec{r}(\sigma,t) = \frac{1}{2}((1-\alpha)sin\sigma_- + \frac{\alpha}{3}sin3\sigma_- + sin\sigma_+, -(1-\alpha)cos\sigma_- - \frac{\alpha}{3}cos3\sigma_- - cos\phi cos\sigma_+, -2\sqrt{\alpha(1-\alpha)}cos\sigma_- - sin\phi cos\sigma_+)$ where $\sigma_\pm = \sigma \pm t$ and σ runs from 0 to 2π. This family of solutions satisfies the equations (52) for all α and ϕ. They are periodic, with period π.

Figure 2.11

A string network in a radiation dominated universe. The cube shown is half a Hubble radius on each side and the scale factor of the universe has doubled since the beginning of the simulation. The picture is from a forthcoming paper by Albrecht and myself.

Figure 2.12

The string density in the latest simulations by Albrecht and myself, for radiation dominated universes. The vertical axis shows the string density in units of the string tension divided by the Hubble radius squared $(\mu/(2t)^2)$. The horizontal axis shows the comoving Hubble radius in units of the initial domain size ξ when the strings are formed. This is proportional to the scale factor. The scaling density appears to be at $200\mu/(2t)^2$. The upper and lower solid lines show simulation results starting at higher and lower than scaling density. The dashed line shows the prediction of the simple model explained in Section 2.12.

Figure 2.13

The conical metric of a long straight string. If an angle $8\pi G\mu$ is 'cut out' of flat space (left) and the two edges identified, a cone is the result (right).

Figure 3.1

The equilibrium loop distribution at low density in flat spacetime found in numerical simulations by Smith and Vilenkin (squares) for different cutoffs in the size (number of points) of the loops allowed, and the prediction from the analysis of Mitchell and myself explained in Section 3.5 for the same quantity. $N \geq E$ is the number of loops with energy greater than E and δ is the energy per point on the string. μ ids the string tension. The two parameters in the predicted distribution were fitted to one curve, then used to predict the second.

Figure 3.2

The results of flat spacetime simulations by Sakellariadou and Vilenkin.
a) The equilibrium loop distribution at high density. The three sets of points show the distribution at $\rho = 0.2$ (open circles), $\rho = 0.45$ (solid circles) and $\rho = 0.8$ (triangles). The solid line is the prediction, explained in Section 3.7, that $n(E) \propto E^{-\frac{5}{2}}$.
b) the total density in loops (total density minus the density in long string) versus the total density. At low density all the string is in loops, with the distribution of Figure 3.1 . As the density is increased, a point is reached beyond which the density in loops saturates and everything goes into the long string. This is explained in Section 3.7 .

Figure 3.3

The energy versus temperature curve discussed in Section 3.8, for spacetime dimension $d \geq 4$. The dashed line shows the canonical prediction, where one fixes T and calculates E. The calculation fails above the Hagedorn temperature T_H. The solid line shows the microcanonical result, where one fixes E and calculates T. In the infinite volume limit this turns vertical beyond T_H.

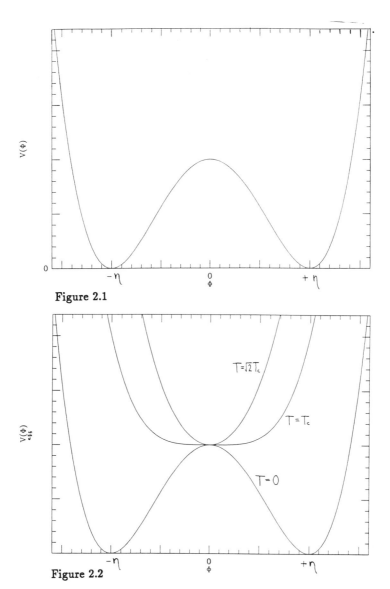

Figure 2.1

Figure 2.2

Figure 2.3

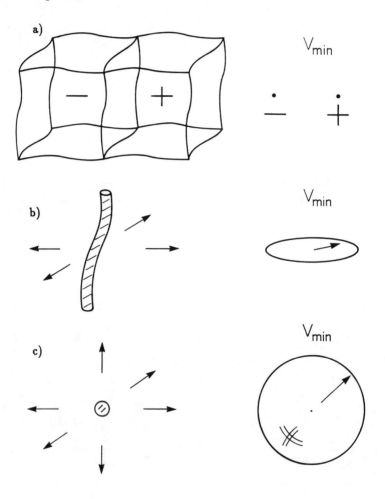

Figure 2.4

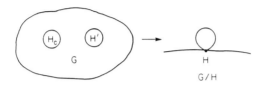

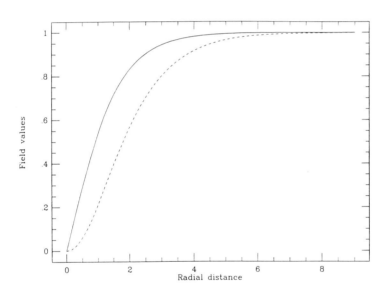

Figure 2.5

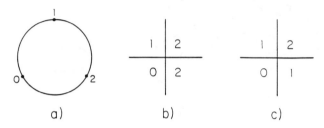

Figure 2.6

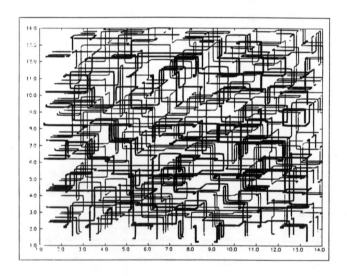

Figure 2.7

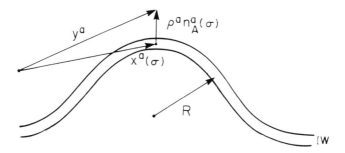

Figure 2.8

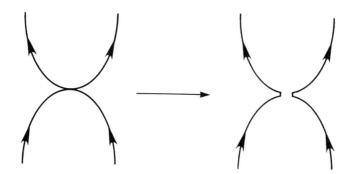

Figure 2.9

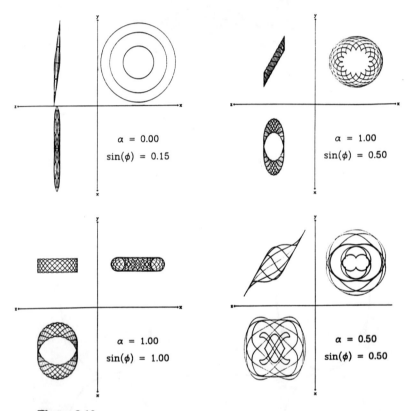

Figure 2.10

Figure 2.11

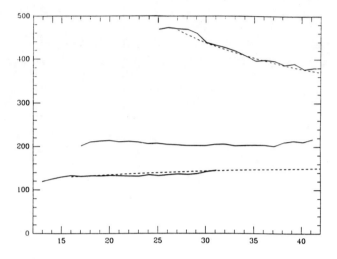

Figure 2.12

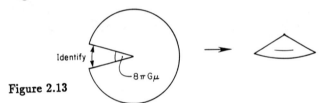

Figure 2.13

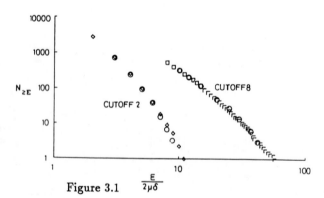

Figure 3.1

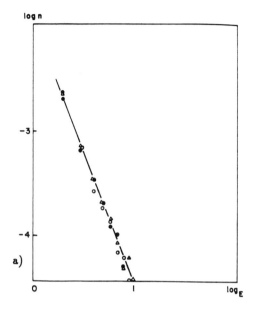

Figure 3.2

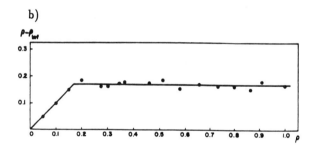

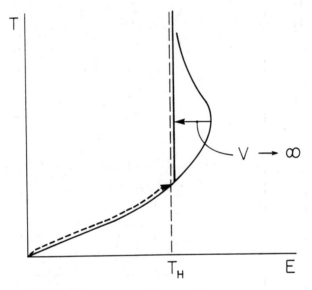

Figure 3.3

Remarks on Thermodynamics of Strings

GREGORY G. ATHANASIU[*] AND JOSEPH J. ATICK[†]

School of Natural Sciences
Institute for Advanced Study
Princeton, NJ 08540

ABSTRACT

Some aspects of string thermodynamics are explored within the context of canonical as well as microcanonical ensembles, with emphasis on exposing the nature of the Hagedorn temperature. We consider several alternative scenarios for what could be happening above T_H. However we argue that available evidence seems to favour a first order phase transition before or at T_H when interactions are included.

---

[*] Work supported by Department of Energy, contract DE-AC02-76ER02220.
[†] Work Supported by National Science Foundation grant number PHYS86-20266.

Despite all the impressive progress of the past few years, our basic notions of what strings are all about have remained at a rather primitive stage. The truth of the matter is that up to now we do not possess any formulation of string theory where its symmetries, its degrees of freedom or the structure of the space of all its vacua could be explored. Not to mention its dynamics or the choice of a unique vacuum (within which contact with reality could be attempted). This has placed strings in the very awkward–though unique–position of being the theory of everything that predicted nothing!

What keeps the dream alive is the hope that all this could change as the amount of "string wisdom" accumulates. In fact in the recent past a lot of valuable insight about strings have started to emerge. Most significant of all in our opinion, is the insight that has been gained, within current formulations of strings, by probing strings under extreme physical conditions in some sort of gedanken experiments. For example in refs. [1–3], one studies string fixed angle scattering at very high energy; a regime in QCD for example, dominated by parton like behaviour and thus very revealing. The philosophy here is to develop some intuition about underlying "stringy phenomena": Phenomena which exhibit marked deviations from corresponding field theoretic behaviour and are more or less insensitive to details of the string model. These should help us summarize the essential features that a fundamental theory of strings has to embody.

Another region that we feel is worth exploring in the same spirit is the region of extreme high temperature. In addition to having the potential of teaching us something new about the fundamental nature of strings, it is a physical region. After all, if one adopts the premise that strings have anything to do with reality then we have to accept that they must have been hot in some earlier era. It is precisely in that era, where the extended nature of strings is most prominent, that one expects deviations from standard cosmology. It is not inconceivable that these deviations are what is needed to cure the problems or explain some of the puzzles of early universe physics. One issue which strings very likely could shed some light on is the question of the initial singularity.

In this talk we explore some pertinent aspects of string thermodynamics. Our interest is mainly in the behaviour of strings at high temperature and high energies. We examine these regions within the context of the canonical as well as the microcanonical ensembles. For an ideal gas we shall see that both ensembles lead to unreasonable physics above the Hagedorn temperature. We will then argue that introducing interactions leads us to the belief that strings undergo a first order phase transition before they get to the Hagedorn temperature and that all the nonsensibility can be attributed to the fact that one is studying strings in the wrong phase. This view has been advocated recently in ref. [25]. Several of our results here, have also been derived from different points of view by several people. In fact the literature on the subject is rather substantial and goes as far back as the work of Hagedorn in the mid sixties. The reader is encouraged to consult refs. [4-31] for further details and different points of view.

Region of validity of string thermodynamics

Before we try to address some of the main issues about string thermodynamics, we have to keep in mind that the physical validity of any analysis of this sort is limited. These limitations stem from the well known fact that, in principle, the concepts and tools of thermodynamics in the presence of gravity are ill defined. One way to see this is to recall that statistical mechanics is only rigorously defined in the large volume limit. But any gravitating system with a nonzero energy density and sufficiently large size suffers even classically from the Jeans instability [32], which is the tendency of the system to collapse into lumps. More precisely the instability happens for length scales $R \geq (\pi v_s/G\rho)^{\frac{1}{2}}$, where v_s is the velocity of sound in the medium, ρ the energy density and G is Newtons constant (in terms of the closed string coupling $\sim g^2$). A closely related point is that thermal energy density provides a source in Einstein's equations $R_{\mu\nu} - \frac{1}{2}g_{\mu\nu}R = GT_{\mu\nu}$, that would drive the evolution of spacetime. A correct description of thermodynamics of strings has to take into account this back reaction. Inevitably, this would require formulating string propagation in time dependent backgrounds, about which we know nothing at the moment.

In passing, we should emphasize that the instability is not just a problem of principle, it actually shows up in practical calculations: At one loop at finite temperature, the graviton and dilaton develop a tadpole, which is essentially an indication that we are in the wrong vacuum. Infinities in higher loop amplitudes and higher point functions at one loop, will therefore arise from regions of moduli space where the amplitude factorizes on the massless tadpoles.* These infinities presumably can be eliminated consistently in some sort of Fischler-Susskind [34] mechanism by allowing for the evolution of the background spacetime.

Despite these limitations there still is a regime where our analysis makes approximate sense [25]. First we have to arrange that the coupling constant or the density is small enough that we can justify ignoring the instability. This requires $g^2 \rho R^2 \ll 1$. Nevertheless, we still have to have a big enough ensemble so that we can do thermodynamics. This can be arranged if $R \gg \sqrt{\alpha'}$, where α' is the string tension which sets a characteristic length scale in the problem. These two conditions are simultaneously satisfied by choosing $g^2 \rho \alpha' \ll 1$. Henceforth we shall assume that this is so.

Let us start our discussion by first examining an ideal gas of strings.

Thermodynamics of an Ideal Gas of Strings

(i) Canonical Ensemble

In most physical systems we know, the canonical ensemble $Z(\beta)$, has proven to be more informative and to some extent easier to calculate than the seemingly more fundamental microcanonical ensemble. It is logical therefore, to examine what it says about string thermodynamics first.

★ For a somewhat detailed discussion of the origin of divergences in string theory see ref.[33].

For strings it is known that we can calculate $Z(\beta)$ by two different ways. The first of course is through its definition

$$Z(\beta) = Tr \ e^{-\beta H} \tag{1}$$

which requires knowledge of the single string spectrum and the degeneracy of single string states. The second is by calculating the string one loop path integral on a target space $R^9 \times S^1$, with β the radius of the compactified time dimension. As was shown in [15] this path integral yields $-lnZ(\beta)$. This equivalence will play a significant role in our analysis later on when we try to think about interactions. At this stage it is probably more illuminating to calculate $Z(\beta)$ explicitly from (1).

In order to examine the dependence of various thermodynamical quantities on the dimensionality of spacetime we shall carry out the calculation of $Z(\beta)$ for strings on backgrounds of the form $M^d \times T^{(10-d)}$ (or $M^d \times T^{(26-d)}$ for the bosonic string). Here M^d is d dimensional uncompactified flat spacetime and $T^{(10-d)}$ is a $(10-d)$ dimensional torus. For simplicity we choose all the radii R of this internal torus to be equal. We could have considered more general possibilities however this is not needed to exhibit the features of thermodynamics that we are interested in here.

The mass spectrum of a string propagating on the space $M^d \times T^{(10-d)}$ is given by [35]:

$$\frac{\alpha'}{2}M^2 = n_- + \frac{\alpha'}{4}\left(\frac{m_i}{R} - \frac{w_i R}{\alpha'}\right)^2 \\ n_+ + \frac{\alpha'}{4}\left(\frac{m_i}{R} + \frac{w_i R}{\alpha'}\right)^2 \tag{2}$$

subject to the level matching contraint:

$$n_- = n_+ + \vec{m} \cdot \vec{w} \tag{3}$$

The notation that we have adopted is as follows: n_-, n_+ are the level numbers for right and left movers respectively; $\{(m_i, w_i); i = 1, \ldots, 10-d\}$ are the momentum and winding numbers of the state in the internal directions. The degeneracy of the levels in (n_+, n_-) is calculable if the generating function $D_\pm(x) \equiv Tr\,(x^{N_\pm})$ is known; where the trace runs over a complete set of states. Here $N_\pm$ is an operator whose eigenvalue on the state is the level number $n_\pm$. $N_-(N_+)$ can be written as an infinite sum over bilinears in the right (left) handed bosonic and fermionic oscillators that create the states. For example for the bosonic string $N = \sum_{n>0} \sum_\mu \alpha^\mu_{-n} \alpha^\mu_n$. More precisely, if we expand $D(x)$ in powers of x; $D_\pm(x) = \sum_{n=1}^\infty d_\pm(n_\pm) x^{n_\pm}$, the coefficient of the $x^{n_\pm}$ term is nothing but the degeneracy of the $n_\pm$'th level. The generating function is explicitly known for all string theories of interest to us. What one actually finds from these expressions is that for large $n_\pm$, $d(n)$ grows exponentially with $n_\pm$. For closed strings the net degeneracy is $d_+(n_+) \cdot d_-(n_-)$ with n_-, n_+ related by appropriate level matching condition. For example for flat space one finds [6,9, 35]:

$$d(n) = d_+(n)d_-(n) \sim (2n)^{-c} exp(b\sqrt{\frac{4n}{\alpha'}}) \qquad (4)$$

where c and b for different string theories are given as follows:

Bosonic String	$c = 27/2$	$b = 4\pi\sqrt{\alpha'}$
Type II superstring	$c = 11/2$	$b = \pi\sqrt{8\alpha'}$
Heterotic string	$c = 11/2$	$b = \pi(2+\sqrt{2})\sqrt{\alpha'}$

Let us see what such exponential growth in the degeneracy leads in thermodynamics. To regulate things we work in a box of finite volume L^{d-1} for M^d with periodic boundary conditions, however no winding in these uncompactified directions will be allowed. The energy levels of a single string of momentum k are $\epsilon_{k\alpha} = \sqrt{(\frac{2\pi}{L})^2 k^2 + M^2(\alpha)}$, where $\alpha = \{n_+, m_i, w_i\}$, is the set of quantum numbers characterizing the state, n_- should always be thought of as given in terms of the set α through eq. (3).

Nothing fundamental in our discussion will depend on the particular string theory we consider. For definiteness we consider the superstring. For the superstring $Z(\beta)$ is:

$$Z(\beta) = \prod_{k,\alpha} \left(\frac{1 + e^{-\beta \epsilon_{k\alpha}}}{1 - e^{-\beta \epsilon_{k\alpha}}}\right)^{d_+(n_+) \cdot d_-(n_-)} \tag{5}$$

or the free energy

$$-\beta F = \ln Z(\beta)$$
$$= \left(\frac{L}{2\pi}\right)^{d-1} \sum_\alpha d_+(n_+) \cdot d_-(n_-) \int d^{d-1}p \left(\ln(1 + e^{-\beta \epsilon_{p\alpha}}) - \ln(1 - e^{-\beta \epsilon_{p\alpha}})\right) \tag{6}$$

where we have taken the continuum limit in the sum over momenta. Expanding the logarithms in a power series in $e^{-\beta \epsilon}$ one can then carry out the integration over the momenta explicitly. The exact answer turns out to be an infinite series involving modified Bessel functions $K_\nu(z)$:

$$\ln Z(\beta) = \pi^{\frac{d-2}{2}} 2^{d/2} \left(\frac{L}{2\pi}\right)^{d-1} \sum_{\{w_i\}=-\infty}^\infty \sum_{\{m_i\}=-\infty}^\infty \sum_{n=1}^\infty \sum_{p=1}^\infty d_+(n) \cdot d_-(n + \vec{m} \cdot \vec{w})$$
$$\times \frac{1}{(2p-1)} \cdot M^{d-1}(\alpha) \cdot \left(\frac{1}{(2p-1)M(\alpha)\beta}\right)^{\frac{d}{2}-1} \cdot K_{\frac{d}{2}}((2p-1)M\beta) \tag{7}$$

This does not look very illuminating. However, its main properties are not hard to exhibit. The first significant fact to observe about this expression is that it diverges if $\beta < b$, i.e if the temperature T is larger than some temperature $\frac{1}{b}$ which has become known as the Hagedorn temperature T_H. To see this we only need to notice that for large n, the degeneracy factor $\sim e^{b\sqrt{n}}$ while the leading term in (7)($p = 1$, $w_i = m_i = 0$ term) has the asymptotic behaviour of a Bessel function that goes like $e^{-\beta\sqrt{n}}$. For $\beta < b$ this leads to an exponentially growing series and hence to the celebrated divergence in $\ln Z(\beta)$.

What does a divergence in the thermal ensemble of an ideal string gas above the Hagedorn temperature mean? Does it signify a limiting temperature, a phase transition or neither? To gain more insight, still within the context of an ideal gas, we can examine the behaviour of various thermodynamical quantities as we approach the Hagedorn temperature from below. In particular we are concerned with the behaviour of the mean energy $<E>$ and the specific heat c_V.

Expression (7) can be put in a much more illuminating form by noting that for the most part the dependence on $\{w_i, m_i, n\}$ is through the mass $M(\alpha)$. We can therefore trade the three sums over these variables by one over the mass with an appropriate density of states that has to be calculated. To do this we insert in expression (7) the identity element expressed as

$$1 = \frac{\alpha'}{2} \int dM M \delta\left(\frac{\alpha'}{2}M^2 - \frac{\alpha'}{4}\left(\frac{m_i}{R} - \frac{w_i R}{\alpha'}\right)^2 - n\right) \tag{8}$$

and then carry out the sum over n. The measure that we arrive at looks something like, $\rho(M)dM \sim MdM \sum_{w_i, m_i} d_-\left(\frac{\alpha'}{2}M^2 - \frac{\alpha'}{4}\left(\frac{m_i}{R} - \frac{w_i R}{\alpha'}\right)^2\right) \cdot d_+\left(\frac{\alpha'}{2}M^2 - \frac{\alpha'}{4}\left(\frac{m_i}{R} + \frac{w_i R}{\alpha'}\right)^2\right)$ Where we have to keep in mind that the summation over (m_i, w_i) is constrained to $\left(\frac{m_i}{R} \pm \frac{w_i R}{\alpha'}\right)^2 < M^2$. Since expressions for thermodynamics near the Hagedorn temperature will be dominated mainly by large masses we need mostly to exhibit this measure asymptotically in M. In that case we can replace the summations in that measure by integrals. Using (4) for d_- and d_+ these integrals can then be carried out in an expansion of $\frac{1}{M}$. The leading term turns out to be:

$$\rho(M) \sim B\frac{e^{bM}}{M^d}\left(1 + O\left(\frac{1}{M}\right)\right) \tag{9}$$

where B is some mass independent numerical factor.

With the asymptotic density of states in (9), the leading contribution to $\ln Z(\beta)$, apart from irrelevant numerical factors is

$$\ln Z(\beta) \sim \int_\Lambda^\infty \frac{dM}{M} \frac{e^{bM}}{(M\beta)^{d/2-1}} K_{d/2}(M\beta) \tag{10}$$

where we have ignored the contribution from low lying states by introducing an infrared cutoff Λ, below which the density of states deviates from the exponential growth in (9). Typically, Λ will be a few times the mass of the first level. We are justified in introducing such a cutoff since the contribution to (7) from low lying states of very low degeneracy can be seen to be negligible near T_H. Note that the smallest the arguement of the Bessel function could be $\Lambda\beta \sim \Lambda b$ is still $\gg 1$ (recall Λ is few times the mass of the first level, so for example for the superstring the first massive state is at $M_1 \sim 2/\sqrt{\alpha'}$ while $b = \pi\sqrt{8\alpha'}$, hence $M_1 b \sim$ .... $\gg 1$ and $\Lambda b \gg 1$.) This fact finally allows us to replace the Bessel function reliably by its asymptotic form $K_\nu(z) \sim \sqrt{\frac{\pi}{2z}}e^{-z}$, leading to:

$$\ln Z(\beta) \sim (\frac{1}{\beta})^{\frac{d-1}{2}} \int_\Lambda^\infty \frac{dM}{M^{(d+1)/2}} e^{M(b-\beta)}$$
$$\sim \frac{(\beta-b)^{\frac{d-1}{2}}}{\beta^{(d-1)/2}} \Gamma(\frac{1-d}{2}, \Lambda(\beta-b)) \tag{11}$$

where $\Gamma(A, B)$ is a generalized gamma function.

From the first line in (11) we see that strings more or less have a Boltzmann distribution at this temperature given by $n(M) \sim \frac{1}{\beta^{(d-1)/2}} \cdot \frac{e^{M(b-\beta)}}{M^{(d+1)/2}}$. Eq. (11) also makes it clear that $<E> = -\frac{\partial}{\partial\beta} \ln Z(\beta)$ goes as $(\beta-b)^{\frac{d-3}{2}}$ in the limit $\beta \to b$, which is always finite for spacetime dimensions of interest to physics. While the specific heat c_V goes like $(\beta-b)^{\frac{d-5}{2}}$ in the same limit. A curious feature is that in $d = 4$ the specific heat does diverge at the Hagedorn temperature. More generally for any given dimensionality, high enough derivatives of $\ln Z(\beta)$ will diverge at T_H. This behaviour cannot be distinguished from that of a sufficiently high order (continuous to this approximation) phase transition happening at T_H. In fact the critical exponents that we get here are the same as those of the mean field theory Gaussian model in d dimensions [36].

From what has been done so far it would be hard to understand why for $d \geq 6$, where the specific heat is finite, we cannot heat up the system beyond T_H. Even when the specific heat diverges but $<E>$ is finite ($d = 4$), T_H is not necessarily a limiting temperature, it could be just the onset of a phase transition. We feel that within the context of the canonical ensemble therefore there is no basis why T_H should be a limiting temperature. However we cannot argue reliably that it corresponds to a phase transition either unless we have a way of exploring the new phase above T_H. Speculations on this will be given later in the talk. Before we submit to such speculations there is one more possibility that has to be considered: Maybe nothing fundamental is happening with strings above the Hagedorn temperature. Perhaps it is just the validity of the thermal ensemble $Z(\beta)$ as an honest description of thermodynamics is breaking down. If this is so, then a more refined analysis using say the microcanonical ensemble should reveal perfectly descent physics above the Hagedorn temperature to the approximation that we are working with. A major portion of the rest of the talk is allocated to checking this serious possibility.

(ii) Microcanonical Ensemble

Given that $Z(\beta)$ does not exist for an ideal string gas above the Hagedorn temperature, one way we can explore the thermodynamics of that region is to attempt to use the microcanonical ensemble. The microcanonical, although harder to calculate and is typically therefore less useful in physics, is admittedly more fundamental. Within this setup one asks for the configuration that dominates phase space and with the assumption of ergodicity the system is expected to end up in this particular configuration. By definition therefore the microcanonical is more or less guaranteed to exist and no embarrasment of the sort found for $Z(\beta)$ above T_H should be expected here. Nevertheless, there still could be embarrasment in the physics that follows from this ensemble and this is what we are trying to check here. Quantitatively, the object we need to calculate in the microcanonical ensemble is the total number of microstates for a gas of strings with a total energy between E and $E + dE$. If we call that number $\Omega(E)dE$ then thermodynamics follows from Boltzmann's equation

$$S = \ln \Omega(E)$$

In particular, $\beta = \frac{\partial}{\partial E}\ln\Omega(E)|_{E=\bar{E}}$ and $c_V = -\frac{1}{T^2}(\frac{\partial}{\partial E}\frac{\partial}{\partial E}\ln\Omega(E))^{-1}$.

Before we calculate $\Omega(E)$ for strings, we would like to digress a bit to recall for future reference, the precise relationship between $Z(\beta)$ and $\Omega(E)$. As is well known, starting with the microcanonical ensemble, one can arrive at $Z(\beta)$ as a derived concept. More formally $Z(\beta)$ is just a Laplace transform of $\Omega(E)$

$$Z(\beta) = \int_0^\infty dE e^{-\beta E}\Omega(E) \qquad (12)$$

This is of course, what arises from considering thermalization of the system of interest with an infinite heat bath. If this transform exists we can recover $\Omega(E)$ through its inverse:

$$\Omega(E) = \int_{\beta_0-i\infty}^{\beta_0+i\infty} \frac{d\beta}{2\pi i}e^{\beta E}Z(\beta) \qquad (13)$$

Nothing apriori guarantees that the transform exists for all β. In ordinary systems $\Omega(E)$ typically grows like $E^{\text{(size of system)}}$ (e.g. $\sim E^{3N/2-1}$ for an ideal gas) with energy while $e^{-\beta E}$ is decaying. Hence one expects to find an energy where $e^{-\beta E}\Omega(E)$ is peaked very sharply at. This is being so, the Laplace transform in (12), can be evaluated reliably using steepest descent. The energy where the integral picks up the most contribution is around E_0 which satisfies the stationary point condition $-\beta + \frac{\partial}{\partial E}\ln\Omega(E)|_{E=E_0} = 0$. But $\beta = \frac{\partial}{\partial E}\ln\Omega(E)|_{E=<E>}$. So in normal systems the most probable and the mean energy agree. In strings the exponential growth in the degeneracy of the single string states leads (as we shall see below) to an exponential growth in $\Omega(E)$ and hence above the Hagedorn temperature $e^{-\beta E}\Omega(E)$ has no peak for any finite E.

Equation (13) is actually one way we could calculate $\Omega(E)$ if we choose β_0 so that we are in a regime where $Z(\beta)$ exists. The resulting integrals though, are very hard to calculate amd most probably will have to be done numerically[*].

---

[*] For low energy it is possible to reliably use steepest descent techniques to calculate these integrals, however this is not the region of interest to us here.

For this reason we resort to first principles in our calculation of $\Omega(E)$.

To start with notice that $\Omega(E)$ could be written as:

$$\Omega(E) = \sum_{N=1}^{\infty} \Omega(N, E) \tag{14}$$

where $\Omega(N, E)dE$ is the number of microstates for N strings with a total energy E. By definition this is given by:

$$\Omega(N, E) = \frac{1}{N!} \int_\Lambda^\infty \prod_{i=1}^N dE_i \rho(E_i) \delta(\sum_{i=1}^N E_i - E) \tag{15}$$

where $\rho(E_i)$ is the degeneracy of states of a single string with energy E_i and we have divided by $N!$ to account for the indistinguishability of strings.

The density of single string states at a given energy $\rho(E)$ can be calculated from $d_+(n_+)$ and $d_-(n_-)$ the same way we have calculated $\rho(M)$ above. The only difference is that since $E^2 = p^2 + M^2(\alpha)$ we have to carry out the integration over the d dimensional spacetime momenta p_i in addition to the integration over the set of quantum numbers α. It is not hard to check that $\rho(E)$ to leading order in $\frac{1}{E}$ goes as

$$\rho(E) \sim K \frac{e^{bE}}{E^{\frac{1+d}{2}}} (1 + O(\frac{1}{E})) \tag{16}$$

where K is some calculable numerical factor.

Substituting (16) into (15) gives an integral representation of $\Omega(N, E)$ of the form:

$$\begin{aligned}\Omega(N, E) &= \frac{K^N}{N!} \int_\Lambda \prod_{i=1}^N \frac{dE_i e^{bE_i}}{E_i^{\frac{1+d}{2}}} \delta(\sum_i E_i - E) \\ &= \frac{K^N e^{bE}}{N!} E^{N(\frac{1-d}{2})-1} \int_{\frac{\Lambda}{E}} \prod_{i=1}^N \frac{dx_i}{x_i^{\frac{(1+d)}{2}}} \delta(\sum_{i=1}^N x_i - 1)\end{aligned} \tag{17}$$

One strategy for evaluating (17) is to eliminate the delta function constraint all together by carrying out say the x_N integral. One should be a bit careful about the limits of integration after this elimination. We just have to keep in mind that the multidimensional integration is always constrained to the hypersurface $\sum_{i=1}^{N} x_i = 1$. A more explicit expression for $\Omega(N, E)$ is then given by:

$$\Omega(N, E) = \frac{K^N E^{N\frac{(1-d)}{2}-1}}{N!} \int_k^{1-k} dx_1 \int_k^{1-x_1-k} dx_2 \int_k^{1-x_1-x_2-k} dx_3 \cdots \int_k^{1-x_1-\cdots x_{N-2}-k} dx_{N-1}$$

$$\times \prod_{i=1}^{N-1} \frac{1}{x_i^{(1+d)/2}} \cdot \frac{1}{(1 - \sum_{i=1}^{N-1} x_i)^{(1+d)/2}}$$

(18)

where $k = \Lambda/E$. One key fact to observe about (18) is that for small k the integrand in the variable x_{N-1} is sharply peaked near the lower and upper limits of integration, so the integral is dominated by those neighborhoods (This domination is better the higher d is). The behaviour is actually symmetric with respect to the two limits. It is sufficient to calculate near k and multiply the answer by 2. For x_{N-1} near k, the relevant part of the integrand simplifies to to $\frac{dx_{N-1}}{x_{N-1}^{(1+d)/2}}$ and thus the contribution to the integral from that neighborhood is $(\frac{2}{d-1})(\frac{1}{k})^{(d-1)/2}$. The $N - 2$ integrals that are left can be recognized, apart from some numerical factors, as the integral representation for $\Omega(N - 1, E)$. What this means is that in the limit of small k (i.e. large energy) we can derive a recursion relation between $\Omega(N, E)$ and $\Omega(N - 1, E)$. The precise form of this recursion is:

$$\Omega(N, E) = 2(\frac{2}{d-1}) \cdot \frac{K}{N} \cdot (\frac{1}{\Lambda^{(d-1)/2}}) \cdot \Omega(N-1, E) \qquad (19)$$

using this recursion it is an easy matter to calculate $\Omega(E)$ by carrying out the sum over N. We get:

$$\Omega(E) = exp(\frac{4}{d-1} \frac{K}{\Lambda^{(d-1)/2}}) \cdot K \cdot \frac{e^{bE}}{E^{(1+d)/2}} \qquad (20)$$

This is a strikingly simple result. It agrees with the answer obtained by [7,8,22,30] , using different techniques.

The result in (20) has a simple interpretation too: What we have seen in arriving at (20), is that to leading order the result is mostly coming from the region of integration near the boundaries. This is true for all the successive integrations dx_i except for the very first one which eliminated the delta function. What this means is that in a configuration of N strings at very high energy E, $N-1$ strings get the least possible energy allowed which is here equal to the cutoff while one string gets virtually all of E. We see this interpretation of what is happening born explicitly in (20), since the factor $e^{bE}/E^{(1+d)/2}$ is nothing but $\rho(E)$ for a single string. This single string dominance was also recently discussed in [30].

Let us check what thermodynamics follow from $\Omega(E)$ in (20). The microcanonical temperature:

$$\beta = \frac{\partial}{\partial E} \ln \Omega(E) = b - \frac{(1+d)}{2}\frac{1}{E} \qquad (21)$$

exceeds the Hagedorn temperature and asymptotes it from above as $E \to \infty$. What is disturbing about this is that strings seem to have negative specific heat in that region.

$$c_V = -\frac{1}{T^2}\frac{2}{d+1}E^2 = -(\frac{2}{d+1})\frac{1}{T^2}\rho V^2 \qquad (22)$$

where ρ is energy density. A second disturbing fact is that $c_V/V \sim V$ and hence no sensible thermodynamical limit exists for this system.

At this juncture it is perhaps worth summing up the situation. We have tried to explain the origin of the divergence in $Z(\beta)$ for strings above T_H. One scenario that we have entertained for a while is the possibility that nothing fundamental is happening above T_H, it is just the validilty of $Z(\beta)$ is breaking down. We feel the results just derived do not support such a view. The microcanonical ensemble for an ideal gas of strings have lead us to very nonsensical thermodynamics. This takes us back to the drawing board to try again and understand what is physically happening to strings above the Hagedorn temperature. In the next section we shall argue that no sensible physics should be expected until one has included interactions (regardless of how weak the coupling is).

The Hagedorn transition as a first order phase transition

So far we have only considered the thermodynamics of an ideal gas of strings. We found, in our opinion, no sensible thermodynamics above T_H regardless of which ensemble we use. In what follows we would like to examine to what extent this nonsensibility is due to having ignored interactions of strings. Our argument here will be rather sketchy we refer the reader to ref. [25] for further detail.

Naturally the first question one asks is how to introduce interactions in any analysis of string thermodynamics. In the microcanonical ensemble this seems, just as is the case in ordinary point particle theories, very difficult to do. The canonical ensemble on the other hand, could be formulated at the interacting level rather easily. The key to this is the existence of the second way to calculate $Z(\beta)$ that we alluded to earlier: One of the lessons that we have learned in the eighties about strings is the fact that $-\ln Z(\beta)$ is given by the path integral of strings propagating on the target space $R^9 \times S^1$ with S^1 of radius β and with the correct periodicity and antiperiodicity on bosons and spacetime fermions under $X^0 \to X^0 + \beta$. Although the equivalence has been established at one loop we expect that if we calculate the path integral on higher genus riemann surfaces, to be calculating higher order corrections to $Z(\beta)$ in some kind of cluster expansion.

Formulating string thermodynamics in this path integral language gives us some insight on why we are finding sensless thermodynamics for an ideal gas of strings above the Hagedorn temperature. To see this, let us consider the bosonic string propagating on a target space $R^9 \times S^1$, similar analysis holds for fermionic strings. The mode expansion has now winding in the S^1 direction and quantized momenta in that direction:

$$X^0 = x^0 + \frac{2\pi m}{\beta}\tau + \frac{n\beta}{\pi}\sigma + \cdots \qquad (23)$$

where $n, m \in Z$. The spectrum on this space is then given by:

$$\frac{M^2}{4} = N + \frac{1}{2}\left(\frac{m\pi}{\beta} + \frac{n\beta}{2\pi}\right)^2 - 1 \\ + \tilde{N} + \frac{1}{2}\left(\frac{m\pi}{\beta} - \frac{n\beta}{2\pi}\right)^2 - 1 \quad (24)$$

subject to the constraint $N - \tilde{N} = mn$ coming from the level matching condition $L_0 = \bar{L}_0$. Consider the following stringy states ($m = 0, n = \pm 1$) which we shall denote by (ϕ, ϕ^*). These are conjugate winding states with a mass that is given by

$$M_\phi^2 = -8 + \frac{1}{\pi^2 T^2} \quad (25)$$

As the temperature increases these states become less massive, until $T = \frac{1}{2\pi\sqrt{2}}$ they become massless. This is nothing but the Hagedorn temperature for bosonic string. This was first noted in [18,19,20,37]. Above T_H these states become tachyonic. In the formulation on $R^9 \times S^1$, this explains the divergence in $Z(\beta)$ as due to thermal tachyons above T_H. After all in carrying out the integration over the moduli space the only infinities that arise in stable vacua are those due to tachyons in the infrared corners of moduli space.

Let us examine this instability in some further detail. When all interactions are turned off the relevant part of the effective potential involving the fields (ϕ, ϕ^*) is given by:

$$V(\phi, \phi^*) = \left(\frac{1}{\pi^2 T^2} - 8\right)\phi^*\phi \quad (26)$$

which for $T > T_H$ leads to an unstable vacuum around the origin. Doing calculations with strings in flat space with $<\phi> = 0$, is doing perturbation about this *unstable* vacuum. If we take interactions into account the effective potential will involve higher powers of (ϕ, ϕ^*) which could lead to a stable vacuum. In general $V(\phi, \phi^*)$ will look something like:

$$V(\phi, \phi^*) = \left(\frac{1}{\pi^2 T^2} - 8\right)\phi^*\phi + ug^2(\phi\phi^*)^2 + \cdots \quad (27)$$

where u is some numerical coefficient whose sign is very important. If $u < 0$, this potential (assuming that terms higher order than $(\phi\phi^*)^2$ stabilize the vacuum) gives a first order phase transition with ϕ acquiring a VEV at a temperature for which the coefficient of $\phi\phi^*$ is still positive [36]. This implies that the critical temperature $T_c < T_H$. On the other hand if $u > 0$ it is consistent to assume that the transition is 2nd order and T_c = temperature at which the mass term vanishes. This will happen for $T_c = T_H$. Since things are continuous in this case one can calculate perturbatively. The VEV acquired by ϕ to leading order is

$$|<\phi>|^2 = \frac{|\frac{1}{\pi^2 T^2} - 8|}{2g^2} \qquad (28)$$

while the free energy is just the value of the potential at its minimum:

$$F = -\frac{(\frac{1}{\pi^2 T^2} - 8)^2}{4g^2} \qquad (29)$$

what is significant about this is the fact that it goes as $O(1/g^2)$. If we recall that genus k contribution to the free energy goes as $O(g^{2(k-1)})$ then we must interpret this contribution to the free energy as coming from genus zero.

The above is very reminiscent of $\frac{1}{N}$ expansion in QCD [38]. In that case $\frac{1}{N}$ plays the analogue of the loop counting parameter: The contribution from the kth loop goes as $(N^2)^{1-k}$. To make the comparison, we would like to think that the deconfinement temperature in QCD plays the same role as T_c in strings. At $T < T_{dec}$ we calculate the free energy of QCD by counting glueballs and mesons. Their spectrum being independent of N leads to $O(1)$ free energy in the confined phase. This means that in perturbation the free energy receives contributions from genus one riemann surfaces and higher just as in strings. Above T_{dec} on the otherhand one works with gluons whose number grows like N^2. So the free energy is expected to be $O(N^2)$, which signals the fact that genus zero surfaces start contributing, similar to strings. In QCD the order of the phase transition is supposed to be first order, what order would it be in string theory?

It is not too difficult to determine the order of this phase transition in string theory. All we have to do is calculate from strings the sign of the quartic term in the effective action for the winding states (ϕ, ϕ^*). Actually one knows what their vertex operator is. It is given by $e^{\pm i\sqrt{2}(X_L - X_R)}$ at $T = T_H$. Therefore we can explicitly calculate any amplitude we want. In ref. [25] u was found to be negative. This leads us to believe that the transition is first order. This is rather unfortunate since just as is the case with all first order transitions the stable vacuum most likely will exist very far from the origin at a place where our perturbative expansion in field strength breaks down. Also we would not be able to tell within perturbation how far T_c is from the Hagedorn temperature.

To sum up, having a first order phase transition means that $g^2 |<\phi>|^2$ is necessarily $O(1)$, and hence, contrary to intuition that one gets from 2nd order transitions, the error one makes by being in the unstable vacuum is $O(1)$. We see therefore no justification for ignoring this effect of interactions: A correct description of thermodynamics of strings above T_H will have to be done on backgrounds with $|<\phi>|^2 \neq 0$. This is what we call the correct phase of string theory above T_H. It is not clear whether it will be possible within our current tools to find the correct (stable) vacuum. However assuming its existence it was argued, among other things, in ref. [25] that in this phase strings have remarkably much less degress of freedom than one expects. The prospects of connection between this point and the work of ref. [3,40, 39] is tantalizing.

We feel this subject is worthy of further investigations. The level of understanding and rigour is admittedly very primitive. For example, it is not clear to us how to translate the insight gained above into the language of strings on R^{10}. The states (ϕ, ϕ^*) are rather mysterious. They are not states in the spectrum on R^{10}, yet their vev plays the role of the order parameter as we approach the critical temperature. Even if we are convinced that sensible thermodynamics needs interactions it is not clear that we will be able to learn anything from perturbation theory. Maybe saddle point techniques like those of [1] will be valuable in this respect.

ACKNOWLEDGEMENTS

We would like to thank J. Harvey, N. Redlich, N. Seiberg, T. Sienko, A. Strominger, C. Tan, H. Tye, N. Turok and especially E. Witten for very useful discussions. J. J. A. is indebted to the organizers of "Strings ' 88 workshop", University of Maryland; and "Theoretical Advanced Study Institute", Brown University for their hospitality. Relevant discussions with the participants of both conferences are gratefully acknowledged. J. J. A. is also indebted to the Aspen center for Physics where some of the early work on this manuscript was done.

REFERENCES

1. D. J. Gross and P. Mende, 'String Theory Above the Planck Mass, Princeton university preprint no PUPT-1067 November (1987).

2. D. Amati, M. Ciafaloni and G. Veneziano, Int. J. of Modern Physics **7** 1988) 1615.

3. D. J. Gross, Phys. Rev. Lett. **60** (1988) 1229.

4. R. Hagedorn, Nuovo Cimento Suppl. **3** (1965) 147.

5. S. Fubini and G. Veneziano, Nuovo Cimento **64A** (1969) 1640.

6. K. Huang and S. Weinberg, Phys. Rev. Lett. **25** (1970) 895.

7. S. Frautschi, Phys. Rev. **D3** (1971) 2821.

8. S. Carlitz, Phys. Rev. **D5** (1972) 3231.

9. D. Gross, J. Harvey, E. Martinec and R. Rohm, Nucl. Phys. **B256** (1985) 253.

10. E. Alvarez, Phys. Rev. **D31** (1985) 418.

11. B. Sundborg, Nucl. Phys. **B254** (1985) 583.

12. M. J. Bowick and L. C. R. Wijewardhana, Phys. Rev. Lett. **54** (1985) 2485.

13. S. H. Tye, Phys. Lett. **158B** (1985) 388.

14. C. Aragao de Carvalho, O. Eboli and G. Marques, Phys. Rev. **D32** (1985) 3256.

15. J. Polchinski, Commun. Math. Phys. **104** (1986) 37.

16. P. Salomonson and B.-S. Skagerstam, Nucl. Phys. **B268** (1986) 349.

17. B. Maclain and B. Roth, Commun. Math. Phys. **111** (1987) 539.

18. B. Sathiapalan, Phys. Rev. **D35** (1987) 3277.

19. A. Kogan, ITEP preprint no. 110 (1987).

20. K. H. O'Brien and C.-I. Tan, Phys. Rev. **D36** (1987) 1184.

21. M. Axenides, S. D. Ellis and C. Kounnas, preprint LBL-2364, UCB-PTH-87/27

22. D. Mitchell and N. Turok, Phys. Rev. Lett. **58** (1987) 1577; Nucl. Phys. **B294** (1987) 1138.

23. H. Okada, Prog. Theor. Phys. **77** (1987) 751.

24. E. Ahmed, Int. J. Theor. Phys. **26** (1987) 1135.

25. J. J. Atick and E. Witten, IASSNS-HEP-88/14 (1988), Nucl. Phys. B, in press

26. M. McGuigan, Rockeffeler report Number DOE/ER/40325-31 Task B, April 1988.

27. Y. Lablanc, MIT preprints, CTP-1588, April (1988), CTP-1647 and CTP-1648 Sept. (1988).

28. R. Brandenberger and C. Vafa, BROWN-HET-673 (HUTP-88/A035), August. 1988

29. P. Ditsas and E. G. Floratos, Phys. Lett. **201B** (1988) 49.

30. N. Turok, "Phase Transitions as the Origin of Large Scale Structure in the Universe", Princeton preprint, Aug. 1988.

31. M. Spiegelglas, preprint, RU-88-45, Oct. 1988.

32. S. Weinberg, *'Gravitation and Cosmology,'* pp. 563 (Wiley-Interscience, 1972).

33. J. J. Atick, G. Moore and A. Sen, Nucl. Phys. **B307** (1987) 221; **308** (1988) 1.

34. W. Fischler and L. Susskind, Phys. Lett. **171B** (1986) 383; **173B** (1986) 262.

35. M. Green, J. Schwarz and E. Witten, *Superstring Theory*, Cambridge University Press, 1987.

36. P. Pfeuty and G. Toulouse, *Introduction to the Renormalization Group and to Critical Phenomena*, John Wiley and Sons, 1975.

37. C. Vafa, unpublished.

38. C. Thorn, Phys. Lett. **99B** (1981) 458.

39. E. Witten, ' Topological Quantum Field Theory,' 'Topological Gravity,' 'Topological Sigma Models,' IAS preprints 87/72, 88/2, 88/7 (1988).

40. I. Klebanov and L. Susskind, 'Continuum Strings From Discrete Space-Time Models,' preprint SLAC-PUB-4602 (1988).

STATISTICAL MECHANICS OF STRINGS AT HIGH ENERGY DENSITIES[*]

Chung-I Tan

Department of Physics
Brown University, Providence, Rhode Island 02912

ABSTRACT

We discuss in this note conditions under which both microcanonical and canonical quantities remain equally profitable for describing strings at high energy densities. We also comment on the nature of "phase transitions" anticipated in string theories.

---

[*] Work supported in part by the Department of Energy under contract DE-AC02-76ERO3130.027-Task A

INTRODUCTION

For statistical systems with a finite number of fundamental degrees of freedom, it is well-known that microcanonical and canonical descriptions are equivalent. For strings, this equivalence is less well understood, especially in the region of high temperature and/or high energy densities.[1-13] The primary objective of this note is to discuss conditions under which both descriptions remain equally profitable and also to comment on the nature of "phase transitions" anticipated in string theories.

The fundamental quantity in a canonical approach is the partition function:

$$Z(\beta, V) \equiv Tr e^{-\beta \hat{H}} = \sum_\alpha e^{-\beta E_\alpha}, \quad (1)$$

where the sum is over all possible multiparticle states of the system. For a microcanonical approach, one works with a density function, which counts the number of microstates,

$$\Omega(E, V) dE \equiv \sum_\alpha \delta(E - E_\alpha) \, dE. \quad (2)$$

Statistical mechanics based on a microcanonical ensemble is more general, even though it is often more convenient to work with $Z(\beta, V)$, e.g., when interactions must be included.

Representing the δ-function in (2) by an integral along an imaginary axis, we find that $\Omega(E,V) = \sum_\alpha \int_{-i\infty}^{+i\infty} \frac{d\beta}{2\pi i} \, e^{\beta(E-E_\alpha)}$. Note that the integral is in the form of an inverse Laplace transform over the complex-temperature plane. If one can move the contour into a region where interchanging the order of sum and integral is allowed, one obtains

$$\Omega(E, V) = \int_{\beta_0 - i\infty}^{\beta_0 + i\infty} \frac{d\beta}{2\pi i} \, Z(\beta, V) \, e^{\beta E}. \quad (3)$$

The allowed region is labelled by the interception of the contour with the real axis, β_0. One can then recover $Z(\beta, V)$ from $\Omega(E, V)$ via a Laplace transform,

$$Z(\beta, V) = \int_0^\infty dE \, \Omega(E, V) \, e^{-\beta E}, \quad (4)$$

which provides an alternative analytic definition for $Z(\beta, V)$. Under what conditions are (3) and (4) valid?

For "normal" systems in a *finite* volume, Eq. (1) defines an analytic function in β. Since $\Omega(E,V)$ can be shown to grow with E at most as $0(E^{const.})$, (4) also yields an analytic function in β for $Re\,\beta > 0$. These two definitions agree with each other, and the contour in (3) can be taken along *any* imaginary axis in the right-half β-plane. Normal thermodynamics, however, also requires taking the large volume limit so that the total degrees of freedom become unbounded. What happens to the analyticity structure of $Z(\beta, V)$? Is it still meaningful to make use of complex-β? The situation is more confusing for a string theory since the degrees of freedom are already infinite even in a finite volume. One normally assumes that the system is weakly interacting so that, at sufficiently low temperatures, $Z(\beta, V)$ is controlled by the free-particle spectrum. When $Re\,\beta$ is large, it can be shown that (1) is analytic in β. As we increase the temperature, will there be singularities in the right-half complex β-plane? If there are, do they depend on interactions, and/or on the volume V? Under what conditions will (3) and (4) hold for strings?

For normal systems, Eq. (3), (or equivalently Eq. (4)), can often be approximated by a saddle contribution. Introducing entropy and free-energy per unit volume by $S \equiv \log \Omega/V$ and $F \equiv -\beta^{-1} \log Z/V$ respectively, and denoting the saddle point by β_1, one has

$$S(E,V) \approx \beta_1 E/V - \beta_1 F(\beta_1, V) + \Delta S/V, \tag{5}$$

where ΔS represents fluctuations about the saddle point. For a closed system where E is fixed, the usual notion of a temperature is given by β_1^{-1} and is related to E by the stationary condition: $E = -\frac{\partial \log Z}{\partial \beta}|_{\beta_1}$. We shall refer to this as the Boltzmann temperature. Alternatively, one can also follow Gibbs by defining a different inverse temperature by $\beta_2 \equiv V \frac{\partial S(E,V)}{\partial E}$. In the thermodynamic limit, both entropy and free energy are extensive, $S \to S(E/V)$ and $F \to F(\beta_1)$, whereas ΔS is proportional to $\log V$. Therefore, for V large, ΔS in (5) can be ignored, and these two notions of temperature agree in this limit.[14]

The equivalence of microcanonical and canonical approaches for strings has heretofore been addressed at the level of whether (5) holds. This kind of discussions clearly misses the more serious issues which we have just raised. For strings, we must first clarify the relation between canonical and microcanonical quantities, *i.e.*, (3) and (4). The question of saddle point analysis plays a peripheral role when compared to our broader issues.

IDEAL GAS AND ANALYTICITY IN INVERSE TEMPERATURE

Let us begin by taking a closer look at statistical mechanics of conventional systems by examining a simple example. Consider a classical gas of N non-interacting identical particles of mass m. Under Maxwell-Boltzmann (MB) statistics, the microcanonical density, $\Omega_N(E,V)$, can be obtained by introducing a generating function, $\Omega(E,V;z) \equiv \sum_{N=1}^{\infty} \Omega_N(E,V) z^N$. It is easy to verify that $\Omega_N(E,V)$ obeys a simple one-term recursion relation, and

$$\partial \Omega(E,V;z)/\partial z = f(E,V;m) + \int_0^E dE_1 f(E_1,V;m) \Omega(E-E_1,V;z), \qquad (6)$$

where $f(E,V;m) \equiv \Omega_1(E,V)$ is the single-particle density. For a relativistic particle in d-spatial dimensions, $f(E,V;m) = V \int \frac{d^d k}{(2\pi)^d} \delta(E - (m^2+k^2)^{1/2})$, which easily leads to $f(E,V;m) \propto V E (E^2 - m^2)^{d/2-1} \theta(E-m)$. This allows one to find $\Omega(E,V;z)$ and to verify that $\Omega_N(E,V)$ grows with E as $0(E^{const.})$.

It is now possible to define, for $\mathrm{Re}\,\beta > 0$, $Z(\beta,V;z)$, $\tilde{f}(\beta,V;m)$, and $Z_N(\beta,V)$ as Laplace transforms of $\Omega(E,V;z)$, $f(E,V;m)$, and $\Omega_N(E,V)$ respectively. Instead of (6), one now has a simple differential equation to work with:

$$\partial Z(\beta,V;z)/\partial z - \tilde{f}(\beta,V;m) Z(\beta,V;z) = \tilde{f}(\beta,V;m). \qquad (7)$$

It follows that $Z(\beta,V;z) = e^{z\tilde{f}(\beta,V;m)} - 1$, $Z_N(\beta,V) = \tilde{f}(\beta,V;m)^N / N!$, where

$$\tilde{f}(\beta,V;m) = \beta(2\pi\beta/m)^{-(d+1)/2} V \int_0^\infty ds\, s^{-(d+3)/2} e^{-(m\beta/2)(s+1/s)}. \qquad (8)$$

For the limit where m is large, (8) reduces to $V(m/2\pi\beta)^{d/2} e^{-\beta m}$, the familiar non-relativistic result.

Note that, other than a singularity at $\beta = 0$, (8) is analytic when $\mathrm{Re}\,\beta > 0$. The expression for $Z_N(\beta,V)$ derived by a Laplace transform agrees with the original definition for $Z_N(\beta,V)$. It follows that (3) and (4) hold when $\mathrm{Re}\,\beta > 0$; and this result is independent of the volume, V. Therefore, in the thermodynamic limit of large V and N with N/V

fixed, the free-energy per unit volume, $F(\beta) \equiv -\beta^{-1} \log Z_N(\beta, V)/V$, approaches a regular function for $Re\beta > 0$. One finds that (3) defines a single-phase theory, and is independent of β_0. The singularity at $\beta = 0$ controls the asymptotic growth of $\Omega_N(E, V)$, it is independent of the volume, and it does not signal for the onset of a phase transition. We also note that (8) can be analytically continued into the entire complex β-plane.

STATISTICAL MECHANICS FOR FREE-STRINGS

Traditional approach to string statistical mechanics suffers several defects which could lead to misleading results. One possible flaw is the assumption of Maxwell-Boltzmann statistics. The use of MB statistics sounds reasonable at first since one normally expects that, at high temperature, the system appraoches a classical limit. However, for strings, the notion of a classical limit remains elusive. Furthermore, since we will be interested in the limit where energy densities are high, one might suspect that properly incorporating quantum statistics is crucial for obtaining the correct analytic property of $Z(\beta, V)$. More seriously, all string statistical studies have been carried out under the assumption that interactions are weak so that it is sufficient to consider the free-string limit. Of course, interactions can never be completely ignored for reaching equilibrium. We shall address the question of interactions in the next two sections, and concentrate here on treating the statistical mechanics of free strings.

For simplicity, we shall first work within MB statistics. We also restrict ourself here to the case of closed bosonic strings with d_c uncompactified spatial dimensions, where d_c is the critical spatial dimension, i.e., $d_c = 25$ for bosonic strings. We shall treat V as a finite but large number, and shall ignore effects of winding modes. Although this theory is strictly speaking ill-defined due to the presence of a tachyon, it does provide a sufficiently simple framework for getting our points across.

Since we are dealing with free strings, our earlier analysis for the noninteracting gas remains valid; the only modifications required are to sum over string masses for (6), (7), and (8), e.g.,

$$\partial \Omega(E, V; z)/\partial z = f(E, V) + \int_0^E dE_1 f(E_1, V) \Omega(E - E_1, V; z), \qquad (9)$$

where $f(E, V) \equiv \Omega_1(E, V)$ is the single-particle density: $f(E, V) = \sum_m d(m) f(E, V; m)$,

$d(m)$ is the mass degeneracy. In particular, we can again solve (9) by an Laplace transform and obtain

$$Z(\beta,V) = e^{\tilde{f}(\beta,V)} - 1 \qquad (10)$$

where $\tilde{f}(\beta,V)$ is the Laplace transofrm of $f(E,V)$. Again, the analyticity of $Z(\beta,V)$ is known once $\tilde{f}(\beta,V)$ is found. [$\Omega(E,V)$ and $Z(\beta,V)$ are $\Omega(E,V;z)$ and $Z(\beta,V;z)$, with $z = 1$.]

The mass spectrum for a closed string is the sum of left- and right-moving pieces, $m^2 = m_l^2 + m_r^2$, subject to a constraint, $m_l^2 - m_r^2 = 0$. At each mass level, the degeneracy, $d(m)$, becomes $d_l(m_l)d_r(m_r)$. Following the analysis of Ref. 4, one finds

$$\tilde{f}(\beta,V) \propto V\beta \int_E (d^2\tau/\tau_2)\tau_2^{-(d_c+1)/2} D^{(l)}(\bar{z}) D^{(r)}(z) \exp(-\beta^2/4\pi\alpha'\tau_2), \qquad (11)$$

where the integration in $\tau \equiv \tau_1 + i\tau_2$ is over a half-strip region E: $-1/2 < \tau_1 < 1/2, \tau_2 > 0$. With $\bar{z} \equiv \exp(-2\pi i\bar{\tau})$ and $z \equiv \exp(2\pi i\tau)$, $D^{(l)}$ and $D^{(r)}$ are generating functions for left- and right-mass degeneracies.

Since β enters analytically in (11), singularities for $\tilde{f}(\beta,V)$ can occur only through endpoint pinchings. An analysis of this type has previously been carried out in Ref. 4. From the known asymptotic behaviors of d_l and d_r, we find that (11) contains a singularity at $\beta_H \equiv 4\pi\alpha'^{1/2}$, which is the inverse Hagedorn temperature for closed bosonic string.

We have considered elsewhere, in a quantum statistical treatment, analytic property of $Z(\beta,V)$ in β for various string theories.[13] We find that, generically, $Z(\beta,V)$ is analytic for $Re\,\beta > 0$ except at isolated points. For each string theory, because of the exponential growth in mass degeneracy, there is always an isolated rightmost singularity at $\beta = \beta_H$, ı.e., the inverse Hagedorn temperature for that theory. There is a finite gap in their real parts between β_H and the next singularity, and this gap is theory-dependent but calculable. By excluding winding modes, the singularity at β_H is bounded.

How does our finding fit in with the earlier analysis? Note that the singularity at β_H is always present for any V. For free strings, one must therefore keep β_0 in (3) larger than β_H so that the integration region overlaps with the analyticity region of (1). The situation is in a way analogous to the gas example, with the singularity moved from $\beta = 0$

to β_H. It follows that $\Omega(E,V) = 0(e^{\beta_H E})$. Conversely, if $\Omega(E,V)$ for strings can grow at most exponentially with E, we conclude that the interchange of the order of sum and integral leading to (3) is allowed, so that canonical and microcanonical quantities are mathematically equivalent. *The totality of physics of microcanonical approach for free strings has been entirely encoded in the analyticity of $Z(\beta,V)$.* Whether one prefers to work with $\Omega(E,V)$ or $Z(\beta,V)$ is a matter of convenience.

At low energies, (3) can be saturated by a saddle point at $\bar{\beta}(E)$, lying to the right of β_H. When this takes place, descriptions based on microcanonical and canonical ensembles are equivalent. However, because of the singularity at β_H, at high energy densities, either the saddle point disappears, or the fluctuations about the saddle point become large; a microcanonical ensemble must be adopted. Nevertheless, the statistical mechanics of free strings is still given unambiguously by (3). One can in fact push the contour in (3) to the left of the branch point by a finite distance η, $\eta > 0$. As one moves past the branch point, one picks up an additional contribution involving the discontinuity across the cut. Denoting the discontinuity by $\Delta Z(\beta,V)$, $\beta < \beta_H$, we have $\Delta Z(\beta_H,V) = 0$. From (3), the large-E behavior of $\Omega(E,V)$ is

$$\Omega(E,V) = -\int_{\beta_H-\eta}^{\beta_H} \frac{d\beta}{2\pi i} \Delta Z(\beta,V) e^{\beta E} + 0(e^{(\beta_H-\eta)E}), \qquad \eta > 0. \tag{12}$$

Once $\Delta Z(\beta,V)$ is known, the dominant behavior of $\Omega(E,V)$ can be found. Therefore, the large-E limit of a free-string theory can best be approached by working first with the canonical quantity, $Z(\beta,V)$. Further discussion on this approach can be found in Ref. 13.

INTERACTIONS AND PHASE TRANSITIONS

Let us next consider the effects of interactions. For instance, in the case of a classical gas, one may ask: What will happen if we include some short-range interactions (an attraction plus a hard core)? For a finite system, interactions cannot alter the fact that $Z_N(\beta,V)$ is analytic for $Re\,\beta > 0$. However, in the thermodynamic limit, one expects that, when the dimension is sufficiently high, "condensation" takes place at low temperature. At some value β_c, (it depends on the particle density), $F(\beta)$ must develop a singularity, though it is continuous there. That is, the large-V limit yields two separate analytic functions for the free-energy, $F_>(\beta)$ and $F_<(\beta)$, depending on whether β is greater or less than β_c! As a first order transition, $F'(\beta)$ will be discontinuous at β_c. How can this come about?

Let us examine a solvable interacting theory–the two-dimensional Ising model. Denote the classical hamiltonian by $H = -\varepsilon \sum_{(i,j)} \mu_i \mu_j$, where the sum is over all nearest neighbors, and $\mu_i = \pm 1$. Consider a square lattice with N^2 sites. Adopting a toroidal boundary condition, $Z(\beta, V)$ can be expressed in terms of 2^N eigenvalues, λ_k, of a "transfer matrix", $Z(\beta, N^2) = \sum_k \lambda_k^N$, where we have replaced the volume by the number of lattice sites, N^2. For a finite system, since the original defining expression for $Z(\beta, N^2)$ is a finite sum of terms, each of which is entire in β, $Z(\beta, N^2)$ is therefore also entire in β. Although it is rather pointless in this case, one can obtain the microcanonical density by (3) with β_0 arbitrary.

Consider next the large-V limit. It can be shown that all λ's are positive; it is convenient to arrange them in a descending order, with λ_0 being the largest. In fact, the free energy per spin is given by $F(\beta) = -\lim_{N \to \infty} \beta^{-1}(\log \lambda_0)/N$, which is known to have the following integral representation when $\beta > 0$:[15]

$$\beta F(\beta) = -\log 2 - (1/2)\log y - \int_0^{2\pi}\int_0^{2\pi} \frac{d\xi d\eta}{(2\pi)^2} \log[y + 1/y - (\cos\xi + \cos\eta)]. \quad (13)$$

In (13), y is related to β by $y \equiv \sinh(2\beta\varepsilon)$. In terms of y, $Z(\beta, V) \approx e^{-\beta F(\beta)V}$, also obeys a duality symmetry between high and low temperatures:[16] $Z(y, V) = y^V Z(1/y, V)$, a phenomenon occurring frequently in "order-disorder" systems.

Because of the self-dual structure, if there is a single phase transition, it has to take place at the self dual point: $\sinh(2\beta_c\varepsilon) = 1$. A careful examination shows that (13) contains a singular piece $\sim (y-1)^2 \log|y-1|$, so that the specific heat diverges logarithmically, when approaching β_c both from above and from below. In fact, (13) yields two analytic functions, $F_>$ and $F_<$, each defined by that branch of the logarithm factor in (13) so that $F_>$ and $F_<$ are real for $y > 1$ and $y < 1$ respectively.

Although (3) holds for any β_0 when the system is finite, depending on whether β_0 is greater or less than β_c, it leads to two different microcanonical densities, $\Omega_>$ and $\Omega_<$, in the thermodynamic limit.

The fact that $F(\beta)$ is piecewise analytic in the thermodynamic limit indicates that a phase transition has taken place. How can this nonanalytic behavior reconcile with the fact that $Z(\beta, V)$ is an entire function when the system is finite? The resolution lies in the recognition that the occurrence of this phase transition can be understood in terms of the phenomenon of "level-crossing" for eigenvalues λ_k. This in general will involve a large number of crossings, accumulating at $y = 1$, in the large-V limit. By keeping all contributions, $Z(\beta, V)$ will always remain analytic no matter how large V gets so long as it remains finite. To illustrate this point, let us pretend that both $F_>$ and $F_<$ are analytic at y=1, and assume that a level-crossing has taken place involving the first two levels only, so that $F_> - F_<$ is negative for $y > 1$ and positive for $y < 1$, (more correctly, $e^{-\beta F_>(y)V}$ and $e^{-\beta F_<(y)V}$ can each have a root-type singularity while their sum is regular). Writing $Z(\beta, V) \approx e^{-\beta F_>(y)V} + e^{-\beta F_<(y)V}$, it is easy to see that this expression produces precisely the desired effect: It assures that $Z(\beta, V)$ remains analytic at β_c and yet leads to a piecewise analytic free energy function.

By examining the ideal gas and the Ising model, we have learned the following: The region of validity for (3) and (4) should be established first before considering the thermodynamic limit. This region is determined by the volume-independent singularities of $Z(\beta, V)$ in β. In the thermodynamic limit, the limiting free energy can be piecewise analytic if a "conventional" phase transition has taken place. This asymptotic singularity is volume dependent, and shows up only in the thermodynamic limit. However, this nonanalytic behavior does not invalidate (3) and (4). Even when a phase transition occurs, it remains in principle meaningful to continue working with $Z(\beta, V)$ as an analytic function of the inverse temperature.

SPECULATIONS FOR INTERACTING-STRINGS

The thermodynamics of free strings at high energy densities has many unusual features. One unsettling fact involves having a specific heat which tends to be negative. However, these nonanalytic behaviors are not volume dependent; they therefore do not correspond to conventional phase transitions.

It has been argued that these strange features suggest that no sensible physics should be expected at high energy densities until interactions are included. Partly based on the analogy to the deconfinement phenomenon in QCD, it has been speculated that a first order phase transition has already taken place before one reaches the Hagedorn temperature.[10] Although solid support for this picture is still lacking, it is certainly an intriguing possibility. The most tantalizing aspect of this scenario is the suggestion that, as one probes the system at smaller and smaller distances, the degrees of freedom become much less than that of a free string. [It has also been argued that the underlying symmetry for the duality relation for heterotic string, $Z_0(\beta, V) = Z_0(1/\beta, V)$, first noted in Ref. 4, must be spontanously broken due to interactions.]

For free strings, the mass degeneracy increases with the mass exponentially: $\sim m^{-c} e^{bm}$. Let us assume that this free-string behavior is an adequent description only for $m \leq l^{-1}$, where l is a very small length scale. However, for $m \geq l^{-1}$, we assume that, through interactions, the effective degrees of freedom become much reduced, (e.g., the widths of excited modes increase sufficiently fast with their masses.) There are of course many possibilities. Without committing oneself to a specific dynamical mechanism, let us concentrate here on the kinematic aspects of this conjecture. Let us continue to represent the degrees of freedom between E and $E + dE$ in a volume V by $f(E, V)$. It could happen that $f(E, V)$ continue to grow exponentially with E, but with a coefficient b' which is reduced to a smaller value. However, this does not represent a drastic change, all it does is to push the Hagedorn temperature to a higher value. A more interesting possibility is that $f(E, V)$ no longer grows with E exponentially. To be specific, let us assume that, when $E \geq l^{-1}$,

$$f(E, V) \sim exp(bE^{1-\epsilon}), \qquad (14)$$

where $0 < \epsilon < 1$. Clearly, if this happens, statistical mechanics of strings looks quite different at different temperature scales. In particular, from (10), the Hagedorn singularity has now been removed and $Z(\beta, V)$ is analytic for $Re\ \beta > 0$. We would like to close by suggesting two possible scenarios in which this phenomenon could come about.

(A) New-Constituents Picture: Since the Hagedorn singularity is no longer present, as one increases the energy density, the temperature of the system will simply continue to rise, as for ordinary systems. The specific heat per unit volume can be used to characterize the effective degrees of freedom of the system at a given energy density. The transition from a free-string description to that characterized by (14) could be like that of "deconfinement", i.e., a QCD-like transition from "confined" strings to yet another set of constituents takes place before reaching the scale l^{-1}. This transition could be first order. The length scale, l, must be of the order or larger than β_H. It represents the beginning of new physics, new space-time structure, *etc.*, definitely of a type unfamilar to most of us.

(B) Finite-Volume Effect: It has been pointed that statistical mechanics of free strings at finite volume can be altered due to the presence of winding modes. Unfortunately, these additional modes have just the opposite effect than what is needed for leading to (14). Nevertheless, let us speculate that, at a finite volume, there is a dynamical effect which suppresses energetic strings so that (14) applies when E/V for a single string is larger than a constant value. That is, we consider the possibility that l is volume-dependent and $l \sim V^{-1}$. At finite V, the Hagedorn singularity is again absent. Therefore, the system can exist either above or below β_H. However, if one could increase V, l would decrease to zero; in a thermodynamic limit, the Hagedorn singularity would reappear for the free energy. The reappearance of this singularity should be interpreted as an ordinary phase transition. This scenario could be realized if an energetic string would always prefer to have an increasingly large spatial extent. [This picture is classically reasonable and has been used for cosmic string studies. However, its validity in a quantum setting could be in doubt.] It is reasonable to assume that, in a finite volume, large strings should be suppressed due to interactions. Unlike the constituent-picture, no dramatically new physics is required, and yet the statistical mechanics of strings at high energy densities could differ significantly from that of free strings.

Finally, it is intriguing to contrast our findings to the case of a blackhole, where one expects a blackbody radiation spectrum with a temperature inversely proportional to its mass. Can one arrive at this picture by treating a blackhole as an ensemble of "stringy" objects? What would be the asymptotic mass degeneracy for this stringy system? Will Eqs. (3) and (4) still hold?

ACKNOWLEDGEMENTS

I would like to thank J. J. Atick, R. H. Brandenberger, N. Deo, S. Jain, A. Jevicki, A. Shapere, A. Strominger, N. Turok, and C. Vafa for useful discussions.

REFERENCES

1. R. Hagedorn, *Nuovo Cimento Suppl.* **3** (1965) 147;
   K. Huang and S. Weinberg, *Phys. Rev. Lett.* **25** (1970) 895.

2. S. Frautschi, *Phys. Rev.* **D3** (1971) 2821; R. D. Carlitz, *Phys. Rev.* **D5** (1972) 3231.

3. S. H. Tye, *Phys. Lett.* **158B** (1985) 388;
   M. J. Bowick and L. C. R. Wijewardhana, *Phys. Rev. Lett.* **54** (1985) 2485.

4. K. H. O'Brien and C-I Tan, *Phys. Rev.* **D36** (1987) 1184.

5. B. Maclain and B. D. B. Roth, *Comm. Math. Phys.* **111** (1987) 539.

6. M. McGuigan, *Phys. Rev.* **D2** (1988) 552; E. Alvarez and M. A. R. Osorio, *Nucl. Phys.* **B304** (1988) 327.

7. B. Sundborg, *Nucl. Phys.* **B254** (1985) 583; P. Salomonson and B.-S. Skagerstam, *Nucl. Phys.* **B268** (1986) 349; D. Mitchell and N. Turok, *Nucl. Phys.* **B294** (1987) 1138; F. Englert, J. Orloff and T. Piran, ULB-TH 88/03 preprint.

8. B. Sathiapalan, *Phys. Rev.* **D35** (1987) 3277; A. Kogan, ITEP -110 (1987).

9. M. Axenides, S. D. Ellis and C. Kounnas, *Phys. Rev.* **D37** (1988) 2964; I. Antoniadis, J. Ellis and D. V. Nanopoulos, *Phys. Lett.* **199B** (1987) 402.

10. J. J. Atick and E. Witten, IAS preprint, IASSNS-HEP-88/14, Nuc. Phys. in press.

11. To appear in *Proceedings for TASI-88*, ed. A. Jevicki and C-I Tan, (World Scientific): N. Turok, "Phase Transition as the Origin of Large Scale Structure in the Universe"; G. G. Athanasiu and J. J. Atick, "Remarks on Thermodynamics of Strings".

12. R. H. Brandenberger and C. Vafa, BROWN-HET-673, August, 1988.

13. N. Deo, S. Jain and C-I Tan, BROWN-HET-692, Phys. Letters (in press).

14. For a recent discussion on "temperature fluctuation", see an Opinion piece by B. Mandelbrot, PHYSICS TODAY, January 1989, page 71, which is a response to an earlier Opinion piece by C. Kittel, May 1988, page 93.

15. See, for instance, *'Statistical Mechanics'*, by K. Huang, second edition, Wiley (1987).

16. A. Shapere and F. Wilczek, IASSNS-HEP-88/36, and references therein.